T0230060

Lecture Notes in Computer Science 903

Ernst W. Mayr   Gunther Schmidt
Gottfried Tinhofer (Eds.)

Graph-Theoretic Concepts in Computer Science

20th International Workshop. WG '94
Herrsching, Germany, June 16-18, 1994
Proceedings

 Springer

Series Editors

Gerhard Goos
Universität Karlsruhe
Vincenz-Priessnitz-Straße 3, D-76128 Karlsruhe, Germany

Juris Hartmanis
Department of Computer Science, Cornell University
4130 Upson Hall, Ithaca, NY 14853, USA

Jan van Leeuwen
Department of Computer Science, Utrecht University
Padualaan 14, 3584 CH Utrecht, The Netherlands

Volume Editors

Ernst W. Mayr
Institut für Informatik, Technische Universität München
Arcisstraße 21, D-80290 München, Germany

Gunther Schmidt
Fakultät für Informatik, Universität der Bundeswehr
D-85577 Neubiberg, Germany

Gottfried Tinhofer
Mathematisches Institut, Technische Universität München
Arcisstraße 21, D-80290 München, Germany

CR Subject Classification (1991): G.2.2, F.2, F.1.2-3, F.3-4, E.1

ISBN 3-540-59071-4 Springer-Verlag Berlin Heidelberg New York

CIP data applied for

© Springer-Verlag Berlin Heidelberg 1995
Printed in Germany

Typesetting: Camera-ready by author
SPIN: 10485464   45/3140-543210 - Printed on acid-free paper

Preface

The 20th International Workshop on Graph-Theoretic Concepts in Computer Science (WG'94) was held in Herrsching (Bavaria), in the conference facilities of the *Bildungsstätte des Bayerischen Bauernverbandes*, June 16–18, 1994. In addition to the wonderful surroundings of Lake Ammersee and a hike to the old monastery of Andechs (once a powerful center of clerical and worldly might, nowadays better known for the potent output of its brewery), it featured sessions on Graph Grammars, Treewidth, Special Graph Classes, Algorithms on Graphs, Broadcasting and Architecture, Planar Graphs and Related Problems, and Special Graph Problems.

The WG'94 represents a cornerstone in the immensely successful WG series, in particular, since it was number 20 in a row. For the historically minded and all those proud of this important annual event for many scientists all around the world interested in the study and application of graph-theoretic concepts in computer science, the table on the next page lists some of the data to remember. At the end of this volume, there is also a bibliography of the proceedings of these twenty workshops.

Sixty-six papers were submitted to WG'94, and the program committee consisting of

G. Engels (Leiden)   P. Spirakis (Patras)
A. Marchetti-Spaccamela (Roma) G. Tinhofer (München, chair)
E.W. Mayr (München)   J. van Leeuwen (Utrecht)
R. Möhring (Berlin)   P. Widmayer (Zürich)
M. Nagl (Aachen)   J. Wiedermann (Praha)
G. Schmidt (München)

selected 32 of these. This volume contains all these contributions, which have been carefully revised based on the comments and suggestions by the program committee and attendants at the workshop. It thus provides an up-to-date snapshot of the research performed in the various areas represented at the workshop, and it can be a starting point for continued work in the computer science oriented parts of graph theory.

The WG'94 meeting took place in a very relaxed atmosphere in the facilities of the *Bildungsstätte des Bayerischen Bauernverbandes* in Herrsching, right next to Lake Ammersee. Its excellent conference facilities provided for an enjoyable stay and many opportunities for discussions, be it on the talk just heard, on a future joint paper, or simply on the destination of the excursion. The organizers would like to thank the Bayerische Bauernverband for hosting the workshop, and they would also like to express their gratitude to the Deutsche Forschungsgemeinschaft (DFG) for sponsoring WG'94.

München, January 1995   Ernst W. Mayr
                                                         Gunther Schmidt
                                                         Gottfried Tinhofer

The first 20 WG's

1	Berlin	1975	U. Pape
2	Göttingen	1976	H. Noltemeier
3	Linz	1977	J. Mühlbacher
4	Castle Feuerstein near Erlangen	1978	M. Nagl, H.-J. Schneider
5	Berlin	1979	U. Pape
6	Bad Honnef	1980	H. Noltemeier
7	Linz	1981	J. Mühlbacher
8	Neunkirchen near Erlangen	1982	H.-J. Schneider, H. Göttler
9	Haus Ohrbeck near Osnabrück	1983	M. Nagl, J. Perl
10	Berlin	1984	U. Pape
11	Castle Schwanberg near Würzburg	1985	H. Noltemeier
12	Monastery Bernried near München	1986	G. Tinhofer, G. Schmidt
13	Castle Banz near Bamberg	1987	H. Göttler, H.-J. Schneider
14	Amsterdam	1988	J. van Leeuwen
15	Castle Rolduc near Aachen	1989	M. Nagl
16	Johannesstift Berlin	1990	R. H. Möhring
17	Richterheim Fischbachau near München	1991	G. Schmidt, R. Berghammer
18	Wilhelm-Kempf-Haus Wiesbaden-Naurod	1992	E.W. Mayr
19	Sports Center Papendal near Utrecht	1993	J. van Leeuwen
20	Herrsching near München	1994	G. Tinhofer, E.W. Mayr, G. Schmidt

Contents

Domino Treewidth
(Extended abstract)

Hans L. Bodlaender[*1] and Joost Engelfriet[**2]

[1] Department of Computer Science, Utrecht University, P.O. Box 80.089, 3508 TB
Utrecht, the Netherlands
[2] Department of Computer Science, Leiden University, P.O. Box 9512, 2300 RA
Leiden, the Netherlands

Abstract. We consider a special variant of tree-decompositions, called
domino tree-decompositions, and the related notion of *domino treewidth*.
In a domino tree-decomposition, each vertex of the graph belongs to at
most two nodes of the tree. We prove that for every k, d, there exists
a constant $c_{k,d}$ such that a graph with treewidth at most k and maximum degree at most d has domino treewidth at most $c_{k,d}$. The domino
treewidth of a tree can be computed in $O(n^2 \log n)$ time. There exist
polynomial time algorithms that — for fixed k — decide whether a given
graph G has domino treewidth at most k. If k is not fixed, this problem is
NP-complete. The domino treewidth problem is hard for the complexity
classes $W[t]$ for all $t \in \mathbf{N}$, and hence the problem for fixed k is unlikely
to be solvable in $O(n^c)$, where c is a constant, not depending on k.

1 Introduction

A topic of much recent research in algorithmic graph theory is the treewidth of
graphs. Applications of this research range from VLSI theory to expert systems
(and many more). (See e.g., [3] for an overview.) In this paper, we introduce a
special variant of treewidth: *domino treewidth*. This notion is derived from the
usual notion of treewidth, by additionally requiring that every vertex $v \in V$
belongs to at most two node sets X_i. (See Section 2 for the precise definitions.)
Our interest in this notion is largely due to a maybe somewhat surprising result,
proved in Section 3: for graphs of bounded degree and bounded treewidth, there
is a uniform upper bound on the domino treewidth. We also investigate the
algorithmic aspects of domino treewidth. We show in Section 5 that the problem
to determine the domino treewidth of a given graph is NP-hard, and also is hard
for the complexity classes $W[t]$, for all $t \in \mathbf{N}$. The latter result tells us that it
is unlikely that the problem, for fixed k, to decide whether a given graph has
domino treewidth $\leq k$, can be solved in $O(n^c)$ time, where c is a constant, not

---

[*] This author was partially supported by the ESPRIT Basic Research Actions of the
EC under contract 7141 (project ALCOM II).
[**] This author was supported by the ESPRIT Basic Research Working Group COM-
PUGRAPH II.

depending on k. Some special cases can be solved in polynomial time (as shown in Section 4): for fixed k, one can check in polynomial time whether the domino treewidth of a given graph is at most k. For trees, the domino treewidth can be computed in $O(n^2 \log n)$ time.

2 Definitions and preliminary results

Definition 1. A *tree-decomposition* of a graph $G = (V, E)$ is a pair $(\{X_i \mid i \in I\}, T = (I, F))$ with $\{X_i \mid i \in I\}$ a collection of subsets of V, and $T = (I, F)$ a tree, such that

(i) $\bigcup_{i \in I} X_i = V$

(ii) for all edges $\{v, w\} \in E$ there is an $i \in I$ with $v, w \in X_i$

(iii) for all $i, j, k \in I$: if j is on the path from i to k in T, then $X_i \cap X_k \subseteq X_j$.

The width of a tree-decomposition $(\{X_i \mid i \in I\}, T = (I, F))$ is $\max_{i \in I} |X_i| - 1$. The treewidth of a graph $G = (V, E)$ is the minimum width over all tree-decompositions of G.

Let $(\{X_i \mid i \in I\}, T = (I, F))$ be a tree-decomposition of $G = (V, E)$. For each vertex $v \in V$, we let T_v be the subgraph of T, induced by $\{i \in I \mid v \in X_i\}$. Condition (iii) above can be rephrased as: for every $v \in V$, T_v is connected (or: a tree).

We use $size(G)$ to denote the number of vertices of G, and $deg(v)$ to denote the degree of vertex v. The degree of G is the maximum degree of its vertices.

Definition 2. A tree-decomposition $(\{X_i \mid i \in I\}, T = (I, F))$ of $G = (V, E)$ is a *domino tree-decomposition*, if for every vertex $v \in V$, $size(T_v) \leq 2$, i.e., every vertex belongs to at most two sets X_i. The domino treewidth of a graph G is the minimum width over all domino tree-decompositions of G.

Equivalently, the definition of a domino tree-decomposition is obtained from Definition 1 by changing condition (iii) into the following requirement: for all $i, j \in I$, if $i \neq j$ and $\{i, j\} \notin F$, then $X_i \cap X_j = \emptyset$. For a connected graph G, this means that T is the intersection graph of the family $\{X_i \mid i \in I\}$. Also, the domino treewidth of a graph G equals the minimum over all chordal dominos H that contain G as a subgraph, of the maximal clique size of H minus 1 (see [11]).

Note that if d is the maximum degree of a graph G and k is its domino treewidth, then $d \leq 2k$. Hence the domino treewidth of a graph is at least half its maximum degree. This shows, e.g., that there is no bound on the domino treewidth of trees.

As an example, the graph G in Figure 1 (and any similar graph) has domino treewidth 3. The dotted lines indicate a domino tree-decomposition of width 3. Note that domino tree-decompositions are easy to visualize. One easily observes that G has treewidth 2: G is outerplanar and all outerplanar graphs have treewidth ≤ 2. The following two lemmas are well known.

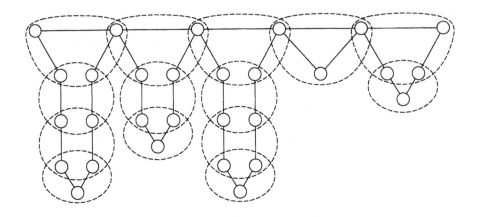

Fig. 1. A domino tree-decomposition

Lemma 3. *Let $(\{X_i \mid i \in I\}, T = (I, F))$ be a tree-decomposition of $G = (V, E)$. Suppose $W \subseteq V$ forms a clique in G. Then there exists a node $i \in I$ with $W \subseteq X_i$.*

Lemma 4. *Let $(\{X_i \mid i \in I\}, T = (I, F))$ be a tree-decomposition of $G = (V, E)$. Let $v_0, v_1, \ldots, v_r$ be a path in G. Suppose $v_0 \in X_i$, $v_r \in X_k$, and j is on the path from i to k in T. Then $X_j \cap \{v_0, \ldots, v_r\} \neq \emptyset$.*

A slightly stronger variant holds for domino tree-decompositions.

Lemma 5. *Let $(\{X_i \mid i \in I\}, T = (I, F))$ be a domino tree-decomposition of $G = (V, E)$. Let $v_0, v_1, \ldots, v_r$ be a path in G. Suppose $v_0 \in X_i$, $v_r \in X_k$, and j is on the path from i to k in T, $j \neq i$, $j \neq k$. Then $|X_j \cap \{v_0, \ldots, v_r\}| \geq 2$.*

3 Bounded treewidth and degree corresponds with bounded domino treewidth

In this section, we give the proof of a structural result (Theorem 10): graphs with bounded degree and bounded treewidth have bounded domino treewidth. A result like this was first shown in the context of graph grammars in [9].

For technical reasons, we will assume that the trees T taken from tree-decompositions are rooted. This induces rootedness of each T_v in a natural way: the root of T_v is the node of T_v that is closest to the root of T. For a node i of a rooted tree, the parent of i and children of i (if they exist) are defined in the usual way. The *rank* of i is the number of children of i. The rank of a tree is the maximum rank of its nodes. A *leaf* is a node of rank 0. Whenever we write down a tree arc $\{i, j\}$ as (i, j), this implicitly means that i is the parent of j. The height of a tree is the maximum distance between the root and the leaves.

If i is a node, and T a tree, we sometimes denote the fact that i is a node of T as $i \in T$.

We will for each tree-decomposition fix a mapping $\psi : I \to \mathcal{P}(E)$, such that
(i) for every $i \in I$, $\{v, w\} \in \psi(i)$: $\{v, w\} \subseteq X_i$.
(ii) for every $\{v, w\} \in E$, there is a unique $i \in I$ with $\{v, w\} \in \psi(i)$.
Such a mapping ψ always exists (but, in general, there may be more than one). It fixes a distribution of the edges of G over the nodes in I.

Let $XT = (\{X_i \mid i \in I\}, T = (I, F))$ be a tree-decomposition of $G = (V, E)$, and let mapping ψ and subtrees T_v be defined as in Section 2 and above. Let $v \in V$, and let $i \in I$ be a node in T_v (i.e., $v \in X_i$). We say that i is a *useful* node of T_v, if there exists an edge $\{v, w\} \in \psi(i)$, otherwise we say that i is a *useless* node of T_v. We say that a tree-decomposition XT has property **(U1)**, if for every vertex $v \in V$, T_v has no useless nodes of degree 1.

Lemma 6. *Let $(\{X_i \mid i \in I\}, T = (I, F))$ be a tree-decomposition of $G = (V, E)$ of width k. There exists a tree-decomposition $(\{Y_i \mid i \in I\}, T = (I, F))$ of $G = (V, E)$ of width $\leq k$ that has property **(U1)**.*

Proof. If i is a useless node of T_v of degree 1, then remove i from T_v, i.e., remove v from X_i. Repeat this procedure as long as possible. The resulting tree-decomposition has property **(U1)**, and width $\leq k$.

Let $XT = (\{X_i \mid i \in I\}, T = (I, F))$ be a tree-decomposition, let the subtrees T_v and mapping ψ be as before. For every node $i \in I$, define

$$chain_{XT}(i) = \{v \in X_i \mid i \text{ is a useless node of rank 1 in } T_v\}$$

For an arc (i, j) of T (recall that the notation means that i is the parent of j), define

$$pass_{XT}(i, j) = chain_{XT}(i) \cap X_j$$

Note: if $j \neq j'$, then $pass_{XT}(i, j) \cap pass_{XT}(i, j') = \emptyset$. We define

$$maxp(XT) = \max\{|pass_{XT}(i, j)| \mid (i, j) \in F\}$$

We say that a tree-decomposition is of *type* (k, r) if its width is $\leq k$ and the rank of its tree is $\leq r$.

Lemma 7. *For every $k, r \in \mathbf{N}$, there exist $k', r' \in \mathbf{N}$, such that for every graph $G = (V, E)$ and for every tree-decomposition XT of G with property **(U1)** of type (k, r), if $maxp(XT) > 0$, then there exists a tree-decomposition $XT' = (\{X_i' \mid i \in I'\}, T' = (I', F'))$ of G with property **(U1)** and of type (k', r'), such that $maxp(XT') < maxp(XT)$. Moreover the height and size of T' are at most the height and size of T, respectively.*

Proof. We begin by first producing an intermediate tree-decomposition $XT'' = (\{X_i'' \mid i \in I''\}, T'' = (I'', F''))$ of G with property **(U1)** and of type (k'', r''), such that for every arc (i, j) in T'', either $|pass_{XT''}(i, j)| < maxp(XT)$, or for every arc (j, h) it holds that $|pass_{XT''}(j, h)| < maxp(XT)$. k'', r'' are constants. The height and size of T'' are at most the height and size of T, respectively.

We say that an arc (i, j) is *maximal*, if $|pass_{XT}(i, j)| = maxp(XT)$, and we say that a node i is maximal, if it has a maximal arc (i, j). The idea of the construction is to fold certain paths of T, that consist of maximal arcs (i, j), all with the same set of vertices $pass_{XT}(i, j)$. These vertices can then be removed from the folded path. We first define these paths. To this aim, we define a marking of the nodes and arcs of T, in a top-down fashion. A node can either stay unmarked, or can be marked as *initial*, *middle*, or *final*. An arc (i, j) can either stay unmarked, or can be marked with the set of vertices $pass_{XT}(i, j)$.

We now describe the marking procedure in detail. Consider a node i of T, and assume that the nodes and arcs on the path from i to the root already have been considered. Let h be the parent of i, if existing. We distinguish two cases.

Case 1. The arc (h, i) is unmarked, or i is the root of T. If i is a maximal node, then mark i as *initial*, and mark each maximal arc (i, j) with $pass_{XT}(i, j)$. After that, consider the children of i.

Case 2. The arc (h, i) is marked with the set of vertices π. If there exists an arc (i, j) with $pass_{XT}(i, j) = \pi$, then mark this arc with π, and mark i as *middle*. Such an arc is necessarily unique. If there is no such arc, then mark i as *final*. After that, consider the children of i.

Note: the marked subgraph of T is a disjoint collection of trees. Each such tree has a root, marked *initial*, of rank at least 1; all other nodes in such a tree either have rank 1 and mark *middle*, or rank 0 and mark *final*. Moreover, all the arcs on one path from the initial node to a final node have the same mark; we will call this the mark of the path. Marks of distinct paths are disjoint.

We now describe the procedure, that computes the desired tree-decomposition XT''. Perform the following steps for each marked subtree of T.

1. Identify the initial node with all the final nodes. The resulting node of T'' is called a *bridge* node.
2. For each path from the initial node to a final node, let $i_1, \ldots, i_\beta$ be the middle nodes on the path. For α, $1 \le \alpha \le \beta$, identify i_α with $i_{\beta-\alpha+1}$. The resulting nodes of T'' are called *folded* nodes.
3. Define the sets X_j'' of folded or bridge nodes as the union of the sets X_i of the corresponding identified nodes. The mapping ψ'' is obtained similarly. (However, see below at point 5!)
4. Unmarked nodes of T will stay the same in XT'', with the same values for X_i and ψ. These will be called the *old* nodes of T''.
5. For each folded node j, remove π from X_j'', where π is the mark of the path to which the corresponding middle nodes belong.

In other words, each marked path from an initial node to a final node is folded in two, and the vertices in the mark of that path are removed from the sets X_i. Note that if in step 2, β is odd, then the node that is half-way the path is not identified with another node.

We first prove that XT'' is a tree-decomposition of G. It is easy to see that after we have applied steps $1 - 4$, we still have a tree-decomposition XT^* of G.

For each path from an initial node to a final node with mark π, the middle nodes are useless nodes of rank 1 in T_v for every $v \in \pi$. Hence, in T_v^*, the corresponding folded nodes form a chain of useless nodes of rank 1, ending in a useless leaf. So, in step 5 we safely remove v from all X_j^* for all folded nodes j.

Since at most $r + 1$ nodes are identified with each other, the width of XT'' is at most $k'' = (r + 1)(k + 1) - 1$, and the rank of T'' is at most $r'' = (r + 1)r$.

It can be shown that XT'' has property (U1) and that for every arc (i, j) in T'', if j is a folded node then $pass_{XT''}(i, j) = \emptyset$, and if i is an old node or a bridge node, then $|pass_{XT''}(i, j)| < maxp(XT)$. We omit the long, detailed proof (mainly based on case analysis) from this extended abstract. This shows that XT'' has the required properties.

Now we apply the following operation to XT'': identify each node i_0 in T'' that is at even distance to the root, with all its children $i_1, \ldots, i_s$. If j is the resulting node of T', then $X_j' = X_{i_0}'' \cup X_{i_1}'' \cup \cdots \cup X_{i_s}''$, and $\psi'(j) = \psi''(i_0) \cup \psi''(i_1) \cup \cdots \cup \psi''(i_s)$. Let XT' be the resulting tree-decomposition. The width k' of XT' is at most $(r'' + 1)(k'' + 1) - 1$, and the rank r' of T' is at most r''^2.

A detailed analysis (that we omit here) shows that XT' has property (U1) and that $maxp(XT') < maxp(XT)$.

Lemma 8. *For every $k, r \in \mathbf{N}$, there exist $k', r' \in \mathbf{N}$, such that for every graph $G = (V, E)$ and for every tree-decomposition XT of G of type (k, r), there exists a tree-decomposition $XT' = (\{X_i' \mid i \in I'\}, T' = (I', F'))$ of G of type (k', r'), such that for every vertex $v \in V$, if $size(T_v') \geq 2$, then T_v' has no useless nodes of rank 1 or 0. Moreover the height and size of T' are at most the height and size of T, respectively.*

Proof. First we apply Lemma 6. Let XT'' be the resulting tree-decomposition. Note that $maxp(XT'') \leq k + 1$. By applying Lemma 7 at most $k + 1$ times, we obtain a tree-decomposition XT' of bounded width and rank, such that $maxp(XT') = 0$. This implies that $chain_{XT'}(i) = \emptyset$, for every node i in T', and hence that for every vertex $v \in V$, if $size(T_v') \geq 2$, then T_v' has no useless nodes of rank 1 or 0.

Lemma 9. *For every $k, r, d \in \mathbf{N}$, there exist $k', r' \in \mathbf{N}$, such that for every graph $G = (V, E)$ with maximum degree $\leq d$ and for every tree-decomposition XT of G of type (k, r), there exists a tree-decomposition $XT' = (\{X_i' \mid i \in I'\}, T' = (I', F'))$ of G of type (k', r'), such that for every vertex $v \in V$, T_v' has height at most one. Moreover the height and size of T' are at most the height and size of T, respectively.*

Proof. First, we apply Lemma 8. Let XT' be the resulting tree-decomposition with width $\leq k'$ and rank of T' at most r'.

Note that every tree T_v' contains at most $deg(v)$ useful nodes, so at most $deg(v)$ nodes of rank 1 or 0. It follows that each T_v' contains at most $2d - 1$ nodes.

Now apply the following operation to XT': let i_0 be the root of T'. Identify all trees T_v' with $v \in X_{i_0}'$, i.e., we have one node i_0', with $X_{i_0}'' = \bigcup_{\{j \in T_v' \mid v \in X_{i_0}'\}} X_j'$.

Repeat this procedure with all subtrees in the forest, obtained by removing the set $\{j \in T'_v \mid v \in X'_{i_0}\}$ from T'. It now holds that for each vertex T''_v has height at most one. The width of the resulting tree-decomposition XT'' is at most $(2d-1)(k'+1)^2 - 1$, and the rank of T'' is at most $(2d-1)(k'+1)r'$, as we identify never more than $(2d-1)(k'+1)$ nodes.

Theorem 10. *For every k, r, $d \in \mathbf{N}$, there exist k', $r' \in \mathbf{N}$, such that for every graph $G = (V, E)$ with maximum degree $\leq d$ and for every tree-decomposition XT of G of width $\leq k$, with the rank of $T \leq r$, there exists a domino tree-decomposition $XT' = (\{X'_i \mid i \in I'\}, T' = (I', F'))$ of G of width $\leq k'$ with the rank of $T' \leq r'$. Moreover the height and size of T' are at most the height and size of T, respectively.*

Proof. By first applying Lemma 9, and then, in the resulting tree-decomposition, identifying all nodes that have a common parent in T'.

Corollary 11. *For every k, $d \in \mathbf{N}$, there exists $k' \in \mathbf{N}$, such that every graph with treewidth at most k and maximum degree at most d has domino treewidth at most k'. Moreover, given a graph $G = (V, E)$ with treewidth at most k and maximum degree at most d, one can find a domino tree-decomposition of width at most k' in $O(|V|)$ time.*

Proof. One can find a tree-decomposition of G with width $\leq k$ and with $O(|V|)$ nodes in $O(|V|)$ time [2]. It is easy to transform this tree-decomposition into one with the same width and of rank 2 in $O(|V|)$ time. The constructions, described in the proofs of Lemmas 6, 7, 8, 9, and Theorem 10 can be carried out in $O(|I|) = O(|V|)$ time in total.

Corollary 12. *For every k, $d \in \mathbf{N}$, there exists $k' \in \mathbf{N}$, such that every graph $G = (V, E)$ with treewidth at most k and maximum degree at most d has a domino tree-decomposition of width at most k' and with the height of the tree $O(\log |V|)$.*

Proof. Applying the algorithm of [12] and then Theorem 10 gives the result.

There is a connection with the notion of *strong treewidth*, introduced by Seese [13].

Definition 13. A *strong tree-decomposition* of a graph $G = (V, E)$ is a pair $(\{X_i \mid i \in I\}, T = (I, F))$ with $\{X_i \mid i \in I\}$ a collection of *disjoint* subsets of V, and $T = (I, F)$ a tree, such that
    (i) $\bigcup_{i \in I} X_i = V$
    (ii) for all edges $\{v, w\} \in E$, either there is an $i \in I$ with $v, w \in X_i$, or there are i, $i' \in I$, that are adjacent in T $((i, i') \in F)$, and $v \in X_i$, $w \in X_{i'}$.
The width of a strong tree-decomposition $(\{X_i \mid i \in I\}, T = (I, F))$ is $\max_{i \in I} |X_i|$. The strong treewidth of a graph $G = (V, E)$ is the minimum width over all strong tree-decompositions of G.

Note that in general, a strong tree-decomposition of a graph G, is *not* a tree-decomposition of G. Every graph of strong treewidth $\leq k$ is of treewidth $\leq 2k - 1$ [13]. However, there are sets of graphs of treewidth 2, that are of unbounded strong treewidth, e.g., the set of all paths with one additional vertex that is adjacent to all vertices of the path. However, if a graph has bounded treewidth and bounded degree, then it is also of bounded strong treewidth. In fact, a domino tree-decomposition of width $\leq k$ can be turned as follows into a strong tree-decomposition of width $\leq k + 1$: for all $v \in V$ and nodes $i, j \in T_v$: if j is the child of i, then remove v from X_j.

Corollary 14. *For every class of graphs $\mathcal{G}$, the following statements are equivalent:*

(i) There exists a constant $c \in \mathbf{N}$, such that every graph in $\mathcal{G}$ has domino treewidth at most c.

(ii) There exist constants k, $d \in \mathbf{N}$, such that every graph in $\mathcal{G}$ has treewidth at most k and maximum degree at most d.

(iii) There exist constants k', $d' \in \mathbf{N}$, such that every graph in $\mathcal{G}$ has strong treewidth at most k' and maximum degree at most d'.

4   Algorithms for determining domino treewidth

4.1   Domino treewidth of trees

It is easy to see that every tree T has domino treewidth $\leq d - 1$, where d is the degree of T; thus, its domino treewidth lies between $\frac{1}{2}d$ and $d - 1$. We now give an algorithm to compute the domino treewidth of a tree, based on dynamic programming.

Let $T = (V, E)$ be a tree. Choose an arbitrary vertex $r \in V$, and consider T as a rooted tree with r as root. Denote the subtree of T, formed by a vertex v and all its descendants by $T(v) = (V_v, E_v)$.

For each vertex $v \in V$, let $W_k(v)$ be defined as the minimum, over all domino tree-decompositions $(\{X_i \mid i \in I\}, (I, F))$ of $T(v)$ of width at most k, of the minimum size of a set X_i with $v \in X_i$. $W_k(v)$ is ∞, if there does not exist a domino tree-decomposition of $T(v)$ with width at most k.

For a leaf vertex v, one can observe directly that $W_k(v) = 1$. The following proposition shows how to compute $W_k(v)$ of a vertex v, given the values of $W_k(w_1), \ldots, W_k(w_p)$, for the children $w_1, \ldots, w_p$ of v.

Proposition 15.

$$W_k(v) = \min \left\{ 1 + \sum_{j \in S} W_k(w_j) \mid S \subseteq \{1, \ldots, p\} , \right.$$

$$\left. \sum_{j \in S} W_k(w_j) \leq k \ \wedge \sum_{j \in \{1, \ldots, p\} - S} W_k(w_j) \leq k \right\}.$$

Proof. We omit the proof. The idea is: we may assume there are i, i', with $v \in X_i \cap X_{i'}$, $i \neq i'$. For each neighbor w_j of v we must choose whether w_j is put into X_i or into $X_{i'}$. When w_j is put into a set, there must be at least $W_k(w_j)$ vertices in V_{w_j} in this set, and this is sufficient. We look for the best partition of the neighbors of v in $T(v)$ over the two sets. The additive factor '1' in the formula is for the vertex v itself.

Finding the set S which achieves the minimum value, as described in Proposition 15 above, when given the values $W_k(w_j)$ for all children of v, corresponds to solving a KNAPSACK problem, and can be solved by a standard dynamic programming algorithm in $O(p \cdot k)$ time. (See e.g., [10], Chapter 5.) So, $W_k(v)$ can be computed in $O(deg(v) \cdot k)$ time, given the values $W_k(w_j)$ for all children of v.

To test whether T has domino treewidth at most k, we compute for all $v \in V$, $W_k(v)$, in a bottom-up order. Clearly, the domino treewidth of T is at most k, if and only if $W_k(r) \neq \infty$. The total time to compute $W_k(r)$ hence is $O(\sum_{v \in V} deg(v) \cdot k) = O(k \cdot |V|)$.

Theorem 16. *There exists an $O(kn)$ algorithm to compute whether the domino treewidth of a tree with n vertices is at most k.*

The algorithm can output a corresponding domino tree-decomposition within the same time bounds. By using this procedure and binary search on the value of k, one can compute the domino treewidth of a given tree in $O(n^2 \log n)$ time.

4.2 Fixed parameter algorithms

For fixed k, the problem to decide whether the domino treewidth of a given graph G is at most k is solvable in polynomial time. A complete algorithm for this problem can be found in the full version. Here, we only remark that our algorithm has a similar structure as the $O(n^{k+2})$ algorithm from Arnborg et al [1] to recognize graphs with treewidth at most k, but that the extra technicalities are involved and tedious.

Theorem 17. *For fixed k, there exists an algorithm, that checks whether a given graph G has domino treewidth at most k, and if so, finds a corresponding domino tree-decomposition in $O(n^{2k+3})$ time.*

5 Hardness results

In this section, we show that the domino treewidth problem is $W[t]$-hard for all $t \in \mathbf{N}$, where the notion of $W[t]$-hardness is taken from the work of Downey and Fellows on fixed parameter (in)tractability (see [6, 7, 8]). Also, we prove the problem to be NP-complete. To be precise, we consider the following problem.

DOMINO TREEWIDTH
Instance: Graph $G = (V, E)$, integer k.
Parameter: k.
Question: Is the domino treewidth of G at most k?

The fact that domino treewidth is $W[t]$-hard for all $t \in \mathbf{N}$ suggests strongly that it is impossible to find algorithms for the domino treewidth $\leq k$ problem that use $f(k)pol(n)$ time, i.e., where only the constant factor in the running time depends on k. (Note that such algorithms do exist for some related problems, like treewidth $\leq k$, pathwidth $\leq k$.) The classes $W[t]$ denote classes of *parameterized problems*: problems $Q \subseteq \{(k, s) \mid k \in \mathbf{N}, s \in \Sigma^*\}$, for some alphabet Σ. To prove $W[t]$-hardness for a problem P, it suffices to find a $W[t]$-hard problem Q and functions $f, h : \mathbf{N} \to \mathbf{N}$, $g : \mathbf{N} \times \Sigma^* \to \Sigma^*$, such that for all $(k, s) \in \mathbf{N} \times \Sigma^*$: $(k, s) \in Q \Leftrightarrow (f(k), g(k, s)) \in P$, and f is any computable function, and g is computable in time $h(k) \cdot |s|^c$ for some constant c. (I.e., the time is polynomial in the length of the non-parameter part of the input, but may depend in any way on the parameter.) Recently, the following problem has been shown to be $W[t]$-hard for all $t \in \mathbf{N}$ [5, 4].

LONGEST COMMON SUBSEQUENCE
Instance: Alphabet Σ, strings $s^1, \ldots, s^r \in \Sigma^*$, integer $m \in \mathbf{N}$.
Parameter: r.
Question: Does there exist a string in Σ^* of length at least m, that is a subsequence of each string $s^1, \ldots, s^r$?

Theorem 18. *For all* $t \in \mathbf{N}$, DOMINO TREEWIDTH *is* $W[t]$-hard.

Proof. We give a transformation from LONGEST COMMON SUBSEQUENCE. Let an instance Σ, $s^1, \ldots, s^r$, m of LONGEST COMMON SUBSEQUENCE be given. We suppose we have a numbering of the characters in Σ, say $\Sigma = \{\sigma_1, \ldots, \sigma_l\}$. Let $l = |\Sigma|$, $k = 20r^2 + 40r - 1$, $q = m(lr^2 + 1)$. We now define a graph $G = (V, E)$, that consists of the following components:

Two anchors. Take two cliques, each with $k + 1$ vertices, with vertex sets $A_1 = \{a_i^1 \mid 1 \leq i \leq k + 1\}$ and $A_2 = \{a_i^2 \mid 1 \leq i \leq k + 1\}$. To ease presentation, we will denote vertices a_i^1 with $1 \leq i \leq \frac{2}{5}(k + 1)$ also as b_i^0, and vertices a_i^2 with $1 \leq i \leq \frac{2}{5}(k + 1)$ also as b_i^{q+1}.

The floor. Take q cliques of $\frac{2}{5}(k+1) = 8r^2 + 16r$ vertices, $\{b_i^\alpha \mid 1 \leq i \leq \frac{2}{5}(k+1), 1 \leq \alpha \leq q\}$. Vertices in the clique B_α with vertex set $\{b_i^\alpha \mid 1 \leq i \leq \frac{2}{5}(k+1)\}$ are made adjacent to all vertices in cliques $B_{\alpha-1}, B_{\alpha+1}$ ($1 \leq \alpha \leq q$). The number $\frac{2}{5}$ is chosen because $2 \cdot \frac{2}{5} < 1$, and $3 \cdot \frac{2}{5} > 1$.

The hills. We have m 'hills'. Take vertices $\{c_{i,j}^\alpha \mid 1 \leq \alpha \leq m, 1 \leq i \leq l \cdot r^2, 1 \leq j \leq \frac{1}{5}(k+1) - (2r+1)\}$. Each such vertex $c_{i,j}^\alpha$ is made adjacent to the two vertices $b_1^{(lr^2+1)\cdot(\alpha-1)+i}$ and $b_1^{(lr^2+1)\cdot(\alpha-1)+i+1}$. Each hill, together with the adjacent floor vertices, represents a character of the subsequence.

The string components. For each string s^i, $1 \leq i \leq r$, we have a string component. Below, we describe the different parts for the ith string component. Suppose $s^i = s_1^i s_2^i \cdots s_{l_i}^i$.

The string path. Take a path with $l_i \cdot (lr^2 + 1) + 2$ vertices, and attach it to a_1^1 and a_1^2: take vertices $\{d_j^i \mid 0 \leq j \leq l_i \cdot (lr^2 + 1) + 1\}$, edges $\{d_j^i, d_{j+1}^i\}$, $\{d_0^i, a_1^1\}$, $\{d_{l_i \cdot (lr^2+1)+1}^i, a_1^2\}$.

The blobs. We take $l_i + 1$ cliques of $2r + 2$ vertices, and attach these to the string paths. Take vertices $\{e^i_{j,j'} \mid 0 \le j \le l_i, 1 \le j' \le 2r + 2\}$. For all j, $0 \le j \le l_i$, each set $\{e^i_{j,j'} \mid 1 \le j' \le 2r + 2\}$ forms a clique, whose vertices are made adjacent to $d^i_{j \cdot (lr^2+1)}$ and $d^i_{j \cdot (lr^2+1)+1}$.

The idea is that blobs cannot come on top of hills. This forces a precise way how the part between blobs falls over a hill (if it does). Each such part represents a character in the string s^i; the $lr^2 + 1$ vertices in such a part are needed for symbol checking.

The symbol checkers. These vertices are used to check that the chosen charac-ters for the different strings are the same. Take vertices $\{f^i_{\alpha,(i-1)rl+(j-1)l+t} \mid 1 \le \alpha \le l_i, 1 \le j \le r, \sigma_t = s^i_\alpha\} \cup \{f^i_{\alpha,(j-1)rl+(i-1)l+t} \mid 1 \le \alpha \le l_i, 1 \le j \le r, \sigma_t \ne s^i_\alpha\}$. If vertex $f^i_{\alpha,\beta}$ exists, make it adjacent to vertices $d^i_{(\alpha-1)(lr^2+1)+\beta}$ and $d^i_{(\alpha-1)(lr^2+1)+\beta+1}$.

Let $G = (V, E)$ be the resulting graph.

Proposition 19. *The domino treewidth of G is at most k, if and only if $s^1, \ldots, s^r$ have a common subsequence of length at least m.*

Proof. $\Rightarrow$: We sketch parts of the (lengthy) proof. Suppose $(\{X_i \mid i \in I\}, T = (I, F))$ is a domino tree-decomposition of G of width $\le k$. First, note that there must be nodes h_0, h_γ with $A_1 \subseteq X_{h_0}, A_2 \subseteq X_{h_\gamma}$ (Lemma 3). Consider the path in T between h_0 and h_γ, and number the nodes on this path consecutively $h_0, h_1, h_2, \ldots, h_{\gamma-1}, h_\gamma$. As $|A_1| = |A_2| = k + 1$, $A_1 = X_{h_0}$ and $A_2 = X_{h_\gamma}$. With induction to α (using Lemma 3, Lemma 2, dominoness, and simple counting arguments), it follows that for all i, $1 \le i \le \frac{2}{5}(k+1)$, b^α_i and $b^{\alpha+1}_i$ are in $X_{h_{\alpha+1}}$. It follows that $\gamma = q + 2$.

For each hill vertex $c^\alpha_{i,j}$, note that there is a unique node containing the neighbors of the vertex, namely $h_{(\alpha-1)(lr^2+1)+i+1}$, so $c^\alpha_{i,j} \in X_{h_{(\alpha-1)(lr^2+1)+i+1}}$. For each blob, there must be a node, containing all vertices in the blob (as it is a clique). However, such a node cannot be a node of the form $h_{(\alpha-1)(lr^2+1)+i+1}$, as the number of hill vertices plus the blob size plus the number of floor vertices that would belong to the node would be larger than $k + 1$. Call nodes of this form *hill nodes*. Nodes of the form $h_{\alpha(lr^2+1)+1}$ are called *valley nodes*.

Note that each hill node also contains at least two vertices per string path (because of Lemma 5). These $2r$ vertices, with the hill vertices and the floor vertices give in total k vertices, so only one extra vertex is possible.

Consider a blob, with vertices $B^i_j = \{e^i_{j,j'} \mid 1 \le j' \le 2r + 2\}$. There must be a non-hill node $i' \in I$, with $B^i_j \cup \{d_{j(lr^2+1)}, d_{j(lr^2+1)+1}\} \subseteq X_{i'}$ (Lemma 3). Let i'' be the first node on the path from i' to h_0 that is also on the path from h_0 to h_γ. One can show that i'' is a valley node. Write $f(i,j) = \alpha$, if the node i'' as above is of the form $h_{\alpha(lr^2+1)+1}$. (α does not depend on the choice of i'.)

Note that $f(i,j) \le f(i, j+1)$; if not, then the string path between blob B^i_j and blob B^i_{j+1} must go back over a hill: that hill then gets too many string path vertices in its node sets. We also must have $f(i,0) = 0$, and $f(i,l_i) = m$. So, for each i, there are exactly m values $\delta_1^i, \delta_2^i, \ldots, \delta_m^i$, with $\delta_1^i < \delta_2^i < \cdots < \delta_m^i$, and for all p, $1 \leq p \leq m$: $f(i, \delta_p^i - 1) = p - 1$ and $f(i, \delta_p^i) = p$.

Finally we can show, that all subsequences $s_{\delta_1^i}^i s_{\delta_2^i}^i \cdots s_{\delta_m^i}^i$ are equal. Showing that two characters $s_{\delta_p^i}^i$ and $s_{\delta_p^j}^j$ are equal follows with the following type of argument: if they are not equal, then there is a hill node in the pth hill, that contains two symbol checker vertices, hence at least $k + 2$ vertices (the symbol checking vertices were created just for this purpose).

$\Leftarrow$: By a rather straightforward, but lengthy construction.

The theorem now follows from the transformation, given above, and the fact that LONGEST COMMON SUBSEQUENCE is $W[t]$ hard for all $t \in \mathbf{N}$ [4].

Theorem 20. DOMINO TREEWIDTH *is NP-complete.*

Proof. Membership in NP is trivial. Observe that the transformation, given in the proof of Theorem 18, is a polynomial time transformation from the NP-complete LONGEST COMMON SUBSEQUENCE problem to DOMINO TREEWIDTH.

6 Conclusions

In this paper, we considered the notion of domino treewidth. We showed a correspondence between bounded domino treewidth, and bounded degree and treewidth, and obtained several results on the complexity of determining the domino treewidth of a given graph.

We believe the notion of domino treewidth can be of use for other investigations in the (algorithmic) theory on the treewidth of graphs. For instance, the notion may be of use for the design of dynamic graph algorithms for several problems on graphs with bounded treewidth and degree.

References

1. S. Arnborg, D. G. Corneil, and A. Proskurowski. Complexity of finding embeddings in a k-tree. *SIAM J. Alg. Disc. Meth.*, 8:277–284, 1987.
2. H. L. Bodlaender. A linear time algorithm for finding tree-decompositions of small treewidth. In *Proceedings of the 25th Annual Symposium on Theory of Computing*, pages 226–234, New York, 1993. ACM Press.
3. H. L. Bodlaender. A tourist guide through treewidth. *Acta Cybernetica*, 11:1–23, 1993.
4. H. L. Bodlaender, R. G. Downey, M. R. Fellows, and H. T. Wareham. The parameterized complexity of sequence alignment and consensus (extended abstract). To appear in: proceedings Conference on Pattern Matching '94, 1993.
5. H. L. Bodlaender, M. R. Fellows, and M. Hallett. Beyond *NP*-completeness for problems of bounded width: Hardness for the W hierarchy. In *Proceedings of the 26th Annual Symposium on Theory of Computing*, pages 449–458, New York, 1994. ACM Press.

6. R. G. Downey and M. R. Fellows. Fixed-parameter tractability and completeness I: Basic results. Manuscript, 1991. To appear in SIAM J. Comput.

7. R. G. Downey and M. R. Fellows. Fixed-parameter tractability and completeness II: On completeness for $W[1]$. Manuscript, 1991. To appear in Theoretical Computer Science, Ser. A.

8. R. G. Downey and M. R. Fellows. Fixed-parameter tractability and completeness III: Some structural aspects of the W hierarchy. Technical Report DCS-191-IR, Department of Computer Science, University of Victoria, Victoria, B.C., Canada, 1992.

9. J. Engelfriet, L. M. Heyker, and G. Leih. Context-free graph languages of bounded degree are generated by Apex graph grammars. *Acta Informatica*, 31:341–378, 1994.

10. E. Horowitz and S. Sahni. *Fundamentals of Computer Algorithms*. Pitman, London, 1978.

11. T. Kloks, D. Kratsch, and H. Müller. Dominoes. *This volume*, pages 106–120, 1995.

12. B. Reed. Finding approximate separators and computing tree-width quickly. In *Proceedings of the 24th Annual Symposium on Theory of Computing*, pages 221–228, New York, 1992. ACM Press.

13. D. Seese. Tree-partite graphs and the complexity of algorithms. In L. Budach, editor, *Proc. 1985 Int. Conf. on Fundamentals of Computation Theory, Lecture Notes in Computer Science 199*, pages 412–421, Berlin, 1985. Springer Verlag.

A Lower Bound for Treewidth and Its Consequences

Siddharthan Ramachandramurthi*

Department of Computer Science
University of Tennessee
Knoxville, TN 37996–1301, USA
siddhart@cs.utk.edu

Abstract. We present a new lower bound for the treewidth (and hence the pathwidth) of a graph and give a linear-time algorithm to compute the bound. With the growing interest in treewidth based methods, this bound has many potential applications.

Our bound helps shed new light on the structure of obstructions for width w. As a result, we are able to characterize completely those treewidth obstructions of order $w + 3$. Unexpectedly, we find that these graphs are exactly the pathwidth obstructions of order $w + 3$. Further, we are also able to enumerate these obstructions.

Surprisingly, while there is only one obstruction of order $w + 2$ for width w, we find that the number of obstructions of order $w + 3$ alone is an asymptotically exponential function of w. Our proof of this is based on the theory of partitions of integers and is the first non-trivial lower bound on the number of obstructions for treewidth.

1 Introduction

Ever since its introduction, interest in the notion of *treewidth* of a graph has been growing (see [B2]). This is mainly because, for many problems that are intractable on general graphs, polynomial-time and even linear-time algorithms can be found for graphs of bounded treewidth (see [AP, ALS] for example).

It is $\mathcal{NP}$-Hard to determine the treewidth of an arbitrary graph [ACP]. For every fixed constant w, there exists an $O(n^{w+2})$ time algorithm to decide whether the treewidth is at most w [ACP]. Robertson and Seymour developed an $O(n^2)$ time algorithm [RS3] for this problem, by showing (non-constructively) that graphs of treewidth at most w can be characterized by a finite number of minimal forbidden minors (called *obstructions*) [RS2]. Subsequently, methods to compute these obstructions were found [FL2, LA]. Besides being impractical, these methods do not reveal much structural information. Algorithms that do not rely on obstructions have also been designed [B1, Re, BK]. However, no practical algorithms are known for treewidth when $w > 4$ [B1, Sa]. Moreover, very few practical upper or lower bounds are known for treewidth.

The *pathwidth* of a graph is a concept akin to treewidth (see [BK, KL] for example). There has been considerable interest in the obstructions for treewidth and pathwidth [APC, FL2, KL, La, ST]. Obstruction based algorithms have been used for integrated circuit design and other applications [DS, GLR, KT]. The reasons for studying the obstructions are two-fold. First, better comprehension of their structure and number can help us design better algorithms for the fixed-parameter problem. Second, being minimal graphs, their structure can give us insights that can help the development of better bounds for treewidth.

Only the obstruction sets for treewidth 1, 2, and 3 have been found so far [APC, ST] and none of the methods used to find them seem to generalize to larger values of w. Lagergren [La] has found a weak upper bound on the number of edges in an obstruction for treewidth w. This bound is triple exponential in w^4 and is purely of theoretical significance. The obstruction sets for pathwidth 1 and 2 have also been computed [KL], but the general constructions only produce relatively large and sparse obstructions (i.e. of order at least $3w+3$ for pathwidth w).

We present a new lower bound for the treewidth of a graph and show how this bound can, in practice, be computed in time linear in the size of the graph. Besides being of independent interest, our bound gives us a new perspective on the obstructions for treewidth and pathwidth. As a result, we obtain the following three important consequences:

A. For every $w \geq 3$, there exists at least one obstruction of order $w + 3$ for treewidth w, and each such obstruction has a simple structural characterization.

B. The graphs in consequence A are exactly the obstructions of order $w + 3$ for pathwidth w.

C. For treewidth w, the number of obstructions of order $w + 3$ alone is an asymptotically exponential function of w.

Consequences A and B are significant in their generality and simplicity. Our characterization shows that these obstructions can be constructed and recognized easily. This is the first direct and general method to construct non-trivial obstructions for treewidth and to construct dense obstructions for pathwidth. In light of the fact that the complete graph is the only obstruction of order $w + 2$ for treewidth w, consequence C is very surprising. Ours is the first proof that there are an exponential number of obstructions for treewidth.

The rest of this paper is organized as follows. Section 2 consists of preliminaries. In section 3 we present a new lower bound for treewidth and a linear time algorithm to compute this bound. The proof of consequence A is spread over Sections 4 and 5. Consequences B and C are proved in sections 4 and 5 respectively. In the conclusions, we discuss the implications of our results and directions for future research.

2   Preliminaries

The graphs we consider are finite, simple, and undirected. If $G = (V, E)$ is a graph then, the *order* of G is $n = |V| = |\{v_1, v_2, \ldots, v_n\}|$. The *size* of G is $e = |E|$. The degree of vertex v_i is $\delta_i = |\{v_j : (v_i, v_j) \in E\}|$, and $\delta = \delta(G) = \min_i\{\delta_i\}$ is the minimum degree of G. K_n denotes the complete graph of order

n. A complete bipartite graph $K_{a,b}$ has vertex set $V = V_1 \cup V_2$ and edge set $E = V_1 \times V_2$ where $|V_1| = a$, $|V_2| = b$, and $|V_1 \cap V_2| = 0$.

Definition[RS1] Given a graph G, a pair (T, Y) is a *tree-decomposition* of G if T is a tree and $Y = \{X_i\}$ is a family of subsets of $V(G)$ indexed by $V(T)$ such that:

(a) $\cup X_i = V(G)$,

(b) for every edge $(v_a, v_b) \in E(G)$, $\exists i \in V(T) \ni \{v_a, v_b\} \subseteq X_i$, and

(c) for $i, j, k \in V(T)$, if j is on the path between i and k in T, then $X_i \cap X_k \subseteq X_j$.

The *width* of a tree-decomposition (T, Y) is $\max_{i \in V(T)}\{|X_i| - 1\}$. Further, $treewidth(G)$ is the minimum width over all possible tree-decompositions of G.

A *path decomposition* of G is just a tree decomposition (T, Y) where T is a simple path. Note that $treewidth(K_n) = pathwidth(K_n) = n - 1$. Therefore, in general $0 \leq treewidth(G) \leq pathwidth(G) \leq n - 1$.

We use a special kind of tree decomposition called a *smooth* decomposition [B1]. A similar notion was independently developed by Yan [Ya] for path decompositions.

Definition [B1] A tree decomposition (T, Y) of width w is *smooth* if

i. for every $i \in V(T)$, $|X_i| = w + 1$, and

ii. if (i, j) is an edge in T then $|X_i - X_j| = |X_j - X_i| = 1$.

As shown in [B1, Ya], any tree-decomposition can be easily transformed into a smooth tree-decomposition without changing the treewidth.

Lemma 2.1 [B1, Ya] If (T, Y) is a smooth tree-decomposition of width w for a graph G, then $|V(T)| = |V(G)| - w$.

If H and G are graphs, then H is a *minor* of G, denoted by $H \leq_m G$, if a graph isomorphic to H can be obtained from a subgraph of G by contracting edges. If $H \leq_m G$ and $H \neq G$, we write $H <_m G$. A family $\mathcal{F}$ of graphs is *closed* in the minor order if $\forall G \in \mathcal{F}$, $H \leq_m G \Rightarrow H \in \mathcal{F}$. The *obstruction set* for a minor-closed family $\mathcal{F}$, written $obs(\mathcal{F})$, is the set of all graphs in the complement of $\mathcal{F}$ that are minimal in the minor order. In other words, $G \in obs(\mathcal{F})$ if and only if $G \in \overline{\mathcal{F}}$ and $H <_m G \Rightarrow H \in \mathcal{F}$. Therefore, if $\mathcal{F}$ is a minor-closed family, then $G \in \mathcal{F}$ if and only if $H \not\leq_m G$, $\forall H \in obs(\mathcal{F})$. If we know all the graphs in $obs(\mathcal{F})$, then we can decide whether $G \in \mathcal{F}$ in polynomial-time using the fact that, for every fixed graph H, there exists a polynomial-time algorithm that when given an input graph G, decides whether $H \leq_m G$ [RS3].

Let $TW(k)$ denote the family of graphs with treewidth at most k. One can easily verify that for any fixed k, $TW(k)$ is minor-closed. It is known that $obs(TW(k))$ is a finite set [RS2]. Unfortunately, the proof of this is non-constructive [FL1]. Hence the need for finding the structure of these obstructions as well as their number.

3 A New Lower Bound for Treewidth

Using smooth decompositions, Yan [Ya] obtained the following lower bound for pathwidth. We observe that it holds for treewidth as well.

Lemma 3.1 [Ya] $treewidth(G) \geq \frac{2n-1-\sqrt{(2n-1)^2-8e}}{2}$, where n is the order of G and e is the size of G.

Proof If $w = treewidth(G)$, then $e \leq nw - \frac{w(w+1)}{2}$. Solve this inequality for w with the condition that $w \leq n - 1$. ∎

The following lemma helps us to find a new lower bound.

Lemma 3.2 If G is not a complete graph and $treewidth(G) = w$, then there exists a pair of non-adjacent vertices, each of degree at most w in G.

Proof $G \subset K_n \Rightarrow w \leq n - 2$. Let (T, Y) be a smooth tree-decomposition of width w for G. In T, let 1 and $n - w$ be two leaves, let 2 be the neighbor of 1, and let $n - w - 1$ be the neighbor of $n - w$. Also, let $\{v_1\} = X_1 - X_2$ and $\{v_2\} = X_{n-w} - X_{n-w-1}$. Since 2 is on every path from 1 to another vertex in T and $v_1 \notin X_2$, $\forall i$, $|X_i \cap \{v_1, v_2\}| \leq 1$. Hence, $(v_1, v_2) \notin E(G)$. Moreover, $|X_i| = w + 1 \Rightarrow \delta(v_1) \leq w$ and $\delta(v_2) \leq w$. ∎

This lemma immediately motivates the definition of a new metric γ of a graph.

Definition For a graph $G = (V, E)$ of order n,

$$\gamma(G) = \min_{1 \leq i < j \leq n} \{n - 1, \max\{\delta_i, \delta_j\} : (v_i, v_j) \notin E\}$$

Lemma 3.3 For every graph G, $treewidth(G) \geq \gamma(G)$.

Proof If G is a complete graph, then $\gamma(G) = n - 1 = treewidth(G)$. Otherwise, let $w = treewidth(G)$ and let v_1, v_2 be a pair of non-adjacent vertices of degree $\leq w$ in G. Then $w \geq \max\{\delta_1, \delta_2\} \geq \gamma(G)$. ∎

There are families of graphs for which γ equals the treewidth whereas the lower bound given by Lemma 3.1 is not. Complete bipartite graphs of the form $K_{m,m}$ with $m \geq 3$ are an example. Thus, γ is a tight lower bound for treewidth. Moreover, Lemma 3.1 is based on the total number of edges in the graph. In contrast, our lower bound γ is based on the neighborhood of individual vertices and hence reveals more useful structural information. Our algorithm GAMMA to compute γ is shown on the next page.

Theorem 3.4 If the input graph G has n vertices and e edges, then algorithm GAMMA computes $\gamma(G)$ in $O(n + e)$ time.

Proof Let $v_1, v_2, \ldots, v_n$ be the vertices of G. Let $x_1, x_2, \ldots, x_n$ be a permutation of $1, 2, \ldots, n$ such that $\delta_{x_1} \leq \delta_{x_2} \leq \ldots \leq \delta_{x_n}$. By definition $\gamma = \delta_{x_i}$ for some i. Algorithm GAMMA is based on the fact that $\gamma = \delta_{x_i}$ for some $2 \leq i \leq \delta + 2$. The proof of this fact is omitted.

Step 1 takes $O(n + e)$ time, step 2 takes $O(n)$ time, and step 3 takes constant time. Each iteration of step 4 takes constant time and there will be $\delta(\delta + 1)/2$

iterations in the worst case. Since there exist at least $\delta + 1$ vertices of degree $\geq \delta$, we have $\delta(\delta + 1) \leq 2e$. Therefore, the total execution time is $O(n + e)$. ∎

Algorithm GAMMA

1. Store the input graph G in uninitialized adjacency matrix form.
   Compute the degree of each vertex while reading the input.
2. Bucket sort the vertices in non-decreasing order of their degree.
   δ = minimum degree;
3. **if** $\delta = n - 1$ **then** $\gamma = n - 1$;
4. **else**
   begin
        Let $v_{x_1}, v_{x_2}, \ldots, v_{x_{\delta+2}}$ be the first $\delta + 2$ vertices in sorted order.
        found = FALSE;
        $i = 2$;
        while $((i \leq \delta + 1)$ and not(found)) **do**
        begin
            $j = 1$;
            while $((j \leq i - 1)$ and not(found)) **do**
                if (edge$(v_{x_j}, v_{x_i}) \notin G$) **then** found = TRUE;
                else $j = j + 1$;
            if (not(found)) **then** $i = i + 1$;
        end
        $\gamma = \delta_{x_i}$;
   end
5. Output γ.

In the next section we use the metric γ to prove consequences A and B.

4   Obstructions of Order $w+3$ for Treewidth w

The complete graph is the smallest obstruction for treewidth (and pathwidth). It is also the only general obstruction known for treewidth. For each $w \geq 0$, $K_{w+2} \in obs(\text{TW}(w))$. What is the next smallest obstruction for treewidth? In this section, we explore obstructions of order $w + 3$ for TW(w).

Let $n = w + 3$ be the order of an obstruction for TW(w). The treewidth of such a graph would be $n - 2$.

Lemma 4.1 *treewidth*$(G) = n - 2$ if and only if $\gamma(G) = n - 2$.
Proof ($\Leftarrow$): $\gamma(G) = n - 2 \Rightarrow G \subset K_n$. Therefore, $n - 2 \leq$ *treewidth*$(G) < n - 1$.
($\Rightarrow$): *treewidth*$(G) = n - 2 \Rightarrow \gamma(G) \leq n - 2$. Suppose $\gamma(G) < n - 2$. Then there exist two vertices v_1 and v_2 in $V(G)$ such that the edge $(v_1, v_2) \notin E(G), \delta_1 < n - 2$, and $\delta_2 < n - 2$. $\delta_1 < n - 2$ implies that there exists $v_x \in V(G)$ such that $(v_1, v_x) \notin E(G)$. Similarly, there exists v_y (x may be equal to y) such that $(v_2, v_y) \notin E(G)$. Then the following is a smooth tree decomposition of width $n - 3$ for G: $X_1 = V - \{v_2, v_x\}$, $X_2 = V - \{v_1, v_2\}$, and $X_3 = V - \{v_1, v_y\}$. This contradicts the fact that *treewidth*$(G) = n - 2$. ∎

Lemma 4.2 *pathwidth*$(G) = n - 2$ if and only if $\gamma(G) = n - 2$.
Proof Notice that the tree decomposition of width $n - 3$ used in the previous proof is also a path decomposition. The rest is trivial. ∎

The following theorem gives the structural characterization promised by consequence A.

Theorem 4.3 For every $n \geq 6$, a graph G of order n is an obstruction for $TW(n - 3)$ if and only if G satisfies the following four conditions:

  i. $\gamma(G) = n - 2$,

 ii. the maximum degree of G is $n - 2$,

iii. $\delta(G) \geq \max\{4, 3q - 3\}$, where

    $q \leq \lfloor \frac{n}{3} \rfloor$ is the number of vertices of degree $< n - 2$, and

iv. G is missing at least three disjoint edges (i.e. $\overline{G}$ has a matching of size ≥ 3).

Proof ($\Rightarrow$): First we show that if $G \in obs(TW(n - 3))$ then it satisfies (i)–(iv). Since $treewidth(G) = n - 2$, (i) follows directly from the previous Lemma. For (ii), suppose there exists v_1 with $\delta_1 = n - 1$. Let v_2 be a vertex of minimum degree in G. Then $\delta_2 = \delta(G) \leq \gamma(G) = n - 2$. Let $H = (V(G), E(G) - \{(v_1, v_2)\})$. $\gamma(H) = \min\{\gamma(G), \max\{\delta_1, \delta_2\}\} = \min\{\gamma(G), \delta_1\} = n - 2$. This implies $treewidth(H) = n - 2$, contradicting the fact that G is minor minimal. Therefore maximum degree of $G = n - 2$.

    The proof of (iii) is by contradiction. Suppose $1 \leq \delta_1 = \delta \leq 3$. Let $v_2, \ldots, v_{\delta+1}$ be the neighbors of v_1. Each vertex in $V_2 = \{v_i : \delta + 2 \leq i \leq n\}$ is non-adjacent to v_1 and has degree $n - 2$. Therefore, each neighbor of v_1 has degree $\geq n - \delta$. If $\delta = 1$ then, $\delta_2 = n - 1$ which contradicts (ii). If $\delta = 2$ then, contracting the edge (v_1, v_2) to v_2 gives us $K_{n-1} <_m G$ and $treewidth(K_{n-1}) = n - 2 = treewidth(G)$, contradicting the minor minimality of G. If $\delta = 3$ then, we claim that at least one of the edges $(v_2, v_3), (v_2, v_4), (v_3, v_4)$ is in G, because otherwise, $\delta_2 = \delta_3 = \delta_4 = n - 3$ and $\gamma(G) < n - 2$. Assume that $(v_2, v_3) \in E$. Contract edge (v_1, v_4) to v_4 to obtain $K_{n-1} \leq_m G$. Therefore $\delta \geq 4$.

    Let $V_1 = \{v_1, \ldots, v_q\}$ be the set of vertices of degree $< n - 2$ in G. Each vertex in V_1 has at least two non-neighbors in G. $\gamma = n - 2 \Rightarrow$ the vertices in V_1 are mutually adjacent and, any vertex in G not adjacent to a vertex in V_1 has degree $= n - 2$. Clearly, there are at least $2q$ such vertices of degree $n - 2$. Therefore, $q \leq \lfloor \frac{n}{3} \rfloor$ and $\delta \geq q - 1 + 2q - 2 = 3q - 3$. Hence, $\delta \geq \max\{4, 3q - 3\}$.

    It is clear from the proof of (iii) that each vertex of degree $< n - 2$ in G contributes an edge to the maximum matching in the complement of G. Thus, when $q \geq 3$, (iv) follows from (iii). Otherwise, we have three different cases.

    When $q = 0$, $\delta_i = n - 2, \forall i$ and, there is a matching of size $\lfloor \frac{n}{2} \rfloor$ in the complement of G. For $n \geq 6$, $\lfloor \frac{n}{2} \rfloor \geq 3$.

    If $q = 1$, let $\delta_1 < n - 2$ and $(v_1, v_2) \notin E$. Then $\delta \geq 4 \Rightarrow \exists \{v_3, v_4, v_5, v_6\} \ni \{(v_1, v_i), (v_2, v_i) : 3 \leq i \leq 6\} \subset E$ and $\{(v_3, v_4), (v_5, v_6)\} \cap E = \phi$.

    If $q = 2$, let $\delta_1 \leq \delta_2 < n - 2$ and $\{(v_1, v_3), (v_2, v_4)\} \cap E = \phi$. Then $\delta \geq 4 \Rightarrow \exists \{v_5, v_6\} \ni \{(v_1, v_i), (v_2, v_i) : 5 \leq i \leq 6\} \subset E$ and $(v_5, v_6) \notin E$. Thus, when $q = 1$ or 2, $(v_1, v_2), (v_3, v_4), (v_5, v_6)$ is a set of three disjoint edges missing from G. This concludes the proof of (iv).

($\Leftarrow$): Suppose G satisfies conditions (i)–(iv). Condition (i) $\Rightarrow$ $treewidth(G) = n - 2$. It is easy to verify that if H is obtained by either deleting or contracting an edge of G, then $\gamma(H) < n - 2$ which implies that $treewidth(H) < n - 2$. Hence G is minor minimal and the proof is complete. ∎

Given a graph G, it is simple to verify in polynomial-time whether G satisfies conditions (i)–(iv) of Theorem 4.3. Hence, the obstructions of order $w + 3$ for TW(w) are easily recognizable.

In the following lemma we prove consequence B.

Lemma 4.4 A graph G of order n is an obstruction for PW($n - 3$) if and only if G is also an obstruction for TW($n - 3$).

Proof ($\Leftarrow$): Let $G \in obs(\text{TW}(n - 3))$. Then $treewidth(G) = n - 2$ and $\gamma(G) = n - 2$, which implies that $pathwidth(G) = n - 2$. If H is a minor of G, then $treewidth(H) < treewidth(G) = n - 2$, and $\gamma(G) < n - 2$. Therefore we have $pathwidth(H) < n - 2$ and $G \in obs(\text{PW}(n - 3))$.

($\Rightarrow$): Let $G \in obs(\text{PW}(n - 3))$. Then $pathwidth(G) = n - 2$ and $\gamma(G) = n - 2$, which implies that $treewidth(G) = n - 2$. Suppose $H <_m G$. Then $pathwidth(H) < pathwidth(G)$. But $treewidth(H) \le pathwidth(H)$. Therefore, for every $H <_m G$, $treewidth(H) < treewidth(G)$, and $G \in obs(\text{TW}(n - 3))$. ∎

The proof of consequence A is not complete, because Theorem 4.3 does not tell us whether there are any graphs that satisfy all four conditions. In the next section, we show that such obstructions do exist for every $n \ge 6$.

5 Enumerating the Obstructions

Given w, we would like to find all the obstructions of order $w + 3$ for TW(w). For this, we use the structural description given by Theorem 4.3 and the theory of partitions of integers.

Definition[An] A *partition* of a positive integer n is a finite non-increasing sequence of positive integers $l_1, l_2, \ldots, l_r$ such that $\sum_{i=1}^{r} l_i = n$. The l_i are called *parts* of the partition. $p_r(n)$ is the number of partitions of n into r parts. The total number of partitions of n is $p(n) = \sum_{r=1}^{n} p_r(n)$.

We adopt the convention that $p_0(0) = 1, p_0(n) = 0$ if $n > 0$, and $p_r(n) = 0$ if $r > n$.

5.1 A Canonical Representation

Let $S(n)$ be the set of obstructions of order n for TW($n - 3$) and let $S(n, q) \subseteq S(n)$ be the set of obstructions with q vertices of degree less than $n - 2$. From Theorem 4.3, it is clear that $|S(n)| = \sum_{q=0}^{\lfloor \frac{n}{3} \rfloor} |S(n, q)|$. We describe a canonical representation for each obstruction in $S(n, q)$ for TW($n - 3$).

Consider an obstruction G of order n for TW($n - 3$) with q vertices of degree less than $n - 2$. Let $V_1 = \{v_i : 1 \le i \le q\}$ where $\forall v_i \in V_1, \delta_i < n - 2$. Let $V_2 = \{v_i : q + 1 \le i \le n\}$ where $\forall v_i \in V_2, \delta_i = n - 2$. Then $V = V_1 \cup V_2$. Since

$\forall v_i \in V_1, \delta_i < n-2$, let $\{(v_i, v_{q+i}), (v_i, v_{2q+i})\} \cap E = \phi$. Then v_{q+i} and v_{2q+i} are adjacent to all other vertices in G. The canonical form is shown in Figure 1(a). For clarity, only the missing edges are shown. All other edges are present.

Let $V_3 = \{v_i : 3q+1 \le i \le n\} \subseteq V_2$. If m is the size of the maximum matching in the complement of the subgraph induced by V_3 then $\forall i, 1 \le i \le m$, let $(v_{3q+i}, v_{3q+m+i}) \notin E$. Therefore, v_{3q+i} and v_{3q+m+i} are adjacent to all other vertices in G. Clearly, $m \le \lfloor \frac{n-3q}{2} \rfloor$.

Let $V_4 = \{v_i : 3q+2m+1 \le i \le n\} \subseteq V_3 \subseteq V_2$. Since every vertex in $V_2 - V_4$ already has degree $n-2$, for each vertex in V_4 there is exactly one non-neighbor in V_1. There exist $p_j(n-3q-2m)$ distinct mappings of V_4 into j elements of V_1. Suppose $l_1, \ldots, l_j$ is a j-partition of $n-3q-2m$. Then, $\forall i, 1 \le i \le j \le q$, and $\forall k, 1 \le k \le l_i$, we let $(v_i, v_{3q+2m+l+k}) \notin E$ where $l = l_1 + \ldots + l_{i-1}$. Therefore, $\forall v_i \in V_1$, if $1 \le i \le j$, then $\delta_i = n-3-l_i$; else $\delta_i = n-3$.

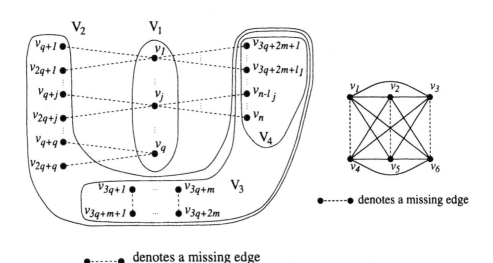

denotes a missing edge

Fig. 1. (a) The canonical form   (b) Obstruction for TW(3)

Observe that each graph that satisfies conditions (i)–(iv) of Theorem 4.3, and hence each obstruction in $S(n)$, has a unique representation in the form of a 5-tuple $(n, q, m, j, (l_1, \ldots, l_j))$. This is stated more formally in the following lemma.

Lemma 5.1 A graph G of order n is an obstruction for TW$(n-3)$ if and only if G can be uniquely represented by a 5-tuple $(n, q, m, j, (l_1, \ldots, l_j))$, where

  i. $n \ge 6$ is the order of the G,

  ii. $0 \le q \le \lfloor \frac{n}{3} \rfloor$ is the number of vertices of degree $< n-2$ (i.e. $q = |V_1|$),

  iii. $\max\{0, 3-q\} \le m \le \lfloor \frac{n-3q}{2} \rfloor$ is the size of the maximum matching in the complement of the subgraph induced by those vertices of degree exactly $n-2$ (i.e. $m = |E(\overline{G}(V_3))|$),

iv. $\min\{1, n - 3q - 2m\} \leq j \leq \min\{q, n - 3q - 2m\}$ is the number of vertices of degree less than $n - 3$ in G, and

v. $l_1 \geq \ldots \geq l_j$ is a j-partition of $n - 3q - 2m$ such that $n - 3 - l_1 \leq \ldots \leq n - 3 - l_j$ is the degree sequence of the set of vertices of degree less than $n - 3$ in G.

Proof Omitted. ∎

5.2 An Exact Formula

Given w, we would like to know the number of obstructions of order $w + 3$ for $TW(w)$. Using Lemma 5.1, we now derive an exact formula for this.

Theorem 5.2 $|S(n)| = \sum_{q=0}^{\lfloor \frac{n}{3} \rfloor} |S(n,q)|$, where $|S(n,q)|$ is as follows:

When $q = 0$:
$$|S(n,0)| = \begin{cases} 1 \text{ for even } n \geq 6; \\ 0 \text{ otherwise.} \end{cases}$$

When $q = 1$:
$$|S(n,1)| = \begin{cases} 0, & \text{for } n \leq 6; \\ \sum_{m=2}^{\lfloor \frac{n-3}{2} \rfloor} 1 \text{ otherwise.} \end{cases}$$

When $q = 2$:
$$|S(n,2)| = \begin{cases} 0, & \text{for } n \leq 7; \\ \sum_{m=1}^{\lfloor \frac{n-6}{2} \rfloor} \sum_{j=0}^{2} p_j(n - 6 - 2m) \text{ otherwise.} \end{cases}$$

$3 \leq q \leq \lfloor \frac{n}{3} \rfloor$:
$$|S(n,q)| = \begin{cases} 0, & \text{for } n \leq 8; \\ \sum_{m=0}^{\lfloor \frac{n-3q}{2} \rfloor} \sum_{j=0}^{q} p_j(n - 3q - 2m) \text{ otherwise.} \end{cases}$$

Proof Each 5-tuple that satisfies the conditions stated in Lemma 5.1 represents a unique graph in $S(n)$. Therefore, for every n, the number of distinct 5-tuples equals the number of obstructions in $S(n)$.

When $q = 0$, every vertex in G has degree $n - 2$, which implies that n should be even, and G is a complete graph of order n with $\frac{n}{2}$ disjoint edges missing. For even $n \geq 6$, G satisfies all four conditions of Theorem 4.3 and is in $S(n,0)$.

We can see from our canonical representation and the proof of Theorem 4.3 that, G is in $S(n)$ whenever $q \geq 1$, $q + m \geq 3$, and $n \geq 3q + 2m$. Therefore, for $q \geq 1$ we have the general expression $|S(n,q)| = \sum_{m=\max\{0,3-q\}}^{\lfloor \frac{n-3q}{2} \rfloor} \sum_{j=0}^{q} p_j(n - 3q - 2m)$. Recall that, $p_0(n - 3q - 2m) \neq 0 \Leftrightarrow n - 3q - 2m = 0$.

If $q = 1$ then $m \geq 2$ and $n \geq 7$. Therefore, $|S(n,1)| = 0$ for $n \leq 6$. For $n \geq 7$, $|S(n,1)| = \sum_{m=2}^{\lfloor \frac{n-3}{2} \rfloor} \sum_{j=0}^{1} p_j(n - 3 - 2m) = p_0(n - 3 - 2\lfloor \frac{n-3}{2} \rfloor) + p_1(n - 3 - 2\lfloor \frac{n-3}{2} \rfloor) + \sum_{m=2}^{\lfloor \frac{n-3}{2} \rfloor - 1} p_1(n - 3 - 2m) = \sum_{m=2}^{\lfloor \frac{n-3}{2} \rfloor} 1$.

If $q = 2$ then $m \geq 1$ and $n \geq 8$. Therefore, $|S(n,2)| = 0$ for $n \leq 7$. For $n \geq 8$, $|S(n,2)| = \sum_{m=1}^{\lfloor \frac{n-6}{2} \rfloor} \sum_{j=0}^{2} p_j(n - 6 - 2m)$.

If $q \geq 3$, then $m \geq 0$ and $n \geq 9$. Therefore, $|S(n,q)| = 0$ when $q \geq 3$ and $n \leq 8$. For $q \geq 3$ and $n \geq 9$, $|S(n,q)| = \sum_{m=0}^{\lfloor \frac{n-3q}{2} \rfloor} \sum_{j=0}^{q} p_j(n - 3q - 2m)$. ∎

This completes the proof of consequence A.

Recall that, $|S(n)| = \sum_{q=0}^{\lfloor \frac{n}{3} \rfloor} |S(n,q)|$. We can compute $|S(n)|$ using Theorem 5.2 and the recurrence relation $p_k(n) = p_k(n-k) + p_{k-1}(n-1)$. The following table lists some representative values:

Table: The number of obstructions of order $w + 3$ for treewidth w

Treewidth $w = n - 3$	1	2	3	4	5	6	7	8	9	10	15	20	25
No. of obstructions $= |S(n)|$	0	0	1	1	3	4	7	9	15	18	79	242	694

For treewidth 3, the graph shown in Figure 1(b) on page 21, is the only obstruction of order six. This graph has been called M_6 in [APC] and $K_{2,2,2}$ in [ST]. Curiously, while the entire obstruction set for TW(3) was previously known, the fact that $S(w+3) \subset obs(\text{TW}(w))$ was not suspected.

Similarly, even though the entire obstruction set for PW(2) was known (see [KL]), the existence of obstructions of order $w + 3$ for PW(w) was previously unknown. This is because $|S(w+3)| = 0$ for $w = 2$, and the general methods in [KL] can only produce obstructions of order at least $3w + 3$.

5.3 An Exponential Number of Obstructions

From Table 1, it is evident that the number of obstructions increases rapidly. In the sequel, we show that $|S(n)|$ grows exponentially.

Lemma 5.3 For $n \geq 0$, $p(n+2) \leq p(n+1) + p(n) \leq 2p(n+1)$.
Proof Using elementary algebra we can show that $p(n+2) \leq p(n+1) + p(n)$. The rest follows from the fact that $p(n) \leq p(n+1)$. We omit the details. ∎

Theorem 5.4 For $n \geq 12$, $|S(n)| \geq \frac{1}{4}p(\lfloor \frac{n}{4} \rfloor)$
Proof For $n \geq 12$, we know that $|S(n)| \geq \sum_{q=3}^{\lfloor \frac{n}{3} \rfloor} |S(n,q)|$. Taking only the $m = 0$ terms of $S(n,q)$ for each $q \geq 3$, and using the fact that $n \geq 12 \Rightarrow \lceil \frac{n}{4} \rceil \geq 3$, we get

$$|S(n)| \geq \sum_{q=3}^{\lfloor \frac{n}{3} \rfloor} \sum_{j=1}^{q} p_j(n-3q) \geq \sum_{q=\lceil \frac{n}{4} \rceil}^{\lfloor \frac{n}{3} \rfloor} \sum_{j=1}^{q} p_j(n-3q)$$

If $q \geq \frac{n}{4}$, then $q \geq n - 3q$ and we have

$$|S(n)| \geq \sum_{q=\lceil \frac{n}{4} \rceil}^{\lfloor \frac{n}{3} \rfloor} \sum_{j=1}^{n-3q} p_j(n-3q) = \sum_{q=\lceil \frac{n}{4} \rceil}^{\lfloor \frac{n}{3} \rfloor} p(n-3q)$$
$$\geq p(n - 3\lceil \tfrac{n}{4} \rceil)$$
$$\geq p(\lfloor \tfrac{n}{4} \rfloor - 2)$$

Since $n \geq 12 \Rightarrow \lfloor \frac{n}{4} \rfloor - 2 \geq 1$, we can use Lemma 5.3 to get

$$|S(n)| \geq \tfrac{1}{2}p(\lfloor \tfrac{n}{4} \rfloor - 1) \geq \tfrac{1}{4}p(\lfloor \tfrac{n}{4} \rfloor). \quad \blacksquare$$

There is no closed form expression known to compute either $p_k(n)$ or $p(n)$. We use the following result to bound the number of obstructions.

Lemma 5.5 (see page 70 of [An]) As $n \to \infty$, $p(n) \sim \frac{e^{c\sqrt{n}}}{4n\sqrt{3}}$ where $c = \pi\sqrt{\frac{2}{3}}$.

From Theorem 5.4 and Lemma 5.5, it is clear that $|S(n)|$ is asymptotically exponential in $c\sqrt{n} - \log n$. This completes the proof of consequence C.

6 Conclusions

We presented a tight lower bound γ for the treewidth of a graph. The role of γ in helping characterize the dense obstructions was significant. The obstructions in $S(w+3)$ were previously not known to be obstructions for $\text{TW}(w)$ or $\text{PW}(w)$. The rapid growth of the number of obstructions of order $w+3$ with increasing w poses a formidable new challenge in the design of practical algorithms to decide membership in $\text{TW}(w)$ based on testing for minor containment of obstructions alone. One way to surmount this potential difficulty would be to develop general tests for entire families of structurally related obstructions rather than testing for each obstruction individually. Better lower and upper bounds will also be useful. We are currently exploring some of these avenues.

Acknowledgements

We thank Rajeev Govindan, Nancy Kinnersley, and Mike Langston for their helpful comments and encouragement during this work.

References

[ACP] S. Arnborg, D. Corneil, and A. Proskurowski, "Complexity of finding embeddings in a k-tree," *SIAM J. Alg. Disc. Meth.* 8 (1987), 277–284.

[ALS] S. Arnborg, J. Lagergren, and D. Seese, "Problems easy for tree-decomposable graphs," *Journal of Algorithms* 12 (1991), 308–340.

[An] G. E. Andrews, "The Theory of Partitions," in Gian-Carlo Rota, Editor, Encyclopedia of Mathematics and its Applications, Vol. 2, Addison-Wesley, 1976.

[AP] S. Arnborg and A. Proskurowski, "Linear time algorithms for $\mathcal{NP}$-hard problems restricted to partial k-trees," *Discrete Applied Math.* 23 (1989), 11–24.

[APC] S. Arnborg, A. Proskurowski, and D. Corneil, "Forbidden minors characterization of partial 3-Trees," *Discrete Mathematics* 80 (1990), 1–19.

[B1] H. L. Bodlaender, "A linear time algorithm for finding tree-decompositions of small treewidth," *Proceedings, 25th ACM Symposium on Theory of Computing* (1993), 226–234.

[B2] H. L. Bodlaender, "A tourist guide through treewidth," *Acta Cybernetica* 11 (1993), 1–23.

[BK] H. L. Bodlaender and T. Kloks, "Better algorithms for the pathwidth and treewidth of graphs," *Proceedings, 18th ICALP*, Lecture Notes in Computer Science 510 (1991), 544–555.

[DS] W. W-M. Dai and M. Sato, "Minimal forbidden minor characterization of planar 3-trees and application to circuit layout," *Proceedings, IEEE International Symposium on Circuits and Systems* (1990), 2677–2681.

[FL1] M. R. Fellows and M. A. Langston, "Nonconstructive tools for proving polynomial time decidability," *Journal of the ACM*, 35:3 (1988), 727–739.

[FL2] M. R. Fellows and M. A. Langston, "An analogue of the Myhill-Nerode theorem and its use in computing finite-basis characterizations," *Proceedings, 30th Symposium on Foundations of Computer Science* (1989), 520–525.

[GLR] R. Govindan, M. Langston, and S. Ramachandramurthi, "A practical approach to layout optimization," *Proceedings, 6th International Conference on VLSI Design* (1993), 222–225.

[KL] N. G. Kinnersley and M. A. Langston, "Obstruction set isolation for the Gate Matrix Layout problem," *Annals of Discrete Mathematics*, to appear.

[KT] A. Kornai and Z. Tuza, "Narrowness, pathwidth, and their application in natural language processing," *Discrete Applied Mathematics* 36 (1992), 87–92.

[La] J. Lagergren, "An upper bound on the size of an obstruction," in Graph Structure Theory, N. Robertson and P. Seymour (editors), *Contemporary Mathematics* 147 (1993), 601–621.

[LA] J. Lagergren and S. Arnborg, "Finding minimal forbidden minors using a finite congruence," *Proceedings, 18th ICALP*, Lecture Notes in Computer Science 510 (1991), 533–543.

[Re] B. Reed, "Finding approximate separators and computing treewidth quickly," *Proceedings, 24th ACM Symposium on Theory of Computing* (1992), 221–228.

[RS1] N. Robertson and P. D. Seymour, "Graph Minors II. Algorithmic aspects of treewidth," *Journal of Algorithms* 7 (1986), 309–322.

[RS2] N. Robertson and P. D. Seymour, "Graph Minors IV. Tree-Width and Well-Quasi-Ordering," *J. of Combinatorial Theory, Series B* 48 (1990), 227–254.

[RS3] N. Robertson and P. D. Seymour, "Graph Minors XIII. The Disjoint Paths Problem," manuscript (1986).

[Sa] D. P. Sanders, "On linear recognition of treewidth at most four," manuscript (1992).

[ST] A. Satyanarayana and L. Tung, "A characterization of partial 3-trees," *Networks* 20 (1990), 299–322.

[Ya] X. Yan, "A relative approximation algorithm for computing the pathwidth," Master's Thesis, Department of Computer Science, Washington State University, Pullman (1989).

Tree-width and Path-width of Comparability Graphs of Interval Orders

Renate Garbe

Department of Applied Mathematics, University of Twente, P.O.Box 217
NL-7500 AE Enschede, The Netherlands, garbe@math.utwente.nl

Abstract. The problem to decide whether the tree-width of a comparability graph is less than k is NP-complete, if k is part of the input. We prove that the tree-width of comparability graphs of interval orders can be determined in linear time and that it equals the path-width of the graph. Our proof is constructive, i.e., we give an explicit path decomposition of the graph.

1 Introduction

The *tree-width* of a graph can be defined as the smallest number k such that the graph can be embedded into a k-tree, where a *k-tree* is a graph that can be reduced to the complete graph on k vertices by a sequence of removals of vertices of degree k whose neighborhood is completely connected. Graphs with tree-width k are also known as *partial k-trees*.

Equivalently to the above definition the tree-width of a graph G can be defined to be the smallest number k such that G can be embedded into a *chordal graph* with *clique number* k (cf. [ACP 87]). If the chordal graph is required to be an *interval graph* we get the definition of the *path-width* of a graph G. Because in general the path-width of a graph exceeds its tree-width an interesting problem is to investigate whether there are classes of graphs whose path-width and tree-width are equal.

The notions of tree-width and path-width were introduced by Robertson and Seymour [RoSe 83] and [RoSe 86] in the framework of graph minor theory. They showed by non-constructive methods that for fixed k the problem of determining whether a graph has tree-width less than k is polynomially solvable.

Their result was improved by Bodlaender and Kloks [BoKl 91] who gave an algorithm of order $O(n \log^2 n)$ for each constant k to determine whether the tree-width or path-width of a given graph is less than k. The algorithm, however, depends exponentially on k.

Arnborg, Corneil and Proskurowski [ACP 87] proved that if k is part of the problem instance then the problems to decide whether a graph G has tree-width or path-width less than k is NP-complete.

Graphs with bounded tree-width or path-width are of special interest because many problems on graphs that are intractable in general can be solved by polynomial algorithms when restricted to graphs with bounded tree-width or path-width.

See for example the *List Coloring* Problem, a generalization of the problem of determining the *chromatic number* of a graph, given by Jansen and Scheffler [JS 1992]. More examples and a survey on tree-width and path-width are given in Proskurowski and Sysło [PrS 90] and Bodlaender [Bo 92].

In [JS 1992] is proved that List Coloring is *NP*-complete when restricted to complete bipartite graphs. Hence it is *NP*-complete for comparability graphs of interval orders. But for graphs with fixed tree-width k there exists an algorithm whose running time is bounded by a polynomial of degree $k + 2$ in the number of vertices of the graph.

In the special case of comparability graphs of *series-parallel orders* Bodlaender and Möhring [BoM 93] showed that the tree-width can be determined by a linear time algorithm and that in this case tree-width equals path-width.

Our main result asserts that also the tree-width of comparability graphs of *interval orders* can be computed by a linear time algorithm and path-width equals tree-width for this class of graphs. "Linear" means linear in the size of the directed acyclic graph inducing the interval order.

In general path-width does not equal tree-width when we restrict to comparability graphs because we give an example of a comparability graph with different path-width and tree-width. This is in remarkable contrast with a result of Habib and Möhring [HM 92] who proved that in the case of *cocomparability graphs*, namely complements of comparability graphs, tree-width equals always path-width.

The result in [HM 92] has been extended to the class of *asteroidal triple free graphs* (cf. Möhring [M 93]). Comparability graphs of interval orders, however, are not necessarily asteroidal triple free.

Our algorithmic results improve a result of Kloks [Kl 93] who proves that the tree-width of *co-chordal graphs*, i.e., complements of chordal graphs, can be determined by an $O(n^2)$ and the path-width of co-chordal graphs by a polynomial algorithm.

2   Definitions and Notations

We treat (partial) orders induced by directed acyclic graphs. Let $G = (V, E)$ be a directed acyclic graph. The adjacency structure of G induces an order $P(G) = (V, <)$ on V by

$$x < y \quad \text{if} \quad (x, y) \in E.$$

Denote for each element $x \in V$ by $N(x) \stackrel{\text{def}}{=} \{y \in V \mid x < y\}$ its successor set and its predecessor set by $N^{-1}(x) \stackrel{\text{def}}{=} \{y \in V \mid y < x\}$.

The order $P(G)$ is called an *interval order* if it allows an interval representation $\{I_x\}_{x \in V}$ with real compact intervals, such that for all $x, y \in V$ we have:

$$x < y \text{ in } P(G) \text{ if and only if } r(I_x) < l(I_y) \text{ in } \mathbb{R},$$

where $r(I)$ resp. $l(I)$ denotes the right resp. left endpoint of an interval I.

It is well known (cf. Papadimitriou and Yannakakis [PY 79]) that each of the following two properties also characterizes interval orders:

1. If $x < u$ and $y < w$, then either $x < w$ or $y < u$.
2. The elements of V can be arranged into an *N-maximal decomposition sequence* $x_1 x_2 \ldots x_n$, such that $N(x_1) \supseteq N(x_2) \supseteq \ldots \supseteq N(x_n)$,

where $n \overset{\text{def}}{=} |V|$ is the number of vertices.

Given a partial order $P = (V, <)$ its *comparability graph* $C(P) = (V, E)$ is the undirected graph that has vertex set V and edge set E given by

$$xy \in E \text{ if and only if either } x < y \text{ or } y < x.$$

Comparability graphs of interval orders are the complements of *interval graphs* (cf. Golumbic [Go 80]). An undirected graph $G = (V, E)$ is an *interval graph* if it allows an interval representation $\{I_x\}_{x \in V}$ of real compact intervals such that

$$xy \in E \text{ if and only if } I_x \cap I_y \neq \emptyset.$$

A combinatorial characterization of interval graphs is given by the following Theorem.

Theorem 1 (Gilmore and Hofman [GiHo 64]). *Let G be an undirected graph. The following statements are equivalent.*

1. *G is an interval graph.*
2. *The maximal cliques of G can be linearly ordered such that, for every vertex x of G, the maximal cliques containing x occur consecutively.*

The following definitions of tree-width and path-width of an undirected graph were introduced by Robertson and Seymour [RoSe 83] and [RoSe 86].

Definition 2. A tree decomposition of an undirected graph $G = (V, E)$ is a pair $(\{X_i \mid i \in I\}, T = (I, F))$ with $\{X_i \mid i \in I\}$ a family of subsets of V and T a tree with vertex set I and edge set $F \subseteq I \times I$, such that

1. $\bigcup_{i \in I} X_i = V$
2. for all edges $(v, w) \in E$ there exists $i \in I$ such that $v \in X_i$ and $w \in X_i$
3. for all $i, j, k \in I$ such that j lies on a path in T from i to k holds $X_i \cap X_k \subseteq X_j$

The *width* of a tree decomposition is $\max_{i \in I} |X_i| - 1$. The *tree-width of a graph* G is the minimum width over all tree decompositions of G. The tree-width of G is denoted by $tw(G)$.

In the special case that the tree T in Definition 2 is a path we get a *path decomposition* of the graph.

Definition 3. A path decomposition of an undirected graph $G = (V, E)$ is a sequence of subsets of vertices $(X_1, X_2, \ldots, X_r)$, such that

1. $\bigcup_{1 \le i \le r} X_i = V$.
2. for all edges $(v, w) \in E$ there exists $i \in \{1, 2, \ldots, r\}$ such that $v \in X_i$ and $w \in X_i$.
3. for all $i, j, k \in \{1, 2, \ldots, r\}$: if $i \le j \le k$ then $X_i \cap X_k \subseteq X_j$.

The *width* of a path decomposition is $\max_{1 \le i \le r} |X_i| - 1$. The *path-width of a graph* G is the minimum path-width over all path decompositions of G. The path-width of G is denoted by $pw(G)$.

Each path decomposition is also a tree decomposition of a graph G. Thus

$$tw(G) \le pw(G). \tag{1}$$

An undirected graph G is *chordal* if every cycle of length strictly greater than 3 possesses a chord, that is, an edge joining two non consecutive vertices of the cycle.

The tree-width respectively the path-width of a graph G is related to the clique number of the smallest chordal respectively interval graph into that the graph can be embedded, as follows (cf. Bodlaender [Bo 92])

$$tw(G) = \min\{\omega(H) \mid H \text{ is a chordal graph and } G \subseteq H\} - 1 \tag{2}$$
$$pw(G) = \min\{\omega(H) \mid H \text{ is an interval graph and } G \subseteq H\} - 1 \tag{3}$$

Where $\omega(G)$ is the *clique number* of G, that is the number of vertices in a maximal subset of pairwise adjacent vertices of G.

Next we state known results that we need for the proof of our results.

Lemma 4 (Clique Containment Lemma). *Let* $(\{X_i \mid i \in I\}, T = (I, F))$ *be a tree-decomposition of* $G = (V, E)$ *and let* $W \subseteq V$ *be a clique in* G. *Then there exists* $i \in I$ *such that* $W \subseteq X_i$.

Lemma 4 implies that $tw(G) \ge \omega(G) - 1$.

Lemma 5 (Complete Bipartite Subgraph Containment Lemma). *Let* $(\{X_i \mid i \in I\}, T = (I, F))$ *be a tree-decomposition of* $G = (V, E)$. *Let* $A, B \subseteq V$ $A \cap B = \emptyset$ *and suppose* $\{(v, w) \mid v \in A, w \in B\} \subseteq E$. *Then there exists* $i \in I$ *such that* $A \subseteq X_i$ *or* $B \subseteq X_i$.

Lemma 5 implies that $tw(G) \ge \min(|A|, |B|) - 1$.

For a proof of Lemmata 4 and 5 see e.g. Bodlaender and Möhring [BoM 93].

3 Tree-width and Path-width of Comparability Graphs of Interval Orders

First we show that in general the tree-width of comparability graphs of interval orders is not bounded by a constant. Consider for example the interval order that has the complete bipartite graph $G = (V = V_1 \cup V_2, V_1 \times V_2)$, with $|V_1| = |V_2|$, as its comparability graph. By Lemma 5 we have $tw(G) \geq |V|/2 - 1$.

Next we give an example that shows that in general $tw(G) < pw(G)$ may hold for a comparability graph G.

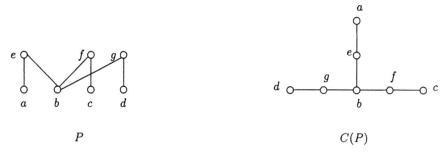

$$P \qquad\qquad\qquad C(P)$$

Fig. 1.

Take the comparability graph $C(P)$ of the order P in Figure 1. It is chordal and has clique number 2. Thus by (2) $tw(G) = 1$.

All maximal cliques of $C(P)$ are also maximal cliques of an interval graph with clique number 2 containing $C(P)$. But the maximal cliques of $C(P)$

$$\{d,g\}, \{g,b\}, \{b,f\}, \{b,e\}, \{a,e\} \text{ and } \{f,c\}$$

can not be linearly ordered such that for each vertex x in $C(P)$ the maximal cliques containing x occur consecutively. Thus by Theorem 1 an interval graph containing $C(P)$ has clique number at least 3.
By (3) we get $pw(C(P)) \geq 2 > 1 = tw(C(P))$.

In the case of comparability graphs of interval orders we have

Theorem 6. Let $G = (V, E)$ be a directed acyclic graph such that $P(G)$ is an interval order. The tree-width of the comparability graph $C(P)$ of $P(G)$ can be determined by an algorithm of order $O(|V| + |E|)$ and

$$tw(C(P)) = pw(C(P)).$$

The proof of Theorem 6 consists of three parts. First we prove Lemma 8 that gives a lower bound, say B, on the tree-width of $C(P)$.

In the second part we show that there exists a path decomposition of $C(P)$ with width equal to B, i.e.,

$$B \leq tw(C(P)) \leq pw(C(P)) \leq B.$$

Thus $tw(C(P)) = pw(C(P)) = B$.

The number B we determine in the first part of the proof provides us with a formula for the tree-width of $C(P)$. In the third part of the proof we show that this formula enables us to compute the tree-width of $C(P)$ by an algorithm of order $O(|V| + |E|)$.

In order to derive a lower bound on the tree-width of $C(P)$ we will use the Complete Bipartite Subgraph Containment Lemma, i.e., we have to look for complete bipartite subgraphs in $C(P)$. Below we show that for each $x \in V$ there exists a complete bipartite subgraph in $C(P)$.

Denote for each $x \in V$ the set of vertices with strictly more successors than x and the set of vertices with exactly the same successors than x by

$$G(x) \stackrel{\text{def}}{=} \{y \in V \mid N(y) \supset N(x)\} \quad \text{and} \quad Z(x) \stackrel{\text{def}}{=} \{y \in V \mid N(y) = N(x)\}.$$

Because for each $y \in G(x) \cup Z(x)$ holds $N(x) \subseteq N(y)$ the sets

$$G(x) \cup Z(x) \quad \text{and} \quad N(x)$$

form a complete bipartite subgraph in $C(P)$ and

$$(Z(x) \cup G(x)) \cap N(x) = \emptyset.$$

Let $(\{X_i \mid i \in I\}, T = (I, F))$ be a tree decomposition of $C(P)$ with minimum width, i.e., $tw(C(P)) = \max_{i \in I} |X_i| - 1$. By the Complete Bipartite Subgraph Containment Lemma 5 for each $x \in V$ there exists $i \in I$ such that

$$N(x) \subseteq X_i \quad \text{or} \quad G(x) \cup Z(x) \subseteq X_i. \tag{4}$$

Let now w be an element with largest number of successors such that there is $i \in I$ with

$$N(w) \subset X_i.$$

Let $z \in V$ such that $N(z) \supset N(x)$ and there is no $y \in V$ such that $N(z) \supset N(y) \supset N(x)$. By the choice of w, we have $N(z) \not\subseteq X_i$ for all $i \in I$. Thus by (4) that there is $j \in I$ such that

$$G(w) = Z(z) \cup G(z) \subseteq X_j.$$

Let H be the minimal graph containing $C(P)$ such that for all $i \in I$ the sets X_i form a clique. Clearly, $tw(H) \leq tw(C(P))$ because the tree decomposition of $C(P)$ is also a tree decomposition of H.

Because $N(w) \subseteq X_i$ for some $i \in I$ and $G(w) \subseteq X_j$ for some $j \in I$ and $G(w)$ and $N(w)$ are disjoint and form a complete bipartite subgraph in $C(P)$ we get

$$G(w) \cup N(w)$$

forms a clique in H. Thus by the Clique Containment Lemma 4 we get

$$tw(C(P)) \geq tw(H) \geq |G(w) \cup N(w)| - 1 = |G(w)| + |N(w)| - 1.$$

If there is $y \in Z(w)$ with $N^{-1}(y) = G(w)$, then the set

$$G(w) \cup N(w) \cup \{y\}$$

forms a clique. Thus by the Clique Containment Lemma 4 we have

$$tw(C(P)) \geq tw(H) \geq |N(w) \cup G(w) \cup \{y\}| - 1 = |N(w)| + |G(w)|.$$

Thus we have to distinguish two types of elements in V.
We say x is of *first type* if

$$N^{-1}(y) \subset G(x) \text{ for all } y \in Z(x) \tag{5}$$

and of *second type* if

$$N^{-1}(y) = G(x) \text{ for some } y \in Z(x). \tag{6}$$

That each element in V is either of first or of second type, i.e., for all $x \in V$ there is no $y \in Z(x)$ with $N^{-1}(y) \not\subseteq G(x)$, is guaranteed by Lemma 7 below.

Lemma 7. *For all $x \in V$ and all $y \in V \setminus (N(x) \cup G(x))$ holds*

1. $N(y) \subseteq N(x)$
2. $N^{-1}(y) \subseteq G(x)$

Proof of Lemma 7 : Ad 1.: If $y \notin G(x)$ then $N(y) \not\supset N(x)$. Thus $N(y) \subseteq N(x)$ because $P(G)$ is an interval order.

Ad 2.: Suppose there is $z \in N^{-1}(y)$ such that $z \notin G(x)$. But then $y \in N(z) \subseteq N(x)$, contradicting $y \in V \setminus (N(x) \cup G(x))$.$\Box$

Define for each $x \in V$, $g(x) \stackrel{\text{def}}{=} |G(x)|$, $n(x) \stackrel{\text{def}}{=} |N(x)|$ and

$$d(x) \stackrel{\text{def}}{=} \begin{cases} g(x) + n(x) - 1 & \text{if } x \text{ is of first type} \\ g(x) + n(x) & \text{if } x \text{ is of second type.} \end{cases}$$

From the above follows

Lemma 8. $tw(C(P)) \geq \min_{x \in V} d(x).$

Next we prove that $\min_{x \in V} d(x)$ is an upper bound on the path-width of $C(P)$.

Lemma 9. $pw(C(P)) \leq \min\limits_{x \in V} d(x).$

Combining Lemma 8 and Lemma 9 we get

$$tw(C(P)) = pw(C(P)) = \min\limits_{x \in V} d(x)$$

which proves the first part of Theorem 6.

Proof of Lemma 9: In order to prove Lemma 9 we show that for each $x \in V$ there exists a path decomposition of $C(P)$ of width $d(x)$.

For each $x \in V$ the family of subsets forming the path decomposition determined by x depends on the vertices in $V \setminus (N(x) \cup G(x))$ as we will see below. Each vertex in $V \setminus (N(x) \cup G(x))$ lies in exactly one of these subsets and if x is of first type then an additional subset is given by $N(x) \cup G(x)$.

First we consider $x \in V$ of first type. Because in this case for all $y \in Z(x)$ we have $N^{-1}(y) \subset G(x)$ there exists v_1 such that for all $y \in Z(x)$ we have

$$N^{-1}(y) \subseteq G(x) \setminus \{v_1\}$$

The set of vertices in V that have strictly less successors than x but are no successors of x is given by

$$U(x) \stackrel{\text{def}}{=} V \setminus (Z(x) \cup G(x) \cup N(x)) = \{y \in V \mid N(y) \subset N(x) \text{ and } y \notin N(x)\}.$$

Because for all $y \in U(x)$ holds $N(y) \subset N(x)$ there exists $v_2 \in N(x)$ such that for all $y \in U(x)$ holds

$$N(y) \subseteq N(x) \setminus \{v_2\}.$$

Define $z(x) \stackrel{\text{def}}{=} |Z(x)|$ and $u(x) \stackrel{\text{def}}{=} |U(x)| + z(x)$ and let

$$Z(x) = \{z_1, z_2, \ldots, z_{z(x)}\} \text{ and } U(x) = \{z_{z(x)+2}, \ldots, z_{u(x)+1}\}.$$

Define for $i = 1, 2, \ldots, z(x)$

$$X_i \stackrel{\text{def}}{=} N(x) \cup (G(x) \setminus \{v_1\}) \cup \{z_i\},$$

for $i = z(x) + 2, \ldots, u(x) + 1$

$$X_i \stackrel{\text{def}}{=} (N(x) \setminus \{v_2\}) \cup G(x) \cup \{z_i\}$$

and

$$X_{z(x)+1} \stackrel{\text{def}}{=} N(x) \cup G(x).$$

We claim that a path decomposition of $C(P)$ is then given by the sequence

$$(X_1, X_2, \ldots, X_{u(x)+1}).$$

Clearly,

$$\bigcup_{i=1}^{u(x)+1} X_i = N(x) \cup G(x) \cup \left(\bigcup_{z \in V \setminus (N(x) \cup G(x))} \{z\} \right) = V$$

which proves that the first condition in Definition 3 holds.

Next we prove that the second condition in Definition 3 holds. Let zy be an edge of $C(P)$, i.e, either $z < y$ or $y < z$. If both $z, y \in G(x) \cup N(x) = X_{z(x)+1}$ then there is nothing to prove. Consider now the case that $z \notin N(x) \cup G(x)$. If $z \in Z(x)$, i.e., $z = z_i$ for some i with $1 \le i \le z(x)$, then, because either $y > z_i$ or $y < z_i$ and by Lemma 7

$$y \in N(z_i) \cup N^{-1}(z_i) \subseteq N(x) \cup (G(x) \setminus \{v_1\}).$$

Thus

$$z_i, y \in N(x) \cup (G(x) \setminus \{v_1\}) \cup \{z_i\} = X_i.$$

If $z \in U(x)$, i.e., $z = z_i$ for some i with $z(x) + 2 \le i \le u(x) + 1$ then

$$y \in N(z_i) \cup N^{-1}(z_i) \subseteq (N(x) \setminus \{v_2\}) \cup G(x)$$

which implies

$$z_i, y \in (N(x) \setminus \{v_2\}) \cup G(x) \cup \{z_i\} = X_i.$$

Thus also the second condition in Definition 3 holds.

Definition 3 holds, if we show that for each $y \in V$ the subsets of the path decomposition containing y occur consecutively.
For each $y \in (N(x) \setminus \{v_2\}) \cup (G(x) \setminus \{v_1\})$ holds

$$y \in X_i \quad \text{if and only if} \quad 1 \le i \le u(x) + 1.$$

$$v_2 \in X_i \quad \text{if and only if} \quad 1 \le i \le z(x) + 1.$$

$$v_1 \in X_i \quad \text{if and only if} \quad z(x) + 1 \le i \le u(x) + 1.$$

And, finally, for each $y \in V \setminus (N(x) \cup G(x))$ there exists exactly one i such that $y \in X_i$. Thus also the third condition in Definition 3 holds.

Because we have

$$\max_{1 \le i \le u(x)+1} |X_i| = n(x) + g(x) = d(x) + 1$$

the width of the path decomposition equals $d(x)$.

Consider now the case that x is of second type. Let

$$Z(x) \cup U(x) = \{z_1, z_2, \ldots, z_{u(x)}\}$$

and define for $i = 1, 2, \ldots, u(x)$

$$X_i \stackrel{\text{def}}{=} N(x) \cup G(x) \cup \{z_i\}.$$

A path decomposition of $C(P)$ is then given by the sequence

$$(X_1, X_2, \ldots, X_{u(x)}).$$

Similarly to the first case we can prove that the conditions in Definition 3 are fulfilled. For $\max_{1 \le i \le u(x)} |X_i| = n(x) + g(x) + 1 = d(x) + 1$ also in this case the width of the tree decomposition equals $d(x)$.$\square$

Given a path decomposition of $C(P)$ of minimum width, an interval graph H such that $H \supseteq C(P)$ and $\omega(H) = pw(C(P)) + 1$ can be constructed by adding edges such that each set in the family of subsets of the path decomposition is a clique. An interval representation of H can be constructed by assigning to each $x \in V$ a real compact interval I_x (cf. Figure 2).

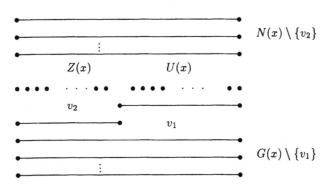

Fig. 2. Interval representation of an interval graph containing $C(P)$

In order to compute the tree-width of $C(P)$ we have to determine for each $x \in V$ the number $d(x)$. That means we have to compute for each $x \in V$ the numbers $n(x)$, $g(x)$ and to decide whether x is of second or of first type.
If we know an N-maximal decomposition sequence

$$x_1 x_2 \ldots x_n$$

of $P(G)$ and its transitive reduction, i.e, for each $x \in V$ the set

$$H(x) \overset{\text{def}}{=} \{y \in V \mid (x,y) \in E \text{ and there is no } z \in V \text{ with } (x,z),(z,y) \in E\},$$

then this can be done by an algorithm of order $O(|V| + |E|)$ as follows. The numbers $n(x)$ and $g(x)$ for each $x \in V$ can then be computed in $O(|V| + |E|)$ time by the following recursions.

$$n(x_n) = 0 \text{ and } n(x_i) = n(x_{i+1}) + |H(x_i) \setminus H(x_{i+1})| \text{ for } i = n-1,\ldots,2,1$$

and $g(x_1) = 0$ and

$$g(x_i) = \begin{cases} g(x_{i-1}) & \text{if } n(x_i) = n(x_{i-1}) \\ g(x_{i-1}) + z(x_{i-1}) & \text{if } n(x_i) < n(x_{i-1}) \end{cases} \quad \text{for } i = 2, 3, \ldots, n.$$

Because x is of first type if and only if $|N^{-1}(y)| < g(x)$ for all $y \in V$ with $N(y) = N(x)$ and the number of predecessors of the elements in V can be computed in $O(|V|+|E|)$ time, we can decide in $O(|V|+|E|)$ time which elements are of first type.

This completes the proof of Theorem 6 because an N-maximal decomposition sequence and the transitive reduction of an interval order $P(G)$ can be computed by an algorithm of order $O(|V| + |E|)$ (cf. [Ga 81] or [Gb 94]).

4   Remark

We are able to generalize Theorem 6 as follows (cf. [Gb 94]).
An order $P = (V, <)$ is a *generalized interval order* if for all $x, y \in V$ holds

$$N(x) \cap N(y) \neq \emptyset \text{ implies } N(x) \subseteq N(y) \text{ or } N(y) \subseteq N(x).$$

The order P in Figure 3 is an example of a generalized interval order. Thus for comparability graphs of generalized interval orders may hold

$$tw(C(P)) \neq pw(C(P)).$$

By similar methods as used in the proof of Theorem 6 we can prove that for a generalized interval order P holds

$$pw(C(P)) \leq tw(C(P)) + 1$$

and that also in this case tree-width and path-width can be computed in linear time. Moreover, we give a necessary and sufficient condition for a generalized interval order to have equal tree-width and path-width.

References

[ACP 87] S. Arnborg, D.G. Corneil, and A. Proskurowski (1987). Complexity of finding embeddings in a k-tree. *SIAM J. Alg. Disc.Meth.*, vol. 8, No.2, 277-284.

[Bo 92] H.L. Bodlaender (1993). A tourist guide through treewidth. *Acta Cybernetica*, Vol. 11, No. 1-2, Szeged.

[BoKl 91] H.L. Bodlaender and T.Kloks (1991). Better algorithms for the pathwidth and treewidth of graphs. In *Proceedings of the 18th Colloquium on Automata, Languages and Programming*, Springer Verlag, Lecture Notes in Computer Sciences, vol. 510, 544-555.

[BoM 93] H.L. Bodlaender and R.H. Möhring (1993). The path-width and tree-width of cographs. *SIAM J. Discrete Math.*, vol. 6, 181-188.

[Ga 81] H.N. Gabow (1981). A linear-time recognition algorithm for interval dags. *Information Processing Letters*, vol. 12, No. 1, 20-22.

[Gb 94] R. Garbe (1994). Algorithmic Aspects of Interval Orders. Ph.D. Thesis, University of Twente, The Netherlands. To appear in October 1994.

[GiHo 64] P.C. Gilmore and A.J. Hoffman (1964). A characterization of comparability graphs and of interval graphs. *Cand. J. Math.*, vol. 16, 539-548.

[Go 80] M.C. Golumbic (1980). *Algorithmic Graph Theory and Perfect Graphs*. Academic Press, New York.

[HM 92] M. Habib and R. H. Möhring (1992). Treewidth of cocomparability graphs and a new order-theoretic parameter. Technical report 336, Technische Universität Berlin. To appear in *Order*.

[JS 1992] K. Jansen and P. Scheffler (1992). Generalized coloring for tree-like graphs. Proceedings of the 18th International Workshop on Graph-Theoretic Concepts in Computer Science, Lecture Notes in Computer Science 657, 50-59.

[Kl 93] T.Kloks (1993). Tree-width, Ph.D. Thesis, Utrecht University, The Netherlands.

[M 93] R.H. Möhring (1993). Triangulating graphs without Asteroidal Triples. Technical report No. 365/1993, Technische Universität Berlin.

[PY 79] C.H. Papadimitriou and M. Yannakakis (1979). Scheduling interval ordered tasks. *SIAM J. Computation* , 8, 405-409.

[PrS 90] A. Proskurowski and M.M. Sysło (1989). Efficient Computations in Tree-Like graphs. In G. Tinhofer, E. Mayr, H. Noltemeier and M. Sysło (eds.), *Computational Graph Theory* Springer Verlag, Wien New York, 1-15.

[RoSe 83] N.Robertson and P.D. Seymour (1983). Graph minors. I. Excluding a forest. *J. Comb. Theory Series B*, vol. 35,39-61.

[RoSe 86] N. Robertson and P.D. Seymour (1986). Graph minors. II. Algorithmic aspects of tree-width. *J. Algorithms,* vol. 7, 309-322.

A Declarative Approach to Graph Based Modeling

Jürgen Ebert[1] and Angelika Franzke[2]

[1] Fachbereich Informatik, Universit"at Koblenz-Landau, Abt. Koblenz,
`ebert@informatik.uni-koblenz.de`
[2] Fachbereich Informatik, Universit"at Koblenz-Landau, Abt. Koblenz,
`franzke@informatik.uni-koblenz.de`

Abstract. The class of TGraphs, i.e. typed, attributed, and ordered directed graphs, is introduced as a general graph class for graph based modeling. TGraphs are suitable for a wide area of applications. A declarative approach to specifying subclasses of TGraphs by a combination of a schematic graphical description and an additional constraint language is given. The implementation of TGraphs by an appropriate software approach is described.

1 Introduction

Modeling parts of the real world is a wide area of applications of graph theoretic concepts in practice. E.g. in the software design process the structure of the information to be kept and handled by the system is often describable by graphs. In order to describe which graphs are correct models and which are not, an approach to describing classes of graphs (graph languages) is necessary. Since graphs are to the same extent

▷ expressive pictures,
▷ formal models, and
▷ efficient data structures,

a well-based approach to working with graphs is necessary, which covers all three aspects in a consistent manner.

In this paper, we introduce the class of TGraphs as a general graph class, which by being very general is suitable for a wide area of applications. We give a declarative approach to specifying subclasses of TGraphs by a combination of a schematic graphical description and an additional constraint language, and we show how TGraphs can be implemented appropriately. The declarative approach for describing graph classes, allows to check the correctness of a graph at any point of time. Here the constraints are forced by the graph module itself and not by the application program.

The paper is organized as follows: In section 2 the class of TGraphs is described and a formal definition is given[3]. Based on this definition, section 3 introduces our approach to graph based modeling. We describe a specification

---

[3] Throughout the text, a $\mathcal{Z}$-like notation is used. For an introduction to $\mathcal{Z}$, the reader is refered to [Diller 90] or [Spivey 92].

language suitable to describe TGraph languages - i.e. subclasses of the general class - from a conceptual point of view, by using extended entity-relationship modeling in section 4 and an appropriate constraint description language in section 5. Section 6 certifies that the formal and conceptual descriptions may be implemented accordingly. Section 7 compares this work with related approaches, especially with graph grammar techniques. The approach is illustrated by an example chosen from a software engineering context, namely by a class of graphs that are models for state charts.

2 Basic Definitions

Graph models are to a certain extent abstractions of an object-oriented view of the world. When using graphs, objects are usually represented by vertices, whereas relationships are modeled by edges. To enhance the modeling power of graphs, vertices and edges may be classified into different types, and (depending on their types) they also may have attributes. To control graph traversal order in the context of graph algorithms an additional order can be put on the edges incident with a given vertex. Thus, a general class for modeling is the class of ordered directed graph with typed and attributed vertices and edges, called **TGraphs** in the following.

An **ordered directed graph** is defined by its set of vertices V and its set of edges E. The incidence relation between vertices and edges is a function Λ, which maps each vertex $v \in V$ to a sequence of all edges $e \in E$ that are incident to v. For each edge e in $\Lambda(v)$ also the direction of e with respect to v (e may be an ingoing or an outgoing edge) is specified. In order for G to be a correct graph, it has to be assured that every edge in E is mapped to exactly one vertex as an outgoing edge and to exactly one vertex as an ingoing edge.

To model **types**, a universe of type identifiers is assumed. An **attribute** (instance) is defined to be an ordered pair consisting of an attribute identifier and a corresponding attribute value. Then, a TGraph is an ordered directed graph extended by two functions. *type* assigns a type identifier to the vertices and edges, whereas *value* is an assignment of finite sets of attribute instances. Here, it is demanded that every vertex and every edge has exactly one type, whereas graph elements may or may not have attribute instances:

$UNIVERSE ::= vertex\langle\!\langle \mathbf{N} \rangle\!\rangle \mid edge\langle\!\langle \mathbf{N} \rangle\!\rangle$
$VERTEX == \mathrm{ran}\ vertex$
$EDGE == \mathrm{ran}\ edge$
$DIR ::= in \mid out$

$[ID, VALUE]$
$typeID == ID$
$attrID == ID$
$attributeInstanceSet == attrID \twoheadrightarrow VALUE$

```
┌─ TGraph ─────────────────────────────────────────────
│  V : F VERTEX
│  E : F EDGE
│  Λ : VERTEX ↛ seq(EDGE × DIR)
│  type : UNIVERSE ↛ typeID
│  value : UNIVERSE ↛ attributeInstanceSet
├──────────────────────────────────────────────────────
│  Λ ∈ V → iseq(E × DIR)
│  ∀ e : E • ∃₁ v, w : V • (e, in) ∈ ran(Λ(v)) ∧ (e, out) ∈ ran(Λ(w))
│  dom type = V ∪ E
│  dom value = V ∪ E
└──────────────────────────────────────────────────────
```

Note, that this definition of a TGraph defines a very general class of graphs. A TGraph may contain loops and multiple edges, it may be interpreted as being undirected or unordered by ignoring the *DIR*-value or the ordering implied by the seq-operator. Edges are first-class objects which allows also edge-oriented algorithms. It is possible to derive most graph definitions used in a practical contexts by loosening the definition of a TGraph.

3 Graph Based Modeling

In the previous section, the general schema for TGraphs has been introduced. These graphs can be used as a modeling structure in many different contexts. We used them e.g. to model many different kinds of languages (string languages and graphical languages) in building software design and software reengineering tools, as well as for representing the subject matter aspects and the learner model inside a tutor system. Furthermore street maps, bus tours and the like were respresented with appropriately defined graph classes inside a tour planning tool.

To model reality by TGraphs, the following principles should be adhered to:

▷ each identifiable and relevant object is represented exactly once by a vertex,
▷ each relationship between objects is represented exactly once by an edge,
▷ similar objects and relationships are assigned a common type,
▷ informations on objects and relationships are stored in those attribute instances that are assigned to the corresponding vertices and edges, and
▷ an ordering of relationships is expressed by edge order.

This approach is very close to object-oriented information modeling of reality. Note, that objects and relationships of a common type will always have the same kind of information assigned and that objects are modeled by vertices whereas their relationships (including occurrences of an object in some context) are modeled by edges.

Example. To illustrate these modeling principles, a TGraph model for state charts is presented in the following. State charts are a graphical language introduced by David Harel ([Harel 87]) for describing reactive systems. They provide

a visual formalism which extends finite state descriptions by introducing hierarchy and orthogonality as additional concepts. Figure 1 shows a sample state
chart.

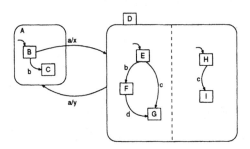

Fig. 1. A Sample State Chart

To model state charts inside a tool, a conceptual analysis shows, that there
are four different kinds of entities, namely *blobs* (rectangles), *transitions* (arcs),
events, and *responses* (actions). *Blobs* may be either *elementary* (like $B, C, E, ...$)
or composite (like A and D). Composite blobs may be *xor-blobs* (like A) containing a set of other (elementary or composite) blobs or they may be *and-blobs* (like
D), which contain at least two unnamed xor-blobs separated by dashed lines.
Each xor-blob (as well as the state chart itself) contains exactly one so-called
start blob, which is marked by a small arrow. Transitions connect blobs and are
annotated by an *event symbol* (like $a, b,$ and c) and an optional *response symbol*
(like x and y), which are separated by a slash character from each other. Some
blobs and all events and responses are described by strings, i.e. they a carry a
string-type attribute.

Following these modeling decisions every concrete state chart may be viewed
as an SCGraph, i.e. as a TGraph, which reflects its abstract syntactical structure.
Figure 2 shows the syntax graph corresponding to the state chart in figure 1.
(For the sake of readability the edge types have only been differentiated coarsely
in this figure.)   ◊

4   Schematic Description

To document the class of legal models according to some modeling decisions, a
suitable specification language for TGraph classes is needed. A graphical specification of the graph class for modeling state charts is shown in figure 3. The
diagram captures all the informations on vertex and edge types gathered so far.
It is an (extended) entity relationship(ER) description. Here a diagram is used
to define a graph class, i.e. the class of TGraphs representing a state chart. Thus,

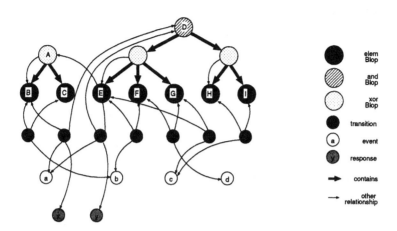

Fig. 2. A TGraph Representation of a State Chart

we use ER-diagrams as a specification language for graph classes[4]. Like EBNF grammars may be used to describe and discuss string languages, ER-diagrams serve as a description of graph languages. Note, that they are not read as a generative scheme, but as a means of declarative description.

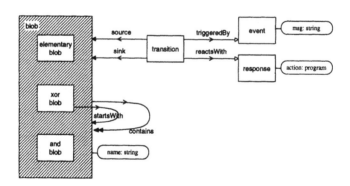

Fig. 3. ER-Description of SCGraphs

The entity types defined in an ER-diagram correspond to vertex types, while relationship types represent edge types. The incidence structure shown in the diagram imposes structural constraints on the class of TGraphs defined thereby.

---

[4] A complete definition of our notation is given in [CaEbWi 94], where also a precise graph based semantics for ER-diagrams is defined, i.e. the class of graphs defined by an ER-diagram is specified.

A TGraph matches this specification if
- ▷ the type of every vertex or edge is defined in the ER-Diagram and
- ▷ for every edge the incidence information defined in the ER-diagram is respected.

4.1 Type Systems

In the following, we will formalize the notion of a TGraph matching a specification given by an ER-diagram. To do this, we introduce the terms **type system** and **incidence system**, which are abstractions of what is described by an ER-diagram. Given these definitions, the relations $\models_{typeSystem}$ and $\models_{incSystem}$ between graphs and type systems are defined. These relations specify whether a given TGraph matches the type restrictions imposed by a type system, as well as the incidence information.

From a modeling point of view, objects of one type share common properties. Thus, when defining a type t, one should also define the attributes that characterize the objects of t. Moreover, it is often useful to build up type hierarchies. Thus, when defining a set of valid types, one has also to specify a subtyping relation. This is reflected by the following definition of the schema *typeSystem*.

$$domID == ID$$
$$attributeSchema == attrID \leftrightarrow domID$$

---
typeSystem _____
$typeDefinitionSet : typeID \twoheadrightarrow attributeSchema$
$isA : typeID \leftrightarrow typeID$

_____
$(\_isA\_) \in \text{dom } typeDefinitionSet \times \text{dom } typeDefinitionSet$

Example. In figure 3 the entity types and the relationship types form the domain of *typeDefinitionSet*. *blobs* have an attribute schema consisting of the attribute identifier *name* paired with the domain identifier *string*. Accordingly *events* and *responses* have non-empty attribute schemes. All other type identifiers have the empty scheme associated. The types *elementaryBlob*, *xorBlob* and *andBlob* are subtypes of *blob*, which is expressed by the inclusion of their respective rectangles.   ◊

A TGraph G matches a given type system, if the following conditions hold:
- ▷ All types that are assigned to vertices or edges are defined in the type system.
- ▷ If a vertex or an edge has type t, then the value of this vertex or edge is correct with respect to the attribute schemes assigned to t and all superclasses of t in the corresponding type system.

If we define t_1 to be a subtype of t_2, we express that every object of type t_1 has also the properties of an object of type t_2, i.e. that the objects of type t_1 inherit the attributes defined for objects of type t_2. The set of all attribute definitions which belong to a type t due to inheritance is described by the following function (here, *union* should be defined appropriately as the union of sets in its argument):

$$allAttributesOfType : typeSystem \rightarrow (typeID \rightarrow attributeSchema)$$

$$
\begin{aligned}
&allAttributesOfType = \\
&\quad \lambda\, typeSystem;\ t : typeID \mid t \in \mathrm{dom}\ typeDefinitionSet\ \bullet \\
&\qquad union(\{\, t' : \mathrm{dom}\ typeDefinitionSet \mid t\ isA^*\ t'\ \bullet\ typeDefinitionSet(t')\,\})
\end{aligned}
$$

The relation $\models_{\mathbf{attrSchema}}$ describes the fact, that the attribute instances of a graph element are correct with respect to an attribute schema. (Note that the definition of $\models_{\mathbf{attrSchema}}$ does *not* imply that there is an instance for every definition in the given schema.)

$$carrier : domID \rightarrow \mathbf{P}\ VALUE$$

$$
\begin{aligned}
&-\models_{\mathbf{attrSchema}} - :\ attributeInstanceSet \leftrightarrow attributeSchema \\
\hline
&\forall I : attributeInstanceSet;\ S : attributeSchema\ \bullet \\
&\quad I \models_{\mathbf{attrSchema}} S \Leftrightarrow \\
&\qquad \forall i : I\ \bullet \\
&\qquad\quad \exists_1 d : S\ \bullet \\
&\qquad\qquad first(i) = first(d)\ \wedge \\
&\qquad\qquad second(i) \in carrier(second(d))
\end{aligned}
$$

This definition implies, that an *attributeInstance* may only have values for attribute identifiers, which appear only once in the *attributeSchema*. Thus, the following definition solves conflicts due to multiple inheritance by forbidding conflicting attributes on the instance level.

$$
\begin{aligned}
&-\models_{\mathbf{typeSystem}} - :\ TGraph \leftrightarrow typeSystem \\
\hline
&\forall TGraph;\ typeSystem\ \bullet \\
&\quad \theta TGraph \models_{\mathbf{typeSystem}} \theta typeSystem \Leftrightarrow \\
&\qquad \mathrm{ran}\ type \subseteq \mathrm{dom}\ typeDefinitionSet \\
&\qquad \forall obj : V \cup E\ \bullet \\
&\qquad\quad value(obj) \models_{\mathbf{attrSchema}} \\
&\qquad\qquad allAttributesOfType(\theta typeSystem)(type(obj))
\end{aligned}
$$

4.2 Incidence Systems

The edges of a TGraph specified by an ER-diagram may be incident only with vertices, whose types fit to the description. The schema *incidenceSystem* specifies restrictions on the incidence structure of a TGraph. Given an edge type t, *incidences*(t) determines the types of the start and goal vertices.

$$
\begin{aligned}
&\_\,incidenceSystem\,\_\_\_ \\
&typeSystem \\
&incidences : typeID \nrightarrow (typeID \times typeID) \\
\hline
&\mathrm{dom}\ incidences \subseteq \mathrm{dom}\ typeDefinitionSet \\
&\mathrm{ran}\ incidences \subseteq \mathrm{dom}\ typeDefinitionSet \times \mathrm{dom}\ typeDefinitionSet
\end{aligned}
$$

The relation $\models_{\text{incSystem}}$ defines whether a given TGraph G satisfies an incidence system. This is the case, if for every edge e of G the typing conditions for its start and its goal vertex are respected. (Here, inheritance has to be taken into account, too.)

$$
\begin{array}{l}
\underline{\quad} \models_{\text{incSystem}} \underline{\quad} : \; TGraph \leftrightarrow incidenceSystem \\
\hline
\forall \; TGraph; \; incidenceSystem \; \bullet \\
\quad \theta TGraph \models_{\text{incSystem}} \theta incidenceSystem \Leftrightarrow \\
\qquad \forall \, e : \; E; \; t : typeID \mid type(e) = t \; \bullet \\
\qquad\quad type(\alpha(e)) \in \{\, t' : typeID \mid t' \; isA^* \; first(incidences(t)) \,\} \wedge \\
\qquad\quad type(\omega(e)) \in \{\, t' : typeID \mid t' \; isA^* \; second(incidences(t)) \,\}
\end{array}
$$

Example. From Figure 3 one may derive that

$$incidences(source) = (transition, blob)$$

holds, which e.g. implies, that the types of a goal vertex v of a *source*-edge e may only be *blob*, *elementaryBlob*, *xorBlob*, or *andBlob*. ◊

4.3 ERSpecifications

The specification given by an ER-diagram consists of a type system and a corresponding incidence system. A TGraph matches a specification if it matches the type and incidence claims specified therein.

$$
\begin{array}{l}
\underline{\quad}\; ERSpecification \; \underline{\hspace{6cm}} \\
typeSystem \\
incidenceSystem \\
\hline
\end{array}
$$

$$
\begin{array}{l}
\underline{\quad} \models_{\text{ERSpecification}} \underline{\quad} : \; TGraph \leftrightarrow ERSpecification \\
\hline
\forall \; TGraph; \; ERSpecification \; \bullet \\
\quad \theta TGraph \models_{\text{ERSpecification}} \theta ERSpecification \Leftrightarrow \\
\qquad \theta TGraph \models_{\text{typeSystem}} \theta typeSystem \wedge \\
\qquad \theta TGraph \models_{\text{incSystem}} \theta IncidenceSystem
\end{array}
$$

Thus, the class of TGraphs given by an ERSpecification *Spec* is the set of those TGraphs G with $G \models_{\text{ERSpecification}} Spec$.

5 Additional Constraints

So far, it has been discussed how to define classes of TGraphs by specifying vertex and edge types and restrictions on possible incidences in an ER-diagram, as well as how to define appropriate attribute schemes for objects and relations. Often, this information alone does not suffice to describe an application domain in a satisfying manner. Hence, the diagram still has to be extended by additional predicates which specifiy additional constraints to complete the specification language.

Some of these predicates, like e.g. degree restrictions might also be expressed graphically in the diagram, but more elaborate (especially global) constraints must be stated explicitly.

Example. In state charts, for instance, the set of *contains*-arcs defines a tree-like hierarchy on blobs. This is not expressed in the diagram. In this sense, the specification of the class of TGraphs modeling state charts is still incomplete. In graph theoretical terms, the dexcribed constraint is given by the following predicate:

isTree(eGraph(edges(*contains*)))

Here, **edges**(*contains*) is the set of all edges e with $type(e) = contains$. eGraph returns the graph induced by a set of edges, and isTree is a predicate on graphs. Further conditions on state charts are: a blob is either elementary, or an xor-blob, or an and-blob, elementary blobs are not refined, xor-blobs contain exactly one start blob, and and-blobs consist of xor-blobs, only:

$\forall\, q : $ nodes(*blob*) •
    $type(q) \in \{elementaryBlob, xorBlob, andBlob\} \wedge$
    if $type(q) = elementaryBlob$ **then** outdegree($q, contains$) = 0 $\wedge$
    if $type(q) = xorBlob$ **then** outdegree($q, startsWith$) = 1 $\wedge$
    if $type(q) = andBlob$ **then** $(q \rightarrow_{contains}) \subseteq$ nodes($xorBlob$)   ◊

To describe additional constraints on graph classes the constraint language GRAL ([EbeFra 92]) has been developed. GRAL is a Z-like notation to describe structural properties of TGraphs, which can be translated into efficient algorithms that test the property described ([CapFra 91]).

A GRAL predicate is a first order predicate logic term containing only elements, which can be tested by polynomial algorithms. Basic elements of GRAL are
▷ a set of predicates to describe basic properties of graph elements (e.g. isIsolated, isSink), or graphs (like isTree, isConnected, isBipartite)
▷ a set of functions to compute sets of vertices (like **nodes**) and sets of edges (like **edges**) of a given type, or induced subgraphs (like **eGraph**).
▷ predicates and functions on sets and numbers,
▷ logical operators (like $\wedge, \vee$) and quantifiers (like $\forall$), which are allowed to range over finite sets, and
▷ (regular) path expressions.

A path expression describes a path in a graph in an abstract manner. A simple path expression consists of a single edge ($\leftarrow, \rightarrow, \rightleftarrows$). Each edge symbol may be annotated with restrictions on its type and attributes (e.g. $\rightarrow_{contains}$). Analogously, restrictions on nodes reached by following an edge can be expressed (e.g. $\rightarrow_{contains} \bullet_{elem}$). Simple path expressions may be combined to form more complex ones. Here, only regular structures (sequence, alternative, iteration) are allowed. Path expressions may be applied in postfix notation to vertices, thus delivering the set vertices which are reachable via a path matching the description, or they may denote predicates, if they are applied to two vertices in infix notation. Sets defined by path expressions may be computed by appropriate search algorithms in $\mathcal{O}(\#E \cdot s)$ time, if s is the number of symbols in the path expression ([CapFra 91]).

Example. To describe the set of all transitions in a state chart that are enabled through a given event e by a blob q, the following expression suffices:

    $\{\, t : $ nodes(*transition*) $\mid q \overset{*}{\leftarrow}_{contains} \leftarrow_{source} t \rightarrow_{triggeredBy} e \,\}$ ◊

Since GRAL - as a constraint description language - allows also to define the type system and the incidence structure of graph classes. It depends on the user, where he puts the borderline between the ER-like description and the GRAL-part. We recommend to use the diagrammatic description as far as possible and to use GRAL only for the addition of further constraints, which are not expressible graphically by the ER-dialect chosen.

6 Programming With Graph Based Models

Up to now, we described the class of graphs used for modeling and gave a an approach to define subclasses as model classes for practical applications which consists of a combination of a graphical and a textual description. In this section we sketch how graph based models specified in this way may be implemented.

TGraphs may be handled efficiently by graph algorithms written in a pseudocode according to the algorithmic interface described in figure 4. Using an implementation technique like the one described in [Ebert 87] a direct implementation of pseudocode algorithms is possible, without any intermediate transformation. If the traversal operations are implemented in optimal time, complexity analysis of algorithms also carries through to the final implementation.

procedure init ()	initializes the graph as empty
for all v **in** V **do** ... **end**	processes all vertices v
for all e **in** E **do** ... **end**	processes all edges e
for all e **in** $A(v)$ **do** ... **end**	processes all edges e incident with v
for all e **in** $A^+(v)$ **do** ... **end**	processes all edges e going out of v
for all e **in** $A^-(v)$ **do** ... **end**	processes all edges e going into v
procedure vertexCount(): nat0	returns the number of vertices
procedure edgeCount(): nat0	returns the number of edges
procedure isEdge (v, w: vertex): edge	returns an edge from v to w
procedure areNeighbours (v, w: vertex): edge	returns an edge between v and w
procedure alpha (e: edge): vertex	returns the start vertex of edge e
procedure omega (e: edge): vertex	returns the end vertex of edge e
procedure this (e: edge): vertex	returns the vertex from whose A-sequence e was taken
procedure that (e: edge): vertex	returns the other vertex of e with respect to this (e)

Fig. 4. Algorithmic Interface for TGraphs

The algorithmic interface has been realised by C++-software package, which may be used for programming with TGraphs ([DaEbLi 94]). This package works on the edge-oriented basis described in this paper, where the data type edge includes the edge object plus its direction. This allows to view a graph as being undirected or directed without any change of its representation. The view on the graph is only determined by the interface operations that are used.

Internally graphs are represented by symmetrically stored forward and backward adjacency lists ([Ebert 87]). The sets of attribute instances belonging to vertices and types are implemented by application dependent C++-objects, thus using the multiple inheritance concept of the language to implement the analogous concept in TGraphs.

Since according to the modeling approach given above ER-diagrams themselves may easily be modeled as ERGraphs (according to the (meta)-ER-schema in figure 5), the graph interface can be realised in such a way that creation and deletion of edges are controled by a constraint checker, which consults the ERGraph of the graph class with every updating operation.

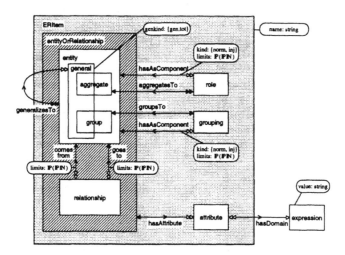

Fig. 5. Schema for ERGraphs

If G is an ERGraph, which describes a graph class, its *entity* vertices correspond to vertex types and its *relationship*, *role*, and *grouping* vertices correspond to edge types. The *generalizesTo* edges give the *isA*-relation. Furthermore, the *hasAsComponent*, *aggregatesTo*, *groupsTo*, *comesFrom*, and *goesTo* edges contain the incidence information, and the subgraphs induced by the sets of *entityOrRelationship*, *attribute*, and *expression* vertices describe the attribute schemes[5].

In addition to the checking via the ER-diagram, the checkings procedure derived from the GRAL predicates may be activated every time, a modification of the graph is performed. Thus, the declarative description of TGraph classes is directly usable in consistency assurance.

7   Conclusion and Related Work

In this paper we gave an overview on our approach of using graph theory and graph algorithms in a consistent software technological manner in practical applications. A

---

[5] [CaEbWi 94] contains a formal description of the type system and the incidence system defined by an ER-diagram via its ERGraph.

formal definition of the class of TGraphs was given together with a declarative approach of specifying subclasses for modeling purposes by giving their type and incidence structure graphically and by supplying a constraint specification language, which is translatable into efficient test procedures. Since this language includes regular path expressions, it already leads quite far. All aspects are implementable accordingly without any gap. Thus TGraphs build a framework which uniformly supports the aspects of expressiveness, formality, and efficiency.

In the KOGGE project ([Ebert 94]) software requirement and design documents are modeled inside a Meta-CASE-tool by TGraphs. Here, graphs are used as internal data structures exactly in the way described above. The modelling classes are specified using ER-diagrams and GRAL, and the repository component of the system assures that all specified integrity constraints are automatically kept by the system.

Since the class of TGraphs is defined in terms of objects, relations, attributes, etc. it fits very well into today's object-oriented way of thinking. Thus, TGraphs models are object-oriented structural models. Their definition shows also a way to formalizing structural aspects of object-oriented description.

The approach of characterizing graph classes by ER-diagrams, which have a widespread use in information modeling in database systems, may as well be read as a definition of a graph based data model for databases, including a powerful specification language for consistency conditions, namely GRAL. Comparable work has been done by e.g. Paredaens and others ([GyPaGu 90]) and Consens and Mendelzon ([ConMen 90]), where the emphasis lies on the definition of suitable query languages while we concentrate on the modeling process.

Foundations of declarative specification for graph languages have been explored by Courcelle ([Courcelle 90]). He showed that graph properties describable in monadic second order logic (MSOL) are efficiently testable for large graph classes. The specification language proposed in this paper is less powerful then MSOL (by offering sufficient expressive power for practical applications). We can guarantee that our declarative specification lead to efficient implementations, which is also shown by our translator in a constructive manner.

The concept of declarative graph class specification contrasts with graph grammar approaches, which assure constraints on graphs by specifying the operations, which build or update the graph. The graph grammar based approach has been used in the IPSEN project by Nagl and others ([EnLeNa 92] where graph grammar based specifications are used in the development of software engineering environments. In this context, the operational specification language PROGRESS ([Schuerr 91]) has been defined that allows users to specify even complex graph transactions. Furthermore, implementation is supported by a powerful graph database system.

In the graph grammar based approach to graph class specification it is quite easy to implement tools that control the construction of graphs matching the specification (e.g. syntax directed editors). On the other hand, given a more declarative specification it is easier to test whether a given graph is correct with respect to the specification. Test procedures for specified graph properties may be embedded in a module managing the graph structure and thus be separated from application programs. Any test may be invoked at any point in time, on system or user demand. Thus, flexible strategies for consistency control might be implemented. It is depends on the goals of graph based modeling in an application context which kind of specification should be used. From a modeling point of view, the declarative approach seems to be more intuitive or natural since the user is able to describe the class of graphs he thinks of without having to think of operations.

References

[CaEbWi 94]   Martin Carstensen, Jürgen Ebert, Andreas Winter. *Entity–Relationship–Diagramme und Graphenklassen.* to appear as: Fachbericht Informatik 1994. Universität Koblenz 1994.

[CapFra 91]   Carla Capellmann, Angelika Franzke. *GRAL: Eine Sprache für die graphbasierte Modellierung.* Universität Koblenz, Diplomarbeit, 1991.

[ConMen 90]   Mariano P. Consens, Alberto O. Mendelzon. *GraphLog: a Visual Formalism for Real Life Recursion.* In: Proc. 9th Symposium on Principles of Database Systems. New York: ACM Press, 1990, pp. 417-424.

[Courcelle 90]   Bruno Courcelle. *Graph rewriting: An algebraic and logic approach.* In: Jan van Leeuwen (ed.): Handbook of Theoretical Computer Science, Vol. B. Amsterdam: Elsevier 1990. pp. 193 - 242.

[DaEbLi 94]   Peter Dahm, Jürgen Ebert, Christoph Litauer. *Das EMS-Graphenlabor.* in preparation.

[Diller 90]   Anthony Diller. *Z: An Introduction to Formal Methods.* Wiley 1990.

[EbeFra 92]   Jürgen Ebert, Angelika Franzke. *Specification of a Graph Based Data Model for a CASE Tool.* Universität Koblenz, Fachbericht Informatik 5/92

[Ebert 87]   Jürgen Ebert. *A Versatile Data Structure For Edge-Oriented Graph Algorithms.* Communications ACM 30 (1987,6), 513-519.

[Ebert 93]   Jürgen Ebert. *Efficient Interpretation of State Charts.* in: Zoltán Ésik (Ed.). *Fundamentals of Computation Theory (FCT '93), Szeged, Hungary.* Berlin: Springer, LNCS 910, 1993, pp. 212-221.

[Ebert 94]   Jürgen Ebert. *KOGGE: A Generator for Graphical Design Environments.* in preparation.

[EnLeNa 92]   Gregor Engels, Claus Lewerentz, Manfred Nagl, Wilhelm Schäfer, Andreas Schürr. *Building Integrated Software Development Environments. Part I: Tool Specification.* ACM Transactions on Software Engineering and Methodology 1 (1992, 2), 135 - 167.

[GyPaGu 90]   M. Gyssens, J. Paredaens, D. van Gucht. *A graph-oriented object database model.* In: Proc. 9th Symposium on Principles of Database Systems. New York: ACM Press, 1990, pp. 417-424.

[Harel 87]   David Harel. *Statecharts: a Visual Formalism for Complex Systems.* Science of Computer Programming 8 (1987,3), 231-274

[Schuerr 91]   Andreas Schürr. *Operationales Spezifizieren mit programmierten Graphersetzungssystemen.* Wiesbaden: Deutscher Universitätsverlag 1991.

[Spivey 92]   J.M. Spivey. *The Z Notation - A Reference Manual.* New York: Prentice Hall, 1992.

Multilevel Graph Grammars

Francesco Parisi-Presicce[1] and Gabriele Piersanti[2]

[1] Dipartimento di Scienze dell'Informazione, Universita' di Roma 'La Sapienza',
00198-Roma (Italy)
[2] Dipartimento di Matematica Pura ed Applicata, Universitá degli Studi de
L'Aquila, 67100-L'Aquila (Italy)

Abstract. The classical double pushout approach to the algebraic theory of graph grammars is extended to multilevel graph representations, where parts of graphs are not visible and the information can be restored via the explicit application of productions. The notions of applicability and derivation are investigated and the compatibility of the representations with the derivations is shown. Production mechanisms for multilevel graph are motivated by problems in Visual Languages and the representation of Iconic Languages in particular.

1 Introduction

The use of graph–like structures for modelling range from the description of software systems to that of biological systems, with applications to concurrency problems, database integrity, logic programming and hardware design. Among the different approaches to the formalization of graph rewriting and generation, the Algebraic approach [1] based on double pushout construction, and subsequent modifications and extensions [6, 15, 7], allows the exploitation of known tools and results from category theory and a more concise description of properties than other approaches at a more operational level. The main motivation for the extension proposed here is the study of a model for the global representation , via rewriting rules and in a unified framework, of the lexical elements, the syntax and the semantics of Visual Language. In the classical approach, graphs are defined as sets of vertices and edges between pairs of vertices, with two 'labelling' functions associating to vertices and edges elements of two distinct 'alphabets'. The representation of graphs is rarely considered and both vertices and edges are always "visible". In many applications, it is very useful to be able to represent graphs in terms of their particular subgraphs and to hide details of structures needed only in certain conditions; repeated hiding of details gives origin to representations on more than one level of visibility. Instead of dealing with graphs, we treat "graph representations", in which we associate to a graph the sequence of productions needed to restore the original graph, making visible again the information hidden at a different level. Different levels of visibility are available, depending on the amount of information stored in the 'restoring' productions. Of course, if the sequence of restoring productions associated to a graph is empty, the development reduces to the algebraic approach to 'totally visible' graph grammars. Starting from this idea, we define an extension of the algebraic approach to graph grammars, characterizing formally the notions of graph production, derivation, grammar and language, and introduce some important properties, some of which generalize well known properties of the algebraic approach, while others are specific to the multilevel representation.

2 Graph Model

We briefly review the basic notions of the algebraic approach to graph grammars necessary to explain our notation.

Definition 1. Let $C = (C_A, C_N)$ be a pair of sets, called color alphabets.

1. A **C-colored graph** G is a six-tuple $(G_A, G_N, s, t, m_A, m_N)$, consisting of:
   - sets G_A and G_N, called the set of arcs and the set of nodes, respectively;
   - *source* and *target* mappings $s : G_A \to G_N, t : G_A \to G_N$;
   - *arcs* and *nodes* coloring mappings $m_A : G_A \to C_A, m_N : G_N \to C_N$.

   For $e \in G_A$, denote $\{s(e), t(e)\}$ by $end(e)$.
   A graph G' is a subgraph of a graph G if $G'_A \subseteq G_A$, $G'_N \subseteq G_N$, and all the mappings s', t', m'_A, m'_N are restrictions of the corresponding ones from G.
2. A **graph morphism** $f : G \to G'$ is a pair $(f_N : G_N \to G'_N, f_A : G_A \to G'_A)$ such that
   - $f_N \circ s_G = s_{G'} \circ f_A$; $f_N \circ t_G = t_{G'} \circ f_A$ (the structure is preserved)
   - $m_{G'} \circ f_N = m_G$; $m_{G'} \circ f_A = m_G$ (the labels are preserved).

The idea, common to every theory on Graph Grammars, is to realize, from an initial graph, a sequence of transformations, following rules allowing a subgraph DEL of G to be replaced by another graph ADD, leaving unchanged the subgraph D of G not involved in the deletion of DEL. Every transformation of this kind requires the specification of how ADD must be connected to the graph D (called *enbedding*).

Definition 2. A graph production p is a pair $(L \leftarrow K \to R)$ of graph morphisms, where L, R and K are called the left side, the right side and the interface of the production, respectively.

A production specifies that a graph L must be replaced by a graph R, using an interface graph K (whose arcs and nodes are the gluing items) for the embedding.

Definition 3 (Direct Derivation). Given a production $p = (L \leftarrow K \to R)$ with $l : K \to L$ injective, a direct derivation from $G1$ to $G2$ via p, denoted by $p : G1 \Rightarrow G2$ or by $G1 \Rightarrow_p G2$, consists of the following double pushout

$$L \xleftarrow{\quad l \quad} K \xrightarrow{\quad r \quad} R$$

$$g1 \downarrow \qquad c \downarrow \qquad \downarrow g2$$

$$G1 \xleftarrow{\quad l' \quad} D \xrightarrow{\quad r' \quad} G2$$

A derivation $G \Rightarrow_P^* H$ is a sequence of direct derivations $G = G1 \Rightarrow_{p_1} \ldots \Rightarrow_{p_{n-1}} Gn = H$ using productions $p_i \in P$.

Notice that the derivation diagram is symmetric and therefore, if $p : G1 \Rightarrow G2$, then $p^{-1} : G2 \Rightarrow G1$ where $p^{-1} = (R \leftarrow K \rightarrow L)$.
The applicability of a production $p = (L \leftarrow K \rightarrow R)$ to a graph G1 is determined by the existence and properties of a total morphism $g1 : L \rightarrow G1$. The existence of a pushout complement C for a given g1 depends on the following Gluing Conditions being satisfied [1].

Theorem 4 (Gluing Conditions). *Given* $p = (L \leftarrow K \rightarrow R)$ *and* $g1 : L \rightarrow G1$, *let*

$$ID_{g1} = \{x \in L : \exists x' \in L, x \neq x', g1(x) = g1(x')\}$$

$$DANG_{g1} = \{n \in N_L : \exists e \in A_{G1} - g1_A(A_L) \text{ such that } g1_N(n) \in end_{G1}(e)\}$$

Then the pushout complement D exists **if and only if** $DANG_{g1} \cup ID_{g1} \subseteq l(K)$

This limitation of the traditional double pushout approach can be overcome either by using single pushouts [7] or by using "restricting derivation sequences" [10]. In the context of visual languages, the violation of the gluing conditions has a meaningful interpretation.

Definition 5. A graph grammar $GG = (C, T, PROD, START)$ consists of color alphabets $C = (C_A, C_N)$, terminal alphabets $T = (T_A, T_N)$, included in C, a finite set $PROD$ of productions , and an initial graph $START$. The Graph Language generated is $L(GG) = \{H \mid START \Rightarrow^* H, \text{ and } H \text{ is } T-colored\}$.

Productions can be compared using triples of graph morphisms, one for each component of a production.

Definition 6. Given a pair of productions $p = (L \leftarrow K \rightarrow R)$ and $p' = (L' \leftarrow K' \rightarrow R')$, a *production morphism* $f : p \rightarrow p'$ is a triple $(f_L : L \rightarrow L', f_K : K \rightarrow K', f_R : R \rightarrow R')$ of graph morphisms as in the following commuting diagram

$$L \xleftarrow{\quad l \quad} K \xrightarrow{\quad r \quad} R$$

$$f_L \downarrow \qquad f_K \downarrow \qquad \downarrow f_R$$

$$L' \xleftarrow{\quad l' \quad} K' \xrightarrow{\quad r' \quad} R'$$

A production p is a *subproduction* of p' (and we write $p \subseteq p'$) if, in addition,
- $BOUNDARY(L \to L') \subseteq l(K)$ and $IDENTIFICATION(R \to R') \subseteq r(K)$
and
- $f_L^{-1}(l'(K')) \subseteq l(K)$ and $f_R^{-1}(r'(K')) \subseteq r(K)$.
(These two conditions guarantee that the applicability of p' to an occurrence also implies the applicability of p to the same occurrence).

Our descriptive model also uses some extensions of the algebraic approach:
- Variables as colors for arcs and nodes [11]. This extension is based on structured color alphabets, which are "ordinary" color alphabets equipped with particular relations. Variables on arcs and nodes may be replaced by legal values using SC-morphisms, which are graph morphisms respecting the structure of alphabets and are consistent with respect to sources and targets of arcs.
- Applications conditions on productions [8], possibly defined in an informal way, to restrict the contexts of applicability for productions;

3   Multilevel Graph Grammars

As mentioned in the introduction, a basic feature of our model is the possibility to represent a graph on several levels. Given a graph G, we want to replace temporarily a subgraph G_H with a subgraph $G_I \subseteq G_H$, called interface of G_H, using a production (which we call hiding production) of the form $p = (G_H \leftarrow G_I \to G_I)$. The graph G_H, which is hidden by its interface G_I, can be restored in its context, when needed, if we keep the information which connects G_H and G_I. Such information is called restoring production, and is defined as the inverse $p^{-1} = (G_I \leftarrow G_I \to G_H)$ of the production $p = (G_H \leftarrow G_I \to G_I)$ used to hide G_H with G_I.

Definition 7. Given a subgraph $G_H \subseteq G$ to be hidden, $G_I \subseteq G_H$ is called *interface* if the nodes of G_I are the sources or the targets of arcs in $G_H - G_I$. Every node of an interface graph is called an *interface item*.

Since a hidden graph can in turn contain interfaces of other previously hidden graphs, and a graph can simultaneously contain many interfaces, it is possible to obtain a graph represented on various levels of visibility. To realize such a representation, we must associate to a graph G the set of all productions required to expand any interface, restoring the corresponding hidden graph. Furthermore, since a graph might hide an interface of another graph, in general, the application of restoring productions will be realizable without unexpected effects, only in inverse order with respect to the order of their construction. This constraint may be imposed in various ways, such as structuring the set of restoring productions as a LIFO list, or using not-terminal numbered labels as indices to prevent the productions from being applied either in an incorrect order or to a different subgraph than the one intended.

Definition 8. A multilevel graph is a triple $G_M = (G, RP_G, HP_G)$, where G is an "ordinary" graph, called base graph, RP_G is a sequence of restoring productions, and HP_G is a sequence of hiding productions. Any multilevel graph of the form $(G, RP_G, \emptyset)$ is said to be in normal form and is simply represented as a pair (G, RP_G).

Example 1. An example of a multilevel graph in normal form is $G_M = (G, RP_G, \emptyset)$ with $RP_G = \{p\}$ and $p = (G_I \leftarrow G_I \rightarrow G_H)$ where

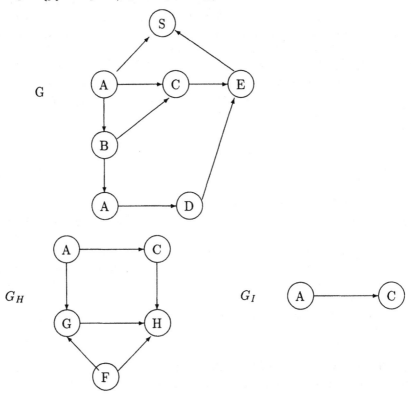

Definition 9. Given a multilevel graph $G_M = (G, RP_G, HP_G)$, an *R-extension* (resp., an *R-restriction*) of G_M is any multilevel graph $G'_M = (G', RP_{G'}, HP_{G'})$ obtained as follows

- the graph G' is derived from G by applying (in the imposed order) all the productions of a subsequence $R = \{p1, \ldots, pk\}$ of RP_G
  (resp. HP_G)
- the new restoring productions are defined by $RP_{G'} = RP_G - \{p1, \ldots, pk\}$
  (resp. $RP_{G'} = RP_G \cup \{(pk)^{-1}, \ldots, (p1)^{-1}\}$)
- the hiding productions are updated by $HP_{G'} = HP_G \cup \{(pk)^{-1}, \ldots, (p1)^{-1}\}$
  (resp. $HP_{G'} = HP_G - \{p1, \ldots, pk\}$)

Every production of RP_G restores a graph previously hidden by an interface, while any production in HP_G hides graphs contained in G. Extensions of a multilevel graph are considered as temporary representations of a graph in normal form and are used to realize derivations.

Definition 10. A *multilevel graph production* is a pair $q = (q_1, q_2)$, where $q_1 = (L \leftarrow K \rightarrow R)$ is called transformation production, and describes a "normal" rewriting of graphs, while $q_2 = (L' \leftarrow K' \rightarrow R')$ is a subproduction (possibly a null production) of q_1, such that $R' = K' \subset L'$. The component q_2 is called hiding production (and its inverse is called restoring production).

Remark. In the general case $q = (q_1, q_2)$, where q_2 is defined as a proper subproduction, any application of q_1 is only partially an ordinary rewriting of graphs, since its subproduction q_2 hides some structure. In this case q_2^{-1} must be added to the set of restoring productions.
If there is no structure to be hidden by q_1, q_2 is not specified as in $q = (q_1, -)$ and the application of q realizes an ordinary rewriting.
When $q = (q_2, q_2)$, the whole production q hides the structure removed from the graph to which it is applied, and therefore q_2^{-1} must be added to the set of restoring productions

As in the classical theory, not all ocuurrence morphisms $L \rightarrow G$ allow the application of the production.

Definition 11. Given a production $q = (q_1, q_2)$, with $q_1 = (L \leftarrow K \rightarrow R)$ and $q_2 = (L' \leftarrow K' \rightarrow R') \subseteq q_1$, an occurrence map $g : L \rightarrow G$ is *safe for a sequence RP* of restoring productions if for each $p = (G_I \leftarrow G_I \rightarrow G_H) \in RP$, $g(L) \cap G_I \subseteq g(l(K)) \cap g(L')$

Note that condition imposed is trivial when $q_1 = q_2$ while it reduces to $g(L) \cap G_I \subseteq g(l(K))$ when q_2 is null.

Definition 12 (Extended Gluing Condition). Given a multilevel graph $G_M = (G, RP_G)$, a production $q = (q_1, q_2)$, with $q_1 = (L \leftarrow K \rightarrow R)$, an occurrence map $g : L \rightarrow G$ satisfies the *extended gluing condition (EGC)* on (G, RP_G) for $q = (q_1, q_2)$ if it satisfies the Gluing Condition and it is safe for the ordered sequence of productions RP_G.

The Extended Gluing Condition guarantees that any rewriting of multilevel graphs will not create inconsistencies within hidden graphs, or loss of links between them and the associated interfaces.

Example 2. Consider again the multilevel graph G_M previously described. Given

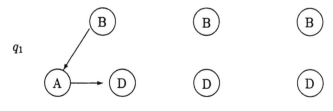

the production $q = (q_1, -)$ can be applied to G_M without restoring the hidden graph and without trasforming the associated restoring productions, because an occurrence map satisfying the Extended Gluing Condition exists for the base graph G. We thus derive from G_M by q the multilevel graph $H_M = (H, RP_G,)$, with

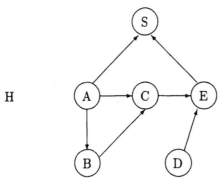

Sometimes it is necessary to restore part of the hidden graph in order to find an appropriate occurrence morphism.

Definition 13 (Minimal Extension). Given $G_M = (G, RP_G, HP_G)$ and a production $q = (L \leftarrow K \rightarrow R)$, a graph $G'_M = (G', RP_{G'}, HP'_G)$ is called a *minimal extension* of G_M if it is obtained from G_M by applying the shortest subsequence $\{p_1, \ldots, p_k\}$ of RP_G such that there exists an occurrence map $g : L \rightarrow G'$ which satisfies the Gluing Condition and is safe for $RP_G - \{p_1, \ldots, p_k\}$

Example 3. Given the production $q = (q_1, q_2)$, where q_1 and q_2 are, respectively,

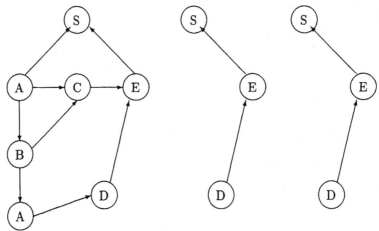

and

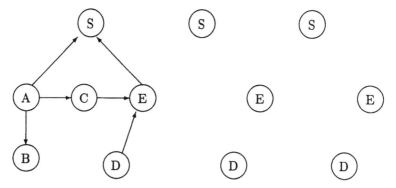

the production q (which simultaneously hides some structures and removes others) can be applied to G_M without either restoring the hidden graph or trasforming the associated restoring production, but it removes some items (including interface items) to put them at a higher level. We thus derive from G_M by q the multilevel graph $H_M = (H, RP_H,)$, where $RP_H = RP_G \cup \{q_2^{-1}\}$ and

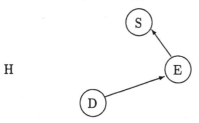

After having restored part of the hidden subgraph in order to find an occurrence morphism which satisfies the EGC, the graph resulting from the derivation should be represented at the same level of details as the original one.

Definition 14 (Maximal Restriction). Given $B_M = (B, RP_B, HP_B)$ such that $q_1 : G \Rightarrow B$, the *maximal restriction* of B_M is the graph $B'_M = (B', RP_{B'}, \emptyset)$ obtained from B by applying (in the imposed order) the productions in HP_B as follows : assuming that the first $r \leq j$ productions to B have been applied to obtain $(B_j, RP_{B_j}, HP_{B_j})$, consider, if $HP_{B_j} \neq \emptyset$, the first production $p \in HP_{B_j}$

- if p is applicable to B_j, then $p : B_j \Rightarrow B_{j+1}$, $RP_{B_{j+1}} = RP_{B_j} \cup \{p^{-1}\}$ and $HP_{B_{j+1}} = HP_{B_j} - \{p\}$
- if $p = (G_H \leftarrow G_I \rightarrow G_I)$ is not applicable to B_j and $q_1 \neq q_2$, then apply the production $p' = (G'_H \leftarrow G'_I \rightarrow G'_I)$ defined as follows
  - if an occurrence map $g' : G_H \rightarrow B_j$ exists but does not satisfy the Gluing Condition, let $G'_I = DANGLING(g')$ and $G'_H = G_H$;
  - if there is no occurrence map $g' : G_H \rightarrow B_j$, let $G'_H = G_H - g(L - l(K))$ (which allows the definition of an occurrence $g' : G'_H \rightarrow B_j$) and $G'_I = DANGLING(G'_H \rightarrow B_j)$

Example 4. Consider again the multilevel graph G_M seen in the previous examples. With the production q_1 given by

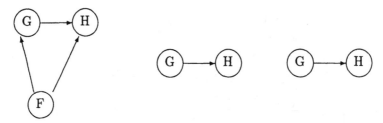

the production $q = (q_1, -)$ can be applied to the extension of G_M obtained by applying the restoring production p. Note that since part of the previously hidden graph is deleted by the application of q_1 to G, the inverse of p cannot be applied to the derived graph H:

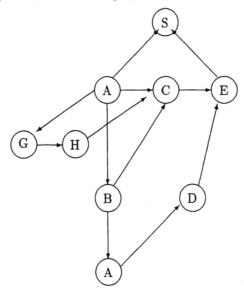

Thus, a production $p' = (G'_I \leftarrow G'_I \rightarrow G'_H)$ where

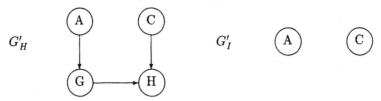

is defined and its inverse applied to H to derive its maximal restriction H' given by

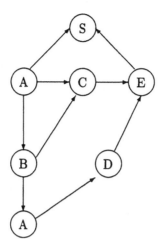

which defines the multilevel graph $H_M = (H', \{p'\},)$ derived from G_M via q.

We can now give the formal definition of derivation.

Definition 15 (Direct Derivation). Given a production $q = (q_1, q_2)$ with $q_1 = (L \leftarrow K \rightarrow R)$ and a multilevel graph $G_M = (G, RP_G, \emptyset)$, a direct derivation is given by the following diagram

$$L \longleftarrow K \longrightarrow R$$

$$\downarrow \qquad \downarrow \qquad \downarrow$$

$$G \underset{ext}{\longrightarrow} G' \longleftarrow D \longrightarrow B \underset{rest}{\longrightarrow} B'$$

where $HP_B = HP_{G'}$, $RP_B = RP_{G'} \cup \{q_2^{-1}\}$,
$(G, RP_G, \emptyset) \Rightarrow (G', RP_{G'}, HP_{G'})$ [minimal extension]
$(B, RP_B, HP_B) \Rightarrow (B', RP_{B'}, \emptyset)$ [maximal restriction]

Interpretation of a Direct Derivation Given an arbitrary multilevel graph (G, RP_G) and a production $q = (q_1, q_2)$ (note that it can be either $q_2 \subseteq q_1$ or $q_2 = -$), the execution of a derivation step $q : (G, RP_G) \Rightarrow (H, RP_H)$, assuming $RP_G = \{p_1, ..., p_n\}$, with $p_i = (G_{H_i} \leftarrow G_{I_i} \rightarrow G_{I_i})$, and $q_1 = (L \leftarrow K \rightarrow R)$, distinguishes two main cases :

Case 1 There exists an occurrence map $g : L \rightarrow G$:
  if the EGC is satisfied on (G, RP_G) by q, then $q : (G, RP_G) \Rightarrow (H, RP_H)$, with $q_1 : G \Rightarrow H$, and $RP_H = RP_G$ if $q_2 = -$ while $RP_H = RP_G \cup \{q_2^{-1}\}$ if $q_2 \subseteq q_1$.
  If the EGC is not satisfied, then q is not applicable via g to (G, RP_G).
Case 2 There is no occurrence map $g : L \rightarrow G$ satisfying the EGC:
  to try to find an occurrence map on a graph derived from G, apply productions of RP_G (in the imposed order) until either an occurrence has been

found or all the productions of RP_G have been used. At the end of this process, having applied productions of a set $R = \{p_1, ..., p_r\} \subseteq RP_G$, and derived a graph G_r, one of two situations is possible:

- $R = RP_G$, and an occurrence map $g : L \to G_r$ satisfying the EGC does not exist, in which case q is not applicable to (G, RP_G);
- an occurrence map $g : L \to G_r$ exists, satisfying the EGC on $(G_r, RP_G - R)$ and allowing the derivation step $q_1 : G_r \Rightarrow H_0$.

  The corresponding set RP_{H_0} of restoring productions is given by $RP_G - R$ if $q_2 = -$ or by $RP_G - R \cup \{q_2^{-1}\}$ if $q_2 \subseteq q_1$

  To complete the derivation, we must now redefine a multilevel organization of the graph, hiding all the previously restored graphs, or their transformations, in inverse order. Starting from $H_{M_0} = (H_0, RP_{H_0}, \{p_r^{-1}, ..., p_1^{-1}\})$, the inverses of all productions of the sequence $p_r, ..., p_1$ are applied to produce its maximal restriction $H_M = (H, RP_H, \emptyset)$.

Example 5. Given the production q_1

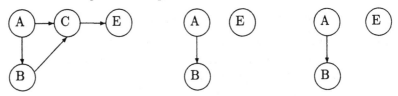

the production $q = (q_1, -)$ cannot be applied to G_M since q_1 removes the node labelled C which is an interface item for the hidden graph G_H.

Definition 16. A multilevel Graph Grammar $GG_H = (C, T, P, S)$ consists of

- a multilevel start graph $S = (START, RP_{START})$ with $START$ an "ordinary" graph in which some subgraphs are interfaces for graphs hidden in origin, restorable through productions in RP_{START};
- a pair of color alphabets $C = (C_A, C_N)$;
- terminal color alphabets for arcs and nodes $T = (T_A, T_N) \subseteq (C_A, C_N)$;
- $P \subseteq \{(q_1, q_2) \mid q_1, q_2 \in PROD$ and $q_2 \subseteq q_1\}$ for some set $PROD$ of "ordinary" productions.

Definition 17. The Language generated by the multilevel Graph Grammar $GG_H = (C, T, P, S)$ is $L(GG_H) = \{(G, RP)|G$ is a $T-colored$ graph, and $S \Rightarrow^* (G, RP)\}$

4 Main Results

We now summarize without proof some of the properties of the model. Details can be found in [13].

Definition 18. Given $G_M = (G, RP_G, HP_G)$ and $H_M = (H, RP_H, HP_H)$ with $RP_G = \{p_1, ..., p_n\}$ where $p_i = (L_i \leftarrow L_i \to R_i)$ $i = 1, ..., n$

$RP_H = \{q_1, \ldots, q_m\}$ where $q_j = (L'_j \leftarrow L'_j \rightarrow R'_j)$ $j = 1, \ldots, m$
$HP_G = \{s_1, \ldots, s_{n'}\}$ where $s_i = (N_i \leftarrow I_i \rightarrow I_i)$ $i = 1, \ldots, n'$
$HP_H = \{t_1, \ldots, t_{m'}\}$ where $t_j = (N'_j \leftarrow I'_j \rightarrow I'_j)$ $j = 1, \ldots, m'$
a *multilevel graph morphism* $f_M : G_M \rightarrow H_M$ is a triple $(f, \{g_1, \ldots, g_n\}, \{k_1, \ldots, k_{n'}\})$ where

- $f : G \rightarrow H$ is a graph morphism
- for $i = 1, \ldots, n$, $g_i : p_i \rightarrow q_{s(i)}$ is a production morphism $(g_{L_i}, g_{L_i}, g_{R_i})$ such that $1 \leq i < j \leq n$ implies $1 \leq s(i) < s(j) \leq m$
- for $i = 1, \ldots, n'$, $k_i : s_i \rightarrow t_{s'(i)}$ is a production morphism $(k_{N_i}, k_{I_i}, k_{I_i})$ such that $1 \leq i < j \leq n'$ implies $1 \leq s'(i) < s'(j) \leq m'$
- the morphisms g_i are consistent with respect to the composition of restoring productions
- the morphisms k_i are consistent with respect to the composition of hiding productions

The first result indicates that multilevel graph morphisms do not depend on the level of representation.

Proposition 19. *Given a multilevel graph morphism $f_M : G_M \rightarrow H_M$ with $G_M = (G, RP_G, HP_G)$ and $H_M = (H, RP_H, HP_H)$, if G'_M is the result of applying one restoring production of RP_G to G and H'_M of applying the corresponding productions of RP_H to H, then there exists a unique multilevel graph morphism $f'_M : G'_M \rightarrow H'_M$ which extends f_M.*

An immediate application of the Proposition allows to extend any multilevel graph morphism to the full extension of the representation.

Theorem 20. *Given a multilevel graph morphism $f_M : G_M \rightarrow H_M$ with $G_M = (G, RP_G)$ and $H_M = (H, RP_H)$, there is always a unique multilevel graph morphism between the RP_G-extension of G and the RP_H-extension of H which extends f_M.*

Theorem 21. *Given a multilevel graph grammar GG_H, a multilevel graph $G_M = (G, RP_G, HP_G)$ and a production $q = (q_1, -)$ with $q_1 = (L \leftarrow K \rightarrow R)$, if the production q is applicable to G_M without using any restoring production, then the graph derivable from (G, RP_G, HP_G) via q can be derivable from an extended graph $(G', RP_{G'}, HP_{G'})$.*

Theorem 22. *Given a multilevel graph $G_M = (G, RP_G, HP_G)$ and a production $q = (q_1, q_2)$ where q_2 is not null and $q_1 = (L \leftarrow K \rightarrow R)$, if the production q is applicable to (G, RP_G, HP_G) without using any restoring production deriving a multilevel graph $H_M = (H, RP_H, HP_H)$, then the application of q to the RP_G-extension of G_M produces the $(RP_H - \{q_2^{-1}\})$-extension of H_M.*

The last two results relate the application of a production to two multilevel graphs connected by a morphism.

5 Concluding Remarks

The definition of graph grammar to generate or, more generally, to rewrite multilevel representations of graphs presented here is motivated by the use of graph grammars to describe the lexical, syntactical and semantical aspects of visual languages within a unified framework. The classical treatment is not adequate for our purposes since it does not deal with representation of graphs, requiring the extension of the notions of production and derivation of the classical approach still in a categorical framework. We have shown, among other things, that if a production is applicable to a representation G to produce a representation H, then any "unfolding" G' of the representation G will produce, using the same production, an "unfolding" H' of the representation H.

Multilevel graph grammars are the basis of our approach to modelling of Visual Languages. The model can be applied to all Visual Languages (not necessarily based on visual constructs, like VPCL), possibly having several views on constructs, like Iconic Languages [13]. Among the advantages of the model are:

1. It's possible to define, in formal and schematical terms, informations useful to any one who wants to study, extend or modify the characteristics of a Visual Language.
2. If the description of a Visual Language also requires the representation of the internal organization of the associated software system, it's possible to analyze the behaviour of the system itself, in reaction to visual programs.
3. Besides visual elements, it's also possible to describe formally rules and constraints not visually defined, giving a clear and global view of all features of a Visual Language.
4. It's possible to model a Visual Language using a graphical notation, more expressive than any textual description.

This research is still at a preliminary stage. Among the topics still under investigation, are the verification of the properties to classify our framework within the HLRn Hierarchy of [2].

References

1. Ehrig, H.: Introduction to the Algebraic Theory of Graph Grammars. Springer Lect. Notes Comp. Sci. 73 (1979) 1–69
2. Ehrig, H., Habel, A., Kreowski, H.J., Parisi–Presicce, F.: From Graph Grammars to High Level Replacement Systems. Springer Lect. Notes Comp. Sci. 532 (1991) 269–287
3. Golin, E.J., Reiss, S.P.: The Specification of Visual Language Syntax. Proc. IEEE Workshop on Visual Languages, Rome (Italy) (1989) 105–110
4. Hesse, W.: Two level graph grammars. Springer Lect. Notes Comp. Sci. 73 (1979) 255–269
5. Helm, R., Marriot, K.: Declarative Specification of Visual Languages. Proc. IEEE Workshop on Visual Languages, Skokie (Illinois, USA) (1990) 98–103

6. Kennaway, J.R.: On On graph rewriting. Theoret. Comp. Sci. **52** (1987) 37–58
7. Löwe, M.: Extended Algebraic Graph Transformation. Technical Report, Technische Universitat Berlin feb 1991, 180 pages
8. Nagl, M.: A tutorial and bibliographical survey on Graph Grammars. Springer Lect. Notes Comp. Sci. 73 (1979) 70–126
9. Ollongren, A.: On multilevel graph grammars. Springer Lect. Notes Comp. Sci. 73 (1979) 341–349
10. Parisi-Presicce, F. : Single vs. Double pushout derivations of Graph. Springer Lect. Notes Comp. Sci. 657 (1993) 248–262
11. Parisi-Presicce, F., Ehrig, H., Montanari, U.: Graph Rewriting with unification and composition. Springer Lect. Notes Comp. Sci. 291 (1987) 496–514
12. Parisi-Presicce, F., Piersanti, G.: Multilevel Graph Grammars. Techn.Rep. 93/33, Dip. Matematica Pura ed Applicata, Universita' di L'Aquila, (1993)
13. Parisi-Presicce, F., Piersanti, G.: Graph Based Modelling of Visual Languages. in preparation
14. Pratt, T.W.: Pair grammars, Graph Languages and string to graph transformations. J. Comput. System Sci. **5** (1971) 560–595
15. Raoult, J.C.: On graph rewriting. Theoret. Comp. Sci. **32** (1984) 1–24

The algorithmic use of hypertree structure and maximum neighbourhood orderings

Andreas Brandstädt[1], Victor D. Chepoi[2] and Feodor F. Dragan[2]

[1] Gerhard-Mercator-Universität –GH– Duisburg FB Mathematik FG Informatik I
D 47048 Duisburg Germany
[2] Department of Mathematics and Cybernetics Moldavian State University
A. Mateevici str. 60 Chişinău 277009 Moldova ***

Abstract. The use of (generalized) tree structure in graphs is one of the main topics in the field of efficient graph algorithms. The well–known partial k–tree (resp. treewidth) approach belongs to this kind of research and bases on a tree structure of constant–size bounded maximal cliques. Without size bound on the cliques this tree structure of maximal cliques characterizes chordal graphs which are known to be important also in connection with relational database schemes where hypergraphs with tree structure (acyclic hypergraphs) and their elimination orderings (perfect elimination orderings for chordal graphs, Graham–reduction for acyclic hypergraphs) are studied.

We consider here graphs with a tree structure which is dual (in the sense of hypergraphs) to that one of chordal graphs (therefore we call these graphs *dually chordal*). The corresponding vertex elimination orderings of these graphs are the *maximum neighbourhood orderings*. These orderings were studied recently in several papers and some of the algorithmic consequences of such orderings were given.

The aim of this paper is a *systematic treatment of the algorithmic use of maximum neighbourhood orderings*. These orderings are useful especially for dominating–like problems (including Steiner tree) and distance problems. Many problems efficiently solvable for strongly chordal graphs remain efficiently solvable for dually chordal graphs too.

Our results on dually chordal graphs not only generalize, but also improve and extend the corresponding results on strongly chordal graphs, since a maximum neighbourhood ordering (if it exists) can be constructed in linear time and we consequently use the underlying structure properties of dually chordal graphs closely connected to hypergraphs.

1 Introduction

The fundamental importance of tree structure for the design of efficient graph algorithms is well-known. Chordal and strongly chordal graphs are good examples for generalizations of trees which have many algorithmic implications. They are characterized by properties of their clique hypergraphs.

*** Second and third author supported by the VW–Stiftung Project No. I/69041
e-mail addresses: ab@marvin.uni-duisburg.de, chepoi@university.moldova.su, dragan@university.moldova.su

Here we investigate the algorithmic use of another generalization of trees (including strongly chordal graphs) namely *dually chordal graphs* which are especially useful for domination-like problems (including Steiner tree) and distance problems. They are also characterized by tree properties of corresponding hypergraphs. Hereby our central notion is that of a hypertree:

A hypergraph $\mathcal{E}$ is a *hypertree* iff there is a tree T with the same vertex set such that the hyperedges of $\mathcal{E}$ induce subtrees in T.

Now to some further notions. Let $G = (V, E)$ be a finite connected simple (i.e. without loops and multiple edges) and undirected graph. $N(v) = \{v : uv \in E\}$ is the *open neighbourhood* and $N[v] = N(v) \cup \{v\}$ is the *closed neighbourhood*. For $Y \subseteq V$ let $G(Y)$ be the *subgraph induced by* Y. For a graph G with the vertex set $V = \{v_1, \ldots, v_n\}$ let $G_i = G(\{v_i, v_{i+1}, \ldots, v_n\})$ and $N_i[v]$ $(N_i(v))$ be the *closed (open) neighbourhood* of v in G_i.

A vertex v is *simplicial* iff $N[v]$ is a clique. The ordering $(v_1, \ldots, v_n)$ of V is a *perfect elimination ordering* iff for all $i \in \{1, \ldots, n\}$ the vertex v_i is simplicial in G_i. The graph G is *chordal* iff G has a perfect elimination ordering.

The ordering $(v_1, \ldots, v_n)$ is a *strong elimination ordering* iff for all $i \in \{1, \ldots, n\}$ $N_i[v_j] \subseteq N_i[v_k]$ when $v_j, v_k \in N_i[v_i]$ and $j < k$. The graph G is *strongly chordal* iff G has a strong elimination ordering.

A vertex $u \in N[v]$ is a *maximum neighbour* of v iff for all $w \in N[v]$ the inclusion $N[w] \subseteq N[u]$ holds (note that $u = v$ is not excluded). A vertex is *doubly simplicial* iff it is simplicial and has a maximum neighbour. The ordering $(v_1, \ldots, v_n)$ is a *maximum neighbourhood ordering* iff for all $i \in \{1, \ldots, n\}$ there is a maximum neighbour $u_i \in N_i[v_i]$:

$$\text{for all } w \in N_i[v_i], N_i[w] \subseteq N_i[u_i] \text{ holds.}$$

$(v_1, \ldots, v_n)$ is a *doubly perfect ordering* iff each vertex v_i is doubly simplicial in G_i.

The graph G is *dually chordal* ([14] – called *HT-graphs* there, cf. [3] for a systematic investigation collecting many characterizations) iff G has a maximum neighbourhood ordering.

The graph G is *doubly chordal* ([23]) iff G has a doubly perfect ordering. There is a close connection between chordal and dually chordal graphs which can be expressed in terms of hypergraphs (for hypergraph notions we follow [2]). Let $\mathcal{N}(G) = \{N[v] : v \in V\}$ be the (*closed*) *neighbourhood hypergraph* of G and let $\mathcal{C}(G) = \{C : C$ is a maximal clique of $G\}$ be the *clique hypergraph* of G. The *distance* $d(u, v)$ between vertices $u, v \in V$ is the length (i.e. number of edges) of a shortest path connecting u and v. The *disk* centered at vertex v with radius k is the set of all vertices having distance at most k to v:

$$N^k[v] = \{u \in V : d(v, u) \leq k\}.$$

By $\mathcal{D}(G) = \{N^k[v] : v \in V, k$ a non–negative integer$\}$ we denote the *disk hypergraph* of the graph G.

Now let $\mathcal{E}$ be a hypergraph with underlying vertex set V i.e. $\mathcal{E}$ is a set of subsets of V. The *dual hypergraph* $\mathcal{E}^*$ has $\mathcal{E}$ as its vertex set and for every

($v \in V$) a hyperedge $\{e \in \mathcal{E} : v \in e\}$. The *line graph* $L(\mathcal{E}) = (\mathcal{E}, E)$ of $\mathcal{E}$ is the intersection graph of $\mathcal{E}$, i.e. $ee' \in E$ iff $e \cap e' \neq \emptyset$. The underlying graph (or two-section graph) of a hypergraph $\mathcal{E}$ has the same vertex set as $\mathcal{E}$ and two vertices form an edge iff they are contained in a common hyperedge of $\mathcal{E}$. A *partial hypergraph* of hypergraph $\mathcal{E}$ has V as the underlying vertex set and some edges of $\mathcal{E}$.

As already mentioned a hypergraph $\mathcal{E}$ is a *hypertree* iff there is a tree T with vertex set V of $\mathcal{E}$ such that every edge $e \in \mathcal{E}$ induces a subtree in T. It is well-known that $\mathcal{E}$ is a hypertree iff the line graph $L(\mathcal{E})$ is chordal and $\mathcal{E}$ has the *Helly* property i.e. any pairwise intersecting subfamily of edges of $\mathcal{E}$ has a common vertex (see [2]). A hypergraph $\mathcal{E}$ is a *dual hypertree* (or α-*acyclic hypergraph*) iff there is a tree T with vertex set $\mathcal{E}$ such that for all vertices $v \in V$ $T_v = \{e \in \mathcal{E} : v \in e\}$ induces a subtree of T, i.e. $\mathcal{E}^*$ is a hypertree.

Theorem 1.1 *[14], [3]*   Let $G = (V, E)$ be a graph. Then the following conditions are equivalent: (i) G is a dually chordal graph; (ii) $\mathcal{N}(G)$ is a hypertree; (iii) $\mathcal{N}(G)$ is a dual hypertree (i.e. α-acyclic); (iv) $\mathcal{D}(G)$ is a hypertree; (v) $\mathcal{C}(G)$ is a hypertree; (vi) G is the underlying graph of a hypertree.

It is well-known ([5]) that G is chordal iff $\mathcal{C}(G)$ is a dual hypertree, i.e. G is the underlying graph of some dual hypertree. Therefore the equivalence of parts (i) and (v) of Theorem 1.1 justifies the name 'dually chordal graphs' for graphs with maximum neighbourhood ordering.

Condition (iii) and the linear-time recognition of α-acyclic hypergraphs [30] give a linear-time recognition of dually chordal graphs.

2 r-Domination and r-packing: duality results

In this section we formulate general domination and packing problems and establish some duality results between them.

Let $G = (V, E)$ be a graph and $r : V \to N \cup \{0\}$ (N the set of positive integers) be a function defined on V. A subset $D \subseteq V$ is an r-*dominating set* if for any vertex $v \in V$ there is a vertex $u \in D$ with $d(u, v) \leq r(v)$. An r-*packing set* is a subset $P \subseteq V$ such that $d(u, v) > r(u) + r(v)$ for all $u, v \in P$. The r-*domination problem* is to find an r-dominating set with minimum size $\gamma_r(G)$ and the r-*packing problem* is to find an r-packing set with maximum size $\pi_r(G)$. (For $r \equiv 1$ these are the usual *domination* or *packing* problems.) Then $\gamma_r(G)$ and $\pi_r(G)$ are called the r-*domination* and r-*packing numbers* of G.

For a graph G and a function $r : V \to N \cup \{0\}$ define the partial hypergraph $\mathcal{D}(G, r)$ of the disk hypergraph $\mathcal{D}(G)$ as follows:

$$\mathcal{D}(G, r) = \{N^{r(v)}[v] : v \in V\}.$$

The r-domination and r-packing problems on G may be formulated as the transversal and matching problems on the hypergraph $\mathcal{D}(G, r)$. Recall that a *transversal* of a hypergraph $\mathcal{E}$ is a subset of vertices which meets all edges of $\mathcal{E}$.

A *matching* of $\mathcal{E}$ is a subset of pairwise disjoint edges of $\mathcal{E}$. For a hypergraph $\mathcal{E}$, the *transversal problem* is to find a transversal with minimum size $\tau(\mathcal{E})$ and the *matching problem* is to find a matching with maximum size $\mu(\mathcal{E})$.
From the definitions we obtain

Lemma 2.1 D *is an* r-*dominating set of a graph* G *iff* D *is a transversal of* $\mathcal{D}(G,r)$.
P *is an* r-*packing of* G *iff* P *is a matching of* $\mathcal{D}(G,r)$.
Thus $\tau(\mathcal{D}(G,r)) = \gamma_r(G)$ *and* $\mu(\mathcal{D}(G,r)) = \pi_r(G)$ *hold for any graph* G *and any function* $r : V \to N \cup \{0\}$.

The parameters $\gamma_r(G)$ and $\pi_r(G)$ are related by a min–max duality inequality: $\gamma_r(G) \geq \pi_r(G)$ for any graph G and any function $r : V \to N \cup \{0\}$. As was shown in [9] for $r(v) \equiv k$, the equality $\gamma_k(G) = \pi_k(G)$ holds on strongly chordal graphs.

Each partial hypergraph of a hypertree is a hypertree again. Thus, by Theorem 1.1, Lemma 2.1 and the well–known equality $\tau(\mathcal{E}) = \mu(\mathcal{E})$ for hypertrees ([2]) we are able to derive a more general duality result on dually chordal graphs.

Theorem 2.2 *[14] For any function* $r : V \to N \cup \{0\}$ *defined on the vertices of a dually chordal graph* G $\gamma_r(G) = \pi_r(G)$ *holds.*

3 Computing maximum neighbourhood orderings and distances

In sections 4 and 5 the special algorithm for determining a maximum neighbourhood ordering is important. For graphs with such an ordering it can be determined in linear time. [11] contains such an algorithm. The algorithm bases on the *maximum cardinality search (MCS)* algorithm of [30] for finding a perfect elimination ordering of chordal graphs. According to MCS the vertices of a graph are numbered from n to 1 in decreasing order. As the next vertex to number, select the vertex adjacent with the largest number of previously numbered vertices, breaking ties arbitrarily [30].

Now we present an algorithm which for dually chordal graphs gives a maximum neighbourhood ordering .

Algorithm 3.1 MNO *(Find a maximum neighbourhood ordering of* G*)*

Input: *A dually chordal graph* $G = (V, E)$ *with* $|V| = n > 1$.
Output: *A maximum neighbourhood ordering of* G.

(0) *Initially all* $v \in V$ *are unnumbered and unmarked;*
(1) *Choose an arbitrary* $v \in V$, *number* v *with* n *i.e.* $v_n := v$ *and* $mn(v_n) := v$;
    repeat
(2)   *among all unmarked vertices select a numbered vertex* u *such that* $N[u]$
      contains a maximum number of numbered vertices;

(3)   *number all unnumbered vertices x from N[u] consecutively by decreasing*
        numbers between n − 1 and 1 and let mn(x) := u for all of them;

(4)   *mark u;*
        until *all vertices are numbered.*

The meaning of $mn(x)$ is that of a maximum neighbour of x. Note that the
algorithm also yields a maximum neighbour for each vertex, and all vertices of
$N[v_n]$ occur consecutively in the ordering on the left of v_n and have v_n as their
maximum neighbour. Furthermore for all v_i with $i \leq n - 1$ $mn(v_i) \neq v_i$ holds.

Recall that the *square* G^2 of a graph $G = (V, E)$ has the same vertex set V
and two distinct vertices $u, v \in V$ are adjacent in G^2 iff $d(u, v) \leq 2$ in G. It is
easy to see that for any graph G the equality $G^2 = L(\mathcal{N}(G))$ holds. Therefore if
G is dually chordal then by Theorem 1.1 the graph G^2 is chordal.

Lemma 3.2 *Let G be a dually chordal graph. Then any numbering of vertices
of G generated by the MNO algorithm can also be generated by a MCS of G^2
and thus this numbering is a maximum neighbourhood ordering of G.*

Proof. Let $(v_1, \ldots, v_n)$ be the ordering obtained by MNO. Assume that the
end vertices $v_{i+1}, \ldots, v_n$ of the ordering can also be generated by a MCS of
G^2 and each $v_k, k \geq i + 1$, has a maximum neighbour in the subgraph $G_k =
G(\{v_k, \ldots, v_n\})$. Let u be the next vertex chosen by MCS. Suppose that u cannot
be replaced by v_i in MCS of G^2. This means that in G^2 the vertex u is adjacent with more numbered vertices than v_i (we call *numbered* only the vertices
$v_{i+1}, \ldots, v_n$). Necessarily, $v_i \notin N[v_n]$, otherwise v_i is selected by both procedures
when the lists of numbered vertices consist of v_n only.

Since G^2 is chordal u is a simplicial vertex of $G^2(\{u, v_{i+1}, \ldots, v_n\})$. Therefore
$d(x, y) \leq 2$ for any two numbered vertices x, y with $d(u, x) \leq 2$ and $d(u, y) \leq 2$.
By the Helly property all closed neighbourhoods of numbered vertices of the disk
$N^2[u]$ have a common vertex w adjacent to u. We claim that w is numbered.
Otherwise, since u is chosen by MCS

$$|N^2_{i+1}[w]| = |N^2_{i+1}[u]|$$

holds. Since $N_{i+1}[w] = N^2_{i+1}[u]$ also $N_{i+1}[w] = N^2_{i+1}[w]$ holds. This implies that
w is adjacent to all numbered vertices i.e. w and v_n are adjacent or coincide. Thus
w is a numbered vertex because all vertices from $N[v_n]$ are already numbered.
Now return to vertex v_i chosen by MNO. By this algorithm we have that v_i is
adjacent to a numbered vertex v such that $N[v]$ contains the maximum number
of numbered vertices. In particular,

$$|N_{i+1}[v]| \geq |N_{i+1}[w]| = |N^2_{i+1}[u]|.$$

Since $N_{i+1}[v] \subseteq N^2_{i+1}[v_i]$ we get a contradiction with the assumption that
$N^2_{i+1}[u] > N^2_{i+1}[v_i]$.
Thus we obtain that the vertex v_i can be chosen by MCS too i.e. we can assume
w.l.o.g. that $v_i = u$ and

$$N_{i+1}[v] = N_{i+1}[w] = N^2_{i+1}[v_i].$$

Therefore the vertex v_i has a maximum neighbour v. □

Theorem 3.3 *[11]*   *The algorithm MNO finds a maximum neighbourhood ordering of a dually chordal graph G in linear time.*

Maximum neighbourhood orderings of graphs immediately lead to optimal algorithms for computing the distance matrix for all graphs having such orderings. Let $(v_1, \ldots, v_n)$ be a maximum neighbourhood ordering of a graph G. The maximum neighbour $mn(v_i)$ of the vertex v_i in $G(\{v_i, v_{i+1}, \ldots, v_n\})$ has an important metric property:

For any vertex $v_j, j > i$, which is non–adjacent to v_i there exists a shortest path between v_i and v_j which passes through $mn(v_i)$. In particular, we obtain that any graph $G(\{v_i, v_{i+1}, \ldots, v_n\})$ is a distance–preserving subgraph of G.

Let $D(G) = (d(v_i, v_j))_{i,j \in \{1,\ldots,n\}}$ denote the distance matrix of the dually chordal graph $G = (V, E)$. By $D_{i+1}(G)$ we denote the submatrix of $D(G)$ which contains the distances between the vertices $v_{i+1}, \ldots, v_n$. The next submatrix $D_i(G)$ is obtained from $D_{i+1}(G)$ by adding the $i - th$ row and $i - th$ column according to the rule:
For any $k > i$ define

$$d(v_i, v_k) = d(v_k, v_i) = \begin{cases} 1 & \text{if } v_i \text{ and } v_k \text{ are adjacent} \\ d(mn(v_i), v_k) + 1 & \text{otherwise} \end{cases}$$

Evidently, this procedure correctly finds the whole matrix $D(G)$ in optimal time $O(n^2)$. Moreover, the maximum neighbourhood ordering of G for any two query vertices u and v allows to find in time $O(c \cdot d(u, v))$ a shortest path between u and v (c is the necessary time to verify the adjacency of two vertices). Let $num(v) = i$ if $v = v_i$ in the maximum neighbourhood ordering of G.

Procedure 3.4 (sh–path(u, v))

> **if** u *and* v *are adjacent* **then return** (u, v)
> **else**
> > **if** $num(u) < num(v)$ **then**
> > > **return** $(u, \text{sh–path}(mn(u), v))$
> > **else**
> > > **return** $(\text{sh–path}(u, mn(v)), v)$

Lemma 3.5 *If $u = x_1, \ldots, x_k = v$ is the shortest path from u to v constructed by $sh - path(u, v)$ then the sequence $x_1, \ldots, x_k$ is unimodal i.e. $num(x_1) < \ldots < num(x_t)$ and $num(x_t) > \ldots > num(x_k)$ for some $1 \le t \le k$.*

4 r–Domination problems

There is a lot of special graph classes where the domination problem remains NP-complete. For the k–domination problem polynomial time algorithms are known for trees ([26]), *sun*–free chordal graphs ([8]), 3–sun–3–anti–sun–free chordal graphs ([9]). For the general r–domination problem polynomial algorithms were developed for sun–free chordal graphs ([7]), 3–sun–3–anti–sun–free chordal graphs ([10]), sun–free and 3–sun–3–anti–sun–free bridged graphs ([12]) and dually chordal graphs ([14]).

In ([14]) we presented $O(n^2)$ time algorithms for the r–domination and r–packing problems, using the fact that the disk hypergraphs of dually chordal graphs are hypertrees. Below we present a linear $O(|E|)$ algorithm for the r–domination and r–packing problems on dually chordal graphs, which avoids the construction of the disk hypergraph. The algorithm uses the maximum neighbourhood ordering and is similar to Chang's algorithm for r–domination on strongly chordal graphs ([7]). The algorithm simultaneously finds an r–dominating set D and an r–packing set P such that $|D| = |P|$. This provides an algorithmic proof of duality results between these two problems on dually chordal graphs.

Let $G = (V, E)$ be a dually chordal graph and $(v_1, \ldots, v_n)$ be the ordering of V generated by the MNO algorithm. The algorithm processes the vertices in the order from v_1 to v_n. In iteration i the algorithm decides whether the vertex v_i has to be put into the r–dominating set D. If v_i is included in D then a certain vertex u_i which is r–dominated by v_i is included in the r–packing set P. Initially both sets D and P are empty. After processing, vertex v_i is deleted from the graph and an information whether or not v_i was included in D is given to its maximum neighbour $mn(v_i)$ and/or to its other neighbours.

For technical reasons, as in [7] we associate to each vertex v_i the domination radius $r(v_i)$ and the non–negative integer $c(v_i)$. Initially $c(v_i) = \infty$ for all i. Both $r(v_i)$ and $c(v_i)$ keep decreasing during the execution of the algorithm. At each step $r(v_i)$ becomes the current radius within which the vertex v_i must be r–dominated in the remaining graph. Opposite to $r(v_i)$, $c(v_i)$ means that in the initial graph G there exists a vertex u from the current set D such that $d(v_i, u) \leq c(v_i)$. $r(v_i)$ decreases in the case where v_i is the maximum neighbour of a vertex v_j, $j < i$, that is not properly r–dominated by a vertex of D within distance $r(v_j)$ in iteration j. In this case, $r(v_i)$ is set to be $r(v_j) - 1$. Similarly, in a previous iteration, $r(v_j)$ is set to be $r(v_k) - 1$, $k < j$. Continuing this argument, we find that there is a smallest vertex v_{i^*} that forces $\ldots, k, j, i$ to decrease their $r(\cdot)$ values, although $r(v_{i^*})$ never changes. We use $fn(v_i)$ (furthest neighbour of v_i) to denote this initial vertex v_{i^*} from which $r(v_i)$ decreases.

In step i the algorithm processes vertex v_i according to the following rules. When $r(v_i) = 0$ then v_i must be in D because no other vertex r–dominates v_i. Moreover, we put $fn(v_i)$ in P and set $c(v_i) = 0$. Otherwise, if $r(v_i) > 0$ then we distinguish two cases. If $c(v_i) > r(v_i)$, $c(v) + 1 > r(v_i)$ and $r(v) > 0$ for any $v \in N_i[v_i]$ then we update $r(mn(v_i))$ by

$$r(mn(v_i)) = \min\{r(mn(v_i)), r(v_i) - 1\}.$$

Finally, if $c(v_i) \leq r(v_i)$ or $c(v) + 1 \leq r(v_i)$ for some $v \in N_i[v_i]$ or $r(v) = 0$ for some $v \in N_i[v_i]$ then we do nothing. At each step we have to update $c(v)$ for all $v \in N_i[v_i]$:

$$c(v) = \min\{c(v), c(v_i) + 1\}.$$

Algorithm 4.1 RDP *(Find a minimum r–dominating set and a maximum r–packing set of a dually chordal graph G)*

Input: *Dually chordal graph $G = (V, E)$ with maximum neighbourhood ordering $(v_1, \ldots, v_n)$ generated by the MNO algorithm and non–negative integers $r(v_1), \ldots, r(v_n)$*

Output: *A minimum r–dominating set D and a maximum r–packing set P of G*

(1) $D := \emptyset; P := \emptyset;$
(2)   **for all** $v \in V$ **do begin** $c(v) := \infty; fn(v) := v$ **end**;
(3)   **for** $i := 1$ **to** $n - 1$ **do**
       begin
(4)   **if** $r(v_i) = 0$ **then**
         begin
(5) $D := D \cup \{v_i\}; P := P \cup \{fn(v_i)\};$
(6) $c(v_i) := 0;$
         end
       else
(7)   **if** $c(v_i) > r(v_i)$ **and** $c(v) + 1 > r(v_i)$ **and** $r(v) > 0$
           for all $v \in N_i(v_i)$ **then**
(8)   **if** $r(m, n(v_i)) \geq r(v_i)$ **then**
            begin
(9) $r(mn(v_i)) := r(v_i) - 1;$
(10) $fn(mn(v_i)) := fn(v_i)$
          end
(11)   **for all** $v \in N_i[v_i]$ **do** $c(v) := \min\{c(v), c(v_i) + 1\}$
       end
(12)   **if** $r(v_n) < c(v_n)$ **then begin** $D := D \cup \{v_n\}; P := P \cup \{fn(v_n)\}$ **end**

Theorem 4.2 *Algorithm RDP is correct and works in linear time $O(|V| + |E|)$.*

Proof. The time bound of the algorithm is obviously linear. Now to the correctness. In order to prove that D is a minimum r–dominating set of G we use the following reformulation of an r–domination problem in terms of $r(v_i)$ and $c(v_i)$:

"Find a minimum size set $D \subseteq V$ such that for any vertex $v \in V$ either $d(u, v) \leq r(v)$ for some $u \in D$ or $d(v, w) + c(w) \leq r(v)$ for some $w \in V$."

We need some auxiliary results; see also [7] for the case of strongly chordal graphs. Because of space limitations we omit here some of the proofs – for a full version see [4].

Lemma 4.3 *If $r(v_i) = 0$ then D is a minimum r–dominating set of G_i iff $D = D' \cup \{v_i\}$ where D' is a minimum r–dominating set of G_{i+1} with $c(v) := 1$ for $v \in N_i(v_i)$ and $c(v) := c(v)$ otherwise, and $r(v) := r(v)$ for all v.*

Lemma 4.4 *Assume that $c(v_i) > r(v_i), c(v) + 1 > r(v_i)$ and $r(v) > 0$ for all $v \in N_i[v_i]$. A subset $D \subseteq (\{v_{i+1}, \ldots, v_n\} \setminus N_i(v_i)) \cup \{mn(v_i)\}$ is a minimum r–dominating set of G_i iff D is a minimum r–dominating set of G_{i+1} with $c(v) :=*

$c(v)$ for $v \notin N_i(v_i)$ and $c(v) := \min\{c(v), c(v_i) + 1\}$ for $v \in N_i(v_i)$ and $r(v) := r(v)$ for all $v \neq mn(v_i)$ and $r(mn(v_i)) := \min\{r(v_i) - 1, r(mn(v_i))\}$.

Lemma 4.5 *Suppose that* $r(v_i) \geq 1$. *Assume* $c(v_i) \leq r(v_i)$ *or* $c(v) + 1 \leq r(v_i)$ *for some* $v \in N_i(v_i)$ *or* $r(v) = 0$ *for some* $v \in N_i(v_i)$. *A subset* $D \subseteq \{v_{i+1}, \ldots, v_n\}$ *is a minimum* r–*dominating set of* G_i *iff* D *is a minimum* r–*dominating set of* G_{i+1} *with* $r(v) := r(v)$ *for all* $v \neq v_i$ *and* $c(v) := c(v)$ *for all* $v \notin N_i(v_i)$ *and* $c(v) := \min\{c(v), c(v_i) + 1\}$ *when* $v \in N_i(v_i)$.

Now we return to the proof of Theorem 4.2. Observe that if $r(v_n) < c(v_n)$ then v_n is not dominated by any vertex of D and thus v_n must be included in D.

Let D be the set determined by the algorithm before step i, and let D' be an arbitrary minimum r–dominating set of the graph G_i with modified functions $r(v)$ and $c(v)$. By Lemmas 4.3, 4.4 and 4.5 we obtain that $D \cup D'$ is a minimum r–dominating set of the initial graph G with initial domination radii. Thus in step n we obtain the required minimum r–dominating set D of G.

Next we concentrate on the proof that the algorithm correctly finds a maximum r–packing set P of G.

For two vertices u, v of a dually chordal graph G by a *maximum neighbour path* we will mean the shortest path between u and v computed by the procedure $sh - path(u, v)$.

Lemma 4.6 $|P| = |D|$.

Proof. By weak duality we know that $|D| \geq |P|$. If $|P| < |D|$ then there are two vertices $u, v \in D$ and a vertex $x \in P$ such that $fn(u) = x = fn(v)$. By the algorithm $fn(u)$ and $fn(v)$ reach vertices u and v along maximum neighbour paths connecting x, u and x, v resp. Let y be the vertex belonging to both paths which is furthest from x. Denote by u' and v' the next neighbours of y in these paths. By the algorithm $fn(mn(y)) = fn(y) = x$ i.e. reaching y the vertex x is certainly transmitted to its maximum neighbour $mn(y)$. So, either $fn(u) = fn(u') \neq x$ or $fn(v) = fn(v') \neq x$, a contradiction. $\square$

Lemma 4.7 P *is an* r–*packing set of* G.

Proof. Assume to the contrary that $d(u, v) \leq r(u) + r(v)$ for some vertices $u, v \in P$. By the algorithm, there exist $x, y \in D$ such that $u = fn(x)$ and $v = fn(y)$. Let P' and P'' be maximum neighbour paths between u, x and v, y. Both of these paths are increasing. By the algorithm, we obtain that P' and P'' are disjoint, otherwise their common vertex is a 'bottleneck' for transmission of vertices $u = fn(x)$ and $v = fn(y)$ (see also Lemma 4.6).

By Lemma 3.5 the maximum neighbour path between vertices u and v consists of a subpath $(u, \ldots, u')$ of P', an edge $u'v'$ and a subpath $(v, \ldots, v')$ of P''. By the algorithm we have $d(u, x) = r(u)$ and $d(v, y) = r(v)$. Since $d(u, v) \leq r(u) + r(v)$ the maximum neighbour path between u and v does not completely contain both paths P' and P'' i.e. $u' \neq x$ or $v' \neq y$ holds. Among adjacent vertices $u' \in P'$ and $v' \in P''$ with $u' \neq x$ or $v' \neq y$ we choose adjacent vertices $u^* \in P'$ and $v^* \in P''$ whose sum $d(u^*, x) + d(v^*, y)$ is minimal. Let $num(v^*) > num(u^*)$. Then either $u^* = x$ or v^* is adjacent to the maximum neighbour $mn(u^*)$ of u^*. In the second case we obtain a contradiction with our choice of the edge u^*v^* except the case when $mn(u^*) = x$ and $v^* = y$. So, in this case we conclude that vertex y is adjacent to x and is its neighbour in P'. Now consider the first case i.e. $u^* = x$. Then $v^* \neq y$. Before step $num(x)$ we have $r(x) = 0$ – thus after this step we obtain $c(v^*) = 1$. Then in step $num(v^*)$ the condition in line (7) of the algorithm does not hold. This contradicts to our assumption that $fn(y) = fn(mn(v^*)) = fn(v^*) = \ldots = v$. Thus, in the assumption that $num(v^*) > num(u^*)$ we obtain that $v^* = y$, $mn(u^*) = x$ and the vertices x and y are adjacent. Let v^+ be the neighbour of y in P'' if P'' contains at least two vertices, otherwise let $v^+ = v^*$. If $num(v^+) < num(u^*)$ then in step $num(v^+)$ we obtain $r(y) = 0$. Therefore in step $num(u^*)$ we already have a neighbour of u^* which violates the condition in line (7) of the algorithm. Hence $num(v^+) > num(u^*)$ i.e. the vertex $x = mn(u^*)$ is adjacent to v^+. Then $r(x) = 0$ in step $num(u^*)$ and again we can apply the condition in line (7) in order to obtain a contradiction with $fn(x) = fn(u^*) = \ldots = u$. □

From Lemmas 4.6 and 4.7 we immediately obtain that the set P computed by the algorithm RDP is a maximum r–packing of G. This completes the proof of the theorem. □

5 p–Center and q–dispersion problems

Let $G = (V, E)$ be a graph and $w : V \to R^+ \cup \{0\}$ be a non–negative weight function defined on V. We define the *radius* $r(S)$ of a set $S \subseteq V$ as $\max\{w(u)d(u, S) : u \in V\}$, where $d(u, S) = \min\{d(u, v) : v \in S\}$. For a given positive integer $p \leq |V|$ we define the p–*radius* of G as

$$r_p(G) = \min\{r(C) : C \subseteq V, |C| = p\}.$$

A p–*center* of G is a set C that realizes the p–radius of G.

The p–center problem is one of the main models in facility location theory; see [29], [19]. For general graphs, the problem of finding p–centers is NP–hard [19]. Moreover, even the 2–approximation variant of this problem remains NP–hard for $\varepsilon < 2$ while the polynomial 2–approximation algorithm is given in [25]. Note also that since the domination problem is a particular instance of the p–center problem [19] the p–center problem is NP–hard in all graph classes where the domination problem is NP–complete.

Polynomial algorithms for the p–center problem are known only in trees [19], [20], [6] and almost–trees [17]. The best known algorithms have time complexity $O(|V|)$ for unweighted trees [16] and $O(|V| \cdot log^2|V|)$ for weighted trees [22].

As we already mentioned the domination problem on a graph G is a particular case of the p–center problem. Conversely, the p–center problem can be reduced to solving a logarithmic number of r–domination problems on G. The p–radius r_p of G may be defined as a minimum p–radius $r(C)$ of a p–vertex set $C \subseteq V$.

Evidently, r_p is an element of the weighted distance matrix $(w(u)d(u,v))_{u,v \in V}$. To compute r_p we search this matrix for the minimum value which is feasible in the following sense. A value r is *feasible* if there exists a p–vertex set $C \subseteq V$ whose radius $r(C)$ is not greater than r. In order to decide whether a given r ist feasible it suffices to solve the r–domination problem on G with $r(u) = \lceil r/w(u) \rceil$ if $w(u) \neq 0$ and $r(u) = \infty$ (a sufficiently large number) otherwise and compare if the obtained solution contains not more than p vertices. Thus the p–center problem can be solved by computing the weighted distance matrix and searching it by repeatedly using linear–time median finding and solving the corresponding r–domination problem. Since for dually chordal graphs the weighted distance matrix is computed in $O(n^2)$ time and the r–domination problem is solved in $O(|E|)$ time the proposed approach leads to an $O(|V|^2 + |E| log|V|)$ algorithm for the p–center problem in these graphs.

Next we consider the q–dispersion problem [6] which is in some sense dual to the p–center problem. For a given subset $X \subseteq V$ find a q–vertex set $S \subseteq X$ such that they are as far apart as possible i.e. maximizing $\min\{d(u,v) : u,v \in S\}$. Then S is called the q–*dispersion set* of S and $\max\{\min\{d(u,v) : u,v \in S\} : |S| = q\} = d_q(X)$ is the q–*dispersion* of X. The duality between the p–center and q–dispersion problems in trees was established in different variants in [20], [6]. As we will show a similar result holds for dually chordal graphs too.

Theorem 5.1 *For each dually chordal graph G and each subset $X \subseteq V$ the following equality holds:*

$$\min\{\max\{d(v,C) : v \in X\} : C \subseteq V, |C| = p\} =$$

$$= \lceil (\max\{\min\{d(v,u) : v,u \in S, v \neq u\} : S \subseteq X, |S| = p+1\} + 1)/2 \rceil.$$

The solution of the q–dispersion problem on a dually chordal graph G may be obtained in the following way. First, solve the p–center problem on G with $p = q - 1$ and $w(u) = 1$ if $u \in X$ and $w(u) = 0$ if $u \in V \setminus X$. Let r_p be the p–radius of G. By Theorem 5.1 either $d_q(X) = 2r_p$ or $d_q(X) = 2r_p - 1$. In order to compute the exact value of $d_q(X)$ we must solve at most two independent set problems in the subgraph induced by X of the k–th power G^k of G, first for $k = 2r_p - 1$ and later for $k = 2r_p - 2$. Recall that the graph G^k has the same vertex set V and two distinct vertices $u,v \in V$ are adjacent in G^k iff $d(u,v) \leq k$ in G. If for $k = 2r_p - 1$ the obtained maximal independent set S of $G^k(X)$ has at least q vertices then we are done: $d_q(X) = 2r_p$ and S is a q–dispersion set. Otherwise, the maximal independent set $S \subseteq X$ obtained for $k = 2r_p - 2$ is the required one and $d_q(X) = 2r_p - 1$. Observe that in the second case the independent set problem in G^k is equivalent to the r–packing problem on G with $r(u) = r_p - 1$ for any $u \in X$ and $r(u) = \infty$ otherwise, that is not true for the first case.

As it was shown in [3] all powers of doubly chordal graphs are doubly chordal, thus they are chordal. The independent set problem in chordal graphs is solvable in time linear in the size of the graph. So, the presented algorithm gives an $O(|V|^2 + |E| \log |V|)$ time bound for computing the q–dispersion set of a doubly chordal graph. Unfortunately, the presented method is not polynomial for dually chordal graphs, because odd powers of dually chordal graphs have no special structure with respect to the independent set problem. Moreover, as we show below the q–dispersion problem for dually chordal graphs is NP-complete.

Let G_0 be an arbitrary graph. Consider the dually chordal graph G, obtained from G_0 by adding a new vertex v adjacent to all vertices of G_0. Evidently, each independent set of G_0 is independent in G and conversely. Let d_q be the q–dispersion of V in G. Then in the graph G, and therefore in the graph G_0 too, there is an independent set with at least q vertices if and only if $d_q = 2$. Thus we reduce the general independent set problem to the q–dispersion problem on dually chordal graphs.

The same construction works for proving that all four classical graph problems (maximum independent set, minimum clique covering, minimum coloring, maximum clique problem) and other problems like e.g. the Hamiltonian circuit problem remain NP–complete for dually chordal graphs. Summarizing the results of this section we obtain

Theorem 5.2

1) The p-center problem on a dually chordal graph $G = (V, E)$ can be solved in $O(|V|^2 + |E| \cdot \log |V|)$ time;

2) the q–dispersion problem on dually chordal graphs is NP–complete;

3) the q–dispersion problem on a doubly chordal graph $G = (V, E)$ can be solved in $O(|V|^2 + |E| \cdot \log |V|)$ time.

Next consider the simplest cases of the q–dispersion and p–center problems when $q = 2$ and $p = 1$. They are called the *diameter* and the *center problems* in G. Recall some necessary definitions. The *eccentricity* of a vertex $v \in V$ is $e(v) = \max\{d(v, u) : u \in V\}$. The *diameter* $d(G)$ of G is the maximum eccentricity, while the *radius* $r(G)$ of G is the minimum ecentricity of vertices of G. A vertex whose eccentricity is equal to $r(G)$ is called a *central vertex*. Vertices $u, v \in V$ form a *diametral pair* of G if $d(u, v) = d(G)$. We present a linear time algorithm for computing a central vertex, the diameter and a diametral pair of vertices of a dually chordal graph G. Another linear time algorithm for finding a central vertex of G is given in [11].

Let $G = (V, E)$ be a dually chordal graph. According to [11] for any vertex $v \in V$ if $d(v, u) = e(v)$ then $e(u) \geq 2r(G) - 2$. The value $k = [(e(u) + 1)/2]$ is an approximation of the radius of G, more precisely either $k = r(G)$ or $k = r(G) - 1$. Next we apply the r–domination algorithm, first with $r(v) \equiv k$ for all $v \in V$ and later with $r(v) \equiv k + 1$ for all $v \in V$. If in the first case the obtained minimum r–dominating set D consists of a single vertex x then we are done: x is central and $r(G) = k$. Otherwise we have $r(G) = k + 1$ and the single vertex of the minimum r–dominating set with $r(v) \equiv k + 1$ for all $v \in V$ must be central. Thus a central vertex x of a dually chordal graph G can be found in linear time, because the eccentricity of any vertex and an r–dominating set are found in linear time.

Next we present a linear time algorithm for computing the diameter $d(G)$ of a dually chordal graph G.

Algorithm 5.3 Diameter *(Find the diameter $d(G)$ and a diametral pair of vertices of a dually chordal graph G).*

Input: *A dually chordal graph $G = (V, E)$*
Output: *The diameter $d(G)$ of G and a diametral pair of vertices*

(1)   find a central vertex v and radius $r = r(G)$ of graph G;
(2)   using the MNO algorithm find a maximum neighbourhood ordering
       $(v_1, \ldots, v_n)$ of G with $v_n = v$ and let $M := \emptyset$;
(3)   **for all** $v \in V$ **do** $a(v) := 0$ and $fn(v) := v$;
(4)   **for** $i := 1$ **to** n **do**
(5)   **if for all** $v \in N_i[v_i]$ $a(v) < r - 1$ **and** $N(v_i) \cap M = \emptyset$ **then**
(6)   **if** $a(mn(v_i)) \leq a(v_i)$ **then**
              begin
(7) $a(mn(v_i)) := a(v_i) + 1$;
(8) $fn(mn(v_i)) := fn(v_i)$
              end
(9)   **else if** $a(v_i) = r - 1$ **then** $M := M \cup \{v_i\}$
(10) **if** M has two nonadjacent vertices u and v
       then $d(fn(u), fn(v)) = d(G) = 2r(G)$
(11) **else** $d(fn(u), fn(v)) = d(G) = 2r(G) - 1$ *for any pair of*
       vertices $u, v \in M$.

Theorem 5.4 *The diameter $d(G)$, the radius $r(G)$, a diametral pair and a central vertex of a dually chordal graph G can be found in linear time.*

Proof. It is sufficient to show the correctness of Algorithm Diameter only. First remark that any maximum neighbour path between v_n and arbitrary vertex $v \in V$ being unimodal must be decreasing; see Lemma 3.5. Comparing the algorithms Diameter and sh–path we conclude that any vertex v_i belongs to the maximum neightbour path between $f_n(v_i)$ and v_n. Thus, if $a(v_i) = r - 1$ then the vertices v_i and v_n must be adjacent if $i < n$. By the algorithm MNO we obtain that all vertices $v_{i+1}, \ldots, v_{n-1}$ are adjacent to the vertex v_n too. Summarizing, we deduce that M is a subset of $N[v_n]$ consisting of vertices v_i with $a(v_i) = r - 1$.

We assert that $|M| \geq 2$. To show this pick any $v \in M, v \neq v_n$. Let $\overline{v}$ be a farthest vertex from v i.e. $d(v, \overline{v}) \geq r(G)$. Then either v_n lies on a shortest path between v and $\overline{v}$ or $d(v_n, \overline{v}) = d(v, \overline{v}) = r$, otherwise v_n is not central. Therefore, either $v_n \in M$ or M contains a neighbour of v_n one step closer to $\overline{v}$.

Next let $\overline{u} = fn(u), \overline{v} = fn(v)$, where u and v are vertices of M selected on steps (10) or (11). Let $P' = (\overline{u}, u_1, \ldots, u)$ and $P'' = (\overline{v}, v_1, \ldots, v)$ be maximum neighbour paths between $\overline{u}, u$ and $\overline{v}, v$, respectively. Both these paths are increasing and may be extended to maximum neighbour paths between v_n and $\overline{u}$ and $\overline{v}$. By Lemma 3.5 the maximum neighbour path between vertices $\overline{u}$ and $\overline{v}$ consists of a subpath $(\overline{u}, \ldots, u')$ of P', an edge (u', v') and a subpath $(v', \ldots, \overline{v})$

of P''. If $u' = u$ and $v' = v$ then we are in conditions of step (11) i.e. the set M induces a complete subgraph. Evidently, then $d(G) < 2r(G)$. Since $d(\overline{u}, \overline{v}) = d(\overline{u}, u') + 1 + d(\overline{v}, v') = 2r(G) - 1$, we obtain that $d(\overline{u}, \overline{v}) = d(G) = 2r(G) - 1$. So, assume that either $u' \neq u$ or $v' \neq v$.

Among adjacent vertices $u' \in P'$ and $v' \in P''$ with $u' \neq u$ or $v' \neq v$ we choose adjacent vertices $u^* \in P'$ and $v^* \in P''$ whose sum $d(u^*, v) + d(v^*, v)$ is minimal. Let $num(v^*) > num(u^*)$. Then evidently $u^* \neq v_n$. If $u^* \neq u$ then $mn(u^*)$ is adjacent to v^*. From the choice of vertices u^* and v^* we get that $mn(u^*) = u$ and $v = v^*$. Then $a(u^*) = r - 2$, otherwise $fn(u) \neq fn(u^*)$. Let v^+ be a vertex of P'' adjacent to v. If $num(v^+) < num(u^*)$ then according to the algorithm the vertex $u = mn(u^*)$ does not receive the value on step $num(v^*)$ and $fn(u) \neq fn(u^*)$. Otherwise, if $num(v^+) > num(u^*)$ then u is adjacent to v^+. In this case the vertex $v = mn(v^+)$ does not receive the value $r - 1$ on step $j = num(v^+)$, because $u \in N_j(v^+)$ and u already has value $r - 1$.

Next assume that $u^* = u$ i.e. $u^* \in N[v_n]$. Since $num(u^*) < num(v^*)$ v_n and v^* are adjacent and $mn(v^*) = v_n$ (by the MNO algorithm). Thus $v_n = v$. Since $u \in N[v^*] \cap M$ according to the algorithm the vertex $v_n = v$ does not get the value $r - 1$ on step $num(v^*)$, a contradiction. □

In [4] we describe the use of maximum neighbourhood ordering for solving efficiently the *Steiner tree* and *Connected r–Domination* problems. This is not done here because of space limitations.

References

1. H. BEHRENDT and A. BRANDSTÄDT, Domination and the use of maximum neighbourhoods, TECHNICAL REPORT SM–DU–204, University of Duisburg 1992

2. C. BERGE, Hypergraphs, *North Holland*, 1989

3. A. BRANDSTÄDT, F.F. DRAGAN, V.D. CHEPOI, and V.I. VOLOSHIN, Dually chordal graphs, *Technical Report SM–DU–225*, University of Duisburg 1993, *Graph-Theoretic Concepts in Computer Science, 19th International Workshop WG'93, Utrecht, The Netherlands, LNCS 790, Springer, Jan van Leeuwen (Ed.)*, 237–251

4. A. BRANDSTÄDT, V.D. CHEPOI, and F.F. DRAGAN, The algorithmic use of hypertree structure and maximum neighbourhood orderings, *Technical Report SM–DU–244*, University of Duisburg 1994, submitted to *Theor. Comp. Science*

5. P. BUNEMAN, A characterization of rigid circuit graphs, *Discr. Math. 9* (1974), 205–212

6. R. CHANDRASEKARAN and A. DOUGHETY, Location on tree networks: p–center and q–dispersion problems, *Math. Oper. Res.* 1981, 6, No. 1, 50–57

7. G.J. CHANG, Labeling algorithms for domination problems in sun–free chordal graphs, *Discrete Applied Mathematics 22* (1988/89), 21–34

8. G.J. CHANG and G.L. NEMHAUSER, The k–domination and k–stability problems on sun–free chordal graphs, *SIAM J. Algebraic and Discrete Methods 5* (1984), 332–345

9. G.J. CHANG and G.L. NEMHAUSER, Covering, Packing and Generalized Perfection, *SIAM J. Algebraic and Discrete Methods 6* (1985), 109–132

10. F.F. DRAGAN, Dominating and packing in triangulated graphs (in Russian), *Meth. of Discr. Analysis (Novosibirsk)* 51 (1991), 17–36

11. F.F. DRAGAN, HT–graphs: centers, connected r—domination and Steiner trees, *Computer Science Journal of Moldova*, 1993, Vol. 1, No. 2, 64–83

12. F.F. DRAGAN, Domination in Helly graphs without quadrangles (in Russian) *Cybernetics and System Analysis (Kiev)* 6 (1993)

13. F. F. DRAGAN, and A. BRANDSTÄDT, r—Dominating cliques in Helly graphs and chordal graphs, *Technical Report SM–DU–228*, University of Duisburg 1993, *Proc. of the 11th STACS, Caen, France, Springer, LNCS 775*, 735–746, 1994

14. F. F. DRAGAN, C. F. PRISACARU, and V. D. CHEPOI, Location problems in graphs and the Helly property (in Russian), *Discrete Mathematics, Moscow*, 4 (1992), 67–73 (the full version appeared as preprint: F.F. Dragan, C.F. Prisacaru, and V.D. Chepoi, r—Domination and p—center problems on graphs: special solution methods and graphs for which this method is usable (in Russian), Kishinev State University, preprint MoldNIINTI, N. 948–M88, 1987)

15. M. FARBER, Domination, Independent Domination and Duality in Strongly Chordal Graphs, *Discr. Appl. Math.* 7 (1984), 115–130

16. G.N. FREDERICKSON, Parametric search and locating supply centers in trees, *Proc. Workshop on Algorithms and Data Structures (WADS'91), Springer, LNCS 519*, 1991, 299–319

17. Y. GUREVICH, L. STOCKMEYER and U. VISHKIN, Solving NP–hard problems on graphs that are almost trees and an application to facility location problems, *J. ACM* 31 (1984), 459–473

18. S.C. HEDETNIEMI and R. LASKAR, (eds.), Topics on Domination, *Annals of Discr. Math.* 48, North–Holland, 1991

19. O. KARIV and S.L. HAKIMI, An algorithmic approach to network location problems, I: the p–centers, *SIAM J. Appl. Math.* 1979, 37, No. 3, 513–538

20. A.W.J. KOLEN, Duality in tree location theory, *Cah. Cent. Etud. Rech. Oper.*, 25 (1983), 201–215

21. A. LUBIW, Doubly lexical orderings of matrices, *SIAM J. Comput.* 16 (1987), 854–879

22. N. MEGIDDO, A. TAMIR, E. ZEMEL and R. CHANDRASEKARAN, An $O(n log^2 n)$ algorithm for the k–th longest path in a tree with applications to location problems, *SIAM J. Comput.* 10 (1981), 328–337

23. M. MOSCARINI, Doubly chordal graphs, Steiner trees and connected domination, *Networks* 23 (1993), 59–69

24. R. PAIGE and R.E. TARJAN, Three partition refinement algorithms, *SIAM J. Comput.* 16 (1987), 973–989

25. J. PLESNIK, A heuristic for the p–center problem in graphs, *Discr. Math.* 17 (1987), 263–268

26. P.J. SLATER, R–domination in graphs, *J. ACM* 23 (1976), 446–450

27. J.P. SPINRAD, Doubly lexical ordering of dense 0–1– matrices, manuscript 1988, to appear in *SIAM J. Comput.*

28. J.L. SZWARCFITER and C.F. BORNSTEIN, Clique graphs of chordal and path graphs, manuscript 1992, to appear in *SIAM J. Discr. Math.*

29. B. TANSEL, R. FRANCIS and T. LOWE, Location on networks: a survey I, II, *Management Sci.* 29 (1983), 482–511

30. R.E. TARJAN and M. YANNAKAKIS, Simple linear time algorithms to test chordality of graphs, test acyclicity of hypergraphs, and selectively reduce acyclic hypergraphs, *SIAM J. Comput.* 13, 3 (1984), 566–579

31. K. WHITE, M. FARBER and W. PULLEYBLANK, Steiner Trees, Connected Domination and Strongly Chordal Graphs, *Networks* 15 (1985), 109–124

On Domination Elimination Orderings and Domination Graphs

(Extended Abstract)

Elias Dahlhaus[1], Peter Hammer[2], Frédéric Maffray[3], Stephan Olariu [*4]

[1] Basser Department of Computer Science,
University of Sydney, Australia
[2] RUTCOR, Rutgers University Center of Operations Research,
New Brunswick, NJ, USA
[3] C.N.R.S., LSD2-IMAG, Grenoble, France
[4] Department of Computer Science, Old Dominion University,
Norfolk, VA, USA

Abstract. Several efficient algorithms have been proposed to construct a perfect elimination ordering of the vertices of a chordal graph. We study a generalization of perfect elimination orderings, so called domination elimination orderings (*deo*). We show that graphs with the property that each induced subgraph has a *deo* (domination graphs) are related to formulas that can be reduced to formulas with a very simple structure. We also show that every brittle graph and every graph with no induced house and no chordless cycle of length at least five (HC-free graphs) are domination graphs. Moreover, every ordering produced by the Maximum Cardinality Search Procedure on an HC-free graph is a *deo*.

1   Introduction

All graphs in this paper are simple: they contain neither self-loops nor multiple edges. In addition to standard graph theoretic terminology compatible with Golumbic [9], we need to define some new terms. We let $G = (V, E)$ denote an arbitrary graph with vertex-set V and edge-set E; for a vertex x, $N(x)$ stands for the set of all the neighbors of x in the graph G. Note that adjacency is assumed to be non-reflexive, and so $x \notin N(x)$. Similarly, we let $N'(x)$ stand for the set of all the vertices adjacent to x in the complement $\overline{G}$ of G. The *closed neighborhood* of x consists x and all its neighbors and is denoted by $N[x]$.

Given a subset C of vertices of G, we let $N(C)$ stand for the set of vertices in G that have a neighbor in C; G_C will denote the subgraph of G induced by C. However, when no confusion is possible, we shall blur the distinction between a set of vertices and the subgraph it induces, using the same for both. Throughout the paper, P_k (C_k) stands for the chordless path (cycle) on k vertices. If $\{a, b, c, d\}$

* Work supported in part by the National Science Foundation under grant CCR-8909996

induces a P_4 in G with edges ab, bc, cd, then we shall refer to b and c as the *midpoints* of this P_4.

A graph G is said to be *chordal* if every cycle in G of length at least four has a chord. Chordal graphs arise naturally in a number of applications including scheduling [17], the solution of sparse systems of linear equations [19], facility location problems [4], and the study of evolutionary trees [3]. The interested reader is referred to [9] where many of these applications are summarized.

Dirac [7] proved that every chordal graph contains a *simplicial* vertex, that is, a vertex whose neighbors are pairwise adjacent. An ordering

$$x_1, \ x_2, \ \ldots, \ x_n$$

of the vertices of G is said to be a *perfect elimination ordering* (*peo*, for short) if for every i $(1 \le i \le n-1)$

$$x_i \text{ is a simplicial vertex in } G_{\{x_i,x_{i+1},\ldots,x_n\}}. \tag{1}$$

Fulkerson and Gross [8] proved that a graph G is chordal if, and only if, it admits a perfect elimination ordering. Later, Rose, Tarjan, and Leuker [19], Tarjan and Yannakakis [22], and Shier [20] proposed efficient algorithms to find perfect elimination orderings in chordal graphs.

Recently, Jamison and Olariu [15] proposed an extension of the class of chordal graphs obtained by relaxing the condition related to the existence of the simplicial vertex. Specifically, they call a vertex x in G *semi-simplicial* if x is midpoint of no P_4 in G.

An ordering x_1, x_2, ..., x_n of the vertices of G is said to be a *semi-perfect elimination ordering* (*speo*, for short) if for every i $(1 \le i \le n-1)$

$$x_i \text{ is a semi} - \text{simplicial vertex in } G_{\{x_i,x_{i+1},\ldots,x_n\}}. \tag{2}$$

It is not hard to see that if a graph has a *peo* then it also has a *speo*. In general, the converse is not true, however.

We are now in a position to state the contribution of this paper. We begin by introducing a new concept of linear elimination ordering of the vertices of a graph. This new elimination ordering, that we call a *deo* turns out to generalize both *peo* and *speo*. We also show that for graph with no induced $\overline{P}_5$ and no induced chordless cycle of length at least five, *every* ordering of the vertices produced by Tarjan and Yannakakis' MCS [22] is a *deo*. This result is surprising and interesting in its own right since it fathoms the power of the MCS. Actually, this result motivated the title the paper.

To define our new elimination ordering, we shall say that a vertex y *dominates* a vertex x whenever $N(x) \subseteq \{y\} \cup N(y)$ holds. An ordering x_1, x_2, ..., x_n of the vertices of G is said to be a *domination elimination ordering* (*deo*, for short) if for every i $(1 \le i \le n-1)$

$$\text{there exists a } j > i \text{ such that } x_j \text{ dominates } x_i \text{ in } G_{\{x_i,x_{i+1},\ldots,x_n\}}. \tag{3}$$

We propose to show that the *deo* is a generalization of both *peo*'s and *speo*'s. The details are spelled out in the following result.

Lemma 1. *If a connected graph G has an* speo *then it also has a* deo.

Proof of Lemma 1 omitted   □

It is worth noting that not every graph that has a *deo* also has an *speo*. An example of a perfect graph that has a *deo* but is not perfectly orderable is featured in Figure 2.

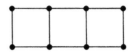

Fig. 1. A graph with a *deo* and no *speo*

As pointed out in [15], every graph featuring an *speo* is perfectly orderable in the sense of Chvátal [6]. This is no longer the case for graphs that have a *deo*. An example, due to Ryan Hayward, is featured in Figure 2.

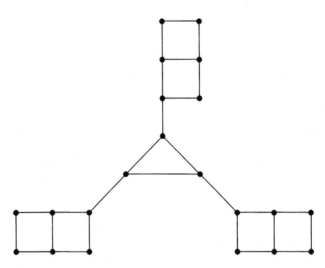

Fig. 2. A graph with a *deo* which is not perfectly orderable

We are particularly interested in graphs with the property that each induced subgraph has a *deo*. It is not difficult to see that these graphs are just those graphs with the property that each induced subgraph has a pair (u, v) of vertices such that u is dominated by v. These graphs are also called *domination graphs*.

We shall see that domination graphs can also be translated into boolean formulas with the vertices as variables with certain properties.

The remainder of this work is organized as follows. Section 2 deals with equivalent characterizations of domination graphs by boolean formulas. In section 3, we show that every brittle graph is a domination graph. Section 4 proves that every ordering of the vertices of an HC-free graph produced by MCS is a *deo*. That also means that every HC-free graph is a domination graph. Finally, section 5 provides concluding remarks and open questions.

2 A Boolean Reduction in Domination Graphs

We recall that an n-dimensional Boolean function is any function from the set $\{0,1\}^n$ to the set $\{0,1\}$. For each 0–1 variable x we write $\bar{x} = 1 - x$; both x and $\bar{x}$ are called *literals*. A *term* is any product of Boolean literals. It is well-known and easy to show that any Boolean function can be written as the Boolean sum of terms in the same variables. This may be done in many ways; each way is called a form of the function. An *implicant* of a Boolean function F is any term P such that $P = 1 \Rightarrow F = 1$. Clearly, any Boolean function is equal to the Boolean sum of its prime implicants. In order to check that two given Boolean functions are equal, one may find the list of all prime implicants of each of them and verify that these two list are identical.

When we have two Boolean terms

$$T_1 = x \prod_{i \in I_1} x_i \prod_{j \in J_1} \overline{x_j} \quad \text{and} \quad T_2 = \bar{x} \prod_{i \in I_2} x_i \prod_{j \in J_2} \overline{x_j}$$

with $I_1 \cap J_2 = \emptyset$ and $I_2 \cap J_1 = \emptyset$, the *consensus* of the two terms over the variable x is the term

$$T' = \prod_{i \in I_1 \cup I_2} x_i \prod_{j \in J_1 \cup J_2} \overline{x_j}.$$

It is easy to see that $T_1 \vee T_2 = T_1 \vee T_2 \vee T'$.

When we have two terms $T = \prod_{i \in I} x_i \prod_{j \in J} \overline{x_j}$ and $T' = \prod_{i \in I'} x_i \prod_{j \in J'} \overline{x_j}$ with $I \subseteq I'$ and $J \subseteq J'$, we say that T' subsumes T.

To find the prime implicants of a function, the following procedure, called the *Consensus Method*, can be applied: Given the Boolean function as a union of terms, do all the possible consensuses between two terms of F, remove any term that is subsumed by another one, and iterate this procedure as long as possible. It can be proved that the terms obtained at the end of this procedure are all the prime implicants of the function (see [10]). This procedure, however, may entail a very large number of steps (in particular one may have an exponential number of steps, as measured with respect to the number of terms given in the initial form of the function).

Now let $G = (V, E)$ be a graph with vertex-set $V = \{v_1, \ldots, v_n\}$. With G we will associate a Boolean function F_G as follows. The variables of F_G are $x_1, \ldots,$

x_n. For every ordered pair of Boolean variables (x_i, x_j) with $i \neq j$, we create a Boolean term $T_G(x_i, x_j)$ defined by:

$$T_G(x_i, x_j) = x_i x_j \cdot \prod_{v_k \in N(x_i) - N[x_j]} \overline{x_k}. \tag{4}$$

Note that for every pair of variables x_i, x_j there is a term $T_G(x_i, x_j)$ and a term $T_G(x_j, x_i)$. Now we write

$$F_G(x_1, \ldots, x_n) = \bigvee_{i,j \leq n,\ i \neq j} T_G(x_i, x_j). \tag{5}$$

This will be called the graph form of function F_G.

Observation 1. If H is an induced subgraph of G, then the function F_H can be obtained from F_G simply by setting to 0 all the variables corresponding to vertices of $G - H$, since the terms $T_G(x_i, x_j)$ with either $v_i \in V(G - H)$ or $v_j \in V(G - H)$ will vanish, and the literals $\overline{x_k}$, for any $v_k \in V(G - H)$, will become equal to 1 in each term containing them.

Observation 2. If v_i and v_j are comparable vertices in G, then one of the inclusions $N(v_i) \subseteq N[v_j]$ or $N(v_j) \subseteq N[v_i]$ holds. It follows that one of the terms $T_G(x_i, x_j)$ or $T_G(x_j, x_i)$ is equal to $x_i x_j$ (and subsumes the other term). Hence $F_G(1, 1, \ldots, 1) = 1$. More generally, we have the following result.

Theorem 2. *Let G be a graph with n vertices and F_G the associated Boolean function as defined above. Then, G is a domination graph if and only if*

$$F_G(x_1, \ldots, x_n) = \bigvee_{i,j \leq n,\ i \neq j} x_i x_j. \tag{6}$$

Proof of Theorem 2 omitted. □

Theorem 2 has an interesting consequence, as we can test whether a given graph G is a domination graph by building the associated function F_G and checking whether equation (6) holds for F_G. It is clear that, if (6) holds, then each term $x_i x_j$ is a prime implicant, and there is no other prime implicant of F_G. However, as pointed out above, finding the prime implicants of a function using the Consensus Method may take a lot of time. We would like to suggest here a simplification of the method, which we will call the Special Quadratic Consensus Method. This is the same as the Consensus Method but with the following restriction: apply the consensus on two terms T_1, T_2 only if $T_1 = x_i x_j$ for some i, j ($1 \leq i < j \leq n$) and T_2 contains either the literals x_i and $\overline{x_j}$ or the literals x_j and $\overline{x_i}$. Notice that in this case the consensus of T_1 and T_2 is the term obtained by "deleting" from T_2 whichever of $\overline{x_i}, \overline{x_j}$ it contains; so the new term subsumes T_2, and this one can be removed. Hence the Special Quadratic Consensus Method reduces the size of the input by one at each step, and it stops after a number of steps which does not exceed the size of the initial form of the function. In general, however, this method may fail to produce all the prime implicants. We do think that it works for domination graphs, precisely:

Conjecture 1 *Let G be a domination graph, and F_G the associated Boolean function given in its initial graph form. Then the Special Quadratic Consensus Method reduces F_G to the form that appears on the right side of (6).*

This conjecture and the precedind remark would imply that domination graphs can be recognized in polynomial time. We do not know any polynomial-time algorithm for this problem.

3 Brittle graphs

It is clear from this definition that any induced subgraph of a domination graph is itself a domination graph. We may also note that the class of domination graphs is self-complementary. Indeed, if two vertices are comparable in a graph G, then they are also comparable in the complement $\overline{G}$ of G. More precisely, if x dominates y in G, then x is dominated by y in $\overline{G}$, and vice-versa. It follows clearly that the complement of every domination graph is a domination graph.

A graph G is said to be *weakly triangulated* ([11]) whenever G and $\overline{G}$ do not contain any induced chordless cycle of length at least five. Since a chordless cycle of length at least five has no pair of comparable vertices, it follows that a domination graph cannot contain such a chordless cycle or its complement. So we obtain that every domination graph is weakly triangulated; consequently every domination graph is perfect. In [11], Hayward shows a graph with 24 vertices which is weakly triangulated and yet has no pair of comparable vertices. Thus we may conclude that the class of domination graphs is strictly included in the class of weakly triangulated graphs.

The class of weakly triangulated graphs contains several well-known classes of perfect graphs. It is worth pointing out that many of these classes actually consist of domination graphs.

A graph G is *brittle* if every induced subgraph H of G contains a vertex which is either not the endpoint or not the midpoint of any P_4 of H. Chvátal ([5]) introduced the concept of brittle graph. Several families of brittle graphs were studied by Hoàng and Khouzam ([14]) and by Preissmann, de Werra and Mahadev ([18]). The fact that brittle graphs are domination graphs is a consequence of Theorem 3.

Theorem 3 [14]. *Let G be a brittle graph. Then either G or $\overline{G}$ has a simplicial vertex or a proper homogeneous set, where a set of vertices M is called homogeneous if all vertices in M have the same neighbors outside M.* □

Theorem 4. *Every brittle graph is a domination graph.*

Proof of Theorem 4. Let G be any brittle graph. We prove that G is a domination graph by induction on the number of its vertices. The fact is clear when G has few vertices. Since the property of being brittle is hereditary, it suffices to prove that G has a pair of comparable vertices. We use the property proved in Theorem 3.

If G has a simplicial vertex x, and if x has any neighbor z in G, then it is clear that z dominates x in G; and if x has no neighbor, then any vertex of $G - x$

weakly dominates x in G. If $\overline{G}$ has a simplicial vertex, we can do the same as above in $\overline{G}$ (recall that comparability between vertices is a self-complementary property). If G or $\overline{G}$ has a proper homogeneous set M, then by the induction hypothesis we know that the subgraph G_M of G induced by M possesses a pair of comparable vertices x, y. Let W be the set of vertices of $G - M$ which are adjacent to (all) the vertices of M. Note that $N_G(x) = N_{G_M}(x) \cup W$ and $N_G(y) = N_{G_M}(y) \cup W$. It follows that x and y are comparable in G. $\quad\square$

The graph in Figure 1 is a domination graph and is not brittle. Thus the class of domination graphs contains the class of brittle graphs strictly.

Weakly triangulated graphs can be recognized by a polynomial time algorithm which simply checks whether the input graph or its complement possesses a cycle of length at least five. In [12] a new characterization of weakly triangulated graphs was given. This characterization yields an efficient polynomial-time algorithm that finds a maximum weighted clique and a maximum weighted stable set in any weakly triangulated graph. The same algorithms can be used for domination graphs.

4   HC-free graphs

Call a graph HC-free if it contains no induced subgraph isomorphic to $\overline{P}_5$ (the *house*) and no induced cycle of length at least five.

The class of HC-free graphs is a generalization of the HHD-free graphs as considered in [15]. It has been shown that HHD-free graphs have a *speo*, and this *speo* can be computed by the lexical breadth-first search procedure of Rose, Tarjan, and Lueker [19]. Note that not every HC-free graph has a *speo*. One example is the *domino* consisting of two quadrangle sharing one edge. It is known that every HHD-free graph is brittle (see for example [14]). But not every HC-graph is brittle. One example is the *triple domino* as in Figure 1.

Tarjan, and Yannakakis [22] proposed a linear-time graph search technique which they called Maximum Cardinality Search (MCS, for short). They also proved that a graph G is chordal if, and only if, any ordering of the vertices of G produced by MCS is a *peo*. To make this paper self-contained we now reproduce the details of MCS.

Procedure MCS(G);
{Input: an arbitrary graph $G = (V, E)$;
Output: an ordering σ of the vertices of G.}
begin
    for $i = n$ **downto** 1 **do begin**
        pick an unnumbered vertex v adjacent to the most numbered vertices;
        $\sigma(v) = i$ {assign to v number i}
    end
end;

Note that we can think of the output of MCS as a linear order on V by placing u before v whenever $\sigma(u) < \sigma(v)$.

Theorem 5. *Any ordering of the vertices of an HC-free graph produced by MCS is a deo.*

Proof of Theorem 5. Let $G = (V, E)$ be an HC-free graph and let

$$x_1, x_2, \ldots, x_n \tag{7}$$

be an arbitrary ordering of the vertices of G produced by MCS.

We only need prove that x_1 is dominated by some vertex x_j with $j \neq 1$, as this extends easily to induced subgraphs. To simplify the notation we write $u = x_1$. Our proof relies on a number of intermediate results that we present next.

Lemma 6. *For each connected component C of $N'(u)$, there is a vertex x_C in C, such that $N(C) \cap N(u) = N(x_C) \cap N(u)$.*

Proof of Lemma 6 First, we show that for adjacent vertices x and y in $N'(u)$ the sets

$$N(x) \cap N(u) \text{ and } N(y) \cap N(u) \text{ are comparable with respect to inclusion.} \tag{8}$$

Otherwise, we find vertices $x' \in N(x) \cap N(u) \setminus N(y)$ and $y' \in N(y) \cap N(u) \setminus N(x)$. In other words, $x'x \in E$, $x'y \notin E$, $yy' \in E$, $xy' \notin E$, $x'u \in E$, $y'u \in E$. However, now $\{x, y, y', u, x'\}$ induces a house or a cycle of length five. Therefore, (11) must hold.

Let vertices x and y belong to the same connected component C of $N'(u)$. We claim that there exists a vertex z in C such that

$$(N(x) \cup N(y)) \cap N(u) \subseteq N(z) \cap N(u). \tag{9}$$

Suppose that x and y in C are joined by a chordless path $x = v_1, v_2, \ldots v_k = y$ with all the internal vertices in C. By induction on k, we show that there is a v_i such that $(N(v_1) \cup N(v_k)) \cap N(u) \subset N(v_i)$, for some v_i.

For $k = 2$, we are done by (11). Suppose that there is an i $(1 \leq i \leq k)$, such that $(N(v_1) \cup N(v_{k-1})) \cap N(u) \subseteq N(v_i)$. We claim that $N(v_k) \cap N(u) \subseteq N(v_i)$ or $N(v_i) \cap N(u) \subseteq N(v_k)$.

Otherwise, we find $x' \in N(v_i) \cap N(u) \setminus N(v_k)$ and $y' \in N(v_k) \cap N(u) \setminus N(v_i)$. Let i' be the largest j such that $x' \in N(x_j)$. Note that $y' \notin N(v_j)$, for any $j = 1, \ldots, k-1$ (since $N(v_j) \cap N(u) \subseteq N(v_i)$). Therefore $x', v_{i'}, v_{i'+1}, \ldots, v_{k-1}, v_k, y'$ induces a chordless path. If $x'y' \in E$ then we have a chordless cycle of length four. If we have a cycle of length four then we get a house together with u.

If $x'y' \notin E$ then together with u we get a cycle of length greater than four. Thus (12) holds and the proof of Lemma 6 is complete. $\qquad\square$

For further reference, we introduce some notation. We write:
$N_i = \{x_{i+1}, \ldots, x_n\}$, and
$M_i = N_i \cap N(u)$.

Lemma 7. *i) If C is a connected component of $N'(u)$ and $C \not\subseteq N_i$ then $M_i \subseteq N(C)$.*

ii) If there exist $y \in V \setminus N_i$ with $|N(y) \cap N_i| > |M_i|$ then there is a connected component C of $N'(u)$ such that all such y are in $C \cup N(C)$.

Proof of Lemma 7. The proof is by backward induction on i. To settle the basis, let $i = n - 1$.

First, let x_n belong to $N'(u)$. Clearly, $M_i = \emptyset \subseteq N(C)$, for any connected component C of $N'(u)$ and part i) is proved.

If $|N(y) \cap N_i| > 0$ then y is a neighbor of x_n, and so y belongs to the same connected component of $N'(u)$ as x_n, or to the neighborhood of this component. Thus, in this case, ii) holds.

Next, let x_n belong to $N(u)$, and let C be a connected component of $N'(u)$. Consider x_i, the vertex in C with the largest index in the ordering (8). Since all greater neighbors of x_i with respect to (8) are in $N(u)$ and $|N(x_i) \cap N_i| = |N(x_i) \cap M_i| \geq |N(u) \cap N_i|$, it follows that all vertices of M_i are neighbors of x_i (otherwise, x_1 would be labeled before x_i).

Therefore, x_n is a neighbor of some vertex in every connected component of $N'(u)$ and i) has been proved. Further, ii) is true since $1 = |N_{n-1}| = |M_{n-1}|$.

For the inductive step, assume that both i) and ii) are true for some $i \leq n-1$.

First, suppose that x_i belongs to $N'(u)$. Since $M_i = M_{i-1}$, part i) follows directly from the induction hypothesis. Note that if there exists a vertex $y \in V \setminus N_i$ with $|N(y) \cap N_i| > |M_i|$, then x_i is such a vertex. If there exists a vertex y in $V \setminus N_{i-1}$ satisfying $|N(y) \cap N_{i-1}| > |M_{i-1}| = |M_i|$ then either $|N(y) \cap N_i| > |M_i|$ is satisfied, or else y is a neighbor of x_i. In the first case, ii) follows by the inductive hypothesis, in the second it follows from the observation that y belongs to the same connected component of $N'(u)$ as x_i, or to the neighborhood of this component. Thus, in this case ii) has been proved.

Next, suppose x_i belongs to $N(u)$. Now, $M_{i-1} = M_i \cup \{x_i\}$ and therefore $|M_{i-1}| = |M_i| + 1$.

Let C be a connected component of $N'(u)$ which is not a subset of N_i and which does not contain vertices $y \in C \setminus N_i$ with $|N(y) \cap N_i| > |M_i|$. Note that, in this case, there must exist a vertex $y \in C \setminus N_i$ with $|M_i|$ neighbors in N_i: otherwise, in all stages $j < i$, the number of neighbors of any $y \in C \setminus N_j$ is smaller than $|M_j|$ and y cannot become an x_j.

Replacing i by $i - 1$, we find a vertex $y \in C \setminus N_{i-1}$ with the property that $|N(y) \cap N_{i-1}| = |M_{i-1}| = |M_i| + 1 = |N(y) \cap M_i| + 1$. This vertex y must be a neighbor of x_i.

Suppose C is the connected component of $N'(u)$ with the property that there are vertices $y \in C$ with $|N(y) \cap N_i| > |M_i|$. Then $|N(x_i) \cap N_i| > |M_i|$ and therefore x_i is in the neighborhood of C. Hereby, i) has been proved.

Next, suppose $|N(y) \cap N_{i-1}| > |M_{i-1}|$. Then $|N(y) \cap N_i| \geq |N(y) \cap N_{i-1}| - 1 > |M_{i-1}| - 1 = |M_i|$. Therefore such a y must be in C or in the neighborhood of C. Therefore also in stage $i - 1$, if there are y with the property $|N(y) \cap N_{i-1}| >$

$|M_{i-1}|$ then such vertices y are in C or in the neighborhood of C and ii) has been proved.

With this the proof of Lemma 7 is complete. □

We now return to the proof of Theorem 5. If some vertex y in $N'(u)$ dominates u, then there is nothing to prove. Therefore, from now on we assume that

$$\text{no vertex in } N'(u) \text{ dominates } u. \tag{10}$$

We shall show that if (13) holds, then u is dominated by a vertex in $N(u)$. More precisely, we shall prove the following result that sheds more light onto the structure of $N(u)$, and that will be exploited later for algorithmic purposes.

Lemma 8. *If no vertex in $N'(u)$ dominates $x_1 = u$, then there exists a subscript l such that the following conditions are satisfied:*

- $x_1, x_2, \ldots, x_l$ *is a clique.*
- *every vertex x_i $(1 \le i \le l)$ is adjacent to all vertices in $N(u) \setminus \{x_2, \ldots, x_l\}$.*
- *There exist a vertex y in $N'(u)$ such that with $N(u) \setminus \{x_2, \ldots, x_l\} \subseteq N(y) \cap N(u)$.*

Clearly, we only need prove Lemma 8, since its conclusion will imply, in particular, that x_l dominates x_1, as desired.

Proof of Lemma 8. Let k is the smallest index in (10) for which x_k belongs to $N'(u)$. Then, clearly, $M_k = M_{k-1}$ and by Lemma 7, $M_{k-1} \subseteq N(C)$ for some component C of $N'(u)$. By Lemma 6, $M_{k-1} \subseteq N(y)$ for some y in $N'(u)$. Note that $k \ge 3$ for otherwise we contradict (11).

By the induction step of Lemma 7, for the case case $x_i \in N'(u)$, if $|N(x_{k-1}) \cap N_{k-1}| > |M_{k-1}|$ then x_{k-1} must belong to the neighborhood of the connected component that contains x_k.

By the induction step of Lemma 7 for the case $x_i \in N(u)$, all vertices x_i with $i < k$ and $|N(x_i) \cap N_i| > |M_i|$ must belong to the neighborhood of the connected component of $N'(u)$ that contains x_k, and must be consecutive in (10).

Let x_l be the vertex in $N(u)$ with the largest index l in (8) such that M_{l-1} is not a subset of the neighborhood of some connected component of $N'(u)$. Then $l < k$ and $|N(x_l) \cap N_l| = |M_l|$. We claim that

$$\text{vertex } x_l \text{ is adjacent to no vertex in } N'(u). \tag{11}$$

Suppose x_l is in the neighborhood of the connected component C_1 of $N'(u)$ and M_l is a subset of $N(C_2)$.

Let x_m be the vertex in C_1 with the smallest index in (8). We may assume that for $x_i \in N'(u)$, $i < m$, $x_i x_l \notin E$. Note that $M_{\bar{m}}$ is a subset of $N(C_1)$ if for all $i, \bar{m} \le i \le m$, $|N(x_i) \cap N_i| > |M_i|$. We consider a minimum $\bar{m}$ with this property. If such $\bar{m}$ does not exist, then we set $\bar{m} = m$. Note that $|N(x_l) \cap N_{\bar{m}}| = |M_{\bar{m}}|$, for otherwise x_l had been numbered in the interval $[\bar{m}, m]$. Therefore we find a vertex $x_q \in M_{\bar{m}-1}$ with $x_l x_q \notin E$.

To preserve the property $|N(x_l) \cap N_i| \geq |M_i|$, for $i \in [l, \bar{m}]$, x_l is a neighbor of x_i if $x_i \in N(u)$. Since $M_{\bar{m}-1}$ is a subset of the neighborhood of C_1 and M_l is not a subset of $N(C_1)$, we find a $p \in [l+1, \ldots, \bar{m}-1]$, such that x_p is not a neighbor of C_1. Note that x_p is a neighbor of x_l. Let $y_1 \in C_1$ such that $N(y_1) \cap N(u) = N(C_1) \cap N(u)$ and $y_2 \in C_2$ such that $N(y_2) \cap N(u) = N(C_2) \cap N(u)$.

Consider the cycle $x_l, y_1, x_q, y_2, x_p, x_l$. The only chord this cycle can have is $x_q x_p$. Therefore we have an induced cycle of length five or a house. This is a contradiction, and so (14) must hold.

To complete the proof of Lemma 8, we claim that

$$x_l \text{ is dominating every } x_i \text{ with } i < l. \tag{12}$$

To see that this must be the case, note that since $|N(x_l) \cap N_l| \geq |M_l|$ and x_l has no neighbors in $N'(u)$, x_l must be adjacent to all vertices in M_l.

Note that for all $i \leq l$, $|N(x_i) \cap N_l| = |M_l|$. To preserve the property $|N(x_i) \cap N_i| \geq |M_i|$, for $i < l$, x_l must be adjacent to all x_i with $i < l$. Therefore, Lemma 8 follows. □

With this, the proof of Theorem 5 is complete. □

5   Concluding remarks

In this paper we have further extended the range of applicability of Maximum Cardinality Search that was used, primarily, for detecting whether a graph is chordal. We have shown that, in fact, in the presence of HC-free graphs, Maximum Cardinality Search *always* returns a *deo* of the graph, regardless of the beginning vertex and regardless of how ties are broken during the execution of the algorithm.

Due to space limitations we have not stated similar results for Lexicographic Breadth-First Search [19] and for Maximum Cardinality in Component [20]. We have shown that the same result holds for all of these. The interested reader is referred to the journal version of this paper for full treatment of these and related issues.

There are a number of open problems, however. We would be happy if we could use domination orderings to get a minimum coloring efficiently.

References

1. C. Berge, Färbung von Graphen, deren sämtliche bzw. deren ungerade Kreise starr sind, Wissenschaftliche Zeitung, Martin Luther Univ. Halle Wittenberg, 1961, 114–115.
2. C. Berge and V. Chvátal, Topics on Perfect Graphs, Annals of Discrete Math 21, North Holland, Amsterdam, 1984.
3. P. Buneman, A characterization of rigid circuit graphs, *Discrete Mathematics* 9 (1974), 205–212.
4. R. Chandrasekharan and A. Tamir, Polynomially bounded algorithms for locating p-centers on a tree, *Mathematical Programming* 22, (1982), 304–315.

5. V. Chvátal Perfect graphs seminar. 1983.

6. V. Chvátal, Perfectly ordered graphs, in Berge and Chvátal [2].

7. G. Dirac, On rigid circuit graphs, Abh. Math. Sem. Univ. Hamburg, 25, 1961, 71–76.

8. D. R. Fulkerson and O. A. Gross, Incidence matrices and interval graphs, *Pacific Journal of Mathematics* 15, (1965), 835–855.

9. M. C. Golumbic, *Algorithmic Graph Theory and Perfect Graphs*, Academic Press, New York, 1980.

10. P.L. Hammer and S. Rudenau *Boolean Methods in Operations Research and Related Areas.* Springer, Berlin, Heidelberg, New York, 1968.

11. R. B. Hayward, Weakly triangulated graphs, *Journal of Combinatorial Theory (B)* 39 (1985), 200–208.

12. R. Hayward, C.T. Hoáng, and F. Maffray Optimizing weakly triangulated graphs. *Graphs and Combinatorics*, 5:339–349, 1989. See erratum in vol. 6, 1990, p. 33–35.

13. A. Hertz, A fast algorithm for coloring Meyniel graphs, *Journal of Combinatorial Theory (B)*, 50, (1990), 231–240.

14. C.T. Hoáng and N. Khouzam On brittle graphs. *J. Graph Theory*, 12:391–404, 1988.

15. B. Jamison and S. Olariu, On the Semi-perfect Elimination, *Advances in Applied Mathematics* 9, (1988) 364–376.

16. H. Meyniel, On the perfect Graph Conjecture, *Discrete Mathematics*, 16, 1976, 339–342.

17. C. Papadimitriou and M. Yannakakis, Scheduling interval-ordered tasks, *SIAM Journal on Computing* 8, (1979), 405–409.

18. M. Preissmann, D. de Werra, and N.V.R. Mahadev A note on superbrittle graphs. *Disc. Math.*, 61:259–267, 1986.

19. D. Rose, R. Tarjan, and G. Leuker, Algorithmic aspects of vertex elimination on graphs, *SIAM Journal on Computing* 5, 1976, 266–283.

20. D. R. Shier, Some Aspects of Perfect Elimination Orderings in Chordal Graphs, *Discrete Applied Mathematics*, (1984), 325–331.

21. Sritaran and J. Spinrad, Personal communication, 1991.

22. R. E. Tarjan and M. Yannakakis, Simple linear-time algorithms to test chordality of graphs, test acyclicity of hypergraphs, and selectively reduce acyclic hypergraphs, *SIAM Journal on Computing*, 13, (1984), 566–579.

Complexity of Graph Covering Problems

Jan Kratochvíl[1], Andrzej Proskurowski[2] and Jan Arne Telle[2]

[1] Charles University, Prague, Czech Republic
[2] University of Oregon, Eugene, Oregon

Abstract. Given a fixed graph H, the H-cover problem asks whether an input graph G allows a degree preserving mapping $f : V(G) \to V(H)$ such that for every $v \in V(G)$, $f(N_G(v)) = N_H(f(v))$. In this paper, we design efficient algorithms for certain graph covering problems according to two basic techniques. The first one is a reduction to the 2-SAT problem. The second technique exploits necessary and sufficient conditions for the existence of regular factors in graphs. For other infinite classes of graph covering problems we derive $\mathcal{NP}$-completeness results by reductions from graph coloring problems. We illustrate this methodology by classifying all graph covering problems defined by simple graphs with at most 6 vertices.

1 Motivation and overview

For a fixed graph H, the H-cover problem admits a graph G as input and asks about the existence of a "local isomorphism": a labeling of vertices of G by vertices of H so that the label set of the neighborhood of every $v \in V(G)$ is equal to the neighborhood (in H) of the label of v. We trace this concept to Biggs' construction of highly symmetric graphs in [4], and to Angluin's discussion of "local knowledge" in distributed computing environments in [2]. More recently, Courcelle and Métivier used graph coverings to show that nontrivial minor closed classes of graphs cannot be recognized by local computations [5]. In [1], Abello *et al.* raised the question of computational complexity of H-cover problems, noting that there are both easy (polynomial-time solvable) and difficult ($\mathcal{NP}$-complete) versions of this problem for different graphs H. A related question of complexity of a "homomorphic mapping" (also parametrized by a fixed graph H and called H-coloring) has been resolved by Hell and Nešetřil [8] who completely classified graphs for which the problem is easy and those for which it is difficult. Our own interest in the subject comes from the view of H-covering as a generalization of domination and perfect code problems.

In this paper, we develop a methodology that is useful in analyzing the complexity of graph covering problems. The paper is organized as follows. First, we introduce our vocabulary by giving the necessary definitions in Section 2. In designing efficient algorithms that solve easy graph covering problems, we reduce those problems to regular factorization problems and/or to the 2-SAT problem. We introduce these tools and present the corresponding results in Section 3. To prove $\mathcal{NP}$-completeness of the difficult graph covering problems, we use polynomial time reductions from known $\mathcal{NP}$-complete restrictions of vertex-, edge-

and H-coloring problems and also reductions between covering problems. These last reductions are based on properties of the automorphism groups of the relevant graphs. We set up a paradigm to construct such reductions and present our findings in Section 4. In the Appendix, we give a catalogue of the complexity of the covering problem for all simple graphs with at most 6 vertices. There are 208 such graphs, with about 100 having non-trivial polytime solution algorithms and 36 being NP-complete (the remaining graphs define trivial covering problems).

2 Definitions

We use standard graph terminology [7], and consider simple, undirected graphs only. For a vertex $v \in V(G)$ of a graph G, let $N_G(v) = \{u : uv \in E(G)\}$ be the set of neighbors of v and $deg_G(v) = |N_G(v)|$ its degree. For $S \subseteq V(G)$ let $G[S]$ denote the graph induced in G by S, let $V(G) \setminus S = \{v \in V(G) : v \notin S\}$ and let $G \setminus S = G[V(G) \setminus S]$. For $F \subseteq E(G)$, we denote by $G \setminus F$ the spanning subgraph of G with edges $\{uv \in E(G) : uv \notin F\}$. If $E(G) = \emptyset$ then G is called a *discrete* graph. The complement of a graph G, $\overline{G}$, has vertices $V(G)$ and edges $\{uv : uv \notin E(G)\}$. $Aut(G)$ is the automorphism group of G.

A graph G is said to *cover* a graph H if there is a function (called *covering projection*) $f : V(G) \to V(H)$ which preserves the identity of the neighborhood of any vertex v of G, $\{f(u)|u \in N_G(v)\} = N_H(f(v))$ with $deg_G(v) = deg_H(f(v))$. Fixing the graph H, and allowing any graph G as the input, one can pose the question: "Does G cover H?" The computational complexity of this problem, called the H-*cover* problem for the particular graph H, is the subject of this paper.

The *degree partition* of a graph is the partition of its vertices into the minimum number of *blocks* $B_1, \ldots, B_t$ for which there are constants r_{ij} such that for each i, j $(1 \leq i, j \leq t)$ each vertex in B_i is adjacent to exactly r_{ij} vertices in B_j. The $t \times t$ matrix R $(R[i, j] = r_{ij})$ is called the *degree refinement*.

The degree partition and degree refinement of a graph are easily computed by a stepwise refinement procedure. Start with vertices partitioned according to degree and keep refining the partition until any two nodes in the same block have the same number of neighbors in any other given block. See Figure 1 for an example. Graph coverings are related to degree partitions and degree refinements (see, for instance, Leighton [12]):

Fact 1. *If f is a covering projection of H by G then H and G have the same degree refinement and have degree partitions $B_1, B_2, ..., B_t$ and $B'_1, B'_2, .., B'_t$ so that for every $v \in B'_i$ we have $f(v) \in B_i$, $i = 1, 2, ..., t$.*

Without loss of generality, we will consider only connected graphs, because of the following observations (whose proofs are left to the reader.)

Fact 2. *(a) A disconnected graph G covers a connected graph H if and only if every connected component of G covers H.*

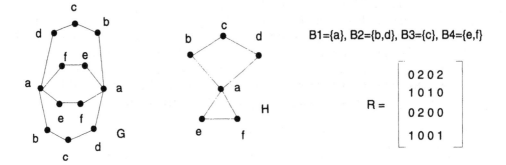

Fig. 1. G labeled by a covering projection of H, their common degree refinement R and the degree partition of H.

(b) For a disconnected graph H, the H-cover problem is polynomially solvable ($\mathcal{NP}$-complete) if and only if the H_i-cover problem is polynomially solvable ($\mathcal{NP}$-complete) for every (for some) connected component of H.

3 Efficient algorithms

For a given graph G and a fixed graph H, it is easy to compare degree partitions and degree refinements in polynomial time. Surprisingly, for many graphs H, the necessary condition for the existence of a covering given by Fact 1 is also sufficient. For many other graphs H (including some infinite classes of graphs), for which those conditions are not sufficient, we are able to design an efficient solution algorithm paradigm by constructing an equivalent instance of the 2-SAT problem, and/or by reducing to a factorization problem in a regular graph. Before we present these results, we note that the only cover of a tree is a graph isomorphic to it, and the only covers of a cycle are cycles with lengths divisible by the length of that cycle. The following observation follows indirectly.

Fact 3. *For a graph H with at most one cycle, the H-cover problem is solvable in polynomial time.*

3.1 2-satisfiability

The 2-SAT problem, that asks about the existence of a truth assignment satisfying a conjuction of clauses with at most two variables, is solvable in polynomial time. We reduce a class of H-covering problems to an instance of the 2-SAT problem.

Theorem 4. *The H-COVER problem is solvable in polynomial time if every block of the degree partition of H contains at most two vertices.*

Proof. Denote the vertices of the i-th block B_i of H by L_i, R_i (or L_i only, if B_i is a singleton). Suppose that G has the same degree refinement as H and its degree partition is $B_1', B_2', \ldots, B_t'$, where the blocks are numbered so that every covering projection sends B_i' onto B_i, $1 \le i \le t$. This structure of G can be checked in polynomial time, and G does not cover H unless it satisfies these assumptions.

The crucial part of the algorithm is to decide which vertices of B_i' should map onto L_i and which onto R_i. This can be done via 2-SAT. For every vertex u of G, introduce a variable x_u. In a truth assignment ϕ, these variables would encode

$$\phi(x_u) = \begin{cases} \text{true} & \text{if } f(u) = L_i \\ \text{false} & \text{if } f(u) = R_i \end{cases} \tag{1}$$

for a corresponding covering projection f (here i is such that $u \in B_i'$). We construct a formula Φ as a conjunction of the following subformulas:

1. (x_u) for every $u \in B_i'$ such that B_i is a singleton;
2. $(x_u \vee x_v) \wedge (\neg x_u \vee \neg x_v)$ for any pair of adjacent vertices u, v which belong to the same block B_i' (i.e., $L_i R_i \in E(H)$);
3. $(x_u \vee \neg x_v) \wedge (\neg x_u \vee x_v)$ if u and v belong to distinct blocks (say $u \in B_i'$ and $v \in B_j'$) and there are exactly the two edges $L_i L_j, R_i R_j$ between B_i and B_j in H;
4. $(x_u \vee x_v) \wedge (\neg x_u \vee \neg x_v)$ if u and v belong to distinct blocks (say $u \in B_i'$ and $v \in B_j'$) and there are exactly the two edges $L_i R_j, R_i L_j$ between B_i and B_j in H;
5. $(x_w \vee x_v) \wedge (\neg x_w \vee \neg x_v)$ if v and w belong to the same block (say B_j') and are both adjacent to u which belongs to a block (say B_i') such that $L_i L_j, L_i R_j \in E(H)$.

Note that in case 2, every $u \in B_i'$ has exactly one neighbor v in the same block, in cases 3 and 4, every $u \in B_i'$ has exactly one neighbor $v \in B_j'$, and in case 5, every $u \in B_i'$ has exactly two neighbors $v, w \in B_j'$.

It is clear that Φ is satisfiable if and only if f defined by (1) is a covering projection from G onto H. The clauses derived from 2 guarantee, if $L_i R_i \in E(H)$, that every vertex mapped on L_i has a neighbor which maps onto R_i and vice versa, the clauses from 3-5 control adjacencies to vertices from different blocks, and the technical clauses from 1 control the singletons. □

3.2 1-factorization

A spanning subgraph H of a graph G is a k-*factor* if all vertices of H have degree k. When $k = 1$, the 1-factor is often referred to as *perfect matching*. The existence of perfect matchings in bipartite graphs is a subject of the celebrated König-Hall theorem. A graph G is k-*factorable* if its edges can be partitioned into k-factors. An application of the König-Hall marriage theorem states that a regular bipartite graph is 1-factorable ([7],[10]). We will use this fact to show that the obvious necessary conditions are also sufficient for a class of graph covering problems.

Theorem 5. *Let H be a graph with all but two vertices of degree 2, all of them lying on paths connecting the two vertices of degree $k > 2$. Then a graph G covers H if and only if H and G have the same degree refinement and the multigraph obtained from G by replacing the paths between vertices of degree k by edges is bipartite. It follows that the H-cover problem is solvable in polynoimal time.*

Proof. The 'only if' part of the statement is obvious. For the 'if' part, note first that since G is connected, the bipartition of its degree k vertices into V_1, V_2 is unique. Denote the vertices of degree k in H by v_1, v_2, and let the paths between them have lengths $n_1 < n_2 < \ldots < n_m$, with exactly k_i paths of length n_i. Number the paths of length n_i from 1 to k_i. For every i, consider an auxiliary multigraph G_i with vertex set $V_1 \cup V_2$ and edges being in one-to-one correspondence with the paths of length n_i between the vertices of $V_1 \cup V_2$ in G. Since degree refinements of G and H are identical, G_i is k_i-regular bipartite. It is therefore 1-factorable, which means that its edges can be colored by k_i colors, say $1, 2, \ldots, k_i$, so that every vertex is incident to exactly one edge of each color. We then define the covering projection by

$f(x) = v_i$ if $x \in V_i, i = 1, 2$, and otherwise

$f(x) = u$ where u is a vertex of degree 2 on the jth path of length n_i which leads from v_1 to v_2 and x is the corresponding vertex on the path in G (from a vertex in V_1 to a vertex in V_2) that is represented by an edge colored by color j in the auxiliary multigraph G_i. □

In a forthcoming paper we will treat in more detail the question of pending trees to vertices of graphs which define polynomially solvable variants of graph covering problems. Presently, to encompass all 6-vertex graphs, we mention that adding a degree 1 vertex to a graph falling under Theorem 5 will not affect the conclusion of the theorem.

3.3 2-factorization

A classical result of Petersen [13] states that any $2k$-regular graph is 2-factorable. We will use this fact to show that the obvious necessary conditions are also sufficient for a class of graph covering problems.

Theorem 6. *Let H be a graph with all but one vertex of degree 2. Then the H-cover problem is solvable in polynomial time, and a graph G covers H if and only if its degree refinement is the same as the degree refinement of H.*

Proof. The structure of H is such that it contains one vertex, say A, of degree $2k$ and all other vertices lie on cycles which pass through A. Let the lengths of the cycles be $n_1 < n_2 < \ldots n_m$ and let there be k_i cycles of length $n_i, i = 1, 2, \ldots, m$.

The obvious necessary condition for a graph G to cover H is for G to contain only vertices of degree 2 and $2k$, and for every $i = 1, 2, \ldots, m$, every vertex of degree $2k$ is an endpoint of exactly $2k_i$ paths of length n_i which contain only vertices of degree 2 and every such path is between degree $2k$ vertices (possibly the same vertex). This is just an explicit reformulation of the fact that G has the same degree refinement as H. We will show that this obvious necessary condition is also sufficient.

It suffices to consider only paths of the same length, say n_i. Consider a multigraph G' whose vertex set are the vertices of degree $2k$ in G, and edges correspond to paths of length n_i. This graph is $2k_i$-regular, and hence 2-factorable [13]. Let $E'_j, j = 1, 2, \ldots, k_i$, be the edge sets of k_i disjoint 2-factors. Each such E'_j is a disjoint union of cycles, which in the original graph G correspond to cycles formed by paths of length n_i. These paths of G must map to paths $P_1, P_2, \ldots, P_{k_i}$ of H, with P_j having vertices $A_{j1}, A_{j2}, \ldots, A_{jn_i}$. In fact, the paths in G represented by a 2-factor E'_j can all be mapped onto the same path P_j in H. If $x_1, x_{11}, \ldots, x_{1n_i}, x_2, x_{21}, \ldots, x_{2n_i}, \ldots, x_r, x_{r1}, \ldots, x_{rn_i}$ is such a cycle (with $x_1, x_2, \ldots, x_r$ being its vertices of degree $2k$), then the vertices x_{ab} will map onto A_{jb}, $1 \leq a \leq r, 1 \leq b \leq n_i$. □

Theorems 5 and 6 can be unified in the following general statement, which is again an example of the 'obvious necessary conditions are also sufficient' scheme. The proof, which we omit, is more or less a confluence of the proofs of Theorems 5 and 6.

Theorem 7. *Let H be a graph with all but two vertices of degree 2 and let these two vertices of higher degree be L and R. Further suppose that for every $i > 1$, L belongs to l_i cycles of length i, R belongs to r_i cycles of length i and there are m_i paths of length i joining L and R. If*
  a) there is an i such that $l_i \neq r_i$, or
  b) $l_i m_i = 0$ for every i,
then H-COVER is solvable in polynomial time, and a graph G covers H if and only if it has the same degree refinement as H and, in case b), if the vertices of degree > 2 in G can be partitioned into classes U and V so that every path of length i such that $m_i \neq 0$ connects a vertex from U to a vertex from V, and every path of length i such that $l_i \neq 0$ either connects vertices from U or vertices from V.

Let us state without proof that in all remaining cases, i.e., when $l_i = r_i$ for all i and there is an i_0 such that $l_0 \neq 0$ and $m_0 \neq 0$, the H-COVER problem is $\mathcal{NP}$-complete.

Fact 3, Theorem 4 and Theorem 6 encompass all but three graphs of at most 6 vertices for which the covering problem is easy. One of these graphs is a particularly easy case of Theorem 7. The covering problems for the two remaining graphs, which will be treated in detail in a forthcoming paper, are solved by a modification of the 2-SAT method.

4 $\mathcal{NP}$-completeness

For any graph H, the H-cover problem is in $\mathcal{NP}$. We will show $\mathcal{NP}$-completeness of H-cover problems for several infinite classes of graphs. We first mention an earlier result proved in [11].

Theorem 8. *For every $k \geq 4$, the K_k-cover problem is $\mathcal{NP}$-complete.*

4.1 Reductions from coloring problems

In this section we reduce from $\mathcal{NP}$-complete problems of edge coloring and vertex coloring. The *vertex k-coloring* problem is well known to be $\mathcal{NP}$-complete for every fixed $k \geq 3$ [6]. The *edge k-coloring* problem asks if each edge of a graph can be assigned one of k colors so that no two edges incident with the same vertex are assigned the same color. Edge 3-coloring of cubic graphs is $\mathcal{NP}$-complete [9]. The following observation is used in our reductions.

Fact 9. *If G covers H by $f : V(G) \rightarrow V(H)$ and $\pi \in Aut(H)$ then $\pi \circ f$ (the composition of f and π) is also a covering projection of H by G.*

Our reductions use vertex and edge gadget construction, providing a graph G' which covers H if and only if a given graph G can be colored appropriately. The general outline of the reductions is as follows:

1. Define vertex gadget for a vertex $v \in V(G)$ by a subgraph of a cover of H, with $deg_G(v)$ 'port's to be used for edge gadgets (edge-coloring requires covers of H with distinct projections for each port and automorphisms of H that allow any permutation of the ports as distinct covers by Fact 9, whereas for k-vertex coloring we need equivalent projections for each port and k distinct covers)
2. Define edge gadget connecting two ports by a subgraph of a cover of H, so that the 'only if' direction of the reduction is met (for edge-coloring (vertex coloring) this amounts to ensuring that the two ports must cover H equivalently (distinctly)).
3. Neighborhoods left unspecified are completed, possibly with added vertices, so that the 'if' direction of the reduction is met (this amounts to extending any partial covering projection defined in step 2 to a cover of H).

Theorem 10. *The $K_{2,2,2}$-cover problem, the $K_{3,3}$-cover problem and the $\overline{C_6}$-cover problem are $\mathcal{NP}$-complete.*

Proof. Let $H = K_{2,2,2}$. For a given cubic graph G we construct a graph G' such that G is edge 3-colorable if and only if G' covers H. The gadget for a vertex $v \in V(G)$ is a 3-cycle, with one vertex for each port. The association of $V(H)$, as labels of the vertex gadget ports, with the 3 edge colors is that each pair of non-adjacent vertices of H corresponds to a unique color.

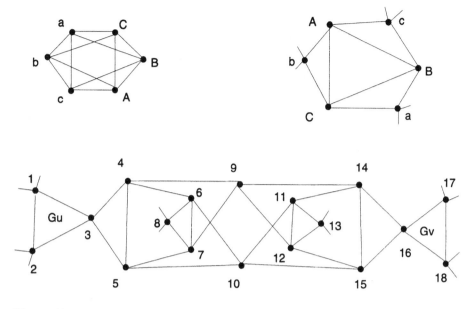

Fig. 2. $K_{2,2,2}$ upper left. Vertex gadgets Gu and Gv connected by an edge gadget at bottom. Lacking neighbors for a, b, c provided by the 3-cycle A, B, C at top right.

For the next step of this reduction, we define the edge gadget for $uv \in E(H)$ as a subgraph of a cover of H, see Figure 2 where vertices 4..15 form an edge gadget. We show that if the graph we are constructing covers H by a projection f then the two ports connected to this edge gadget, called 3 and 16 in the figure, must be labelled by the same color. Since the vertex gadget is a complete graph it cannot have two vertices labelled by two non-adjacent vertices of H. Assume wlog that $f(1) = b, f(2) = c, f(3) = a$, see Figure 2. We show that $f(16) = f(3) = a$. We have $N_H(a) = \{b, c, B, C\}$ so wlog let $f(4) = B, f(5) = C$. Since $N_H(C) = \{a, b, A, B\}$ we have $f(7)$ equal to A or b, and similarly $f(6)$ equal to A or c. But if $f(7) = A$ then $f(10) = b$ so $f(6) = c$ and $f(9) = A$ which cannot be since vertices 7 and 9 are adjacent. We conclude $f(7) = b, f(10) = A$ and similarly $f(6) = c, f(9) = A$. This forces $f(8) = a$. Now, $f(11)$ is b or B and similarly $f(12)$ is c or C. Any of the 4 possible pairs for $f(11), f(12)$ have as common neighbors only a and A, but 11 and 12 already have a neighbor labelled A, so $f(13) = a$. It is not hard to check that for all 4 cases we have $f(3) = f(16) = a$.

Hence if the constructed graph covers H, we color each edge of G by one of the 3 colors $a = A, b = B$ or $c = C$ according to the label of the edge gadget's port connections.

For the other direction of the proof, we complete the neighborhoods left unspecified. Note that we can now assume that G is 3-edge colorable and freely specify the covering projection. For each $u \in V(G)$, we have three vertices, e.g., vertex 8 for u in Figure 2, each lacking two neighbors. Following the projection given above, let these three vertices be labelled a, b, c, lacking neighbors $\{B, C\}, \{A, C\}$ and $\{A, B\}$, respectively. We add a 3-cycle for each $u \in V(G)$, label its vertices A, B, C and use them as the lacking neighbors, see Figure 2. The constructed graph G' thus covers H whenever G is edge 3-colorable.

The reductions for $\overline{C_6}$ and $K_{3,3}$, which we leave out, are again from edge 3-coloring of cubic graphs. □

Theorem 11. *Let Q be a block in the degree-partition of a graph H and let $S \subseteq V(H)$ be a set of vertices such that for every $v \in Q$, $N(v) \setminus Q = S$. Then the H-cover problem is $\mathcal{NP}$-complete in each of the following cases.*
   (a) $H[Q]$ is a discrete graph, $|Q| = 3$ and $|S| \geq 3$;
   (b) $H[Q]$ is a k-cycle ($k \geq 3$) and $|S| \geq 1$;
   (c) $H[Q]$ is a perfect matching ($|Q| = 4$) and $|S| \geq 2$.

Proof. By reduction from vertex and edge coloring problems. See the full version of this paper. □

The k-*starfish* graph has k vertices of degree two and k vertices of degree four with the vertices of degree four inducing a cycle and any two consecutive vertices of this cycle sharing a neighbor of degree two.

Theorem 12. *For every $i \geq 1$, the $(2i + 1)$-starfish-cover problem is $\mathcal{NP}$-complete.*

Proof. By reduction from C_{2i+1}-homomorphism. See the full version of this paper. □

4.2   Reductions from covering problems

A graph H may have an induced subgraph H' for which the H'-cover problem is $\mathcal{NP}$-complete. In general, the H-cover problem could itself be easy. Our next theorem shows $\mathcal{NP}$-completeness in a restricted case by reducing the H-cover problem to the H'-cover problem.

Theorem 13. *The H-cover problem is $\mathcal{NP}$-complete if for some block $Q = \{v_1, v_2, ..., v_k\}$ in the degree partition of H the $H[Q]$-cover problem is $\mathcal{NP}$-complete and there exists an order k latin square L over Q whose columns are elements of $Aut(H[Q])$, and whose rows are elements of $Aut(H \setminus E(H[Q]))|_Q$ (projections onto Q of automorphisms that fix Q setwise)*

Proof. We reduce from the $H[Q]$-cover problem. Given a graph G, we construct a graph G' such that G covers $H[Q]$ if and only if G' covers H. Let $V(G) = \{x_1, ..., x_n\}$ and $V(H) = Q \cup R$. G' will contain k copies of G ($G_1, ..., G_k$) and n copies of $H[R]$ ($R_1, ..., R_n$). A vertex $x_i \in V(G)$ thus has k copies $x_i^1, x_i^2, ..., x_i^k$ in G' ($x_i^j \in V(G_j)$), which will be used as the remaining neighbors for vertices of R_i. We let the vertex x_i^j play the role of vertex $v_j \in Q$ and connect vertices of R_i to its remaining neighbors, as specified by H, thereby completing the construction of the graph G'.

Suppose G' covers H. Since Q is a block in the degree-partition of H the vertices of the n copies of $H[R]$ in G' cannot map to Q, so we have an n-fold cover. The vertices of each of the k copies of G must then map to Q and thus G covers $H[Q]$.

For the other direction, suppose $f : V(G) \to Q$ is a covering projection of $H[Q]$ by G. Let $\Delta_1, \Delta_2, ..., \Delta_k$ be the columns of the latin square L and let $\pi_1, \pi_2, ..., \pi_k$ be its rows. Since $\forall i : \Delta_i \in Aut(H[Q])$, we have by Fact 9 that $\Delta_1 \circ f, ..., \Delta_k \circ f$ are also covering projections of $H[Q]$ by G and we label the vertices of the copy G_j of G by $\Delta_j \circ f$. By construction we have that R_i is connected to vertices $x_i^1, x_i^2, ..., x_i^k$. Assuming that $f(x_i) = v_j$ we label these vertices by the respective labels $\Delta_1(v_j), \Delta_2(v_j), ..., \Delta_k(v_j)$, corresponding to a row π_r of L, when taken in this order. Since $\pi_r \in Aut(H \setminus E(H[Q]))|_Q$, we can send $V(R_i)$ to R by an element of $Aut(H \setminus E(H[Q]))$ which has projection π_r on Q, locally getting a covering projection from R_i to $H[R]$ by Fact 9, and with correct labels for remaining neighbors of R_i as well. The same is done for all n copies of $H[R]$ resulting in a mapping of $V(G')$ to $V(H)$ where each copy of G covers $H[Q]$ and each copy of $H[R]$ covers $H[R]$ and the remaining neighbors of copies of both G and $H[R]$ have correct labels, hence we have a covering projection of H by G'. □

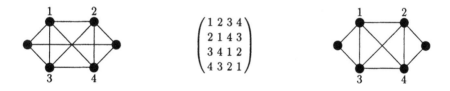

Fig. 3. Application of Theorem 13

As an example of application of this result, consider the graphs depicted in Figure 4.2. For both of the graphs, $Q = \{1, 2, 3, 4\}$ is a block in the degree partition inducing a complete graph and the K_4-cover problem is $\mathcal{NP}$-complete by Theorem 8. Moreover, for the 4 by 4 matrix displayed in the middle of Figure 4.2, both its rows and columns are in $Aut(H)|_Q$. Hence by Theorem 13 the H-cover problem is $\mathcal{NP}$-complete (note $Aut(H)|_Q \subseteq (Aut(H \setminus E(H[Q])))|_Q \cap Aut(H[Q]))$, here H stands for any of the two graphs depicted in Figure 4.2).

Theorem 14. *The H-cover problem is $\mathcal{NP}$-complete if for some block Q ($|Q| \geq$ 4) in the degree-partition of H, $H[Q]$ is a complete graph and $\exists S \subseteq V(H)$, such that $\forall v \in Q, N(v) \setminus Q = S$.*

Proof. By Theorem 8 the $H[Q]$-cover problem is $\mathcal{NP}$-complete and $Aut(H)|_Q$ is the symmetric group on $|Q|$ points, so the conditions in Theorem 13 are easily satisfied. $\square$

Acknowledgments

This research has been initiated under the auspices of the scientific exchange program between the National Research Council and the Czechoslovak Academy of Sciences and completed with partial support of the National Science Foundation through grant NSF-CCR-9213439. The first author acknowledges partial support of EC Cooperative Action, IC-1000 (Project ALTEC) and Charles University Research grant GAUK 351.

References

1. J. Abello, M.R. Fellows and J.C. Stillwell, On the complexity and combinatorics of covering finite complexes, *Australasian Journal of Combinatorics 4* (1991), 103-112;
2. D. Angluin, Local and global properties in networks of processors, in *Proceedings of the 12th STOC* (1980), 82-93;
3. D. Angluin and A. Gardner, Finite common coverings of pairs of regular graphs, *Journal of Combinatorial Theory B 30* (1981), 184-187;
4. N. Biggs, *Algebraic Graph Theory*, Cambridge University Press, 1974;
5. B. Courcelle and Y. Métivier, Coverings and minors: Applications to local computations in graphs, *European Journal of Combinatorics 15* (1994), 127-138;
6. M.R. Garey and D.S. Johnson, *Computers and Intractability*, W.H.Freeman and Co., 1978;
7. F.Harary, *Graph Theory*, Addison-Wesley, 1969;
8. P. Hell and J. Nešetřil, On the complexity of H-colouring, *Journal of Combinatorial Theory B 48* (1990), 92-110;
9. I. Holyer, The $\mathcal{NP}$-completeness of edge-coloring, *SIAM J. Computing 4* (1981), 718-720;
10. D. König, Über graphen und ihre Andwendung auf Determinantentheorie und Mengenlehre, *Math. Ann.* 77, 1916, 453-465;
11. J.Kratochvíl, *Perfect codes in general graphs*, monograph, Academia Praha (1991);
12. F.T. Leighton, Finite common coverings of graphs, *Journal of Combinatorial Theory B 33* (1982), 231-238;
13. J. Petersen, Die Theorie der regulären Graphen, *Acta Mathematica 15* (1891), 193-220;

Appendix

We list every connected, simple graph H on at most six vertices and at least two cycles, showing the complexity of the H-covering problem. Covering of simple graphs with at most one cycle is easy by Fact 3. By Fact 2 this resolves also the complexity of disconnected graphs having components on at most six vertices. The listing thus completes pages $1 \leq p \leq 6$ of the catalogue of the complexity of the covering problem for simple graphs on p vertices.

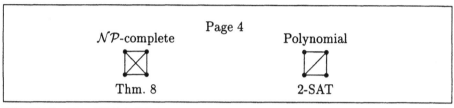

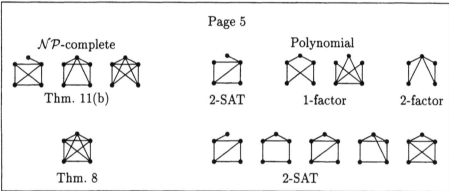

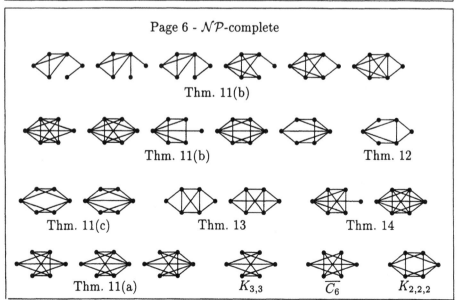

Page 6 - Polynomial

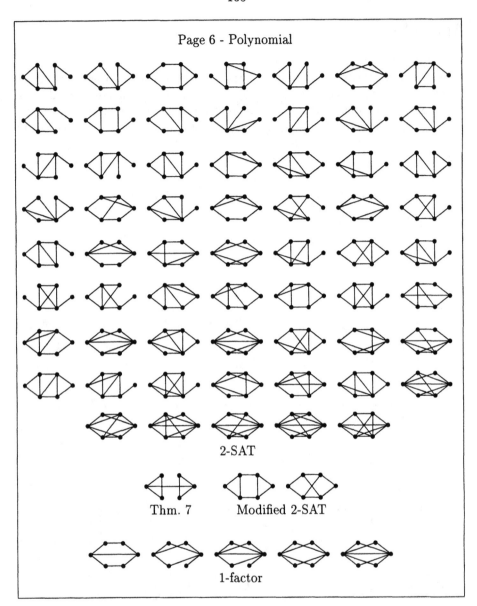

2-SAT

Thm. 7   Modified 2-SAT

1-factor

Dominoes

T. Kloks[1] *, D. Kratsch[2] ** and H. Müller[3]

[1] Department of Mathematics and Computing Science
Eindhoven University of Technology
P.O.Box 513, 5600 MB Eindhoven
The Netherlands
[2] IRISA
Campus de Beaulieu
35042 Rennes Cedex
France
[3] Fakultät für Mathematik und Informatik
Friedrich-Schiller-Universität
07740 Jena
Germany

Abstract. A graph is called a domino if every vertex is contained in at most two maximal cliques. The class of dominoes properly contains the class of line graphs of bipartite graphs, and is in turn properly contained in the class of claw-free graphs. We give some characterizations of this class of graphs, show that they can be recognized in linear time, give a linear time algorithm for listing all maximal cliques (which implies a linear time algorithm computing a maximum clique of a domino) and show that the PATHWIDTH problem remains NP-complete when restricted to the class of chordal dominoes.

1 Introduction

A domino is a graph in which every vertex is contained in at most two maximal cliques. The class of dominoes properly contains the line graphs of bipartite graphs. Every domino is the line graph of a multigraph, and hence claw-free. In the next section we give some characterizations of dominoes. Then we show that dominoes can be recognized in $O(n+m)$ time (where n is the number of vertices and m the number of edges), we also give a linear time clique listing algorithm (and hence a linear time algorithm for MAXIMUM CLIQUE) and, finally, we show that the PATHWIDTH problem 'Given a graph G and a positive integer k, decide whether the pathwidth of G is at most k' remains NP-complete when restricted to chordal dominoes.

---

* Email: ton@win.tue.nl
** On leave from Friedrich-Schiller-Universität Jena, Germany.

2 Characterizations

Unless stated otherwise, all graphs in this paper are finite, undirected, and without loops or multiple edges. If $G = (V, E)$ is a graph then we denote by $N(x)$ the set of neighbors of x. We use $N[x] = \{x\} \cup N(x)$ to denote the closed neighborhood. If S is a subset of vertices, then $G[S]$ is the subgraph of G induced by S. The *line graph* $L(G)$ of G is constructed as follows. The vertex set of $L(G)$ is E (the set of edges of G) and two vertices of $L(G)$ are adjacent if and only if the corresponding edges in G share an endpoint.

Definition 1. A graph is called a *domino* if every vertex occurs in at most two maximal cliques.

Lemma 2. *A domino with n vertices can have at most n different maximal cliques.*

Proof. Isolated vertices account for exactly one maximal clique. For larger connected components, every maximal clique contains at least two vertices. This proves the lemma. ☐

One of the main reasons for studying this class of graphs is, in our opinion, the fact that they generalize in a natural way the class of line graphs of bipartite graphs.

Lemma 3. *Every line graph of a bipartite graph is a domino.*

Proof. Let G be a line graph of a bipartite graph $H = (X, Y, E)$ with color classes X and Y. If $(x, y) \in E$ is an edge of H then $N(x) \cap N(y) = \emptyset$, since H is bipartite. Hence no two vertices (x, a) and (y, b) can be adjacent in G. It follows that the vertices adjacent to (x, y) in G are partitioned in two disjoint cliques. ☐

Clearly not every line graph is domino. For example the 4-wheel W_4 is the line graph of the diamond (i.e., $K_4 - e$), but clearly W_4 is not a domino (Figure 1). The converse is also not true; not every domino graph is a line graph. For example, in the list of the nine forbidden induced subgraphs for line graphs [1, 13] appear already five dominoes. It follows that dominoes and line graphs are incomparable classes, which both properly contain the class of line graphs of bipartite graphs and, as we shall soon see, which are both properly contained in the class of line graphs of multigraphs and hence also in the class of claw-free graphs.

It is not hard to see that a graph is a line graph of a bipartite graph if and only if it does not have a claw, a diamond or an odd hole as induced subgraph (see also [26, 21]). Clearly, all cycles and also the diamond are dominoes and line graphs. Hence not every graph from the intersection of line graphs and dominoes is the line graph of a bipartite graph.

Lemma 4. *The class of dominoes is hereditary; i.e., if G is a domino then also every induced subgraph of G is a domino.*

Fig. 1. The 4-wheel (left) is the line graph of the diamond (right) but not a domino

Proof. Let H be an induced subgraph of G. Assume z is a vertex of H and assume that z is in three maximal cliques S_1, S_2 and S_3 in H. Since these are maximal cliques, there must be pairs od adjacent or equal vertices $a, b \in S_1$, $x, y \in S_2$ and $p, q \in S_3$ such that a, x, y, p and q, b are pairs of different nonadjacent vertices. Since H is an induced subgraph of G these are also non edges in G. Each S_i is contained in a maximal clique D_i in G containing z and no two of these can be equal because of the three non edges. □

A graph is called *chordal* if it does not have a chordless cycle of length at least four as an induced subgraph. Every chordal graph has a *simplicial vertex*; i.e., a vertex whose neighborhood induces a clique [10, 11]. We start with a characterization of chordal dominoes. A *claw* and a *gem* are depicted in Figure 2.

Fig. 2. A chordal graph is domino iff it has no 'claw' (left) or 'gem' (right)

Theorem 5. *A graph is chordal and domino if and only if it does not have C_n $(n \geq 4)$, a claw or a gem as an induced subgraph (Figure 2).*

Proof. First assume G is chordal and domino. Then it cannot contain a chordless cycle C_n $(n \geq 4)$ since G is chordal. But G can also not contain a claw or gem since these contain vertices in more than two maximal cliques and this is forbidden by Lemma 4.

Now assume the converse is not true and let $G = (V, E)$ be a minimal counterexample, i.e., G has a minimal number of vertices among all chordal graphs without gem or claw which are not domino. Then G has some vertex y which is in three different maximal cliques S_1, S_2 and S_3 of G and by minimality $G = G[S_1 \cup S_2 \cup S_3]$. Hence, y is adjacent to all other vertices of G.

G is not a complete graph. Hence G contains two nonadjacent simplicial vertices (see Lemma 4.2 in [11]). Say $x_1 \in S_1$ and $x_3 \in S_3$ are simplicials. If x_1 is adjacent to all vertices of S_2, then $S_1 \cup S_2$ is a clique (since x_1 is simplicial). But then S_1 and S_2 could not be different maximal cliques. Let $p \in S_2$ be not adjacent to x_1. Then p must be adjacent to x_3, otherwise we obtain a claw. In a similar way we obtain a vertex $q \in S_2$ which is adjacent to x_1 but not to x_3. Then we obtain a gem $G[\{p, q, x_1, x_3, y\}]$. □

In the following theorem we give a characterization of dominoes by forbidden induced subgraphs.

Theorem 6. *A graph G is domino if and only if it has no induced claw, gem or 4-wheel.*

Proof. By Lemma 4 it follows immediately that a domino cannot have a 4-wheel, gem or claw as an induced subgraph.

Let G be a counterexample for the converse with a minimal number of vertices.

Consider some vertex y in G which is in the maximal cliques $S_1, S_2, \ldots, S_r$ for some $r \geq 3$. Then clearly, $G[S_1 \cup S_2 \cup S_3]$ has no claw, gem or 4-wheel, and is also not a domino. Since G has a minimal number of vertices, it follows that $G = G[S_1 \cup S_2 \cup S_3]$.

By Theorem 5 G is not chordal. Let C be a chordless cycle of length at least four in G. Each vertex of C is either equal to y or adjacent to y since $G = G[S_1 \cup S_2 \cup S_3]$. If y is a vertex of C then C cannot be chordless. In case y is not a vertex of C we obtain a wheel W_k for some $k \geq 4$ and hence G has a 4-wheel or a gem. □

A *multigraph* $G = (V, E)$ is a graph in which the edges may occur several times. Edges joining the same pair of vertices are called multiple edges. The line graph $L(G)$ of a multigraph $G = (V, E)$ has as vertex set E and two vertices of $L(G)$ are adjacent if they have at least one endpoint in common.

Remark. Notice that if e_1 and e_2 are multiple edges then they have the same closed neighborhood in the line graph. The converse needs of course not be true as a claw and its line graph for example show.

There exists a characterization by a finite set of forbidden induced subgraphs for line graphs of multigraphs [14, 3, 20]. Another characterization, found by Krausz [17, 20], is the following. A graph G is a line graph of a multigraph if and only if G can be written as the union of complete subgraphs such that no vertex of G belongs to more than two of these complete subgraphs. (If in addition the complete graphs have to *partition* the set of edges, we obtain a characterization of line graphs of simple graphs [17, 13]). From both characterizations of line graphs of multigraphs mentioned above immediately follows:

Theorem 7. *Every domino is the line graph of some multigraph.*

Proof. Consider the list of all maximal cliques of G. Clearly, their union is G (every vertex and every edge is in at least one of the cliques). Since G is a domino, every vertex is in at most two of the maximal cliques. □

In Figure 3 we illustrate the relations between the different classes.

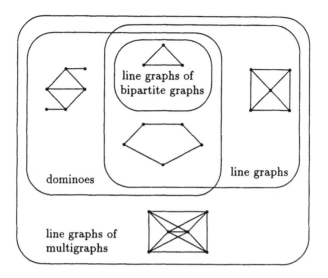

Fig. 3. Relations between the different graph classes

3 Representatives

In this section we show a method to reduce dominoes to certain induced subgraphs which we call representatives. This reduction will enable us to reduce the recognition of dominoes to an elementary graph problem.

Definition 8. Let $G = (V, E)$ be a graph. Two vertices x and y are *equivalent*, $x \equiv y$, if $N[x] = N[y]$.

The relation $\equiv$ is an equivalence relation.

The following lemma is useful for computing equivalence classes.

Lemma 9. *Two vertices x and y are equivalent if and only if for each closed neighborhood $N[z]$, both x and y are in $N[z]$ or both are not in $N[z]$.*

Proof. Notice that for each vertex p, $N[p] = \{q \mid p \in N[q]\}$. □

Definition 10. Let G be a graph. The *representative* of G, $\mathcal{R}(G)$, is the graph constructed as follows. The vertices of $\mathcal{R}(G)$ are the equivalence classes under $\equiv$, and two of these vertices are adjacent if a vertex from one class is adjacent to a vertex of the other class in G.

Remark. For each graph G, the representative $\mathcal{R}(G)$ is isomorphic to an induced subgraph of G.

Lemma 11. *A graph $G = (V, E)$ is a domino if and only if $\mathcal{R}(G)$ is a domino.*

Proof. Clearly, if G is a domino, since $\mathcal{R}(G)$ is isomorphic to an induced subgraph of G, by Lemma 4 also $\mathcal{R}(G)$ is a domino.

Now assume $\mathcal{R}(G)$ is a domino. Suppose, by way of contradiction, that G is not a domino and let z be a vertex which is in three maximal cliques S_1, S_2 and S_3. There are vertices $a, b \in S_1$, $x, y \in S_2$ and $p, q \in S_3$ such that $(a, x), (y, p), (b, q) \notin E$. But then no vertex of the equivalence class of a can be adjacent to a vertex of the equivalence class of x. Hence, in $\mathcal{R}(G)$ these equivalence classes are not adjacent. It follows that the equivalence class of z is also in three different maximal cliques in $\mathcal{R}(G)$. $\qquad\square$

We first exhibit some more useful lemmas concerning representatives.

Lemma 12. *Let G be a domino. Then every edge of $\mathcal{R}(G)$ is in exactly one maximal clique.*

Proof. Assume (x, y) is an edge in $\mathcal{R}(G)$ which is in two maximal cliques S_1 and S_2. Since $\mathcal{R}(G)$ is domino, the closed neighborhood of x and y is $S_1 \cup S_2$. Hence x and y are equivalent in $\mathcal{R}(G)$, which is a contradiction. $\qquad\square$

The following result is an immediate consequence of Theorem 6 and Lemma 12.

Corollary 13. *A graph H is a representative of a domino if and only if H has no induced claw or diamond and $\mathcal{R}(H) = H$ (i.e., H has no equivalent vertices).*

Notice that the list of forbidden subgraphs for line graphs consists of the claw, and graphs which contain a diamond as an induced subgraph.

Corollary 14. *Let G be a domino. Then $\mathcal{R}(G)$ is a line graph.*

Another proof of Corollary 14 is the following. By Lemma 11, $\mathcal{R}(G)$ is a domino, and hence, by Theorem 7, the line graph of some multigraph. Let H be such a multigraph. If e_1 and e_2 would be multiple edges in H, then clearly, they would have the same closed neighborhood in $\mathcal{R}(G)$. This is a contradiction. Hence H does not have multiple edges.

Remark. A characterization similar to Corollary 14 (but somewhat weaker) can also be obtained as follows. Let $H = \mathcal{R}(G)$ be the representative of a domino G and assume H has r vertices that are in two maximal cliques. Let $\mathbf{X}$ be the vertex-clique incidence matrix of H. Then by Lemma 12 $\mathbf{XX}^T = \mathbf{A} + \mathbf{I} + \mathbf{I(r)}$, where $\mathbf{A}$ is the 0/1-adjacency matrix of H and $\mathbf{I(r)}$ is a matrix with r diagonal entries equal to 1 and all other entries equal to zero. It follows that all eigenvalues of $\mathbf{A}$ are at least -2. In [5, 8] it is shown that such a graph is either a line graph of a bipartite graph, a generalized line graph (see, e.g., [8]) or one of a finite class of graphs.

Lemma 15. *Let G be a domino and $\mathcal{R}(G)$ its representative. Then for each vertex x of $\mathcal{R}(G)$, the neighborhood of x in $\mathcal{R}(G)$ is either one clique or the disjoint union of two cliques, (i.e., the neighborhood consists of two cliques, without edges going from one to the other).*

Proof. Let x be a vertex of $\mathcal{R}(G)$. If x is contained in at most one maximal clique, the statement holds. Assume x is contained in two maximal cliques S_1 and S_2. Since $\mathcal{R}(G)$ is a domino, $N(x) = (S_1 \cup S_2) \setminus \{x\}$. By Lemma 12, $S_1 \cap S_2 = \{x\}$. Now let $u \in S_1 \setminus \{x\}$ be adjacent to some $v \in S_2 \setminus \{x\}$. Then, clearly, the triangle $\{x, u, v\}$ is contained in some maximal clique S, which is different from S_1 and S_2. This contradicts the fact that $\mathcal{R}(G)$ is a domino. □

A somewhat different version of the following theorem appeared as an example in [2] (page 32, example 3). A *root graph* K of a line graph H is a graph such that H is the line graph of K. Unless H is K_3, K is unique (see [13]).

Theorem 16. *A graph H is a representative of a domino if and only if H is a line graph and its root graph K is a triangle-free graph in which every vertex is adjacent to at most one pendant vertex.*

Proof. Let H be the representative of a domino. By Corollary 14 H is a line graph. Let K be the root graph. Since H is diamond-free, K cannot have a 3-pan (or paw; i.e., a triangle with a pendant vertex) as a subgraph. Hence every triangle of K is a connected component of K. It follows that every neighborhood in K with at least three vertices induces an independent set. Hence every connected component of K is either a triangle or a triangle-free graph. The fact that H has no equivalent vertices implies that no component of K is a triangle and no vertex is adjacent to two vertices of degree one.

Now assume H is the line graph of a graph K and assume that K is triangle-free and every vertex of K is adjacent to at most one pendant vertex. Since H is a line graph, H is claw-free, and since K is triangle-free, K has no 3-pan as a subgraph and hence H has no induced diamond. It remains to show that H has no equivalent vertices (by Corollary 13). Equivalent vertices a and b in H correspond to edges in K sharing an end vertex x. Since a and b are equivalent every edge incident with a and b in K must be incident with both. Since K is triangle-free, the other end vertices of a and b in K must be pendant vertices. Hence x is adjacent to two pendant vertices. □

4 Computing the representative of a graph

In this section, we describe an algorithm which, given a graph G, computes its representative $\mathcal{R}(G)$ in linear time. In the next section, we describe an algorithm to test whether this representative is a domino.

The first step is a labeling procedure, which assigns labels to vertices such that two vertices get the same label if and only if they are equivalent. The procedure CLASSES is displayed below. We assume the graph G has n vertices, numbered $1, \ldots, n$ and that the closed neighborhood of each vertex i is given in a list $N[i]$. During the algorithm, the number of different labels that is assigned is given by t. Each label $1, \ldots, t$ is used at least once. The number of times a certain label x is used is stored in $a(x)$.

Procedure CLASSES(G); [a]
begin
   Comment: initialization.
   for $i := 1$ **to** n **do begin** $L(i) := 1$; $a(i) := 0$; $r(i) := 0$ **end**;
   $a(1) := n$; $t := 1$;
   for $i := 1$ **to** n **do**
      begin
         Comment: $b(x)$ is the number of occurrences of label x,
           vertices with label x get new label $r(x)$ if necessary.
         for $j \leftarrow N[i]$ **do begin** $b(L(j)) := 0$; $r(L(j)) := 0$ **end**;
         for $j \leftarrow N[i]$ **do** $b(L(j)) := b(L(j)) + 1$;
         for $j \leftarrow N[i]$ **do**
           Comment: If a label occurs in and out the neighborhood,
           then reserve a new label.
           if $b(L(j)) < a(L(j))$ **and** $r(L(j)) = 0$ **then**
              begin $t := t + 1$; $r(L(j)) := t$ **end**;
         for $j \leftarrow N[i]$ **do**
           if $r(L(j)) \neq 0$ **then**
              begin
                  $L(j) := r(L(j))$; $a(L(j)) := a(L(j)) - 1$;
                  $a(r(L(j))) := a(r(L(j))) + 1$
              end
      end
end

---

[a] procedure CLASSES computes equivalence classes

Vertex k is given label $L(k)$. Within the outermost loop we have as an invariant that two vertices p and q have the same label if and only if they have the same neighbors in $\{1, 2, \ldots, i - 1\}$.

For each i, the algorithm goes through the closed neighborhood $N[i]$ four times. First, if some label x occurs as a label of a vertex in $N[i]$, then the number of times it occurs is counted in $b(x)$. This is done the first two times the algorithm passes $N[i]$. If this number is less than $a(x)$, all vertices in $N[i]$ with label x get a unique new label, $r(x)$. Notice that if $b(x) = a(x)$, then all vertices with label x are in $N[i]$ hence no new label needs to be assigned to these vertices. The third time the algorithm runs through $N[i]$, a new label is reserved in $r(x)$ if necessary. Finally, in the fourth loop the actual assignment of new labels is done. Correctness follows from Lemma 9.

The next step of the algorithm to compute the representative, is given by procedure REPRESENTATIVE given below. The input of this procedure is the graph G, the labeling L and the number of labels t. It computes, for vertices $i = 1, \ldots, t$, adjacency list $N_r(i)$. To avoid duplicates in the adjacency lists, for each label x, a boolean variable $B(x)$ is introduced. A label $L(j)$ is added to the adjacency list of $N_r(L(k))$ if j is in the adjacency list $N(k)$ and $B(L(j))$ is true. In that case $B(L(j))$ is made false, such that each label can be added to the list at most once.

Procedure REPRESENTATIVE(G, L, t); [a]
begin
    Comment: initialization.
    for $i := 1$ **to** t **do** $N_r(i) := \emptyset$;
    Comment: make adjacency lists.
    for $k := 1$ **to** n **do**
        for $j \leftarrow N(k)$ **do** $B(L(j)) :=$ TRUE;
        for $j \leftarrow N(k)$ **do**
            if $B(L(j))$ **then**
                begin $N_r(L(k)) := N_r(L(k)) \cup \{L(j)\}$; $B(L(j)) :=$ FALSE **end**
    end

---

[a] procedure REPRESENTATIVE computes the representative graph

The discussion above proves the following theorem.

Theorem 17. *There exists an algorithm which computes, given a graph G, its representative $\mathcal{R}(G)$. This algorithm can be implemented such that it takes linear time.*

5 Recognition of dominoes

In this section we give a very simple linear time recognition algorithm for dominoes. In fact we show a linear time algorithm to test whether H is the representative of a domino. Together with Theorem 17 this gives a linear time recognition for dominoes. We use the characterization of Theorem 16.

Notice that Theorem 16 leads to the following recognition algorithm. The representative H of a given graph G can be computed in linear time which was shown in section 4. Determining whether H is a line graph and, if so, computing the root graph K of H can be done in linear time [19, 24, 25].

Now assume K is known, and has p vertices and q edges. Clearly, the restriction on the pendant vertices can easily be checked in linear time.

Of course, a graph K is triangle-free if and only if no two adjacent vertices have a common neighbor. The following well known (but unfortunately not linear time) algorithms check if a graph is triangle free. For example this can be checked by computing the square of the $0/1$-adjacency matrix. By using fast matrix multiplication techniques this can be done in $O(p^\alpha)$ where $\alpha = 2.37...$ (see, e.g. [18]). (See also [6] for a parallel version of this algorithm.)

In [15] (see also [18]) another algorithm is presented to find a triangle in a graph which gives, in our case, in general, a better worst case timebound. This algorithm runs in time $O(q^{3/2})$. Since $q = n$, we find a domino recognition algorithm which runs in time $O(m + n^{3/2})$.

Our linear time recognition algorithm works as follows. First use a linear time algorithm to test whether H is a line graph and, if so, outputs its root graph K (see, e.g., [19, 24, 25]). Assume K has p vertices and q edges. We assume that vertices of K are numbered $1, 2, \ldots, p$ and that for each vertex i a list of neighbors is given in $NK(i)$.

The next step is to check if every vertex is adjacent to at most one pendant vertex. Obviously, this can be done in linear time.

Now, order the neighbors of each vertex in increasing order. Most easily this is done by making new adjacency lists. Initialize these as empty. Then for each neighbor of vertex 1 insert 1 in the new adjacency list. Next insert 2 in each adjacency list of its neighbors and so on. For reasons of simplicity we call the ordered neighborhoods again $NK(i)$, $i = 1, \ldots, p$.

We use the following simple algorithm to test if K is triangle free. For each edge (x, y) in K, check if $NK(x) \cap NK(y) = \emptyset$. Since the neighborhoods are ordered this takes time $O(|NK(x)| + |NK(y)|)$ for each edge (x, y). Hence the number of steps required by the algorithm is at most:

$$\sum_{x=1}^{p} \sum_{y \in NK(x)} (|NK(x)| + |NK(y)|) = 2 \sum_{x=1}^{p} |NK(x)|^2$$

Notice that, since H is the line graph of K, the number of vertices of H is $n = q$ and the number of edges of H is $m = -q + \frac{1}{2} \sum_{x=1}^{p} |NK(x)|^2$ (see, e.g., [13]). Hence the test if K is triangle free can be performed in time proportional to $2 \sum_{x=1}^{p} |NK(x)|^2 = 4(m + n)$.

6 Listing all cliques of dominoes

In this section we describe an algorithm which lists all maximal cliques of a domino G in linear time.

First compute the representative H of G. Then, compute all maximal cliques of H, and finally, replace every vertex of H by the corresponding equivalence class. Obviously, this gives exactly the list of all maximal cliques of G.

The following procedure finds all maximal cliques of H.

Step 1 For each vertex i sort its adjacency list in increasing order.

Step 2 For each vertex i we seek a partition of its neighborhood into two sets $S_1(i)$ and $S_2(i)$. This partition is such that the sets $\{i\} \cup S_1(i)$ and $\{i\} \cup S_2(i)$ are the maximal cliques containing i, $i = 1, \ldots, n$. For each vertex i let the boolean variable $P(i)$ indicate whether or not the neighborhood of i is partitioned. Initially $P(i) = \text{FALSE}$ for each i. We keep a list of vertices of which the neighborhood is not yet partitioned.

Step 3 Take a vertex i of which the neighborhood is not yet partitioned. Let j be the first neighbor of i (i.e., j is the smallest neighbor of i). Put j in $S_2(i)$. We compute $S_2(i) = \{j\} \cup (N(i) \cap N(j))$. We consider two cases.

Case 1 The neighborhood of j is not yet partitioned. Partition the neighborhoods of i and j at the same time as follows. Go through the neighborhoods, each time comparing the next smallest elements of $N(i)$ and $N(j)$. If the next smallest element of $N(j)$ is i then put i in $S_2(j)$. Otherwise, if the next smallest elements are different then put the smallest of them in $S_1(i)$ (if it is a neighbor of i) or in $S_1(j)$ (if it is a neighbor of j). If the smallest elements are equal then put it in $S_2(i)$ and in $S_2(j)$. Notice that this step can be completed in time $O(|N(i)| + |N(j)|)$.

Case 2 The neighborhood of j is already partitioned. In that case go through the neighborhood of i. Put elements of $N(i)$ into $S_1(i)$ until the first element of $S_1(j)$ or of $S_2(j)$ which is not equal to i is encountered. Assume this is an element of $S_1(j)$. Then put this element in $S_2(i)$. Continue with scanning through $N(i)$. Each time when an element of $S_1(j)$ is encountered, this is put in $S_2(i)$ and otherwise the element is put in $S_1(i)$. Notice that this step takes time $O(|N(i)|)$.

Step 4 In case all neighborhoods are partitioned, a list $\mathcal{P}$ of all maximal cliques is made. In order to avoid duplicates, a set $\{i\} \cup S_k(i)$ ($i = 1, \ldots, n$, $k = 1, 2$) is put in the list $\mathcal{P}$ if and only if i is smaller than the first element of $S_k(i)$.

The description above and Lemmas 12 and 15 prove the following result.

Lemma 18. *The list of all maximal cliques of a domino G with n vertices and m edges can be found in $O(n + m)$ time.*

Corollary 19. *There is a linear time algorithm to solve* MAXIMUM CLIQUE *on dominoes.*

7 Pathwidth

In this section we show the NP-completeness of the PATHWIDTH problem when restricted to chordal dominoes. For applications and further references regarding pathwidth and treewidth of graphs we refer to [4, 16].

Definition 20. A *path decomposition* of a graph $G = (V, E)$ is a sequence $\mathcal{S} = (S_1, S_2, \ldots S_r)$ of subsets of vertices, such that

- $\bigcup_{i=1}^{r} S_i = V$,
- for each edge $\{v, w\} \in E$ there exist an i, $1 \leq i \leq r$, with $v \in S_i$ and $w \in S_i$,
- for $1 \leq i \leq j \leq k \leq r$ holds $S_i \cap S_k \subseteq S_j$.

The pathwidth of the path decomposition $\mathcal{S}$ is $\max_{1 \leq i \leq r} |S_i| - 1$. The *pathwidth* $pw(G)$ of G is the minimum pathwidth over all possible path decompositions of G.

The PATHWIDTH problem, which is: 'Given graph G and a positive integer k, decide whether $pw(G) \leq k$' is NP-complete when restricted to starlike graphs, a subclass of the chordal graphs [12]. We will prove the NP-completeness of PATHWIDTH when restricted to chordal dominoes by a reduction from the MINIMUM CUT LINEAR ARRANGEMENT problem for trees with polynomial edge weights (abbr. MIN CUT TP).

The reduction has two steps. In a first step we reduce MIN CUT TP to the MODIFIED MINIMUM CUT LINEAR ARRANGEMENT problem for trees with polynomial edge weights (abbr. MOD MIN CUT TP). The second step reduces MOD MIN CUT TP to PATHWIDTH for chordal dominoes.

A *linear arrangement* of a graph $G = (V, E)$ is a one-to-one mapping $a : V \to \{1, 2, \ldots |V|\}$, and a *weight function* is a mapping $w : E \to \{0, 1, \ldots\}$. For a class of graphs equipped with weight functions the weights are polynomially bounded if there is a polynomial p such that for each graph G in the class we have $\max(\{w(\{u, v\}) : \{u, v\} \in E\}) \leq p(|V|)$. For simplicity we define $w(\{u, v\}) = 0$ for $\{u, v\} \in \binom{V}{2} \setminus E$.

The MIN CUT problem is the question whether there is a linear arrangement a such that $\max(\{\sum\{w(\{u, v\}) : a(u) < i \leq a(v)\} : 1 < i \leq n\}) \leq k$ for a given graph G with weight function w and a given integer bound k. The MODIFIED MIN CUT problem asks on the same input whether there is a linear arrangement a such that $\max(\{\sum\{w(\{u, v\}) : a(u) < i < a(v)\} : 1 < i < n\}) \leq k$ In [23] the NP-completeness of MIN CUT TP is shown.

Notice that MOD MIN CUT TP belongs to NP. Now we describe a (polynomial time many-one) reduction from MIN CUT TP to MOD MIN CUT TP. For a graph $G = (V, E)$ with weight function $w : E \to \{0, 1, \ldots\}$ and a positive integer $k \leq \sum\{w(e) : e \in E\}$ we define $G' = (V', E')$ and $w' : E' \to \{0, 1, \ldots\}$ by

$$V' = V \times \{0, 1, 2\},$$
$$E' = \{\{(u, 1), (v, 1)\} : \{u, v\} \in E\} \cup$$
$$\{\{(v, 0), (v, 1)\}, \{(v, 1), (v, 2)\} : v \in V\},$$
$$w'(\{(u, i), (v, j)\}) = \begin{cases} w(\{u, v\}), & \text{if } i = j = 1, \\ k + 1, & \text{if } u = v, i = 1, j \neq 1, \\ 0, & \text{if } \{(u, i), (v, j)\} \notin E'. \end{cases}$$

Since G is a tree with polynomial edge weights G' is also a tree and w' is polynomially bounded.

It not hard to check that G with weights w has a linear arrangement a with $\max(\{\sum\{w(\{u, v\}) : a(u) < i \leq a(v)\} : 1 < i \leq n\}) \leq k$ if and only if G' with weights w' has a linear arrangement a' with $\max(\{\sum\{w'(\{(u, i'), (v, j)\}) : a'((u, i')) < i < a'((v, j))\} : 1 < i < 3n\}) \leq k$. For verifying this observe that a' fulfills this condition only if for every $u \in V$ and the edges $\{(u, 1), (u, 0)\}$ and $\{(u, 1), (u, 2)\}$ with weight $k + 1$ holds

$$|a'((u, 1)) - a'((u, 0))| = |a'((u, 1)) - a'((u, 2))| = 1.$$

Hence MOD MIN CUT TP is also NP-complete.

We now continue with the reduction from MOD MIN CUT TP to PATHWIDTH for chordal dominoes. We consider a tree $T = (V, E)$ and a weight function $w : E \to \{0, 1, \ldots\}$. We define $n = |V|$ and

$$m = \max\left(k + 1, \max\left(\left\{\sum\{w(e) : e \in E \wedge v \in e\} : v \in V\right\}\right)\right)$$

We construct a graph G as follows: For each edge $e \in E$ we have $w(e)$ vertices $(e, 1), \ldots (e, w(e))$. For each vertex $v \in V$ we have $w(v) = m - \sum (e : e \in E \wedge v \in e : w(e))$ vertices $(v, 1), \ldots (v, w(v))$. The edges of G are defined such that each maximal clique is of the form

$$C(v) = \{(e, i) : 1 \leq i \leq w(e) \wedge e \ni v\} \cup \{(v, i) : 1 \leq i \leq w(v)\}$$

for a suitable $v \in V$. Then G is a chordal domino.

Theorem 21. *The problem 'Given a chordal domino G and a positive integer k, decide whether $pw(G) \leq k$' is NP-complete.*

Proof. We consider a linear arrangement a of T. We define the path decomposition $S(a) = (S_1, S_2, \ldots S_r)$ of G by

$$S_i = C(a^{-1}(i)) \cup \bigcup_{a(u)<i<a(v)} \{((\{u,v\}, j) : 1 \leq j \leq w(\{u,v\})\}.$$

If a is a linear arrangement of T with $\max(\{\sum\{w(\{u,v\}) : a(u) < i < a(v)\} : 1 < i < n\}) \leq k$ then for $1 \leq i \leq r$ we have $|S_i| \leq m + k$.

Suppose there is a path decomposition $S' = (S'_1, S'_2, \ldots S'_{r'})$ of G such that $|S'_i| \leq m + k$ for all i, $1 \leq i \leq r'$. Then for each $v \in V$ there is an index i, with $C(v) \subseteq S'_i$, and for $C(v) \subseteq S'_i \cap S'_{i'}$ with $i < i'$ we have $C(v) \subseteq S'_{i+1}$ (see, e.g., [16]). Moreover $C(u) \cup C(v) \subseteq S'_i$ implies $u = v$ since $m > k$. Hence S' induces a linear ordering $C(v_1), C(v_2), \ldots, C(v_n)$, $v_i \in V$, of the maximal cliques of G. This defines a linear arrangement a of T (by $a(v_i) = i$ for every $v_i \in V$) with $\max(\{\sum(\{w(\{u,v\} : a(u) < i \leq a(v)\} : 1 < i < n\}) \leq k$. Let $S(a) = (S_1, S_2, \ldots S_n)$. Then $S_{a(v)} = \bigcap\{S'_i : C(v) \subseteq S'_i\}$ and $|S_i| \leq m + k$ for all i with $1 \leq i \leq r$. □

8 Conclusions

In this paper we considered a class of graphs, called dominoes, which is a natural extension of the line graphs of bipartite graphs. We have given some characterizations and a linear time recognition algorithm. We have shown that the PATH-WIDTH problem remains NP-complete when restricted to chordal dominoes. It is an open problem whether the TREEWIDTH problem for dominoes is NP-complete.

A clique listing algorithm for dominoes is given in section 6 that runs in linear time. Clearly, polynomial time algorithms to solve problems for claw-free graphs can be used for dominoes as well. For example, from [22, 20] it follows immediately that there is a polynomial time algorithm for MAXIMUM INDEPENDENT SET on dominoes.

Since dominoes are line graphs of multigraphs we can use the following to obtain an approximate coloring algorithm. Shannon's theorem (see, e.g., [9]) says that if M is a multigraph with maximum valency ρ then $\chi'(M) \leq \lfloor \frac{3}{2}\rho \rfloor$. It follows that a vertex coloring for a domino graph G can be found with at most $\frac{3}{2}\omega(G) \leq \frac{3}{2}\chi(G)$ colors. It is open question whether the coloring problem remains NP-complete when restricted to dominoes.

9 Acknowledgements

We thank J. Spinrad for useful discussions on the problem of obtaining the representative of a graph (section 4) and A. Jacobs for doing a lot of the bibliographical research. Finally, we thank one anonymous referee for making many useful comments.

References

1. Beineke, L. W., On derived graphs and digraphs, In: H. Sachs, H. J. Voss and H. Walther, eds., *Beiträge zur Graphentheorie*, Teubner, Leipzig, (1968), pp. 17–23.

2. Berge, C., *Hypergraphs*, North Holland, 1989.

3. Bermond, J. C. and J. C. Meyer, Graphe représentatif des arêtes d'un multigraphe, *J. Math. Pures Appl.* **52**, (1973), pp. 299–308.

4. Bodlaender, H. L., A tourist guide through treewidth, *Acta Cybernetica* **11**, (1993), pp. 226–234.

5. Cameron, P. J., J. M. Goethals, J. J. Seidel and E. E. Shult, Linegraphs, root systems and elliptic geometry, *J. Algebra* **43**, (1976), pp. 305–327.

6. Chaudhuri, P., An algorithm for finding all triangles in a digraph in parallel, *Pure and Applied Mathematika Sciences*, Vol. XXV, (1987), pp. 27–35.

7. Chiba, N. and T. Nishizeki, Arboricity and subgraph listing algorithms, *SIAM J. Comput.* **14**, (1985), pp. 210–223.

8. Biggs, N., *Algebraic graph theory*, Cambridge University Press, (1993).

9. Fiorini, S. and R. J. Wilson, *Edge colorings of graphs*, Pitman, London, (1977).

10. Fulkerson, D. R. and O. A. Gross, Incidence matrices and interval graphs, *Pacific J. Math.* **15**, (1965), pp. 835–855.

11. Golumbic, M. C., *Algorithmic Graph Theory and Perfect Graphs*, Academic Press, New York, (1980).

12. Gustedt, J., On the pathwidth of chordal graphs, *Disc. Appl. Math.* **45**, (1993), pp. 233–248.

13. Harary, F., *Graph Theory*, Addison-Wesley Publ. Comp., Reading, Massachusetts, (1969).

14. Hemminger, R. L., Characterization of the line graph of a multigraph, *Notices Amer. Math. Soc.* **18**, (1971), pp. 934.

15. Itai, A. and M. Rodeh, Finding a minimal circuit in a graph, *SIAM J. Comput.* **7**, (1978), pp. 413–423.

16. Kloks, T., Treewidth, Ph.D. Thesis, Utrecht University, Utrecht, The Netherlands, 1993.

17. Krausz, J., Démonstration nouvelle d'un théorème de Whitney sur les réseaux, *Mat. Fiz. Lapok* **50**, (1943), pp. 75–85.

18. Leeuwen, J. van, Graph Algorithms. In: J. van Leeuwen, ed., *Handbook of Theoretical Computer Science, A: Algorithms and Complexity*, Elsevier Science Publ., Amsterdam, 1990, pp. 527–631.

19. Lehot, P. G. H., An optimal algorithm to detect a line graph and output its root graph, *J. of the ACM* **21**, (1974), pp. 569–575.

20. Lovász, L. and M. D. Plummer, *Matching Theory*, Ann. Disc. Math. **29**, (1986).

21. Maffray, F., Kernels in perfect line-graphs, *Journal of Combinatorial Theory, Series B* **55**, (1992), pp. 1–8.

22. Minty, G. J., On maximal independent sets of vertices in claw-free graphs, *Journal on Combinatorial Theory, Series B* **28**, (1980), pp. 284–304.

23. Monien, B. and I. H. Sudborough, Min Cut is NP-complete for Edge Weighted Trees, *Theoretical Computer Science* **58**, (1988), pp. 209-229.

24. Roussopoulos, N. D., A max$\{m, n\}$ algorithm for determining the graph H from its line graph G, *Information Processing Letters* **2**, (1973), pp. 108–112.

25. Sysło, M. M., A labeling algorithm to recognize a line digraph and output its root graph, *Information Processing Letters* **15**, (1982), pp. 28–30.

26. Trotter, L. E., Line perfect graphs, *Mathematical Programming* **12**, (1977), pp. 255–259.

GLB-Closures in Directed Acyclic Graphs and Their Applications

Volker Turau[*1] and Weimin Chen[2]

[1] FH Gießen-Friedberg, Wilhelm-Leuschner-Str. 13, 61169 Friedberg, Germany,
turau@courbet.fh-friedberg.de
[2] GMD - Integrated Publication and Information Systems Institute, Dolivostr.15
64293 Darmstadt, Germany, chen@darmstadt.gmd.de

Abstract. A subset S of the vertices of a directed acyclic graph is called glb-closed, if it contains the greatest lower bounds of all pairs of vertices of S. The glb-closure of S is the smallest glb-closed subset containing S. An efficient output sensitive algorithm for computing glb-closures is presented and two applications in the field of object-oriented programming languages are discussed.

1 The problem in general

Directed acyclic graphs are widely used in different areas of computer science. In compiler construction they can be used to identify common subexpressions or to represent type lattices in programming languages supporting subtypes. In the latter case lattice operations for determining least upper bounds or greatest lower bounds are frequently needed. In this paper the notion of closure of a subset of the vertices of a directed acyclic graph is introduced. A subset S of the vertices is called closed if it contains the greatest lower bounds of all pairs of vertices in S, i.e. S is closed with respect to taking greatest lower bounds. In order to avoid confusion with other already existing usages of the notion closure in graph theory (e.g. transitive closure) in this paper the notion *glb-closed* is used. The *glb-closure* of a subset S is the intersection of all *glb-closed* subsets containing S, i.e. the smallest *glb-closed* subset containing S. *Glb-closed* subsets have some interesting properties. In this paper an efficient algorithm for computing the *glb-closure* of a set of vertices is presented. With some preprocessing the *glb-closure* of a subset S can be computed in time $O(e + n)$ where e is the number of edges in G and n is the number of vertices.

Let $G = (V, E)$ be a directed acyclic graph, V denotes the set of vertices and E the set of edges. The following notation is used throughout the paper. Let $v_1, v_2 \in V$ then $v_1 \preceq v_2$ if $v_1 = v_2$ or if there exists a path in G from v_1 to v_2. Furthermore $v_1 \prec v_2$ if $v_1 \preceq v_2$ and $v_1 \neq v_2$. In the following the lattice operations GLB and LUB are defined for directed graphs. The definitions differ from that for lattices. GLB and LUB are sets of vertices

---

* This work was done while the author was visiting the International Computer Science Institute in Berkeley.

(possibly empty), where in lattices GLB and LUB are precisely one vertex. The set of *greatest lower bounds* for $v_1, v_2 \in V$ is defined by $GLB(v_1, v_2) = \{v \in V \mid v \preceq v_1, v \preceq v_2$ and there exists no $u \in V$ such that $v \prec u$ and $u \preceq v_1, u \preceq v_2\}$. Let $v \in V$ and S a subset of V. The set of *least upper bounds* of v in S is defined by $LUB(v, S) = \{u \in S \mid v \preceq u$ and there exists no $w \in S$ such that $v \preceq w \prec u\}$.

Definition 1. A subset S of V is called *glb-closed* if $GLB(s_1, s_2) \subseteq S$ for all $s_1, s_2 \in S$. Let *glb-closure*(S) be the intersection of all *glb-closed* subsets of V which contain S.

Clearly *glb-closure(S)* is *glb-closed* for all subsets S of V and *glb-closure(V)* $= V$. If $v_1 \preceq v_2$ then $GLB(v_1, v_2) = \{v_1\}$. To give examples illustrating these definitions we refer to the Graph in Fig. 1(a):

$$GLB(v_2, v_3) = \{v_8, v_6\}$$
$$GLB(v_1, v_4) = \{v_4\}$$
$$LUB(v_9, \{v_2, v_4\}) = \{v_2, v_4\}$$
$$LUB(v_9, \{v_2, v_4, v_6\}) = \{v_6\}$$
$$glb\text{-}closure\ (\{v_2, v_3\}) = \{v_2, v_3, v_8, v_6, v_{11}\}$$
$$glb\text{-}closure(\{v_6, v_4\}) = \{v_6, v_4\}$$
$$glb\text{-}closure(\{v_2, v_3, v_4\}) = \{v_2, v_3, v_4, v_6, v_7, v_8, v_{11}\}$$

Intersections of *glb-closed* subsets are *glb-closed* and in case G is a DAG where each vertex has outdegree at most 1 then all subsets of V are *glb-closed*.

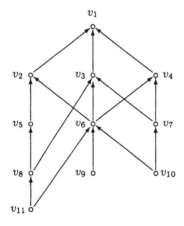
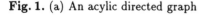
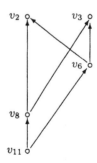

Fig. 1. (a) An acylic directed graph   (b) The graph G_{11}

The following lemmas state some properties of GLB and LUB. The easy proofs are omitted.

Lemma 2. Let $u, v_1, v_2 \in V$ such that $u \preceq v_1$ and $u \preceq v_2$. Then there exists $w \in GLB(v_1, v_2)$ such that $u \preceq w$.

Lemma 3. *Let* $v \in V$, $S \subseteq V$ *and* $u \in S$ *such that* $v \preceq u$. *Then there exists* $w \in LUB(v, S)$ *such that* $v \preceq w \preceq u$.

Lemma 4. *Let* $G_1, \ldots, G_l$ *be the subgraphs of* G *corresponding to the connected components of the underlying undirected graph. Let* $G_i = (V_i, E_i)$. *Then*

$$glb\text{-}closure_G(S) = \bigcup_{i=1}^{l} glb\text{-}closure_{G_i}(V_i \cap S)$$

The index of $glb - closure$ denotes the graph in which the $glb\text{-}closure$ is taken.

2 Applications for the general problem

The following lemma gives a characterization of *glb-closed* subsets in terms of the LUB operation.

Lemma 5. *Let* $S \subseteq V$. *Then* S *is glb-closed if and only if for all* $v \in V$ *the set* $LUB(v, S)$ *contains at most one element.*

Proof. Firstly, suppose that for all $v \in V$ the set $LUB(v, S)$ contains at most one element. Let $v_1, v_2 \in S$. It suffices to prove that $GLB(v_1, v_2) \subseteq S$. Let $u \in GLB(v_1, v_2)$. By Lemma 3 there exists $v_1', v_2' \in LUB(u, S)$ such that $u \preceq v_1' \preceq v_1$ and $u \preceq v_2' \preceq v_2$. Since $LUB(u, S)$ contains at most one element $v_1' = v_2'$. Now $u \in GLB(v_1, v_2)$ yields that $u = v_1' \in LUB(u, S) \subseteq S$.

Secondly, suppose that S is *glb-closed*. Let $v \in V$. If $v \in S$ then $LUB(v, S) = \{v\}$. Hence we can assume that $v \notin S$. Let $v_1, v_2 \in LUB(v, S)$. Then $v \notin GLB(v_1, v_2)$ since S is *glb-closed*. By Lemma 2 there exists $u \in GLB(v_1, v_2)$ such that $v \prec u$. Hence by the definition of LUB, $u = v_1$ and $u = v_2$. Therefore, $LUB(v, S)$ contains at most one element. □

Based on Lemma 5 $LUB(v, S)$ can be calculated efficiently for *glb-closed* subsets S. For this purpose we assume that the vertices of G are labeled $v_1, \ldots, v_n$ in such a way, that if there is a directed edge from v_i to v_j then $j < i$. Topological sorting can be used to produce such an ordering. Then $LUB(v, S) = v_l$ where l is the largest index such that $v_l \in S$ and $v \preceq v_l$. If no such vertex exists then $LUB(v, S) = \emptyset$. Hence $l = \max \{i \mid v_i \in S, v_k \preceq v_i\}$ where $v = v_k$. Clearly $l > k$. The most expensive part is to test the relationship $v_i \preceq v_k$.

In cases where least upper bounds are calculated repeatedly, some preprocessing is very useful. The transitive closure of G and a topological ordering of the vertices of G are needed. Then a test $v_i \preceq v_k$ can be performed in constant time. Then $LUB(v, S)$ can be calculated in time $O(s)$, where s is the number of vertices in S. In the next section, we'll see that topological sorting is also necessary for the computation of *glb-closure*.

Glb-closures are a useful tool in the field of object-oriented programming languages with multiple inheritance, where a typical problem is *conflict resolution*. For example, given a class C which inherits a method m. The conflict arises in case several superclasses of C have defined a method m of the same arity. It is necessary to select one implementation of m. Most object-oriented programming languages provide a mechanism to automaticly solve this conflict [6, 8]. From the point of view of the language implementation, the compiler must determine, for each class, what methods are valid and which are inherited from which super-class, if those methods are not defined in the class itself. This is the prerequisite for the well-known *method dispatching* problem [4, 9]. In the following, we indi-cate a way to answer this question. Suppose that a method m is defined in a set of classes, C_m. Now a specific question is what is the set of classes in which the conflicts with respect to the method m have to be solved. By Lemma 5, this set of classes is glb-$closure(C_m) \setminus C_m$. Generally, glb-$closure(C_m)$ is a small subset compared to the set of all classes. Narrowing the scope where the conflicts have to be solved is significant for the time-efficiency. On the other hand, for each class C' not in glb-$closure(C_m)$, by Lemma 5, we can directly answer whether C' inherits m or not, and if yes which m is inherited by using the function LUB.

In many object-oriented languages, a message is sent to a receiver object, and the runtime "type" of the receiver determines the method that is actually invoked by the message. The arguments of the message are passed on to the invoked method but do not participate in the method dispatching. To surmount these limitations, some object-oriented languages include a more powerful form of function invocation in which all arguments of a method can participate in the method dispatching, i.e. a method is dynamically dispatched based on the types of all arguments. These methods are called multi-methods. The dispatching for multi-methods is called multi-dispatching. Perhaps the most-known languages that support multi-methods are CLOS [6] and its ancestor CommonLisp [5]. One fundamental issue for multi-methods is an efficient mechanism for method lookup. Based on a result of Agrawal, et al. [3] on static type checking of multi-methods, a new mechanism for multi-method dispatching based on *glb-closures* is presented in [7]. The central idea is the introduction of a lookup automaton (LUA) to simulate the dynamic dispatching in a given set of methods. The *glb-closure* is the central operation for the construction of the LUA. The actual dispatching utilizes the function LUB. The total time complexity for the LUA simulation is $O(n)$ where n is the arity of the method. The advantage of this approach is that the memory requirements are very small compared to other approaches with the same time complexity, [4].

3   The Algorithm

Let S be a subset of V. Then glb-*closure(S)* can be calculated recursively as follows. Let $S_0 = S$ and for $i > 0$ let $S_i = \{v \in GLB(v_1, v_2) \mid v_1, v_2 \in S_{i-1}\}$. Clearly, $S_{i-1} \subseteq S_i \subseteq$ glb-*closure(S)* for all $i > 0$. Since V is finite, there exists a k such that $S_{k-1} = S_k$. Clearly $S_k =$ glb-*closure(S)*. If $\overline{n}$ is the number of the vertices in glb-*closure(S)*, then there are $O(\overline{n}^2)$ calls of the function GLB. In the lattice structures, the GLB function can be implemented nearly in constant time [1]. In the general case of a DAG, however, no efficient algorithm for GLB is known. If the time complexity of GLB function on the DAG structure is $g(n,e)$, then the time complexity of the above described algorithm for $glb\text{-}closure(S)$ will be $g(n,e) * O(\overline{n}^2)$.

In the following algorithm the calculation of GLB is avoided. The central idea is to reduce the graph G into a new graph, in which the set of vertices is equal to $glb\text{-}closure(S)$. The complexity is $O(n + e + \overline{n}\,\overline{e})$, where $\overline{e}$ the number of edges incident with the vertices in $glb\text{-}closure(S)$.

In the following it is assumed that the vertices of G are labeled $v_1, \ldots, v_n$ in such a way, that if there is a directed edge from v_i to v_j then $j < i$. Topological sorting can be used to produce such a labeling. Let $SUCC_G(v_i)$ denote the set of vertices reachable by an edge in G from v_i and $PRED_G(v_i)$ the set of vertices having an edge connecting to v_i in G. Furthermore, let $\overline{SUCC_G}(v_i)$ be the set of vertices reachable from v_i in G.

The following algorithm computes $glb\text{-}closure(S)$ for a subset S of V. At the end the variable M contains the $glb\text{-}closure$ of S. During the algorithm the graph changes because vertices are removed. This automatically includes the removal of all edges incident with these vertices.

$M := S;$
$G_0 := G;$
FOR $i := 1$ TO n DO
    IF $v_i \notin S$ THEN BEGIN
(1) $P := SUCC_{G_{i-1}}(v_i);$
        IF $P = \emptyset$ THEN
            remove v_i from G_{i-1} and let G_i be the resulting graph
        ELSE BEGIN
(2)   let v_l be the vertex from P with maximal index;
(3)   IF $\exists\, v_j \in P \setminus \{v_l\}$ such that $v_j \notin \overline{SUCC}_{G_{i-1}}(v_l)$ THEN BEGIN
                $M := M \cup \{v_i\};$
                $G_i := G_{i-1};$
            END
            ELSE BEGIN
(4)   FOR each $v_k \in PRED_{G_{i-1}}(v_i)$ DO
                    insert an edge from v_k to v_l into $G_{i-1};$
                remove v_i from G_{i-1} and let G_i be the resulting graph
            END;
        END;
    END
    ELSE
        $G_i := G_{i-1};$

Fig. 1(b) shows the graph G_{11} when the algorithm is applied to the graph in Fig. 1(a) for computing $glb\text{-}closure\,(\{v_2, v_3\})$.

4 Proof of the algorithm

In the following the correctness of the algorithm is proved in two steps. Note that the set P at line (1) is always a subset of M: since the vertices are numbered with respect to a topological sort, the indices of vertices in P are less than i. But all v_j with $j < i$ not belonging to M are removed from G. Furthermore, if P at line (1) contains only one vertex, then line (4) is executed.

Lemma 6. *M is glb-closed.*

Proof. Assume false. Let $M_1 = $ glb-closure$(M)\backslash M$. Then $M_1 \neq \emptyset$. Let $v_i \in M_1$ such that i is minimal. Let

$$M_2 = \{(v_p, v_k) \mid v_p, v_k \in glb\text{-}closure(M), v_i \in GLB(v_p, v_k), p > k\}$$

Clearly $M_2 \neq \emptyset$. Let $(v_j, v_m) \in M_2$ such that j is maximal. Let

$$M_3 = \{(v_j, v_k) \mid (v_j, v_k) \in M_2\}.$$

Clearly $(v_j, v_m) \in M_3$. Let $(v_j, v_t) \in M_3$ such that t is maximal. Hence if $(v_p, v_k) \in M_2$ then $p \leq j$ and if $p = j$ then $k \leq t$. Furthermore $i > j$ and $i > t$ which implies that $v_j, v_t \in M$.

Assume that $v_j, v_t \in SUCC_{G_{i-1}}(v_i)$. Let v_l as in line (2). Then v_j and v_t cannot both be in $\overline{SUCC}_{G_{i-1}}(v_l)$, because $v_i \in GLB(v_j, v_t)$. Hence v_i is inserted into M in the line following line (3). This contradicts the choice of v_i.

Therefore, first assume that $v_j \notin SUCC_{G_{i-1}}(v_i)$. Let v_s be a vertex on the path in G_{i-1} from v_i to v_j such that $i > s > j$ (see Fig. 2). Then $s > t$ because $j > t$.

Assume that $v_s \in$ glb-closure(M). Then $v_i \notin GLB(v_s, v_t)$, since $s > t$ and $s > j$ and the choice of j. Then Lemma 2 yields that there exists $v_r \in \overline{SUCC}(v_i)$ and $v_r \in GLB(v_s, v_t)$. Since $v_i \in GLB(v_j, v_t)$, either $v_r = v_s$ or $v_r = v_t$. Both possibilities are impossible. This is a contradiction and therefore $v_s \notin$ glb-closure(M).

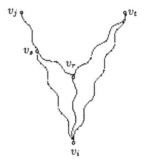

Fig. 2. The case $v_j, v_t \in SUCC_{G_{i-1}}(v_i)$

This yields that none of the vertices on the path from v_i to v_j are in glb-closure(M). Therefore, they cannot be in M and are removed from G. Thus, in

G_{i-1} there exists an edge from v_i to v_j. Similarly there exists an edge from v_i to v_t.

Thus, $v_j, v_t \in SUCC_{G_{i-1}}(v_i)$ which leads to a contradiction as shown above. This completes the proof. □

Theorem 7. $M = glb\text{-}closure(S)$.

Proof. By Lemma 6 $glb\text{-}closure(S) \subseteq M$ since $S \subseteq M$. Hence it suffices to prove that M is a subset of $glb\text{-}closure(S)$. Assume false. Let $M_1 = M \setminus glb\text{-}closure(S)$. Then $M_1 \neq \emptyset$. Let $v_i \in M_1$ such that i is maximal. Clearly $v_i \notin S$. Thus the algorithm inserted v_i into M. The choice of v_i implies that at the point v_i was inserted into M, the set P was a subset of $glb\text{-}closure(S)$. Therefore, there exist $v_j, v_l \in SUCC_{G_{i-1}}(v_i) \cap P \subseteq glb\text{-}closure(S)$ such that $v_i \in GLB(v_j, v_l)$. Hence $v_i \in glb\text{-}closure(S)$. This contradiction completes the proof. □

5 Analysis of the Algorithm

Let e the number of edges and n the number of vertices in G. Then during the algorithm there will be at most e edges inserted into the Graph (for every edge inserted there will be at least one edge removed). Furthermore, in line (3) the test IF $\exists\, v_j \in P \setminus \{v_l\}$ such that $v_j \notin \overline{SUCC}_{G_{i-1}}(v_l)$ can be done by a depth-first search of the reversed subgraph induced by M starting at v_j. Let $\overline{n}$ be the number of vertices in $glb\text{-}closure(S)$ and let $\overline{e}$ the number of edges incident with the vertices in $glb\text{-}closure(S)$. Then the time complexity of the algorithm is $O(n + e + \overline{n}\, \overline{e})$.

An implementation to achieve this time complexity will be discussed in the following.

Lemma 8. *The algorithm to compute* glb-closure(S) *has a time complexity of* $O(n + e + \overline{n}\, \overline{e})$.

Proof. At the end the set of vertices of G_n equals M. But G_n is not necessarily the subgraph of G induced by $glb\text{-}closure(S)$ (because of (4)).

To calculate $SUCC_{G_{i-1}}(v_i)$ in line (1), the adjacency list representation is used for G. Note that P is always a subset of M. There is an array to keep track of which vertex belongs to M. Every time a vertex is inserted into M the corresponding entry in the array is updated. This marking process can be achieved in $O(\overline{n})$ time in total. Thus, the cost of line (1) in total is $O(e + n)$. The adjacency list must be updated during the algorithm. In case a vertex is removed no update is performed. The update in line (4) has a total cost of $O(e)$.

Note that if $v_j \in \overline{SUCC}_{G_{i-1}}(v_l)$ in line (3), then the path from v_l to v_j uses vertices with an index between j and l and all vertices belong to M. Furthermore in case v_l happens to be the vertex with maximal index in P more than once, then $\overline{SUCC}_{G_{i-1}}(v_l)$ remains the same (i.e. independent of the index of G). Hence, this set is stored and can be used again. It is obtained by a depth first search on the subgraph of G_{i-1} induced by M. This subgraph has at most $\overline{e}$ edges. During this depth first search the condition at line (3) can be tested without any extra cost. This has to be done at most $\overline{n}$ times. This yields a cost of $O(\overline{n}\,\overline{e})$. The test of $v_j \notin \overline{SUCC}_{G_{i-1}}(v_l)$ in the case that $\overline{SUCC}_{G_{i-1}}(v_l)$ is stored, is accomplished in constant time and is done for each edge at most once. This yields a total cost of $O(e)$.

To have $PRED_{G_{i-1}}(v_i)$ in line (4) an adjacency list of the inverted graph G is kept. The set $PRED_{G_{i-1}}(v_i)$ contains only vertices v_j with $j > i$. Note that the updates of G have no influence on the predecessors of v_i. □

The most expensive part is the repeated calculation of $\overline{SUCC}_{G_{i-1}}(v_l)$. In case glb-closures have to be computed repeatedly for different subsets S, it is useful to precompute the transitive closure $\overline{G}$ of G. An efficient method for storing the transitive closure is presented in [2]. Then $\overline{SUCC}_{G_{i-1}}(v_l)$ can be directly determined from $\overline{G}$. Note that for all i the relation $\preceq$ in G_i is consistent with that in $\overline{G}$. The cost of executing line (3) is $|P|$, this leads to a total cost of $O(e)$ for this line. Summarizing the following theorem is a consequence of Lemma 8.

Theorem 9. *In case the transitive closure of G is available* glb-closure(S) *can be calculated in time $O(e + n)$, where e is the number of edges in G and n the number of vertices.*

Lemma 4 can be utilized in an implementation. All calculations are performed for the individual connected components. Furthermore, the main loop of the algorithm can be started at $j = \min\{i \,|\, v_i \in S\}$.

6 Conclusion

We have introduced the notion of *glb-closure* of subsets of vertices of directed acyclic graphs. *Glb-closures* have some nice properties which make them useful in several applications. An efficient algorithm for calculating *glb-closures* was presented. Based on some preprocessing the time complexity is $O(e+s)$ where e is the number of edges in G and s the number if vertices in S. We have implemented the presented algorithm in an implementation for dispatching multi-methods and the experience is positive.

An interesting open problem which arose in our application is the dynamic behavior of *glb-closed* sets. Is it possible to compute glb-*closure*$(S \cup \{x\})$ or glb-*closure*$(S \setminus \{x\})$ from glb-*closure*(S)? I.e. is there an efficient incremental algorithm to compute glb-*closure*(S)?

References

1. Aït-Kaci, H., Boyer, R., Lincoln, P., and Nasr, R. *Efficient Implementation of Lattice Operations.* ACM Trans. on Prog. Lang. and Syst. Vol. 11. No. 1, January 1989, 115-146.

2. Agrawal, R., Borgida, A., and Jagadish, H.V. *Efficient Management of Transitive Relationships in Large Data and Knowledge Bases.* Proc. ACM-SIGMOD Int'l Conf. on Management of Data, 1989.

3. Agrawal, R., DeMichiel, L. G., and Lindsay, B. G. *Static Type Checking of Multi-Methods.* Proc. Conf. on Object-Oriented Prog. Sys., Lang., 1991.

4. André, P., and Royer, J.-C. *Optimizing Method Search with Lookup Caches and Incremental Coloring.* In Proc. Conf. on Object-Oriented Prog. Sys., Lang., 1992.

5. Bobrow, D. G., Kahn, K., Kiczales, G., Masinter, L., Stefik, M., and Zdybel, F. *CommonLoops: Merging Lisp and Object-Oriented Programming.* In Proc. Conf. on Object-Oriented Prog. Sys., Lang., 1986.

6. Bobrow, D. G., DeMichiel, L. G., Gabriel, R. P., Keene, S. E., Kiczales, G., and Moon, D. A. *Common Lisp Object System Specification X3J13.* In SIGPLAN Notice 23, special issue, Sept. 1988.

7. Chen, W., Turau, V., and Klas, W. *Efficient Dynamic Look-up Strategy for Multi-Methods.* Proc. European Conf. on Object-Oriented Prog., Bologna, 1994.

8. Ducournau, R., Habib, M., Huchard, M., and Mugnier, M. L. *Monotonic Conflict Resolution Mechanisms for Inheritance.* In Proc. Conf. on Object-Oriented Prog. Sys., Lang., and Appl., 1992, 16-24.

9. Karel, D. *Selector Table Indexing & Sparse Arrays.* In Proc. Conf. on Object-Oriented Prog. Syst., Lang., 1993.

Minimum Vertex Cover, Distributed Decision-Making, and Communication Complexity*

Extended abstract

Pierluigi Crescenzi and Luca Trevisan

Dipartimento di Scienze dell'Informazione
Università degli Studi di Roma "La Sapienza"
Via Salaria 113, 00198 Roma, Italy
E-mail: {piluc,trevisan}@dsi.uniroma1.it.

Abstract. In this paper we study the problem of computing *approximate vertex covers* of a graph on the basis of partial information and we consider the *distributed decision-making* and the *communication complexity* frameworks. In the first framework we do not allow communication among the processors: in this case, we show an optimal algorithm whose performance ratio is equal to p where p is the number of processors. In the second framework two processors are allowed to communicate in order to find an approximate solution: in this latter case, we show a linear lower bound on the communication complexity of the problem.

1 Introduction

The minimum vertex cover (MVC) problem consists of finding, given a graph G, a minimum cardinality set of nodes V' such that, for any edge (u, v), either $u \in V'$ or $v \in V'$. This is a well-studied problem which appeared in the first list of NP-complete problems presented by Karp [10]. A straightforward approximation algorithm, based on the idea of a maximal matching, was successively developed by Gavril (according to [5]) with a performance ratio no greater than 2. Several other approximation algorithms are presented in the lecture notes of Motwani [13]. In this paper we analyse the complexity of finding approximate solutions for the MVC problem under two basic frameworks.

In the first one, we study this problem as one of *distributed decision-making with incomplete information* [4, 15, 16, 17], that is, we assume that the vertex cover is chosen by independent processors, each knowing only a part of the graph and acting in isolation. In particular, we assume that the adjacency list of each node of the graph is known by only one processor which has to decide whether the node should belong to the vertex cover. We then want to develop

---

* Research partially supported by the MURST project *Algoritmi, Modelli di Calcolo, Strutture Informative* and by EEC Human Capital Mobility Program *Efficient Use of Parallel Computers*.

distributed algorithms that always produce feasible solutions (that is, vertex covers) and achieve, in the worst case, a *reasonable* performance ratio (that is, the ratio of the cardinality of the solution computed by the algorithm to the optimum cardinality should be as small as possible). In Sect. 2 we show that a simple *double-matching algorithm* which essentially performs Gavril's algorithm first on the 'bridge' edges and then on the 'inner' edges achieves a performance ratio of p where p is the number of processors. We also show, by means of a quite involved counting technique, that this algorithm is *optimal*, that is, no distributed algorithm can achieve a ratio smaller than p. These results fit into a more general context in which an optimization graph problem has to be solved in a distributed fashion and neither a centralized control nor a complete information are available. Moreover, it has been argued that this kind of results 'can be seen as part of a larger project aiming at an algorithmic theory of the value of information' [17]. Intuitively, this theory should allow to compare in terms of performance ratios two different *information regimes*, that is, two different ways of distributing the input among the processors.

In the second framework, instead, we study the communication complexity of the MVC problem. In standard communication complexity, two processors interact in order to compute a function f of their respective inputs x and y. The question is: how much information do the two processors need to exchange to correctly compute the value $f(x, y)$? The minimum number of bits that must be communicated is the *deterministic communication complexity* of f. This complexity measure was introduced by Yao [22] and has been shown to be tightly related to time-area tradeoffs in VLSI [1, 7, 8, 11, 12, 19, 21], circuit complexity [9], and combinatorial optimization [20]. It has also been studied for its own sake as an interesting model of computation. Indeed, non-deterministic, probabilistic, and alternating variants have been considered and several complexity classes analogous to the more notorius ones in Turing machine complexity have been defined [2, 14, 18]. In this paper we consider the communication complexity of computing, for any graph G and for any rational $\epsilon > 1$, a vertex cover for G whose performance ratio is smaller than ϵ. In particular, we assume that ϵ is known by both processors and the adiacency list of each node of G is known by only one processor. In Sect. 3 we prove that the communication complexity of this function is at least $(1 - H(8(\epsilon - 1)))n/8$ where $H(x) = -(x \log x + (1 - x) \log(1 - x))$ is the entropy of x and n denotes the number of nodes of the graph. The proof is based on an interesting relation with the area of error-correcting codes. We also observe that in the case of planar graphs $n \log n$ bits are sufficient to compute an optimum solution so that, for these graphs, our lower bound is not far from being tight.

2 MVC and distributed decision-making

Suppose that a graph G is described by the n adjacency lists of its n nodes. Given p processors with $p \geq 2$, the ith processor knows the adjacency lists of a subset V_i of nodes (without loss of generality, we shall assume that, for any i

and j, $V_i \cap V_j = \emptyset$). Let G_i be the subgraph of G induced by the set of edges (u, v) such that either u or v belongs to V_i.

A *distributed decision algorithm* A is an algorithm which, for any graph G and for any i, on the basis of the subgraph G_i produces a subset $A(G_i)$ of V_i such that $A(G) = \bigcup_{i=1}^{p} A(G_i)$ is a vertex cover of G. Moreover, let $opt(G)$ be the cardinality of a minimum vertex cover for G. The *performance ratio* of A is

$$R(A) = \max_G \frac{|A(G)|}{opt(G)}.$$

Theorem 1. *A distributed decision algorithm A exists whose performance ratio is at most p.*

Proof. Consider the following algorithm where the edges are supposed to be lexicographically ordered.

begin
  $A(G_i) := \emptyset$; $B_i := \emptyset$;
  for each edge (u, v) such that $u \in V_i$ and $v \notin V_i$ **do**
    if $u \notin A(G_i)$ **and** $v \notin B_i$ **then**
    begin
      $A(G_i) := A(G_i) \cup \{u\}$;
      $B_i := B_i \cup \{v\}$;
    end;
  for each edge (u, v) such that $u, v \in V_i$ **do**
    if $u \notin A(G_i)$ **and** $v \notin A(G_i)$ **then**
      $A(G_i) := A(G_i) \cup \{u, v\}$;
end.

Let $A_1(G_i)$ and $A_2(G_i)$ be the set of nodes included in $A(G_i)$ during the first and the second **for** instruction, respectively. Clearly, all edges 'seen' by processor P_i are covered by the set $A_1(G_i) \cup A_2(G_i) \cup B_i$. Then, in order to prove that $A(G)$ is a vertex cover for G it suffices to show that, for any i, $B_i \subseteq A_1(G) = \bigcup_{k=1}^{p} A_1(G_k)$. The proof is by induction on the number b of *bridge* edges, that is, edges whose extremes are known by different processors (observe that each B_i contains only extremes of bridge edges). If $b = 0$, then the proof is trivial. Suppose that we have $b + 1$ bridge edges and that (u, v) is the last of these edges in the lexicographic order with $u \in V_i$ and $v \in V_j$. Let $A_1(G_i')$, B_i', $A_1(G_j')$, and B_j' be the sets computed by the algorithm on input $G' = (V, E - (u, v))$. By induction hypothesis, $B_i', B_j' \subseteq A_1(G') = \bigcup_{k=1}^{p} A_1(G_k')$. We shall now prove that $B_i \subseteq A_1(G)$ (the proof for B_j is similar). To this aim, we distinguish the following two cases.

1. $u \in A_1(G_i') \vee v \in B_i'$: in this case $B_i = B_i' \subseteq A_1(G') \subseteq A_1(G)$.
2. $u \notin A_1(G_i') \wedge v \notin B_i'$: in this case $B_i = B_i' \cup \{v\}$ and $u \notin B_j'$ (since $u \notin A_1(G_i')$ and $B_j' \subseteq A_1(G')$). If $v \in A_1(G_j')$ then, clearly, $B_i \subseteq A_1(G') \subseteq A_1(G)$, otherwise v will be put into $A_1(G_j)$ when considering edge (u, v) so that $B_i \subseteq A_1(G)$.

We have thus shown that $A(G)$ is a feasible solution. In order to prove that its performance ratio is at most p, let $n_k = \sum_{i=1}^{p} |A_k(G_i)|$ for $k = 1, 2$. Clearly, an index i must exist such that $|A_1(G_i)| \geq n_1/p$. This set $A_1(G_i)$ then corresponds to a set of at least n_1/p disjoint edges. Moreover, the set $\bigcup_{i=1}^{p} A_2(G_i)$ corresponds to another set of $n_2/2$ disjoint edges. It is also clear that the union of these two sets is still a set of disjoint edges. That is, G contains a matching of at least $n_1/p + n_2/2$ edges. Thus, any vertex cover for G must contain at least $n_1/p + n_2/2 \geq (n_1 + n_2)/p$ nodes, that is,

$$\frac{|A(G)|}{opt(G)} \leq \frac{n_1 + n_2}{(n_1 + n_2)/p} = p.$$

We can conclude that the performance ratio of A is at most p. $\qquad\square$

In order to prove that the result of the previous theorem is tight, let us first show that, for any distributed decision algorithm A, $R(A) \geq 2$. Let $K_{i,j}^{m,n}$ denote the instance in which the complete bipartite graph $K^{m,n}$ with vertex classes V_1 and V_2 is distributed in the following way: V_1 is assigned to processor P_i, V_2 is assigned to processor P_j, and all other processors know nothing. Then, for any algorithm A, either P_i or P_j has to choose all its nodes when running algorithm A with input $K_{i,j}^{m,n}$ (otherwise, an uncovered edge exists). Without loss of generality, we can assume that P_i chooses all its nodes. Let us then consider the new instance in which vertices in V_2 are pairwise connected, thus forming a clique of order n. Clearly, P_i still chooses all its nodes since its subinstance is not changed. Moreover, P_j is also forced to choose at least $n - 1$ of its nodes. If $m = n$, then the optimum solution contains n nodes while the solution computed by the algorithm contains at least $2n - 1$ nodes. That is, the performance ratio is at least 2.

In order to improve the above bound, we will show in the next theorem how to find, for any distributed decision algorithm A, an instance G_A in which a processor P_j knows n nodes and the other processors P_i share at least $(p - 1)(n - 1)$ nodes which are all connected to the n nodes of P_j. Moreover, each P_i with $i \neq j$ chooses all its nodes when running algorithm A with input G_A. We can then modify the instance by pairwise connecting all nodes of P_j. The optimum thus contains n nodes while the solution computed by the algorithm contains at least $p(n - 1)$ nodes. That is, the performance ratio is at least p.

Lemma 2. *For any distributed decision algorithm A and for any integer N_0, a graph G, an index j, and an integer $n_0 > N_0$ exist such that*

1. $|V_j| = n_0$.
2. $\sum_{i \neq j} |V_i| \geq (p - 1)(n_0 - 1)$.
3. *For any $i \neq j$, V_i and V_j are the two vertex classes of a complete bipartite graph.*
4. *For any $i \neq j$, $A(G_i) = V_i$.*

Proof. Recall that, for any i, j, m, and n, $K_{i,j}^{m,n}$ denotes the instance in which the complete bipartite graph $K^{m,n}$ with vertex classes V_1 and V_2 is distributed in the following way: V_1 is assigned to processor P_i, V_2 is assigned to processor P_j, and all other processors know nothing. Given a distributed decision algorithm A and an integer n, let $c_{i,j}^n$ be the maximum m such that $A(G_i) = V_1$ (see Fig. 1 where the black nodes have been chosen and the white nodes may or may not have been chosen).

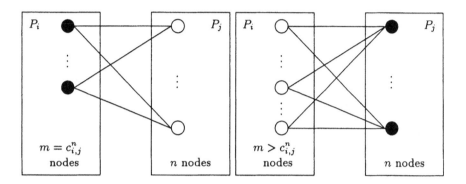

Fig. 1. The definition of $c_{i,j}^n$

Observe that, for any i, j, and n, if $c_{i,j}^n = m < n - 1$ then $c_{j,i}^k \geq n$ for $k = m + 1, \ldots, n - 1$. This observation intuitively suggests that the sum of all these quantities is large enough. Indeed, in the Appendix we prove the following result.

Proposition 3. *For any i, j, and N, the following inequality holds:*

$$\sum_{n=1}^{N}(c_{i,j}^n + c_{j,i}^n) \geq 2\sum_{n=1}^{N}(n-1).$$

From the above proposition it then follows that

$$\sum_{n=1}^{N}\sum_{j=1}^{p}\sum_{\substack{i=1 \\ i\neq j}}^{p} c_{i,j}^n = \sum_{\substack{i,j=1 \\ i<j}}^{p}\sum_{n=1}^{N}(c_{i,j}^n + c_{j,i}^n) \geq p(p-1)\sum_{n=1}^{N}(n-1). \qquad (1)$$

Assume now that an N_0 exists such that, for any $n > N_0$ and for any j,

$$\sum_{\substack{i=1 \\ i\neq j}}^{p} c_{i,j}^n < (p-1)(n-1)$$

and let

$$\sigma = \sum_{n=1}^{N_0} \sum_{j=1}^{p} \sum_{\substack{i=1 \\ i \neq j}}^{p} c_{i,j}^n.$$

Then, for any $N > N_0$, we have that

$$\sum_{n=1}^{N} \sum_{j=1}^{p} \sum_{\substack{i=1 \\ i \neq j}}^{p} c_{i,j}^n = \sigma + \sum_{n=N_0+1}^{N} \sum_{j=1}^{p} \sum_{\substack{i=1 \\ i \neq j}}^{p} c_{i,j}^n \leq \sigma + p(p-1) \sum_{n=N_0+1}^{N} (n-1) - (N-N_0)p$$

which, for N sufficiently large, contradicts (1).

Thus, for any integer N_0, an index j and an integer $n_0 > N_0$ exist such that

$$\sum_{\substack{i=1 \\ i \neq j}}^{p} c_{i,j}^{n_0} \geq (p-1)(n_0 - 1).$$

The graph G is then defined as a star of bipartite graphs in which processor P_j knows n_0 nodes and each processor P_i with $i \neq j$ knows $c_{i,j}^{n_0}$ nodes which are all connected to each node of P_j. Clearly, this graph satisfies Conditions 1-4. □

As a consequence of the above lemma, we then have the following result.

Theorem 4. *No distributed decision algorithm has a performance ratio smaller than p.*

3 MVC and communication complexity

Let V_1 and V_2 be two set of nodes, $E_i \subseteq (V_1 \times V_2) \cup V_i^2$, for $i = 1, 2$, be two set of edges, and $G = (V_1 \cup V_2, E_1 \cup E_2)$ be a graph. For any ϵ, let $V(G, \epsilon)$ be the set of vertex covers V' of G such that

$$\frac{|V'|}{opt(G)} \leq \epsilon$$

where $opt(G)$ denotes the cardinality of a minimum vertex cover. Roughly speaking, $V(G, \epsilon)$ is the set of *approximate solutions within performance ratio ϵ* with respect to the instance G of the MVC problem.

The *MVC communication problem with error ϵ* is then as follows. Two processors P_1 and P_2 get inputs $G_1 = (V_1 \cup V_2, E_1)$ and $G_2 = (V_1 \cup V_2, E_2)$, respectively. Their task is to choose two sets $V_i' \subseteq V_i$ for $i = 1, 2$ such that $V_1' \cup V_2' \in V(G, \epsilon)$. The *communication complexity* of this problem is the minimum number of bits exchanged by the best protocol on a worst case input.

Let us consider the two instances shown in Fig. 2, that is, two rings of eight nodes such that the subinstance held by P_1 is the same in the two cases. The main difference is the 'parity' of the nodes seen by P_1 so that, in the first case,

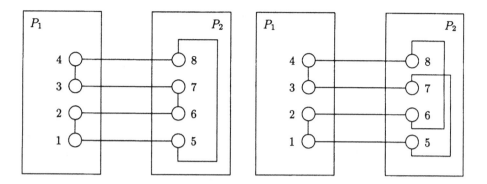

Fig. 2. Two rings of 8 nodes

node 2 belongs to the optimum solution if and only if node 3 does not belong to it while, in the second case, node 2 belongs to the optimum solution if and only if node 3 belongs to it. This means that if the processors execute a protocol that always yields an optimum solution, than P_1 will compute different subsolutions in the two cases, even if its subinstance is the same. This implies that at least one bit of communication is required. This reasoning can be generalized as follows: let $\mathcal{G}_n$ be the set of graphs G containing n disjoint rings $R_1, \ldots, R_n$ where each of these rings can be of one of the two types shown in Fig. 2. Clearly, $|\mathcal{G}_n| = 2^n$ and all these graphs are equal from P_1's point of view. On the other hand, an optimum solution for a graph $G \in \mathcal{G}_n$ cannot be an optimum solution for any other graph in $\mathcal{G}_n$. Thus, while executing a protocol yielding optimum solutions, P_1 will find different subsolutions for any instance in $\mathcal{G}_n$, that is, 2^n different computations are performed with respect to the same P_1's input. This, in turn, implies that different sequences of messages must be exchanged every time and that a sequence of messages exists whose length is at least n.

Let us now consider the case of ϵ-approximate solutions. Since the optimum solutions of any instance in $\mathcal{G}_n$ contains $4n$ nodes, an ϵ-approximate solution will contain at most $4n\epsilon = 4n + 4n(\epsilon - 1)$ nodes. Thus, the number of rings that are not optimally covered is at most $4n(\epsilon - 1)$. It is also easy to see that if $V(G, \epsilon) \cap V(G', \epsilon) \neq \emptyset$ where $G, G' \in \mathcal{G}_n$, then G differs from G' in at most $8n(\epsilon - 1)$ rings. Let $\mathcal{G}'_n$ be a subset of $\mathcal{G}_n$ such that any two instances in $\mathcal{G}'_n$ differ in less than $8n(\epsilon - 1)$ rings. Then $\log |\mathcal{G}'_n|$ is a lower bound on the communication complexity of approximating the MVC problem within factor ϵ on graphs with $8n$ nodes. The problem of finding a large set $\mathcal{G}'_n$ with the required property is equivalent to the problem of finding a large set $\mathcal{C} \subset \{0, 1\}^n$ such that for any two strings $s_1, s_2 \in \mathcal{C}$, the Hamming distance $d(s_1, s_2)$ (i.e., the number of bits where s_1 differs from s_2) is smaller than $8n(\epsilon - 1)$. Such a set $\mathcal{C}$ is indeed an *error correcting code* and a classical result of Gilbert [3, 6] allows us to state that, for any $\epsilon < 17/16$, an error correcting code $\mathcal{C}$ exists with $\log |\mathcal{C}| > (1 - H(8(\epsilon - 1)))n/8$ where H is the entropy function defined as $H(x) = -x \log x - (1 - x) \log(1 - x)$, for $0 < x < 1/2$.

We have thus shown the following result

Theorem 5. *For any $\epsilon < 17/16$, the communication complexity of the MVC communication problem with error ϵ is at least $(1 - H(8(\epsilon - 1)))n/8$.*

For any fixed $\epsilon < 17/16$ we thus have a *linear* lower bound. This result clearly holds even if we restrict ourselves to planar graphs. Since any planar graph with n nodes contains $O(n)$ edges, a trivial upper bound on the communication complexity of exactly solving the MVC problem is $O(n \log n)$, so that our lower bound is not far from being tight.

4 Conclusion and open problems

We studied the problem of computing approximate vertex covers of a graph on the basis of partial information and we analysed two basic frameworks. In the first one we do not allow communication among the processors: in this case, we showed an optimal algorithm whose performance ratio is equal to the number of processors. In the second framework two processors are allowed to communicate in order to find an approximate solution: in this latter case, we showed a linear lower bound on the communication complexity of the problem.

Some problems are left open by this work. From the distributed decision-making point of view, it would be interesting to find tradeoffs between the quantity of information available to each processor and the performance ratio, that is, to compare two different information regimes. From the communication complexity point of view, it would be interesting to pin down the precise communication complexity of approximating the MVC problem. For example, is the linear lower bound optimal? What about $\epsilon > 17/16$?

References

1. Aho, A. V., J.D. Ullman and M. Yannakakis. "On notions of information transfer in VLSI circuits". In *Proceedings of 15th ACM Symposium on Theory of Computing*, pages 133–136, 1983.
2. Babai, L., P. Frankl and J. Simon. "Complexity classes in communication complexity theory". In *Proceedings of 27th IEEE Symposium on Foundations of Computer Science*, pages 337–347, 1986.
3. Berlekamp, E.R. "Algebraic Coding Theory". McGraw-Hill, 1968.
4. Deng, X. and C. H. Papadimitriou. "Distributed decision-making with incomplete information". In *Proceedings of 12th IFIP*, 1992.
5. Garey, M. R. and D. S. Johnson. "Computers and Intractability. A Guide to the Theory of NP-completeness". Freeman, 1979.
6. Gilbert, E.N. "A comparison of signaling alphabets". *Bell System Tech. J.*, 31:504–522, 1952.
7. Hambrusch, S. E. and J. Simon. "Solving undirected graph problems on VLSI". *SIAM Journal of Computing*, 14:527–544, 1985.
8. Ja'Ja', J. "The VLSI complexity of selected graph problems". *Journal of the ACM*, 31:840–849, 1984.

9. Karchmer, M. and A. Wigderson. "Monotone circuits for connectivity require super-logarithmic depth". In *Proceedings of 20th ACM Symposium on Theory of Computing*, pages 539–550, 1988.

10. Karp, R. M. "Reducibility among combinatorial problems". In *Complexity of Computer Computations*. Plenum Press, 1972.

11. Lipton, R. J. and R. Sedgewick. "Lower bounds for VLSI". In *Proceedings of 13th ACM Symposium on Theory of Computing*, pages 300–307, 1981.

12. Melhorn, K. and E. M. Schmidt. "Las Vegas is better than determinism in VLSI and distributed computing". In *Proceedings of 14th ACM Symposium on Theory of Computing*, pages 330–337, 1982.

13. Motwani, R. "Lecture notes on approximation algorithms". 1992.

14. Papadimitriou, C. H. and M. Sipser. "Communication complexity". In *Proceedings of 14th ACM Symposium on Theory of Computing*, pages 196–200, 1982.

15. Papadimitriou, C.H. "The value of information". In *Proceedings of World Congress of Economics*, 1992.

16. Papadimitriou, C.H. and M. Yannakakis. "On the value of information in distributed decision making". In *Proceedings of 10th ACM Symposium on Principles of Distributed Computing*, pages 61–64, 1991.

17. Papadimitriou, C.H. and M. Yannakakis. "Linear programming without the matrix". In *Proceedings of 25th ACM Symposium on Theory of Computing*, 1993.

18. Paturi, R. and J. Simon. "Probabilistic communication complexity". *Journal of Computer and System Sciences*, 33:106–123, 1986.

19. Thompson, C. D. "Area-time complexity for VLSI". In *Proceedings of 11th ACM Symposium on Theory of Computing*, pages 81–88, 1979.

20. Yannakakis, M. "Expressing combinatorial optimization problems by linear programs". In *Proceedings of 29th IEEE Symposium on Foundations of Computer Science*, pages 223–228, 1988.

21. Yao, A. "The entropic limitations of VLSI computation". In *Proceedings of 13th ACM Symposium on Theory of Computing*, pages 308–311, 1981.

22. Yao, A. C. "Some complexity questions related to distributive computing". In *Proceedings of 11th ACM Symposium on Theory of Computing*, pages 209–213, 1979.

Appendix: Proof of Proposition 1

We shall prove the following. Let $a_1, \ldots, a_N, b_1, \ldots, b_N$ be $2N$ nonnegative numbers such that, for any n,

1. $0 \leq a_n, b_n \leq N$.
2. If $a_n < n - 1$, then $b_k \geq n$ for $k = a_n + 1, \ldots, n - 1$.
3. If $b_n < n - 1$, then $a_k \geq n$ for $k = b_n + 1, \ldots, n - 1$.

Then

$$\sum_{n=1}^{N}(a_n + b_n) \geq 2\sum_{n=1}^{N}(n - 1). \tag{2}$$

The proof is by induction on N. For $N = 1$ the proof is trivial since, by property 1,

$$a_1 + b_1 \geq 0.$$

Assume that (2) has been proven for any $N' < N + 1$ and let $a_1, \ldots, a_N, a_{N+1},$ $b_1, \ldots, b_N, b_{N+1}$ be $2(N+1)$ nonnegative numbers satisfying conditions 1-3. Let us consider the case in which both a_{N+1} and b_{N+1} are smaller than N (the other cases are proved similarly). Then $a_{N+1} = N - h$ and $b_{N+1} = N - k$ with $h, k > 0$. From properties 1-3 it follows that

$$a_{N-k+1}, \ldots, a_N = N + 1 \quad \text{and} \quad b_{N-h+1}, \ldots, a_N = N + 1.$$

For any n with $N - k + 1 \leq n \leq N$ and for any m with $N - h + 1 \leq m \leq N$, let us define $a'_n = N$ and $b'_m = N$. The $2N$ numbers $a_1, \ldots, a_{N-k}, a'_{N-k+1}, \ldots, a'_N,$ $b_1, \ldots, b_{N-h}, b'_{N-h+1}, \ldots, b'_N$ clearly satisfy conditions 1-3. This, in turn, implies that

$$\sum_{n=1}^{N+1} (a_n + b_n) \geq 2 \sum_{n=1}^{N} (n - 1) + (h + k) + (a_{N+1} + b_{N+1}) = 2 \sum_{n=1}^{N+1} (n - 1).$$

The proposition thus follows. $\qquad\qquad\qquad\qquad\qquad\qquad\qquad\qquad\qquad$ $\square$

Cartesian products of graphs as spanning subgraphs of de Bruijn graphs* (Extended Abstract)

Thomas Andreae[1], Michael Nölle[2], Gerald Schreiber[2]

[1] Mathematisches Seminar
Universität Hamburg
Bundesstraße 55
D-20146 Hamburg
Germany

[2] Technische Universität Hamburg-Harburg
Technische Informatik I
Harburger Schloßstraße 20
D-21071 Hamburg
Germany

Abstract. For Cartesian products $G = G_1 \times \ldots \times G_m$ ($m \geq 2$) of nontrivial connected graphs G_i and the n-dimensional base B de Bruijn graph $D = D_B(n)$, we investigate whether or not there exists a spanning subgraph of D which is isomorphic to G. We show that G is never a spanning subgraph of D when n is greater than three or when n equals three and m is greater than two. For $n = 3$ and $m = 2$, we can show for wide classes of graphs that G cannot be a spanning subgraph of D. In particular, these non-existence results imply that $D_B(n)$ never contains a torus (i.e., the Cartesian product of $m \geq 2$ cycles) as a spanning subgraph when n is greater than two. For $n = 2$ the situation is quite different: we present a sufficient condition for a Cartesian product G to be a spanning subgraph of $D = D_B(2)$. As one of the corollaries we obtain that a torus $G = G_1 \times \ldots \times G_m$ is a spanning subgraph of $D = D_B(2)$ provided that $|G| = |D|$ and that the G_i are even cycles of length ≥ 4. In addition we apply our results to obtain embeddings of relatively small dilation of popular processor networks (as tori, meshes and hypercubes) into de Bruijn graphs of fixed small base.

Keywords: de Bruijn graphs, Cartesian product, graph embeddings, dilation, processor networks, parallel image processing and pattern recognition, massively parallel computers

---

* This research was partially done as part of the project *Paralleles Rechnen in der digitalen Bildverarbeitung und Mustererkennung* at the Technical University of Hamburg-Harburg supported by the Deutsche Forschungsgemeinschaft (DFG).

1 Introduction

In the context of parallel and distributed computation the problem of embedding one interconnection network into another is of fundamental importance and has gained considerable attention; see, e.g., the textbook of Leighton [3], the survey article of Monien and Sudborough [4], and the literature mentioned there. Nevertheless the field still offers various interesting open problems and sometimes even largely unexplored areas: for example, embeddings into de Bruijn networks have rarely been investigated until, recently, the topic was picked up by Heydemann et al [7] and, independently and with emphasis on applications in digital image processing, by two of the present authors [6]. Either of these papers discuss embeddings of certain Cartesian product networks into de Bruijn graphs: the paper of Heydemann et al [7] contains a variety of results on embeddings of hypercubes and 2–dimensional meshes into their optimal de Bruijn graphs, while meshes of higher dimension and tori are considered only briefly. In contrast, in the paper [6] the emphasis is on tori. (For definitions, see below.) For information on the role of de Bruijn networks in parallel image processing and pattern recognition, see e.g. [2].

In the present paper, for Cartesian products $G = G_1 \times \ldots \times G_m$ $(m \geq 2)$ of non-trivial connected graphs G_i and the n-dimensional base B de Bruijn graph $D = D_B(n)$, we investigate whether or not there exists a spanning subgraph of D which is isomorphic to G and apply our results to obtain embeddings of relatively small dilation of common processor networks (as tori, meshes and hypercubes) into de Bruijn graphs of small fixed base. Our results generalize and improve some of the results previously obtained in [6, 7].

The paper is organized as follows. In the remainder of this section, we collect some definitions and notational conventions. Thereafter, in Section 2, we prove some lemmas and in Section 3, we present our main results (Theorems 7 and 9). We close with a remark on embeddings into de Bruijn graphs of small fixed base and briefly report on experimental results.

Unless stated otherwise, graphs will be undirected, multiple edges and loops are admitted. Our terminology is standard and (essentially) in accordance with the terminology of [1]; graph-theoretic notions not defined here can be found in [1]. For a graph G, we denote by $V(G)$ and $E(G)$ the set of vertices and edges, respectively. The complete graph with n vertices is denoted K_n. The complete bipartite graph with color classes of cardinality n and m, respectively, is denoted $K(n, m)$. In directed graphs, we write $\overrightarrow{ab}$ to denote an arc from a to b. A *directed trail* (of length t) in a digraph $\overrightarrow{D}$ is a sequence $a_0, \ldots, a_t$ of vertices such that $\overrightarrow{a_i a_{i+1}}$ is an arc of $\overrightarrow{D}$ $(i = 0, \ldots, t-1)$ and such that all arcs $\overrightarrow{a_i a_{i+1}}$ are distinct. A graph is *non-trivial* if it has at least two vertices.

For graphs G_i $(i = 1, \ldots, m)$, the *Cartesian product* $G_1 \times \ldots \times G_m$ is the graph with vertex set $V(G_1) \times \ldots \times V(G_m)$ where two distinct vertices $(a_1, \ldots, a_m)$, $(b_1, \ldots, b_m)$ are joined by t (multiple) edges if and only if there exists an $i \in \{1, \ldots, m\}$ such that $a_j = b_j$ for all $j \neq i$ and such that there are t edges of G_i joining a_i with b_i; further, there are $\sum_{i=1}^{m} t_i$ loops at a vertex $(a_1, \ldots, a_m)$,

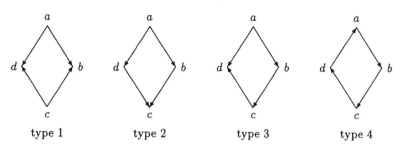

Fig. 1. The different types of 4-cycles in $\overrightarrow{D}_B(n)$.

where t_i is the number of loops at a_i in G_i.

Popular interconnection networks in the area of parallel computing are the tori, meshes and hypercubes which, in terms of the Cartesian product, can be defined as follows. A graph G is an *m-dimensional mesh* (*m-dimensional torus*) if G is the Cartesian product $G_1 \times \ldots \times G_m$ of m non-trivial paths (cycles). The *m-dimensional binary hypercube* is the graph $G = G_1 \times \ldots \times G_m$ where the G_i are complete graphs on two vertices ($i = 1, \ldots, m$). (For $m = 3$, G is called *cube*.)

For integers B, $n \geq 2$, the *directed base B de Bruijn graph of dimension n* (denoted $\overrightarrow{D}_B(n)$) has as its vertex set the set of n-tuples $a = \langle a_{n-1}, \ldots, a_0 \rangle$ where the entries a_i are integers with $0 \leq a_i \leq B-1$, and there exists an arc from $a = \langle a_{n-1}, \ldots, a_0 \rangle$ to $b = \langle b_{n-1}, \ldots, b_0 \rangle$ if and only if $b_i = a_{i-1}$ for $i = 1, \ldots, n-1$. The corresponding *undirected base B de Bruijn graph of dimension n* (denoted $D_B(n)$) results from $\overrightarrow{D}_B(n)$ by ignoring the orientation of the arcs. In other words, $D_B(n)$ has the same vertex set as $\overrightarrow{D}_B(n)$ and distinct vertices $a = \langle a_{n-1}, \ldots, a_0 \rangle$ and $b = \langle b_{n-1}, \ldots, b_0 \rangle$ are joined by a single edge if either $b_i = a_{i-1}$ for $i = 1, \ldots, n-1$ or $a_i = b_{i-1}$ for $i = 1, \ldots, n-1$, and by exactly two edges if both of these conditions hold; moreover, there is a loop at $a = \langle a_{n-1}, \ldots, a_0 \rangle$ if $a_0 = a_1 = \ldots = a_{n-1}$.

2 Some Lemmas on 4-cycles in de Bruijn graphs and Cartesian products

In this section we present a series of lemmas which prepare the proof of Theorem 7. The proof of Theorem 9 is independent of these lemmas. The reader who is mainly interested in results which are relevant for practical applications (Theorem 9 and its corollaries) may skip Section 2.

For a 4-cycle $C \subseteq D_B(n)$, let $\overrightarrow{C}$ be the corresponding directed subgraph of $\overrightarrow{D}_B(n)$. Then C (as well as $\overrightarrow{C}$) is called a *cycle of type t* if t is the maximum length of a directed trail contained in $\overrightarrow{C}$. Figure 1 displays the different types of 4-cycles.

In order to denote the vertices of a 4-cycle of $\vec{D}_B(n)$ and to indicate arc directions, we use notations like $a \to b \to c \to d \leftarrow a$ the meaning of which is clear without explanation.

Lemma 1. $\vec{D}_B(n)$ *contains no cycle of type 2.*

Proof. Suppose that $\vec{D}_B(n)$ contains a cycle of type 2, say, $a \to b \to c \leftarrow d \leftarrow a$ with $a = \langle a_{n-1}, \ldots, a_0 \rangle$, $b = \langle b_{n-1}, \ldots, b_0 \rangle$, $c = \langle c_{n-1}, \ldots, c_0 \rangle$, $d = \langle d_{n-1}, \ldots, d_0 \rangle$. Then $b_i = a_{i-1} = d_i$ $(i = 1, \ldots, n-1)$ and $b_0 = c_1 = d_0$, contradicting $b \neq d$. □

If $a \to b \to c \to d \leftarrow a$ is a cycle of type 3, then $a \to b \leftarrow c \to d \leftarrow a$ is called *the corresponding cycle of type 1*

Lemma 2. *If $\vec{D}_B(n)$ contains a cycle of type 3, then it also contains the corresponding cycle of type 1.*

Proof. Let $a \to b \to c \to d \leftarrow a$ be a cycle of type 3 contained in $\vec{D}_B(n)$ with $a = \langle a_{n-1}, \ldots, a_0 \rangle$, $b = \langle b_{n-1}, \ldots, b_0 \rangle$, $c = \langle c_{n-1}, \ldots, c_0 \rangle$, $d = \langle d_{n-1}, \ldots, d_0 \rangle$. Then $b_i = a_{i-1}$ $(i = 1, \ldots, n-1)$, $c_i = a_{i-2}$ $(i = 2, \ldots, n-1)$, $c_1 = b_0$, $d_i = a_{i-3}$ $(i = 3, \ldots, n-1)$, $d_2 = b_0$, $d_1 = c_0$, $d_i = a_{i-1}$ $(i = 1, \ldots, n-1)$. It follows that $b_i = d_i$ $(i = 1, \ldots, n-1)$ and consequently

$$
\begin{aligned}
b_i = d_i &= a_{i-3} = c_{i-1} \quad (i = 3, \ldots, n-1) \\
b_2 = d_2 &= b_0 = c_1 \\
b_1 = d_1 &= c_0
\end{aligned}
$$

which means that $\vec{D}_B(n)$ contains an arc from c to b. □

The next lemma is an immediate consequence of the definitions.

Lemma 3. *Let $a \to b \leftarrow c \to d \leftarrow a$ be a cycle of type 1 contained in $\vec{D}_B(n)$ with $a = \langle a_{n-1}, \ldots, a_0 \rangle$, $b = \langle b_{n-1}, \ldots, b_0 \rangle$, $c = \langle c_{n-1}, \ldots, c_0 \rangle$, $d = \langle d_{n-1}, \ldots, d_0 \rangle$. Then $a_i = c_i$ $(i = 0, \ldots, n-2)$ and $b_i = d_i$ $(i = 1, \ldots, n-1)$.*

Lemma 4. *If $n \geq 4$ then every vertex of $\vec{D}_B(n)$ is contained in at most one cycle of type 4, and if $n = 3$, then every edge of $\vec{D}_B(n)$ is contained in at most one cycle of type 4.*

Proof. Let $a \to b \to c \to d \to a$ be a cycle of type 4 contained in $\vec{D}_B(n)$ with $a = \langle a_{n-1}, \ldots, a_0 \rangle$, $b = \langle b_{n-1}, \ldots, b_0 \rangle$, $c = \langle c_{n-1}, \ldots, c_0 \rangle$, $d = \langle d_{n-1}, \ldots, d_0 \rangle$. Put $a_{-1} := b_0$, $a_{-2} := c_0$, $a_{-3} := d_0$. Then $b_i = a_{i-1}$, $c_i = a_{i-2}$, $d_i = a_{i-3}$ $(i = 0, \ldots, n-1)$ and $a_i = a_{i-4}$ $(i = 1, \ldots, n-1)$. In particular we have $a_{-3} = a_1$. Further, if $n \geq 3$ then $a_{-2} = a_2$ and if $n \geq 4$ then $a_{-1} = a_3$. Consequently, if $n \geq 4$ then the vertices b, c and d are uniquely determined by $a = \langle a_{n-1}, \ldots, a_0 \rangle$. Further if $n = 3$ then c and d are uniquely determined by a and b. □

Lemma 5. *For $n \geq 4$ let $a = \langle a_{n-1}, \ldots, a_0 \rangle$, $c = \langle c_{n-1}, \ldots, c_0 \rangle$ be a pair of opposite vertices of a cycle of type 4 contained in $\overrightarrow{D}_B(n)$. Then, for at least one i with $1 \leq i \leq n - 2$, we have $a_i \neq c_i$.*

Proof. Choose the notations as in the proof of Lemma 4 and, in addition, define $a_{-4} := a_0$. Then $a_i = a_{i-4}$ ($i = 0, \ldots, n - 1$). Suppose that $a_i = c_i$ for $i = 1, \ldots, n - 2$. Then (because $c_i = a_{i-2}$ for $i = 0, \ldots, n - 1$) we have $a_i = a_{i-2}$ ($i = 1, \ldots, n - 2$) and thus we conclude from $n \geq 4$ that $a_{n-3} = a_{n-5}$ and $a_2 = a_0$. Hence $a_{n-1} = a_{n-5} = a_{n-3} = c_{n-1}$, $a_0 = a_2 = a_{-2} = c_0$ and thus we have $a_i = c_i$ for all $i = 0, \ldots, n - 1$ contradicting $a \neq c$. $\square$

For a Cartesian product $G_1 \times \ldots \times G_m$, a subgraph H of $G_1 \times \ldots \times G_m$ is called *1-dimensional* if there exists an i such that, for any pair of vertices $a = (a_1, \ldots, a_m)$, $b = (b_1, \ldots, b_m)$ of H, $a_j = b_j$ for all $j \neq i$. A 4-cycle $C \subseteq G_1 \times \ldots \times G_m$ which is not 1-dimensional is called *2-dimensional*.

Lemma 6. *Let $G = G_1 \times \ldots \times G_m$ be the Cartesian product of $m \geq 2$ non-trivial connected graphs and let $\sim$ be an equivalence relation on the vertex set of G. Assume that $a \sim c$ for any two vertices a, c which form a pair of opposite vertices on some 2-dimensional 4-cycle of G. Then the partition of $V(G)$ corresponding to $\sim$ consists of at most two classes.*

Proof. Since every connected graph contains a spanning tree, it clearly suffices to prove the lemma for the case that the graphs G_i are trees. Then the graph G is bipartite since it is the Cartesian product of bipartite graphs. Let S be a color-class of G and pick $a, b \in S$. We claim that $a \sim b$ (which implies the lemma). Note that, in order to prove $a \sim b$, it suffices to consider the case when $d_G(a, b) = 2$ since the general case can be settled by iterated application of the distance-two case. Let $P = (a, x, b)$ be a path of G. If a, b is a pair of opposite vertices of a (2-dimensional) 4-cycle of G then we are done; otherwise P is a 1-dimensional subpath of G and (since the graphs G_i are non-trivial) there exists a 1-dimensional subpath $P' = (a', x', b')$ of G such that $P \cap P' = \emptyset$ and $aa', xx', bb' \in E(G)$. Hence $a \sim x' \sim b$. $\square$

3 The main results

We now treat the question whether or not a Cartesian product of graphs is a spanning subgraph of $D_B(n)$. Our first theorem establishes a non-existence result for $n \geq 3$.

Theorem 7. *Let $G = G_1 \times \ldots \times G_m$ be the Cartesian product of $m \geq 2$ non-trivial connected graphs and, for some $n \geq 3$, let $D = D_B(n)$. If $m = 2$ and $n = 3$ then assume that G_1, G_2 are loopless, G_1 is not a path and G_2 has no vertices of degree one. Then G is not a spanning subgraph of D.*

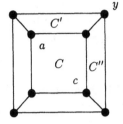

Fig. 2. Cube H contained in G in the case $m \geq 3$.

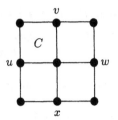

Fig. 3. 4-cycle C of type 4 in the case $m = 2$ and $n \geq 4$.

Let G_1 and G_2 be paths of length one and three, respectively. Then the mesh $G_1 \times G_2$ is a spanning subgraph of $D_2(3)$ which shows that the additional assumption made in Theorem 7 for the case $m = 2$, $n = 3$ cannot be dropped (but, possibly, relaxed). Due to space limitations, in this preliminary report we do not include the proof of Theorem 7 for the case $m = 2$, $n = 3$, the details of which are more involved; this part of the proof will be contained in the full paper.

Proof of Theorem 7 (under the additional hypothesis that $n \geq 4$ when $m = 2$). It suffices to settle the case that the graphs G_i are trees, which will be assumed henceforth. Suppose that G is a spanning subgraph of D.

First, let us assume that $m \geq 3$ or $|G_1|, |G_2| \geq 3$. We claim that the following holds.

$$\text{If } a = \langle a_{n-1}, \ldots, a_0 \rangle, \; c = \langle c_{n-1}, \ldots, c_0 \rangle \text{ is a pair of opposite} \tag{1}$$
$$\text{vertices of a 4-cycle } C \subseteq G, \text{ then } a_i = c_i \; (i = 1, \ldots, n-2).$$

If C is not of type 4, then statement (1) is an immediate consequence of the Lemmas 1–3. Assume that C is of type 4. If $m \geq 3$ then C is contained in a cube $H \subseteq G$. Thus we can find 4-cycles C', C'' of H as shown in Figure 2. By Lemma 4 C' and C'' are not of type 4. Let $y = \langle y_{n-1}, \ldots, y_0 \rangle$ be the vertex of $C' \cap C''$ which is opposite of a in C' and opposite of c in C'' (cf. Figure 2). It follows that $a_i = y_i$ and $c_i = y_i$ $(i = 1, \ldots, n-2)$, which settles the case $m \geq 3$.

It remains to consider the case that $m = 2$. Then $|G_1|, |G_2| \geq 3$ and, therefore, C is contained in a mesh $H = P_1 \times P_2 \subseteq G$ where P_1 and P_2 are paths of length 2. Choose the notations as in Figure 3 and let $u = \langle u_{n-1}, \ldots, u_0 \rangle$, $v = \langle v_{n-1}, \ldots, v_0 \rangle$, $w = \langle w_{n-1}, \ldots, w_0 \rangle$, $x = \langle x_{n-1}, \ldots, x_0 \rangle$. Note that $n \geq 4$ by the additional hypothesis made at the beginning of the proof of Theorem 7. Thus we conclude from Lemma 4 that, with the exception of C, all 4-cycles of H are of type $\neq 4$. Hence (by the Lemmas 1 - 3) $v_i = w_i = x_i = u_i (i = 1, \ldots, n-2)$. On the other hand, by Lemma 5, we have $v_i \neq u_i$ for at least one $i \in \{1, \ldots, n-2\}$. This contradiction completes the proof of (1).

From (1) and Lemma 6 we conclude that the vertices of G can be partitioned into two disjoint classes such that $a_i = b_i (i = 1, \ldots, n - 2)$ for any two vertices $a = \langle a_{n-1}, \ldots, a_0 \rangle$, $b = \langle b_{n-1}, \ldots, b_0 \rangle$ that are members of the same class. Hence

$$2B^2 \geq |G| = |D| = B^n$$

which implies $n = 3$, $B = 2$. Hence (by the assumption that $m \geq 3$ or $|G_1|, |G_2| \geq 3$) G is a cube, and thus we are done since $D_2(3)$ does not contain a cube as a spanning subgraph.

It remains to consider the case that $m = 2$ and, say, $|G_1| = 2$. Then G consists of two disjoint trees T_1, T_2 (which are isomorphic to G_2) together with edges e_v ($v \in V(T_1)$) where e_v connects v with a vertex of T_2 corresponding to v under a fixed isomorphism $T_1 \to T_2$. We refer to an edge e_v as a *rung* of G.

We first assume that $n \geq 5$. Let e be a rung of G connecting vertices $a = \langle a_{n-1}, \ldots, a_0 \rangle$ and $b = \langle b_{n-1}, \ldots, b_0 \rangle$ with $a_0 = \ldots = a_{n-2} = b_1 = \ldots = b_{n-1} = 0$; note that there exists at least one rung with these properties, namely, the rung containing the vertex $\langle 0, 0, \ldots, 0 \rangle$. Let C be a 4-cycle of G passing through e and let e' be the rung of C with $e' \neq e$. Denote by $a' = \langle a'_{n-1}, \ldots, a'_0 \rangle$ and $b' = \langle b'_{n-1}, \ldots, b'_0 \rangle$ the vertices incident with e' where notations are chosen such that a, a' is a pair of opposite vertices of C. Because $a_0 = \ldots = a_{n-2} = 0$ and $n \geq 5$, C cannot be a cycle of type 4; moreover one finds from $a_0 = \ldots = a_{n-2} = b_1 = \ldots = b_{n-1} = 0$ that $\overrightarrow{ba} \notin \overrightarrow{D} = \overrightarrow{D}_B(n)$. Hence $\overrightarrow{a'b}, \overrightarrow{a'b'}, \overrightarrow{ab'} \in \overrightarrow{D}$ by Lemma 1 and 2, and thus we conclude from Lemma 3 that $a'_0 = \ldots a'_{n-2} = b'_1 = \ldots = b'_{n-1} = 0$. Iterating this argument, we conclude from the connectedness of T_1 and T_2 that $v_1 = \ldots = v_{n-2} = 0$ for all vertices $v = \langle v_{n-1}, \ldots, v_0 \rangle$ of G, which is a contradiction. This settles the case $n \geq 5$.

Finally let $n = 4$. If a and b are vertices joined by a rung, then a and b are called *partners*. We claim that there exists a vertex $a = \langle a_3, a_2, a_1, a_0 \rangle$ with the following properties:

$$a_0 \neq a_2 \tag{2}$$

$$\overrightarrow{ab} \in \overrightarrow{D}, \text{where } b \text{ is the partner of } a. \tag{3}$$

Indeed, pick a vertex $v = \langle v_3, v_2, v_1, v_0 \rangle$ with $v_1 \neq v_3$, $v_0 \neq v_2$ and let w be the partner of v. If $\overrightarrow{vw} \in \overrightarrow{D}$, then we are done; otherwise $\overrightarrow{wv} \in \overrightarrow{D}$, and thus $w = \langle v_4, v_3, v_2, v_1 \rangle$ for some $v_4 \in \{0, \ldots, B - 1\}$ which (because $v_1 \neq v_3$) implies that w has the desired properties.

Now let $a =< a_3, a_2, a_1, a_0 >$ be a vertex with partner b such that (2) and (3) hold. Note that $\overrightarrow{ba} \notin \overrightarrow{D}$ since this, together with (3), would imply $a_0 = a_2$. Let $C = (a, b, c, d)$ be a 4-cycle contained in G. If C is of type 4, then $c = \langle a_1, a_0, a_3, a_2 \rangle$ and $\overrightarrow{cd} \in \overrightarrow{D}$, and thus c is (beside a) another vertex having the properties (2) and (3). If C is not of type 4, then one obtains from the Lemmas 1–3 (together with the fact that $\overrightarrow{ba} \notin \overrightarrow{D}$) that $c = \langle a'_3, a_2, a_1, a_0 \rangle$ and

$\overrightarrow{cd} \in \overrightarrow{D}$. Hence, in either case, c is a vertex having the properties (2) and (3). Iterating this argument, we conclude from the connectedness of T_1 and T_2 that every rung contains a vertex with the properties (2) and (3). On the other hand, this is clearly impossible since this cannot hold for the rung containing the vertex $\langle 0,0,0,0 \rangle$. This contradiction settles the case $n = 4$. $\qquad\square$

Note that Theorem 7 leaves open the case that $n = 3$ and G is a 2–dimensional mesh. (For results on embeddings of 2–dimensional meshes, we refer to [7].) For tori, however, Theorem 7 immediately implies the following.

Corollary 8. *A torus $G = G_1 \times \ldots \times G_m$ ($m \geq 2$) cannot be a spanning subgraph of a de Bruijn graph $D_B(n)$ of dimension $n \geq 3$.*

For $n = 2$ the situation is quite different. In order to present our result for $n = 2$, we need some preparation. Let $G = G_1 \times \ldots \times G_m$ be a Cartesian product of $m \geq 2$ nontrivial graphs and $D = D_B(2)$. Let D and G have the same number of vertices, i.e.,

$$\prod_{i=1}^{m} |G_i| = B^2. \tag{4}$$

For each $i \in \{1, \ldots, m\}$, assume that $q_{i,1}, \ldots, q_{i,r_i} (r_i \geq 1)$ are integers with $q_{i,j} \geq 2$ ($j = 1, \ldots, r_i$) such that $|G_i| = q_{i,1} \cdot \ldots \cdot q_{i,r_i}$. Assume further that the set $M = \{(i,j) : 1 \leq i \leq m, 1 \leq j \leq r_i\}$ can be partitioned into disjoint subsets M_1, M_2 such that, for every integer $q \geq 2, |\{(i,j) \in M_1 : q_{i,j} = q\}| = |\{(i,j) \in M_2 : q_{i,j} = q\}|$. Moreover, for each $(i,j) \in M$, let $a_{i,j}, b_{i,j}$ be positive integers with $q_{i,j} = a_{i,j} + b_{i,j}$, where $a_{i,j} = a_{k,l}, b_{i,j} = b_{k,l}$ whenever $q_{i,j} = q_{k,l}$. Put $t_{i,j} = (q_{i,j}, a_{i,j}, b_{i,j})$. We call a family of triples $t_{i,j} (1 \leq i \leq m, 1 \leq j \leq r_i)$ having the described properties a *family of triples associated to $G_1 \times \ldots \times G_m$*. Note that it follows from (4) that there exists at least one family of triples associated to $G_1 \times \ldots \times G_m$: just consider the factorization of $|G_i|$ into primes $q_{i,j}$ and choose the $a_{i,j}, b_{i,j}$ accordingly. One easily concludes from (4) that this results into a family of triples associated to $G_1 \times \ldots \times G_m$. Now the announced theorem reads as follows.

Theorem 9. *Let $G = G_1 \times \ldots \times G_m$ ($m \geq 2$) be a Cartesian product of nontrivial graphs G_i and let $D = D_B(2)$. Assume that $|G| = |D|$ and let $t_{i,j} = (q_{i,j}, a_{i,j}, b_{i,j})$ ($1 \leq i \leq m, 1 \leq j \leq r_i$) be a family of triples associated to $G_1 \ldots \times G_m$. Assume further that*

$$G_i \text{ is a spanning subgraph of } K(a_{i,1}, b_{i,1}) \times \ldots \times K(a_{i,r_i}, b_{i,r_i}), \tag{5}$$
$$i = 1, \ldots, m.$$

Then G is a spanning subgraph of D.

Proof. Let M, M_1, M_2 be as in the paragraph before Theorem 9. Then we conclude that

$$\underset{(i,j) \in M_1}{\times} K(a_{i,j}, b_{i,j}) \cong \underset{(i,j) \in M_2}{\times} K(a_{i,j}, b_{i,j}). \tag{6}$$

Let $H := \underset{(i,j) \in M_1}{\times} K(a_{i,j}, b_{i,j})$. Then it follows from (5) and (6) that G is a spanning subgraph of $H \times H$. Note that $H \times H$ is bipartite and connected; denote by C_0 and C_1 the color–classes of $H \times H$.

Let $\alpha : V(H \times H) \longrightarrow V(H \times H)$ be the function that maps (u, v) onto (v, u) for $u, v \in H$. Clearly α is an automorphism of $H \times H$. Since there exist vertices of $H \times H$ which are fixed by α, we conclude from the connectedness of $H \times H$ that α maps the vertices of C_i onto $C_i (i = 0, 1)$. Let $u, v \in V(H)$. We define a mapping $\Psi : V(H \times H) \longrightarrow V(D)$ by

$$\Psi((u, v)) = \begin{cases} \langle u, v \rangle & \text{if } (u, v) \in C_0 \\ \langle v, u \rangle & \text{otherwise} . \end{cases}$$

Because α maps C_1 onto C_1, we conclude that Ψ is a bijection. Let $(u, v), (x, y)$ be vertices of $H \times H$ which are joined by an edge. We may assume $(u, v) \in C_0, (x, y) \in C_1$. Then $\Psi((u, v)) = \langle u, v \rangle$ and $\Psi((x, y)) = \langle y, x \rangle$. Further $u = x$ or $v = y$. Hence $\langle u, v \rangle$ and $\langle y, x \rangle$ are joined by an edge of D and thus $H \times H$ is a spanning subgraph of D. Hence G is a spanning subgraph of D. $\quad\square$

We now present some corollaries of Theorem 9.

Corollary 10. *Let $G = G_1 \times \ldots \times G_m (m \geq 2)$ be a Cartesian product of even cycles of length ≥ 4 and $D = D_B(2)$. Assume that $|G| = |D|$. Then G is a spanning subgraph of D.*

Proof. Let $|G_i| = q_{i,1} \cdot \ldots \cdot q_{i,r_i}$ be the factorization of $|G_i|$ into primes and let $a_{i,j} = \lfloor \frac{q_{i,j}}{2} \rfloor, b_{i,j} = \lceil \frac{q_{i,j}}{2} \rceil$. Then (as mentioned above) $t_{i,j} = (q_{i,j}, a_{i,j}, b_{i,j})$ is a family of triples associated to $G_1 \times \ldots \times G_m$. Note also that each $K(a_{i,j}, b_{i,j})$ has a Hamiltonian path. Further, for each $i \in \{1, \ldots, m\}$ there exists a $j \in \{1, \ldots, r_i\}$ such that $K(a_{i,j}, b_{i,j}) \cong K_2$ (since $|G_i|$ is even). From this one easily concludes that the graph $K(a_{i,1}, b_{i,1}) \times \ldots \times K(a_{i,r_i}, b_{i,r_i})$ has a Hamiltonian cycle $(i = 1, \ldots, m)$, which means that (5) holds. Thus the assertion follows from Theorem 9. $\quad\square$

In a similar way one obtains

Corollary 11. *If $G = G_1 \times \ldots \times G_m (m \geq 2)$ is the Cartesian product of nontrivial paths G_i and if $|G| = |D_B(2)|$, then G is a spanning subgraph of $D_B(2)$.*

Note that the Corollaries 10 and 11 cover the cases of bipartite tori, arbitrary meshes and, as a special case of Corollary 11, binary hypercubes. The above results improve some of the results of Heydemann et al [7] who obtained the result of Corollary 10 under the additional assumption that the cycle lengths are powers of two (see also [6]); similarly the paper [7] contains the result of Corollary 11 under additional assumptions on the path length.

further the authors of [7] present a restricted version of Corollary 11 covering the hypercube case. We mention that, in a sense, Corollary 10 is sharp since the Cartesian product $C_3 \times C_3$ is not a spanning subgraph of $D_3(2)$, as can easily be checked. (Here C_t denotes the cycle of length t.) In addition, we found that $C_5 \times C_5$ *is* a spanning subgraph of $D_5(2)$ but do not know whether or not $C_t \times C_t$ is a spanning subgraph of $D_t(2)$ for odd $t \geq 7$.

Other corollaries of Theorem 9 are possible. For example, the following can be derived from Theorem 9. (The proof is left for the reader.)

Corollary 12. *Let $G = G_1 \times \ldots \times G_m (m \geq 2)$ where the G_i are simple nontrivial bipartite graphs with color classes A_i, B_i and assume that $|A_i| = |A_j|, |B_i| = |B_j|$ whenever $|G_i| = |G_j|$. Assume further that $|G| = |D_B(2)|$ and that one of the following conditions (i) or (ii) holds:*
  (i) $|G_i|$ is a prime for $i = 1, \ldots, m$
  (ii) $|G_1| = \ldots = |G_m|$ and m is even.
Then G is a spanning subgraph of $D_B(2)$.

We close with a remark on embeddings into base C de Bruijn graphs where C is a fixed constant. An *embedding* of a graph G into a graph H is a pair of mappings $\{\phi_1, \phi_2\}$ where ϕ_1 maps each vertex of G onto a vertex of H and ϕ_2 assigns to every edge ab of G a path $\phi_2(ab)$ of H which connects $\phi_1(a)$ with $\phi_1(b)$. The *dilation* of an embedding is the maximum length of the paths $\phi_2(ab)$. If ϕ_1 is injective, then the embedding is said to have *load 1*.

Proposition 13. *For $n = 2\lceil \log_C B \rceil$ there exists an embedding $\{\phi_1, \phi_2\}$ of $D_B(2)$ into $D_C(n)$ with load 1 and dilation $\frac{n}{2}$.*

Proof. (cf. [5, 7]) Let $r = \lceil \log_C B \rceil$. For a vertex $\langle a, b \rangle$ of $D_B(2)$, $0 \leq a, b < B$, let

$$\phi_1(\langle a, b \rangle) = \langle a_{r-1}, \ldots, a_0, b_{r-1}, \ldots, b_0 \rangle \in D_C(n)$$

with $a = \sum_{i=0}^{r-1} a_i C^i$ and $b = \sum_{i=0}^{r-1} b_i C^i$ where $a_i, b_i \in \{0, \ldots, C-1\}$ for $i = 0, \ldots, r-1$. Clearly ϕ_1 is injective. Then for any edge $\langle a, b \rangle \langle b, c \rangle \in E(D_B(2))$ there exists a path $(x_0, \ldots, x_r)$ in $D_C(n)$ with

$$x_0 = \langle a_{r-1}, \ldots, a_0, b_{r-1}, \ldots, b_0 \rangle$$
$$x_i = \langle a_{r-i-1}, \ldots, a_0, b_{r-1}, \ldots, b_0, c_{r-1}, \ldots, c_{r-i} \rangle$$
$$x_r = \langle b_{r-1}, \ldots, b_0, c_{r-1}, \ldots, c_0 \rangle$$

of length r from $\phi_1(\langle a, b \rangle)$ to $\phi_1(\langle b, c \rangle)$. Here $c = \sum_{i=0}^{r-1} c_i C^i$ with $c_i \in \{0, \ldots, C-1\}$. Thus the dilation of the embedding is $r = \frac{n}{2}$. $\square$

In applications in the field of parallel algorithms, embeddings of relatively small dilation of common processor networks (as tori, meshes and hypercubes) into de Bruijn graphs of small fixed base (typically, 2, 3 or 4) are of significant importance (cf. [2, 6]). Note that Theorem 9 (and its corollaries) in conjunction with the above proposition provides us with embeddings of this kind, where "relatively small dilation" means that the dilation is at most half the diameter n of $D_C(n)$. We refer to [6] for experimental results showing that this gain of a factor $\frac{1}{2}$ for the dilation leads to considerable gains in efficiency of parallel algorithms.

References

1. J.A. Bondy and U.S.R. Murty. *Graph Theory with Applications*. Macmillian, London, Great Britan, 1977.
2. H. Burkhardt, B. Lang, and M. Nölle. Aspects of Parallel Image Processing Algorithms and Structures. pages 65–84. ESPRIT BRA 3035 Workshop, Bonas (F), August 1990, North-Holland, 1991.
3. F.T. Leighton. *Introduction to Parallel Algorithms and Architectures*. Morgan Kaufman Publishers, San Matio, California, 1992.
4. B. Monien and H. Sudborough. Embedding one interconnection network in another. *Computing Suppl., Springer Verlag*, 7:257–282, 1990.
5. M. Nölle. *Konzepte zur Entwicklung paralleler Algorithmen der digitalen Bildverarbeitung*. PhD thesis, Technische Universität Hamburg-Harburg, 1994. In preparation.
6. M. Nölle and G. Schreiber. Einbettung von Gitter-Algorithmen in de Bruijn-Graphen. *3. PASA Workshop Bonn, in Mitteilungen-Gesellschaft für Informatik e.V.*, pages 12–19, April 1993.
7. D. Sotteau M.C. Heydemann, J. Opatrny. Embeddings of hypercubes and grids into de Bruijn graphs. *Proceedings of the 1992 International Conference on Parallel Processing*, 3:28–37, August 1992.

Specification of Graph Translators
with Triple Graph Grammars

Andy Schürr[1]

Lehrstuhl für Informatik III, RWTH Aachen
Ahornstr. 55, D-52074 Aachen, Germany
e-mail: andy@i3.informatik.rwth-aachen.de

Abstract. Data integration is a key issue for any integrated set of soft-ware tools. A typical CASE environment, for instance, offers tools for the manipulation of requirements and software design documents, and it provides more or less sophisticated assistance for keeping these documents in a consistent state. Up to now, almost all data consistency observing or preserving integration tools are hand-crafted due to the lack of generic implementation frameworks and the absence of adequate specification formalisms. Triple graph grammars are intended to fill this gap and to support the specification of interdependencies between graph-like data structures on a very high level. Furthermore, they are the fundamentals of a new machinery for the production of batch-oriented as well as incrementally working data integration tools.

1 Introduction

Graphs play an important role within many application areas of computer science, as e.g. in the form of data flow or control flow graphs in compiler construction or structured analysis and entity relationship diagrams in software engineering. Furthermore, rewriting systems, especially in the form of *graph rewriting systems*, are well-suited for the description of complex transformation or inference processes on complex data structures. They were, for instance, successfully used for the development of an Integrated Project Support ENvironment, called IPSEN [EL92].

Nevertheless, graph rewriting systems have one important draw-back. They are usually restricted to the specification of processes which perform *in-place graph modifications* and transform one instance of a class of graphs into another instance of the same class. Therefore, they are not well-suited for the specification of integration or traceability tools, which either take a complex data structure (source graph) as input and translate it into a new related data structure (target graph), or check consistency of simultaneously existing related data structures.

Examples of related graph-like data structures are for instance the requirements and design documents of a piece of software or the syntax tree and control flow diagram for a program. Figure 1 shows a cutout of a program's syntax tree and its corresponding control flow diagram. The latter one consists of an if-statement with a single assignment in its then-branch and a while-statement in its else-branch. Labels of edges are omitted in order to keep diagrams legible with

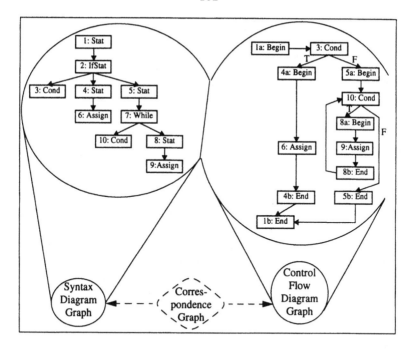

Fig. 1. Related Graphs: Syntax Tree and Control Flow Diagram.

the exception of those control flow edges that have conditions as their sources: "T(rue)" and "F(alse)" are their labels and have the usual meaning. The example demonstrates our needs for fine-grained *m-to-n inter-graph relationships* with the following characteristics:

- Elements of one graph are related to distinct elements of another graph: in figure 1, a vertex with number x of the left-hand side graph is related to a vertex with the same number (if existent) in the right-hand side graph.
- Correspondences between vertices and edges of related graphs are at least 1-to-n: in figure 1, a vertex x of the left-hand side graph corresponds to vertices xa and xb (if existent) of the right-hand side graph.
- Related graphs contain referenced as well as private elements: in our running example, all vertices of the right-hand side graph are referenced, whereas all its edges are private, and the referenced part of the left-hand side graph are all its vertices with exception of 2 and 7.

In general, inter-graph relationships themselves have annotations about ongoing translation or analysis processes. Sometimes, we have even dependencies between inter-graph relationships like "this part of the source graph was translated before that part of the source graph" or "this inter-graph relationship is only valid as long as that inter-graph relationship is existent". These additional annotations and dependencies are of particular importance in the case of incrementally working data integrators, the description of which is outside the scope of this paper (cf. [Le93, Le94]). Therefore, inter-graph relationships will be modeled as separate *correspondence graphs* with references to related source and target graph elements, and data integrators will be specified by means of productions, which rewrite three graphs in parallel, in the sequel.

2 Triple Graph Grammars

Triple graph grammars are a refinement of the old idea of *pair (graph) grammars* already suggested by Pratt more than 20 years ago [Pr71]. Pair graph grammars as well as (almost) all EBNF-oriented approaches [No87, We92] for tree-to-tree translations are restricted to context-free productions and one-to-one correspondences between objects in related data structures. Therefore, these formalisms are far too limited with respect to their expressiveness (see example of figure 1).

To compensate for these deficiencies *triple graph grammars* extend the original pair graph grammar approach to the case of context-sensitive productions with rather complex left- and right-hand sides. They offer even separate correspondence rules and graphs for modeling m-to-n-relationships between related graphs. The name "triple graph grammars" has been chosen in order to emphasize the important role of these additional correspondence rules and graphs.

Before going into details, we have to discuss an important "design decision", the chosen *representation of inter-graph relationships* between corresponding source or target graph elements. One approach is to embed all three graph components as substructures into a common "superstructure" and to use (higher order) relations for modeling correspondences between vertices and edges of these substructures. This solution has the advantage that simple rewrite rules of high level replacement systems with well-known properties [EH91] are sufficient for the specification of graph-to-graph translation processes. Nevertheless, it has also a number of significant drawbacks:

- Relationships between a production's left- and right-hand side are modeled different than relationships between productions themselves, although both kind of relationships play similar roles within the following definitions.
- The wide-spread data model of directed graphs must be abandoned in favor of a new data model which allows relationships between relationships.
- Using one production of a high level replacement system instead of three related productions destroys the "modularity" of the new approach, and makes local modifications of source or target graph structures more difficult.

Therefore, it seems to be more appropriate to use *morphisms* from correspondence graphs to separate source and target graphs for modeling inter-graph relationships.

Returning to the example of section 1, we will now discuss its specification in figure 2. The first triple production defines the correspondences between an if-statement's syntax tree and its control flow (sub-)diagram, whereas following triple productions deal with while-statements and assignments. A dashed vertex (x,y) of a correspondence production cp_i models a relation between a vertex x in the left-hand side syntax tree and a vertex y in the right-hand side control flow diagram of figure 1. Due to lack of space, productions dealing with statement sequences are omitted, and branches of if-statements as well as bodies of while-statements are confined to be single statements only. Last but not least pairs of new "Begin" and "End" vertices within right-hand side productions are intentionally left unconnected. As a consequence the language of all control flow

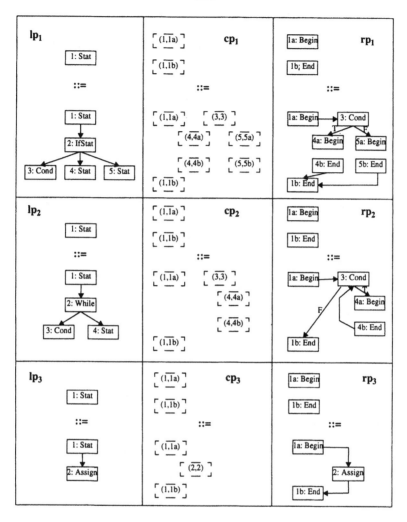

Fig. 2. Triple Production for Syntax Tree ↔ Control Flow Diagram Translations.

diagrams generated by isolated right-hand side productions is a proper superset of the language of all diagrams generated by the triple graph grammar. It contains entangled control flow diagrams, where single statements connect wrong pairs of "Begin" and "End" vertices. This reflects the fact that, in general, we cannot expect that all possible inputs of graph-to-graph translations are also legal inputs which have a defined result.

The presented specification may be used to develop tools of rather different functionality:

- an *LR-translator*, which takes any left-hand side syntax tree as input and returns a corresponding right-hand side control flow diagram,
- an *RL-translator*, which analyzes a right-hand side control flow diagram and

produces the corresponding syntax tree if possible, and

- a *correspondence analyzer*, which monitors the relationships between a given syntax tree and a given control flow diagram.

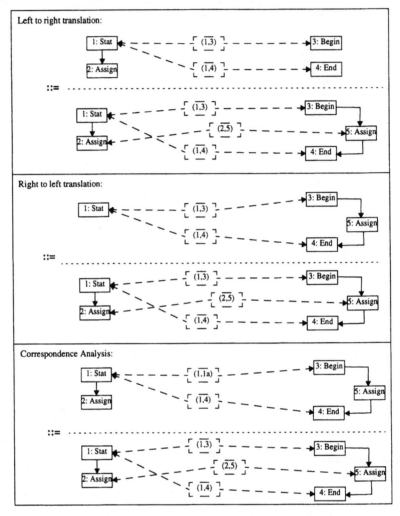

Fig. 3. Conventional Productions for Syntax Tree $\leftrightarrow$ Control Flow Diagram.

The advantage of triple graph grammars in comparison to previously used graph grammar approaches for the specification of inter-graph relationships (cf. [We92]) is clearly visible, when we compare triple production (lp_3, cp_3, rp_3) of figure 2 with its corresponding "conventional" productions of figure 3. In the latter case, we have three productions with a single "flat" and often unintelligible graph on their left- and right-hand sides[1].The first production of figure

---

[1] with correspondences represented as dashed nodes and edges and with left-hand sides above and right-hand sides below dotted lines

3 specifies the corresponding LR-translation step and the second one the corresponding RL-translation step. The last production specifies a correspondence analysis step, which takes a syntax diagram and a control flow diagram as input and tries to establish correct correspondences between these graphs.

That means that *one* triple production replaces *three* rather similar but nevertheless different conventional productions. Furthermore, there exists even an algorithm which takes a set of triple productions as input and generates corresponding sets of (triple/conventional) productions for LR-/RL-translations as well as correspondence analysis as output. The next section presents this algorithm and proves its correctness.

3 Simple Triple Graph Grammars and LR-Translators

Triple graph grammars may be used to specify rather complex graph-to-graph translations as languages of graph triples. Elements (LG, CG, RG) belonging to these languages represent related graph structures LG and RG, respectively, which are linked to each other by means of an additional correspondence graph CG. The grammar for such a graph triple language consists of triples of productions (lp, cp, rp), where each production component is responsible for generating or extending the corresponding graph component.

In principle, any graph model and *any graph grammar approach* may be used as the underlying basic formalism of triple graph grammars. To emphasize this, we will use a very simple class of graphs and rather straightforward rewriting rules, such that we are able to explain the principles of the new formalism both within the framework of the algebraic and the algorithmic graph grammar approach [Eh79, Na79] without getting stuck into technical details. As the reader may imagine, the development of tools which extend related graphs in parallel by simultaneously applying related productions to related vertices and edges is a straightforward task. But the development of tools which translate an existing left-hand (right-hand) side graph into a new right-hand (left-hand) side graph is a rather difficult task. In general, it requires an efficiently working graph parser for context-sensitive productions, which is able to recover a sequence of production applications yielding the given source graph. Provided with this sequence we are then able to apply the related sequence of productions to a target side start graph and to produce thereby the required target graph.

In the presented version of triple graph grammars we simplify the problem of graph parsing by regarding *monotonic productions* only. This means that any production's left-hand side must be part of its right-hand side, i.e. productions do not delete vertices and edges. In this case, a given graph directly contains all necessary information about its derivation history, and graph parsing simply means covering a given graph with right-hand sides of productions (for further details cf. [Sch94]). This simplifies the development of LR- or RL-translators considerably. Requiring monotonicity is not as restrictive as it seems to be at a first glance, since triple graph grammars are not intended to model editing processes on related graphs (with insertions as well as deletions and modifications of graph elements), but are a generative description of graph languages and their relationships.

Following this line, we start with the definition of simple graphs, graph morphisms, and monotonic productions:

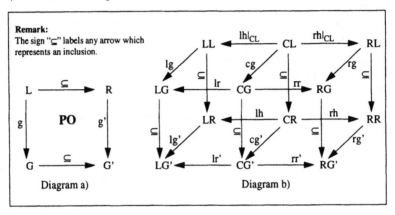

Fig. 4. Application of Simple Productions and Triple Productions.

Definition 1. A quadruple $G := (V, E, s, t)$ is a **graph**
- V being a finite set of vertices,
- E being a finite set of edges, and
- $s, t : E \to V$ as source and target vertex assigning functions. □

Definition 2. Let $G := (V, E, s, t)$, $G' := (V', E', s', t')$ be two graphs. A pair of functions $h := (h_V, h_E)$ with $h'_V : V \to V'$ and $h_E : E \to E'$ is a **graph morphism** from G to G', i.e. $h : G \to G'$, iff:

$$\forall e \in E : h_V(s(e)) = s(h_E(e)) \wedge h_V(t(e)) = t(h_E(e)).$$

Furthermore, we will assume that **operators** $\subset, \subseteq, \cup, \cap$, and $\setminus$ are defined as usual for graphs. □

Definition 3. Any tuple of graphs $p := (L, R)$ with $L \subseteq R$ is a **monotonic production** and p **applied** to a given graph G produces another graph $G' \supseteq G$, denoted by: $G \sim p \rightsquigarrow G'$, with respect to redex selecting morphisms $g : L \to G$ and $g' : R \to G'$, iff:
- $g'|_L = g$, i.e. g and g' are identical w.r.t. the left-hand side graph L.
- g' maps new vertices and edges of $R \setminus L$ onto unique new vertices and edges of $G' \setminus G$. □

Using the categorical framework [Eh79], the two conditions of def. 3 may be replaced by requiring the existence of the pushout diagram a) of figure 4. Based on this fundamental terminology we are now able to define graph triples as well as triple productions and their application to graph triples:

Definition 4. Let LG, RG, and CG be three graphs, and $lr : CG \to LG$, $rr : CG \to RG$ are those morphisms which represent m-to-n relationships between the left-hand side graph LG and the right-hand side graph RG via the correspondence graph CG in the following way:

$$x \in LG \text{ is \textbf{related} to } y :\Leftrightarrow \exists z \in CG : x = lr(z) \cap rr(z) = y.$$

The resulting **graph triple** is denoted as follows:

$$GT := (LG \leftarrow lr - CG - rr \rightarrow RG). \qquad \square$$

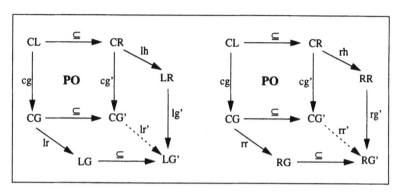

Fig. 5. Existence and Uniqueness of lr' and rr

Definition 5. Let $lp := (LL, LR)$, $rp := (RL, RR)$, and $cp := (CL, CR)$ be monotonic productions. Furthermore, $lh : CR \rightarrow LR$ and $rh : CR \rightarrow RR$ are graph morphisms such that $lh|_{CL} : CL \rightarrow LL$ and $rh|_{CL} : CL \rightarrow RL$ are morphisms, too, which relate the left- and right- hand sides of productions lp and rp via cp to each other. The resulting **triple production** is:

$$p := (lp \leftarrow lh - cp - rh \rightarrow rp).$$

And the **application** of such a triple production to a graph triple

$$GT := (LG \leftarrow lr - CG - rr \rightarrow RG)$$

with redex selecting morphisms (lg, cg, rg) produces another graph triple

$$GT' := (LG' \leftarrow lr' - CG' - rr' \rightarrow RG'),$$

i.e.: $GT \sim p \rightsquigarrow GT'$, which is uniquely defined (up to isomorphism) by the existence of the "pair of cubes" in diagram b) of figure 4. Its new morphisms (lg', cg', rg') are already determined by def. 3. Furthermore, the left-hand side diagram of figure 5 proofs the existence and uniqueness of

$$lr' : CG' \rightarrow LG' \text{ with } lr = lr'|_{CG} \text{ and } lh \circ lg' = cg' \circ lr'.$$

This is a direct consequence of the pushout property for the square with corners CL, CG, CG', and CR. In the same way, the existence and uniqueness of rr' can be shown (an algorithmic version of the proof may be found in [Sch94]). $\square$

In the sequel, we often have to deal with triple production applications, where the redex or result for their left- or right-hand side production application is already known in the form of a morphism g. We denote these *restrictions for rewriting* GT into GT' by $GT \sim p(g) \rightsquigarrow GT'$ in the sequel.

Having defined the application of triple productions to graph triples we are now able to model processes which extend related graphs (and their interrelationships) synchronously. But how can we handle the case, where a left-hand side graph is given and we have to construct the missing right-hand side graph including all inter-graph relationships or vice versa? For symmetry reasons, the solution is the same in both directions. Therefore, the construction of *LR-translators* will be discussed in detail and the solution for *RL-translators* may be obtained by simply exchanging the roles of "left"- and "right"-hand side components.

Informally speaking we have to split a triple production p into a pair of triple productions p_L and p_{LR}. p_L is a *left-local triple production* which rewrites the left-hand side graph only. p_{LR} is a *left-to-right translating triple production* which keeps the new left-hand side graph unmodified but adjusts its correspondence and right-hand side graph. Within the following propositions we will show how to split a triple production into a left-local production and a left-to-right transformation. Furthermore, we will prove that the application of a sequence of triple productions is equivalent to the application of the corresponding sequence of left-local productions followed by the sequence of left-to-right transformations.

Proposition 6. A given triple production

$$p := ((LL, LR) \leftarrow lh - (CL, CR) - rh \rightarrow (RL, RR))$$

may be split into the following pair of equivalent triple productions:

$$p_L := ((LL, LR) \leftarrow \epsilon - (\oslash, \oslash) - \epsilon \rightarrow (\oslash, \oslash))$$

is the **left-local production** for p, where $\oslash$ is the empty graph and ϵ is an inclusion of the empty graph $\oslash$ into any graph.

$$p_{LR} := ((LR, LR) \leftarrow lh - (CL, CR) - rh \rightarrow (RL, RR))$$

is the **left-to-right translating production** for p. For these triple productions and any graph triples GT and GT' (as in def. 5), and a morphism $lg' : LR \rightarrow LG'$ the following proposition holds:

$$GT \sim p(lg') \rightsquigarrow GT' \Leftrightarrow \exists HT : GT \sim p_L(lg') \rightsquigarrow HT \wedge HT \sim p_{LR}(lg') \rightsquigarrow GT'.$$

Proof. The following equivalences prove that the vertical sides of the cubes of figure 4 b) and figure 6 imply each other if all production applications use the same morphism lg' to select an image of graph LR in LG' (and thereby of LL in LG):

$$LG \sim (LL, LR) \rightsquigarrow LG' \Leftrightarrow LG \sim (LL, LR) \rightsquigarrow LG' \wedge LG' \sim (LR, LR) \rightsquigarrow LG'.$$
$$CG \sim (CL, CR) \rightsquigarrow CG' \Leftrightarrow CG \sim (CL, CL) \rightsquigarrow CG \wedge CG \sim (CL, CR) \rightsquigarrow CG'$$
$$\Leftrightarrow CG \sim (\oslash, \oslash) \rightsquigarrow CG \quad \wedge CG \sim (CL, CR) \rightsquigarrow CG'.$$
$$RG \sim (RL, RR) \rightsquigarrow RG' \Leftrightarrow RG \sim (RL, RL) \rightsquigarrow RG \wedge RG \sim (RL, RR) \rightsquigarrow RG'$$
$$\Leftrightarrow RG \sim (\oslash, \oslash) \rightsquigarrow RG \quad \wedge RG \sim (RL, RR) \rightsquigarrow RG'.$$

And def. 5 guarantees existence and uniqueness of all horizontal arrows. Furthermore, diagram 4 b) is equivalent to $GT \sim p(lg') \rightsquigarrow GT'$. We can even merge the two rows of cubes in the upper part of diagram 6 to a single row of cubes. Then, the new upper part of diagram 6 is equivalent to $GT \sim p_L(lg') \rightsquigarrow HT$. Finally, the lower part of diagram 6 is equivalent to $HT \sim p_{LR}(lg') \rightsquigarrow GT'$. $\square$

Please note that we used the name lr in figure 6 to denote a morphism from CG to LG as well as its range extension to a morphism from CG to $LG' \supseteq LG$.

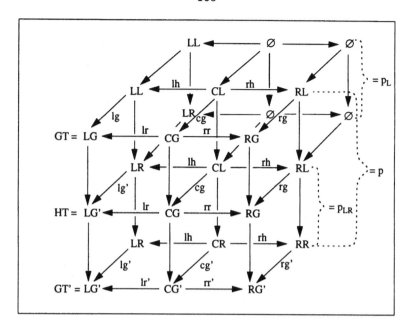

Fig. 6. Splitting of Triple Production Application

Furthermore, all arrows without any label denote inclusions and the domain restrictions of lh and rh from CR to CL have been omitted in order to keep the diagram as legible as possible.

In a similar way the splitting of a triple production into a right-local production followed by a right-to-left translating production may be defined, but we have still to show that we can use these locally equivalent splittings for the definition of graph transformations which create first a left-hand side graph completely and add a corresponding right-hand side graph and the accompanying correspondence graph afterwards or vice versa, i.e. we have to prove:

Proposition 7. Given n triple productions p^1 through p^n and morphisms lg^1 to lg^n, which determine the application results of left-hand side production components of p^1 through p^n, we can prove that

$$p^1(lg^1) \circ \ldots \circ p^n(lg^n) = (p_L^1(lg^1) \circ \ldots \circ p_L^n(lg^n)) \circ (p_{LR}^1(lg^1) \circ \ldots p_{LR}^n(lg^n)).$$

Proof. This follows directly from proposition 3.6 that ensures

$$p^1(lg^1) \circ \ldots \circ p^n(lg^n) = (p_L^1(lg^1) \circ p_{LR}^1(lg^1)) \circ \ldots \circ (p_L^n(lg^n) \circ p_{LR}^n(lg^n))$$

and the fact that

- a triple production $p_L^k := ((LL, LR) \leftarrow \epsilon - (\varnothing, \varnothing) - \epsilon \to (\varnothing, \varnothing))$ modifies left-hand side graph components only and has no requirements with respect to correspondence or right-hand side graphs,
- a simple production (LR, LR) may be applied to a graph LG' without causing any modifications, whenever LG' is the result of applying first a monotonic production (LL, LR) followed by an arbitrary number of different monotonic productions,

- and $p_{LR}^k := ((LR, LR) \leftarrow lh - (CL, CR) - rh \rightarrow (RL, RR))$ keeps its left-hand side graph unmodified.

Therefore, we are allowed to exchange the application order of triple productions freely as long as for any natural numbers $i \leq k$ the application of p_L^i precedes the application of p_L^k for $i \neq k$, the application of p_{LR}^i precedes the application of p_{LR}^k for $i \neq k$, and the application of p_L^i precedes the application of p_{LR}^k. $\square$

In a similar way, we can proof that a sequence of triple productions may be replaced by an equivalent sequence of corresponding right-local and right-to-left translating productions. Therefore, the problem of constructing LR- or RL-translations is solved in principle. The *realization* of such a translation process is divided into two steps:

- The given source graph is analyzed and a sequence of left-local (right-local) productions is computed, which creates the given source graph (if possible).
- Afterwards, the corresponding sequence of LR-translating (RL-translating) productions is applied to the initial (empty) target graph.

For further details concerning the first step of this algorithm the reader is referred to [Sch94]. Furthermore, the topic of correspondence analysis is discussed on an informal level in [Le94].

4 Conclusions

To summarize, *triple graph grammars* are a new formalism for the specification of complex interdependencies between separate and, in general, quite different graph-like data structures. Comparing them with previously suggested formalisms for the specification of data transformations they have the following advantages:

- The underlying data model are directed graphs instead of trees as in [AC89, LM89, No87].
- A specification describes a bidirectional transformation and analysis process in contrast to [AC89, LM89] with their specifications of unidirectional transformation processes.
- Correspondences between source and target elements are not restricted to the case of 1-to-1 relationships between rather similar data structures as in [Pr71, No87, We92].

Note that, due to lack of space, we had to focus our interest on the case of batch-oriented transformation tools, and we had to omit the discussion of *incrementally working* tools and consistency observing analyzers.Furthermore, we have to admit that the presented graph model and underlying conventional graph rewriting approach is rather primitive. There is a significant gap between definitions 1 and 3 of this paper and the graph model and rewriting approach supported by application-oriented graph rewriting language like PROGRES [EL92]. In practice, nodes and edges are labeled and attributed items, and productions (triple productions) are parameterized and contain additional application conditions. For further details concerning incrementally working translation processes and more general forms of rewrite rules, the reader is referred to [Le93, Le94].

Finally, we would like to get rid of the restriction to monotonic productions and to use triple graph grammars with arbitrary graphs as left- and right-hand sides of production components. Unfortunately, this requires the existence of a more or less efficiently working *parsing algorithm* for a very general class of graph grammars. The development of such an algorithm is very difficult and the subject of ongoing research activities (cf. [RS94]).

References

[AC89]   Aho A.V., Ganapathi M., Tijang S.W.K.: *Code Generation Using Tree Matching and Dynamic Programming*, in: acm Transactions on Programming Languages and Systems, vol. 11, no. 4, acm Press (1989), 491–516

[Eh79]   Ehrig H.: *Introduction to the Algebraic Theory of Graph Grammars (a Survey)*, in: Proc. Int. Workshop on Graph-Grammars and Their Application to Computer Science and Biology, LNCS 73, Springer Verlag (1979), 1–69

[EH91]   Ehrig H., Habel A., Kreowski H.J., Parisi-Presicce F.: *From Graph Grammars to High Level Replacement Systems*, in: Proc. 4th Int. Workshop on Graph Grammars and Their Application to Computer Science, LNCS 532, Springer Verlag (1991), 269–291

[EL92]   Engels G., Lewerentz C., Nagl M., Schäfer W., Schürr A.: *Building Integrated Software Development Environments Part I: Tool Specification*, in: acm Transactions on Software Engineering and Methodology, vol. 1, no. 2, acm Press (1992), 135–167

[Le93]   Lefering M.: *Tools to Support Life Cycle Integration*, in: Proc. 6th Software Engineering Environments Conference 1993 (SEE 93), IEEE Computer Society Press (1993), 2–16

[Le94]   Lefering M.: *Development of Incremental Integration Tools Using Formal Specifications*, Technical Report AIB-94-2, RWTH Aachen, Fachgruppe Informatik Germany (1994)

[LM89]   Lipps P., Möncke U., Wilhelm R.: *OPTRAN - A Language/System for the Specification of Program Transformations, System Overview and Experiences*, LNCS 371, Springer Verlag (1989), 52–65

[Na79]   Nagl M.: *Graph-Grammatiken*, Vieweg Press (1979)

[NS91]   Nagl M., Schürr A.: *A Specification Environment for Graph Grammars*, in: Proc. 4th Int. Workshop on Graph Grammars and Their Application to Computer Science, LNCS 532, Springer Verlag (1991), 599–609

[No87]   Normark K.: *Transformations and Abstract Presentations in a Language Development Environment*, Ph.D. Thesis, University of Aarhus, Denmark (1987)

[Pr71]   Pratt T.W.: *Pair Grammars, Graph Languages and String-to-Graph Translations*, in: Journal of Computer and System Sciences, vol 5, Academic Press (1971), 560–595

[RS94]   Rekers J., Schürr A.: *Graph(ical) Parsers and Graph Translators*, appears in: Proc. 5th Int. Workshop on Graph Grammars and Their Application to Computer Science, Williamsburg, Nov. 1994

[Sch94] Schürr A.: *Specification of Graph Translators with Triple Graph Grammars (extended version)*, Technical Report AIB-94-?, RWTH Aachen, Fachgruppe Informatik Germany (1994)

[We92]   Westfechtel B.: *A Graph-Based Approach to the Construction of Tools for the Life Cycle Integration between Software Documents*, in: Proc. 5th International Workshop on Computer-Aided Software Engineering, IEEE Computer Society Press (1992), 2–13

Using Programmed Graph Rewriting for the Formal Specification of a Configuration Management System*

Bernhard Westfechtel

Lehrstuhl für Informatik III, RWTH Aachen, Ahornstr. 55, D-52074 Aachen,
bernhard@i3.informatik.rwth-aachen.de

Abstract. Due to increasing complexity of hardware and software systems, configuration management has been receiving more and more attention in nearly all engineering domains (e.g. electrical, mechanical, and software engineering). This observation has driven us to develop a configuration management model (called CoMa) for managing systems of engineering design documents. The CoMa model integrates composition hierarchies, dependencies, and versions into a coherent framework based on a sparse set of essential configuration management concepts. In order to give a clear and comprehensible specification, the CoMa model is defined in a high-level, multi-paradigm specification language (PROGRES) which combines concepts from various disciplines (database systems, knowledge-based systems, graph rewriting systems, programming languages).

1 Introduction

Due to increasing complexity of hardware and software systems, *configuration management* [2] has been receiving more and more attention in nearly all engineering domains (e.g. electrical, mechanical, and software engineering). Configuration management has been defined as the discipline of controlling the evolution of complex systems [17]. In particular, configuration management is concerned with managing system components, their versions, and their interrelations. In this way, configuration management aids in maintaining system consistency.

In this paper, we present a *configuration management model* (called *CoMa* [4]) for managing systems of engineering design documents. The CoMa model integrates composition hierarchies, dependencies, and versions into a coherent framework based on a sparse set of essential configuration management concepts. The model is generic because it is based on principles common to diverse application domains. On the other hand, it can be adapted to a specific scenario (e.g. in a software engineering scenario, documents such as requirements, software architectures or module implementations have to be managed).

* This work was partially supported by Deutsche Forschungsgemeinschaft under the project title SUKITS ('software and communication structures in technical systems').

The CoMa model is defined in a high-level, multi-paradigm specification language (*PROGRES* [15]) which combines concepts from various disciplines (database systems, knowledge-based systems, graph rewriting systems, programming languages). A PROGRES specification is useful in many respects. First, it precisely defines object and relation types by means of a database schema. Second, it describes complex operations on a high level of abstraction. Third, the specification serves as starting point for developing an efficient implementation.

This paper primarily demonstrates an application of the PROGRES specification language; the CoMa model is explained only briefly (see [21] for a more comprehensive description). Section 2 gives a concise informal survey of the CoMa model. Sections 3-5 are devoted to its formal specification. Section 6 describes its implementation, and section 7 compares it to other work. Section 8 concludes the paper.

2 Survey of the Model

The CoMa model represents the following structures which we consider essential for configuration management: The (recursive) *composition hierarchy* describes which components a (sub-)system consists of. Leaf and non-leaf nodes of the hierarchy are called *documents* and *document groups*, respectively. Components of a document group are related by various kinds of *dependencies*. Both documents and document groups evolve into multipe *versions*.

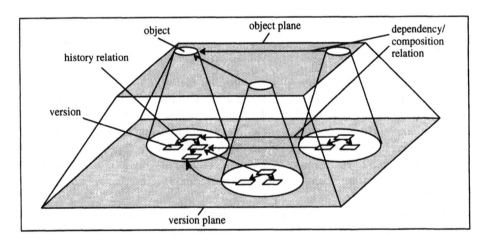

Fig. 1. Object and Version Plane

The CoMa model distinguishes between an *object plane* and a *version plane* (fig. 1). An object represents a set of versions. The version plane refines the object plane: each object is refined into its versions, and each relation between two objects is refined into relations between corresponding versions. Both objects and versions are connected by (hierarchical) composition and (non-hierarchical) dependency relations. Furthermore, history relations between versions of one object represent its evolution.

A database for configuration management is represented by a *CoMa graph* consisting of three kinds of subgraphs which are illustrated in fig. 2. The examples refer to a small program system written in Modula-2. Modula-2 distinguishes between three types of compilation units: program, definition, and implementation modules (denoted by extensions .p, .d, and .i, respectively). Each system contains exactly one program module acting as main program. A module interface is specified in a definition module and realized in the corresponding implementation module. A definition module exports resources (e.g. types or procedures) which may be imported by modules of any type.

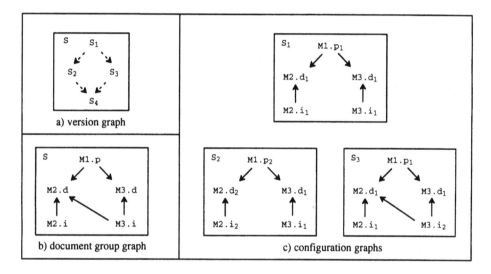

Fig. 2. Subgraphs of CoMa graphs

A *version graph* (fig. 2a) - whose root belongs to the object plane and whose contents is situated in the version plane - represents the evolution history of one object. A history relation from v1 to v2 indicates that v2 is a successor of v1. In general, the evolution history is not required to be linear. Rather, it may contain branches (e.g. a bug fix may have to be applied to an old version having already been delivered to a customer) and merges which combine changes performed on different branches.

With respect to the composition hierarchy, versions are classified into revisions and configurations. A *revision* acts as leaf of the composition hierarchy. The internal structure of a revision (e.g. the declarations and statements a module consists of) is not represented within the CoMa graph; rather, a revision is considered an atomic unit (coarse-grained approach). Inner and root nodes of the composition hierarchy are denoted as *configurations*.

A *configuration graph* - which belongs to the version plane - contains version components and their mutual dependencies. Fig. 2c) shows some examples. S1 (initial configuration of S) contains initial revisions of all components. M1.p1 imports from M2.d1 and M3.d1 which are realized by M2.i1 and M3.i1, respectively. The transition to S2 involves changes to the interface of M2, resulting in new revisions of M2.d, M2.i, and M1.p. On the other branch of the evolution history (i.e. in S3), the body of M3 is modified, yielding a new revision M3.i2 which imports from M2.d1.

A *document group graph* - which belongs to the object plane - represents version-independent structural information. In general, the relation between a document group graph and its versions is constrained by the following condition: For each version component (dependency) contained in a configuration graph, a corresponding object component (dependency) must exist in the document group graph. Loosely speaking, a graph monomorphism must exist for each configuration graph into the corresponding document group graph. Fig. 2b) shows a document group graph for S which satisfies this condition. Note that for almost all elements of S, corresponding elements occur in all configurations of S (the only exception consists in the import from M2.d into M3.i).

3   Schema

After having motivated the CoMa model, we will now turn to its formal specification. We will use the language *PROGRES* [15] which has been developed within the IPSEN project [10]. PROGRES integrates concepts from various disciplines (database systems, knowledge-based systems, graph rewriting systems, programming languages) into one coherent language. A formal semantics for PROGRES based on a logic calculus is described in [14]. In this paper, we will rather be concerned with the application of PROGRES to configuration management problems. In the following sections, we will demonstrate the use of PROGRES by studying a non-trivial, 'real' application ('real' in the sense that a prototypical configuration management system has been built according to the CoMa model).

Let us start with a classical database issue: Each PROGRES specification includes a *schema* which declares types of graph elements. PROGRES is based on attributed graphs consisting of attributed nodes which are connected by binary, directed edges which don't carry attributes. A schema declares node classes and edge types. A *node class* declares attributes, and an *edge type* declares source and target class, as well as its cardinality. Node classes are organized into a (multiple) inheritance hierarchy. A subclass inherits from its superclasses all attributes, and all incoming or outgoing edge types.

Fig. 3 displays a *schema diagram* for the CoMa model. Boxes, dashed and solid arrows denote node classes, inheritance relations, and edge types, respectively. Note that relations which were represented as arrows in fig. 2 are modeled as nodes and adjacent edges (e.g. a history relation is modeled as a HISTORY node and **Predecessor** and **Successor** edges). This solution allows for attaching at-

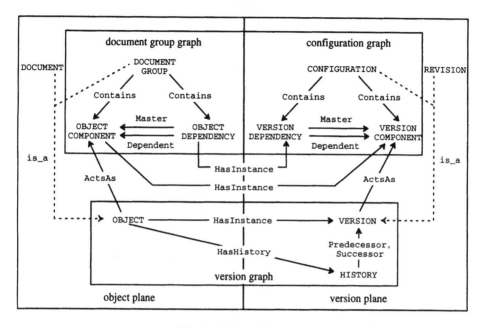

Fig. 3. Schema Diagram

tributes to relations, and establishing relations between relations. In general, relations can be modeled either directly as edges, or as nodes and connecting edges.

Each *subgraph* of the CoMa graph is represented by a root node which is connected to all nodes belonging to this subgraph (e.g. a configuration graph is represented by a CONFIGURATION node which is connected to VERSION_COMPONENT and VERSION_DEPENDENCY nodes by Contains edges). The graph model is constructed such that subgraphs are mutually disjoint.

Apart from HISTORY nodes, each node of the version plane is connected to the corresponding node of the object plane by an (incoming) HasInstance edge. Such a node of the version plane may be regarded as an *instance* of exactly one node of the object plane.

Fig. 4 shows a part of the *textual schema* which refines the overview diagram displayed in fig. 3 by definitions of attributes and cardinalities. Each OBJECT node carries an intrinsic Name attribute which serves as a unique key. MaxVersionNo denotes the number of the next version to be created. Edge type HasInstance connects OBJECT to VERSION nodes. Each object may have any number of versions; conversely, each version is attached to exactly one object (lower and upper bounds of cardinality are enclosed in square brackets). A VERSION node carries a number which identifies it uniquely among all versions of one object, a Stable attribute indicating whether the version is frozen or may be modified, and two date attributes (CreationDate and LastModificationDate). Finally, history relations are represented by HISTORY nodes and Predecessor/Successor edges, and they are connected to OBJECT nodes by incoming HasHistory edges.

```
section VersionGraphs
    node class OBJECT
        intrinsic
            key Name : string;
            MaxVersionNo : integer := 1;
    end;
    edge type HasInstance : OBJECT [1:1] -> VERSION [0:n];
    node class VERSION
        intrinsic
            VersionNo : integer;
            Stable : boolean := false;
            CreationDate : string := CurrentDate;
            LastModificationDate : string := CurrentDate;
    end;
    edge type HasHistory : OBJECT [1:1] -> HISTORY [0:n];
    node class HISTORY end;
    edge type Predecessor : HISTORY [0:n] -> VERSION [1:1];
    edge type Successor : HISTORY [0:n] -> VERSION [1:1];
end;
```

Fig. 4. Textual Schema for Version Graphs

The schema is not powerful enough to express all kinds of consistency constraints. Fig. 5 summarizes some important *structural constraints* which have (at least partially) already been mentioned in passing. These structural constraints have to be preserved by all operations on the CoMa database.

```
1. Each configuration may contain at most one version of a given object
   (version consistency).
2. Versions of one object have to be numbered in a unique way.
3. A version which has a successor must be stable.
4. All components of a stable configuration must be stable, as well.
5. Cycles in history, dependency, and composition relations are not allowed.
6. Each version component (dependency) has to be mapped "monomorphically" to
   the corresponding object component (dependency).
```

Fig. 5. Consistency Constraints

4 Operations

While many data manipulation languages rely on rather low-level operations such as creating/deleting single entities/relationships or modifying single attributes, PROGRES provides *graph rewrite rules* for specifying complex graph transformations in a declarative and graphical way. In PROGRES, it is possible to specify operations formally and precisely on a high level of abstraction; in many other approaches, you either have to content yourself with some informal, imprecise comments, or you have to deal with lengthy, low-level programs. All graph transformations specified in PROGRES are checked for consistency with the schema at specification time. Thus, PROGRES integrates the database world with the world of graph rewriting systems.

A graph rewrite rule (also called production) consists of the following parts: The *header* is composed of an identifier and a list of formal parameters. The *left-*

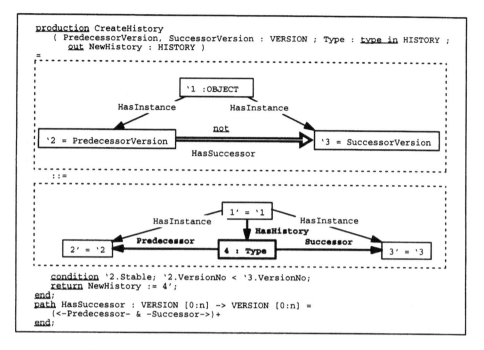

Fig. 6. Graph Rewrite Rule for Creating a History Relation

hand side describes the subgraph to be replaced. The *right-hand side* specifies the subgraph to be inserted. The *condition part* lists conditions on attributes of nodes belonging to the left-hand side. The *transfer part* assigns values to attributes of nodes belonging to the right-hand side. Finally, result parameters receive values in the *return part*.

The graph rewrite rule `CreateHistory` (fig. 6) receives two *node-valued parameters* (`PredecessorVersion` and `SuccessorVersion`) identifying the versions to be connected by a history relation. Furthermore, it is supplied with a *type parameter* indicating the actual type of history relation to be created. Finally, it returns the new history node as an <u>out</u> parameter.

The *left-hand side* contains two `VERSION` nodes which are associated to the same `OBJECT` node by `HasInstance` edges (i.e. both versions must belong to the same object). Nodes '2 and '3 of the left-hand side are identified with parameters `PredecessorVersion` and `SuccessorVersion`, respectively. Thus, the left-hand side need not be searched globally. The double arrow between node '2 and node '3 denotes a path condition. The keyword <u>not</u> means that the rule is applicable only if there is no path of the specified structure.

The *path declaration* is given below the rule declaration. <-.- and -.-> indicate traversal of an edge in negative/positive direction, respectively; & and (.)+ denote concatenation and transitive closure, respectively. Thus, the <u>not</u> condition excludes duplicates of (sequences of) history relations. In general, path conditions are a very powerful and flexible way to specify complex graph patterns.

In the *condition part*, two constraints are checked which were listed in fig. 5: The predecessor version must be stable, and the successor must carry a greater number than the predecessor in order to prevent cycles (constraints 3 and 5 in fig. 5, respectively).

`CreateHistory` is a protective rule, i.e. all nodes and edges of the left-hand side are not affected by its application. Nodes which are replaced identically carry labels of the form `i'='j`. The effect of applying `CreateHistory` consists in creating a new `HISTORY` node (the identifier of which is returned as result parameter) and connecting it to predecessor, successor, and object node, respectively. These insertions are emphasized in bold face on the *right-hand side*.

```
transaction DeleteVersionAndReorganizeHistory
   ( Version : VERSION ; Type : type in HISTORY )
=
     not ( Version is with -ActsAs-> )
   &
   for all PredecessorVersion : VERSION := elem ( Version.HasPredecessor );
           SuccessorVersion : VERSION := elem ( Version.HasSuccessor )
   do
      use NewHistory : HISTORY
      do
         CreateHistory
            ( PredecessorVersion, SuccessorVersion, Type, out NewHistory )
      end
   end
   & DeleteVersion ( Version )
end;
```

Fig. 7. Transaction for a Complex Operation

Although graph rewrite rules may be used to specify rather complex graph transformations, we are convinced that the rule-based specification paradigm alone suffers from severe limitations. PROGRES exceeds the rule-based paradigm by providing *control structures* for the composition of graph rewrite rules [22]. These control structures are similar to those found in procedural programming languages; however, they are designed such that they take atomicity and non-determinism of graph rewrite rules into account.

Fig. 7 shows a *transaction* which makes use of the graph rewrite rule presented in fig. 6. The sample transaction deletes a version and reorganizes the evolution history by connecting all predecessors to all successors. On the top level, its body consists of a *sequence* of statements which are separated by the operator &. The first statement asserts that there is no applied occurrence of the version to be deleted. If this assertion is violated, the sequence fails and leaves the host graph unaffected. Note that each control structure preserves *atomicity* of graph rewrite rules, i.e. in case of failure its execution does not affect the host graph. The next statement consists of a *loop* iterating over all predecessors and successors of the current version (the operator elem is used to iterate through all elements of a set). Each pair is connected by a history relation (the use statement introduces a local variable). Finally, the current version is deleted.

It is beyond the scope of this paper to give a comprehensive description of CoMa operations. Typical examples of *primitive operations* are: create/delete an object; change the name of an object; create/delete/copy a version; create/delete a history relation; ... Based on these primitives, we have also defined *complex operations* which are more convenient to use (e.g. freeze configuration recursively, including all transitive components).

5 Adaptations

So far, the CoMa specification has been independent of a specific application domain. The domain-independent part of the specification is called *generic model*. In the following, we will discuss how the generic model is adapted to a specific domain. The result of such an adaptation is denoted as *concrete model*. As running example, we will use configuration management for Modula-2 programs (see section 2).

The PROGRES type system supports a clear separation between generic model and concrete model. PROGRES has a *stratified type system* which distinguishes between node classes, node types (instances of classes), and nodes (instances of types). Node classes and types are used to specify generic and concrete model, respectively. Nodes are actual instances manipulated at runtime. Due to the stratified type system, types are first order objects which may be supplied as typed parameters, and may be stored as typed values of node attributes.

In order to adapt the generic model, concrete types of documents, document groups, dependencies, etc. have to be defined. Furthermore, operations have to be adapted such that they enforce consistency constraints imposed by the concrete model. For example, in the Modula-2 scenario dependencies from definition modules (dependent) to program modules (master) are prohibited. To achieve this, generic operations are extended such that they access scenario-specific type information. To this end, the schema is enriched by defining *meta attributes* some of which are assigned node types as values. Since meta attributes may only be assigned (i.e. initialized) in node class or node type declarations, their values are type- rather than instance-specific. On the level of the generic model, meta attributes are declared, but not initialized; operations access these attributes. On the level of the concrete model, meaningful values are assigned to meta attributes in node type declarations. In this way, CoMa operations are adapted to a concrete scenario by merely extending the schema and leaving the 'code' of operations unchanged.

To illustrate the approach sketched above, let us describe adaptation of CreateVersionDependency to the *Modula-2 scenario*. Fig. 11 displays an excerpt of the specification of this scenario. The composition of the document group type **Program** is given by an ER-like diagram whose boxes and arrows denote component and dependency types, respectively. To both kinds of elements, cardinalities are attached (cardinalities attached to component types actually refer to the composition relations between document groups and their components). Note that the ER diagram applies to both object and version plane.

Let us describe now how such an *ER diagram* is *transformed* into a *PRO-GRES schema*. We will confine our discussion to version dependencies. In order to perform scenario-specific type checking, meta attributes are attached to nodes of class VERSION_DEPENDENCY (fig. 8). MasterType and DependentType are type-valued attributes (keyword type in) which denote the type of master and dependent component, respectively. Boolean attributes MasterAtMostOnce and DependentAtMostOnce represent upper bounds of cardinality; they are assigned true if a given component may play the master or dependent role at most once, respectively. Lower bounds may be defined analogously.

```
node class VERSION_DEPENDENCY
   meta
      MasterType, DependentType : type in VERSION_COMPONENT;
      MasterAtMostOnce, DependentAtMostOnce : boolean;
end;
   ...
node type ProgModRevisionComponent : VERSION_COMPONENT
   ...
end;
node type DefModRevisionComponent : VERSION_COMPONENT
   ...
end;
   ...
node type ProgDefImportDependency : VERSION_DEPENDENCY
   redef meta
      MasterType := DefModRevisionComponent;
      DependentType := ProgModRevisionComponent;
      MasterAtMostOnce := true;
      DependentAtMostOnce := false;
end;
```

Fig. 8. Meta Attributes for Specifying Scenario-Specific Constraints

Fig. 8 also shows how these *meta attributes* are *redefined* in case of import dependencies between program and definition modules. DependentType and MasterType are defined such that components representing program and definition modules may act as dependents and masters, respectively. DependentAtMostOnce is assigned false because a program module may import from multiple definition modules; conversely, each definition module may act as master at most once because each configuration contains at most one program module component.

Fig. 9 presents the graph rewrite rule CreateVersionDependency which receives master and dependent component, and the dependency type as parameters. The rule has to check a lot of constraints enforced by the generic model: master and dependent must belong to the same configuration, the configuration must not be stable, a corresponding object dependency must exist, and there must not yet exist any dependency between master and dependent. As in fig. 6, changes performed by the rule are emphasized in bold face. Furthermore, all elements of the rule concerning *checks of scenario-specific constraints* are printed in bold face and italics. The dependency must be legal with respect to both master and dependent type (see condition part which accesses values of meta attributes associated to dependency type **Type**), and no cardinality overflow must occur (see restrictions applying to nodes '5 and '6 of the left-hand side). Note that OutgoingDependency (IncomingDependency) is a *restriction* (not specified in the figure), i.e. a unary relation which is fulfilled if a component already participates in a dependency of a given type with upper bound 1.

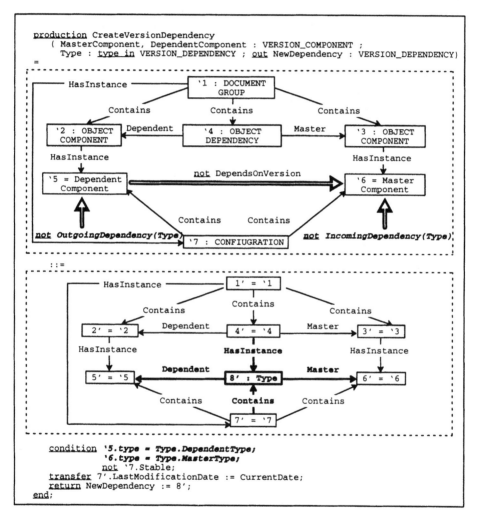

Fig. 9. Graph Rewrite Rule for Creating a Version Dependency

To conclude this section, let us go into another feature of PROGRES which has a nice application in configuration management, namely maintenance of derived information. A significant portion of configuration management research is devoted to *system building*. In large software systems, it is a rather complex task to build (the executable of) a system correctly, i.e. to trigger compile and link steps in the right order, with correct options, and with minimal effort.

In the sequel, we will sketch how these tasks are supported by means of *derived attributes*. In contrast to intrinsic attributes which are assigned values explicitly, derived attributes are calculated from other attributes attached to the same node or to nodes in the neighborhood. Note that an analogous distinction applies to relations (edges and paths, respectively). Neighbor nodes need not belong to the 1-context; rather, they need only be connected via some path of arbitrary length. The PROGRES runtime system evaluates derived attributes in a lazy fashion, i.e. values are calculated on demand only.

To maintain *compiled code* in the Modula-2 scenario, derived attributes are used in the following way (fig. 10): To each module revision, its source code is attached as an intrinsic **Contents** attribute of type **File**. Object code attributes are attached to all component nodes contained in program configurations (class **PROGRAM_COMPONENT**). Note that **ObjectCode** is declared as an optional attribute (cardinality enclosed in square brackets) because its evaluation will not always succeed.

```
node class REVISION is a VERSION
    intrinsic
        Contents : File;
end;
...
node class PROGRAM_COMPONENT is a VERSION_COMPONENT
    intrinsic
        ObjectCode : File [0:1];
end;
...
node type ProgModRevisionComponent : PROGRAM_COMPONENT
    redef derived
        ...
        ObjectCode =
            [MastersCompiled(self) and def (SourceCode(self))
            ?
            CompileProgMod
            ( SourceCode(self), self.Imports.ObjectCode )
            | nil ];
end;
...
transaction Make
    ( Component : PROGRAM_COMPONENT ; out CompiledComponent : File )
=
    CompiledComponent := Component.ObjectCode
end;
```

Fig. 10. Using Derived Attributes to Specify Compilations

For program modules (node type **ProgModRevisionComponent**), a conditional expression (denoted by [. | .]) is given as *evaluation rule*. Since its second alternative evaluates to nil, it will yield a defined value only if the first alternative is selected. This alternative is a guarded expression (denoted by .?.) which is selected when the guard evaluates to true. The guard states that the source of the current component must exist, and that all components on which it depends must have been compiled successfully. In this case, the function **CompileProgMod** is called with two parameters, namely the source code and the set of object codes of all imported components. Within the body of this function, the Modula-2 compiler is called. The function returns <u>nil</u> if compilation fails, and the compiled code otherwise.

After all, it is an easy task to simulate the functionality of the well-known *Make tool* [3]. A call to the function Make triggers all necessary compilations in the correct order with minimal effort and delivers the requested object code, if possible. Linking may be handled in an analogous way (attach attribute Executable to program module components, define attribute evaluation rules, and provide a function MakeExecutable).

6   Implementation

Since PROGRES specifications are executable, the CoMa specification itself may be regarded as a rapid prototype. The CoMa specification covers about 30 pages; it includes all primitive operations and a subset of complex operations offered by the CoMa system (see below). The specification has been developed with the help of the *PROGRES development environment* [11] consisting of editor, analyzer, browser, compiler, and interpreter tools. The PROGRES environment is available as free software.

Starting from the CoMa specification, a configuration management system has been developed within the *SUKITS project* [4] which is dedicated to a posteriori integration of heterogeneous CIM application systems (CAD systems, CAD systems, NC systems, etc.). The *CoMa system* developed in the SUKITS project consists of a schema editor for adapting the system to a specific scenario (a screen dump of which is shown in fig. 11), tools for editing, browsing, and analyzing configuration management data on the instance level, interfaces to CIM application systems to be integrated, and an OSI-based communication system gluing all components together in a heterogeneous environment (multiple types of machines, operating systems, and data management systems). The implementation of the CoMa system comprises more than 60,000 loc written in Modula-2 and C; about 50 % of the code are dedicated to the implementation of operations on the CoMa database. The implementation of the CoMa system was made considerably easier by reuse of components developed within the IPSEN project [10], which is concerned with the construction of integrated software development environments.

7   Relation to Other Work

To the best of our knowledge, the work presented in this paper (and previous work performed by the author [19, 20]) is *unique* in *applying* a specification language based on *programmed graph rewriting* to *configuration management* of engineering design documents. We hope to have convinced the reader that the PROGRES specification language is well-suited to describe and manipulate complex graph structures occurring in this application domain.

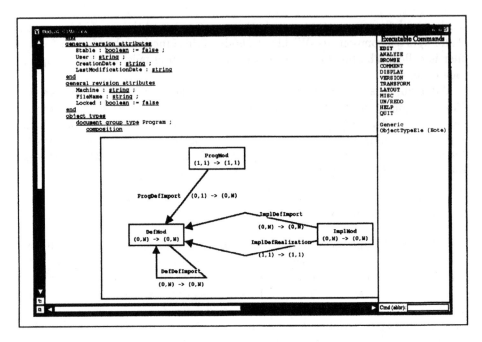

Fig. 11. Screen Dump from the CoMa Schema Editor

Comparing the *CoMa model* and the *functionality* of the CoMa system to other configuration management systems, we observe the following: Traditional configuration management tools such as Make [3], which supports consistent production of derived objects, or SCCS [13] and RCS [16], which both efficiently store revisions of text files, have been very successful. However, each of these tools solves just one configuration management problem, and integration is possible only to a limited extent. Furthermore, they are file-based and do not rely on a formal data model.

More recently, the advantages of using *databases* have been acknowledged [7], and configuration management tools and systems have been developed either on top or as built-in components of database systems. This approach opens the door for configuration management which benefits from the semantic expressiveness of their underlying data models. DAMOKLES [6] and PCTE [12], which both support versions of complex objects based on some extended ER data model, may be quoted as examples. More comprehensive approaches to configuration management, which integrate versioning, composition, and dependencies into a coherent framework, have been realized e.g. in the Nelsis CAD framework [18] and the DEC Cohesion environment [1].

To conclude this section, let us briefly compare PROGRES to other approaches based on *graph rewriting*. Research conducted in this area has mainly been driven by theoretical computer scientists who put a strong emphasis on developing a sound theory. Only recently, this situation has begun to change (e.g. recent extensions to the categorical approach which increase expressive power of graph rewrite rules at the expense of loosing some theoretical properties [8, 9]). On the other hand, the design of PROGRES has strongly been driven by application domains such as software engineering or database systems from its very beginning. By designing a specification language and implementing an integrated development environment, we have moved away from the 'paper and pencil mode' of applying graph rewriting systems. In these respects, the intentions of PROGRES are similar to those followed by Göttler [5] which has designed and implemented tools for editing and executing programmed graph rewriting systems. However, his approach is different in many respects (e.g. type system, control structures, attributes).

8 Conclusion

We have presented a configuration management model (called CoMa) for managing systems of engineering design documents. The CoMa model integrates composition hierarchies, dependencies, and versions into a coherent framework based on a sparse set of essential configuration management concepts. In order to give a clear and comprehensible formal specification, the CoMa model has been defined in the PROGRES language. With the exception of non-determinism and backtracking (which play only a minor role in the CoMa specification), we have exploited more or less the full range of constructs provided by the PROGRES specification language (schema definition, derived attributes and relations, stratified type system, graph rewrite rules, control structures, transactions). We are convinced that PROGRES is superior to other approaches relying on rather low-level data manipulation primitives; furthermore, we believe that we actually need the expressiveness of a multi-paradigm specification language. Our experiences we have gained in configuration management have strongly confirmed these attitudes.

References

1. *A Tool Integration Standard*, ANSI Draft, Digital Equipment Corporation (1990)
2. Feiler, P. *Configuration Management Models in Commercial Environments*, Technical Report, Software Engineering Institute, Carnegie Mellon University, Pittsburgh (1991)
3. Feldman, S.I. *Make - A Program for Maintaining Computer Programs*, Software - Practice and Experience, vol. 9, 255-265 (1979)
4. Große-Wienker, R., Hermanns, O., Menzenbach, D., Pollack, A., Repetzki, S., Schwartz, J., Sonnenschein, K., Westfechtel, B. *Das SUKITS-Projekt: A-posteriori-Integration heterogener CIM-Anwendungssysteme*, Aachener Informatik-Berichte 93-11, RWTH Aachen (1993)
5. Göttler, H. *Graphgrammatiken in der Softwaretechnik - Theorie und Anwendungen*, Informatik Fachberichte 178, Springer Verlag (1987)
6. Gotthard, W. *Datenbanksysteme für Software-Produktionsumgebungen*, Informatik Fachberichte 193, Springer Verlag (1988)

7. Katz, R.H. *Toward a Unified Framework for Version Modeling in Engineering Databases*, ACM Computing Surveys, vol. 22-4, 375-408 (December 1990)

8. Löwe, M. *Extended Algebraic Graph Transformation*, Dissertation, Technical University Berlin (1991)

9. Löwe, M. *Single-Pushout Transformation of Attributed Graphs: A Link between Graph Grammars and Abstract Data Types*, Proc. SEMAGRAPH Symposium 1991, John Wiley & Sons, 359-379 (1991)

10. Nagl, M. *Characterization of the IPSEN Project*, in: Madhavji, Schäfer, Weber (Eds.): Proc. of the 1st Int. Conf. on Systems Development Environments & Factories 1989, Pitman Press, 141-150 (1990)

11. Nagl M., Schürr A. *A Specification Environment for Graph Grammars*, in: Ehrig, Kreowski, Rozenberg (Eds.): Graph Grammars and Their Application to Computer Science, Proc. of the 4th Int. Workshop 1990, LNCS 532, Springer Verlag, 599-609 (1991)

12. Oquendo, F. et al. *Version Management in the PACT Integrated Software Engineering Environment*, Proc. of the 2nd European Software Engineering Conference, 222-242 (1989)

13. Rochkind, M.J. *The Source Code Control System*, IEEE Transactions on Software Engineering, vol. 1-4, 364-370 (December 1975)

14. Schürr, A. *Operationale Spezifikation mit programmierten Graphersetzungen: Formale Definitionen, Anwendungen und Werkzeuge*, Deutscher Universitäts Verlag (1991)

15. Schürr, A. *PROGRES: A VHL-Language Based on Graph Grammars*, in: Ehrig, Kreowski, Rozenberg (Eds.): Graph Grammars and Their Application to Computer Science, Proc. of the 4th Int. Workshop 1990, LNCS 532, Springer Verlag, 641-659 (1991)

16. Tichy, W.F. *RCS - A System for Version Control*, Software : Practice and Experience, vol. 15-7, 637-654 (July 1985)

17. Tichy, W.F. *Tools for Software Configuration Management*, Proc. International Workshop on Software Version and Configuration Control, Teubner Verlag, Stuttgart, 1-20 (1988)

18. van der Wolf, P., Bingley, P., Dewilde, P. *On the Architecture of a CAD Framework: The Nelsis Approach*, Proc. 1st European Design Automation Conference, IEEE Computer Society Press, 29-33 (1990)

19. Westfechtel, B. *Revision Control in an Integrated Software Development Environment*, Proc. of the 2nd International Workshop on Software Configuration Management, ACM SIGSOFT Software Engineering Notes, vol. 14-7, 96-105 (November 1989)

20. Westfechtel, B. *Revisions- und Konsistenzkontrolle in einer integrierten Softwareentwicklungsumgebung*, Informatik Fachberichte 280, Springer Verlag (1991)

21. Westfechtel, B. *A Graph-Based System for Managing Configurations of Engineering Design Documents*, Aachener Informatik-Berichte AIB 94-15, Technical University of Aachen (1994)

22. Zündorf, A., Schürr, A. *Nondeterministic Control Structures for Graph Rewriting Systems*, Proc. 17th Int. Workshop on Graph-Theoretic Concepts in Computer Science WG '91, LNCS 570, Springer Verlag, 48-62 (1991)

Exponential time analysis of confluent and boundary eNCE graph languages [*]

K. Skodinis[1] and E. Wanke[2]

[1] Department of Computer Science, University of Passau, D-94032 Passau, Germany
[2] Institute for Algorithms and Scientific Computing, German National Research Center for Computer Science, D-53757 Sankt Augustin, Germany

Abstract. eNCE (edge label neighborhood controlled) graph grammars belong to the most powerful graph rewriting systems with single-node graphs on the left-hand side of the productions. From an algorithmic point of view, confluent and boundary eNCE graph grammars are the most interesting subclasses of eNCE graph grammars. In confluent eNCE graph grammars, the order in which nonterminal nodes are substituted is irrelevant for the resulting graph. In boundary eNCE graph grammars, nonterminal nodes are never adjacent. In this paper, we show that given a confluent or boundary eNCE graph grammar $\mathcal{G}$, the problem whether the language $L(\mathcal{G})$ defined by $\mathcal{G}$ is empty, is DEXPTIME-complete.

1 Introduction

The theory of graph grammars constitutes a well-motivated and well-developed area within theoretical computer science. The area of graph grammars has grown quite impressively in recent years. This growth was motivated by applications in pattern recognition, software specification and development, VLSI design, data bases, analysis of concurrent systems, and many other areas. For an overview of different application areas and types of graph grammars; see [3, 6, 7, 8].

In this paper, we analyze so-called eNCE *(edge label neighborhood controlled)* graph grammars [1, 9, 14, 15]. In such graph grammars all nodes and additionally all edges of the graphs are associated with labels, in contrast with ordinary NCE graph grammars [13] which do not use edge labels. There are terminal and nonterminal node labels as well as terminal and nonterminal edge labels. Thus, the graphs involved have terminal and nonterminal nodes and edges.

In general, graph grammars are specified by a set of *productions*. A production is usually a tuple (H_1, H_2), where H_1, H_2 are special graphs. Such a production is applied to an occurrence $\hat{H}_1$ of H_1 in some graph by replacing $\hat{H}_1$ with a copy of H_2 with respect to the embedding mechanism of the graph grammar. In eNCE graph grammars, all H_1 are single-node graphs without edges and all H_2 are associated with some embedding information C. The embedding mechanism of eNCE graph grammars allows the substitution to establish terminal as well as nonterminal edges between nodes of H_2 and former neighbors of $\hat{H}_1$. The

---

[*] The work of the first author was supported by the German Research Association (DFG) grant Br-835-3/2

embedding mechanism works as follows: An edge labeled by l_2 is established between a node w from H_2 and a former neighbor v of $\hat{H}_1$, which is labeled by a and was connected with $\hat{H}_1$ by an edge labeled by l_1, if and only if the tuple (a, l_1, l_2, w) is a member of C. Here the nodes in H_2 can be treated separately, whereas all nodes in the former neighborhood of $\hat{H}_1$ with the same label are treated identically. The language of a graph grammar is the set of all terminal labeled graphs derivable from some axiom.

We are interested in the question whether or not a given eNCE graph grammar generates at least one terminal graph. For many types of graph grammars studied in literature, the emptiness problem can be solved in linear time. This holds because in many graph grammars the left hand sides of all productions are atomic (single-node or single-edge graphs) and the embedding mechanism does not produce nonterminal objects. Such graph grammars are for example HR (hyper-edge replacement) systems [10], NLC (node label controlled) graph grammars [11, 12], or NCE (neighborhood controlled embedding) graph grammars [13]. For these types of graph grammars the emptiness problem is equivalent to the emptiness problem of context-free string grammars, and thus P-complete.

The property that makes eNCE graph grammars difficult to analyze is the existence of so-called *blocking edges*. These are nonterminal edges incident to two terminal nodes. Since only nonterminal nodes and their adjacent edges disappear in a substitution step, graphs containing a blocking edge and all graphs derivable from them have at least one nonterminal edge, and thus are not in the language of the grammar. The feature of blocking edges can, for example, be used to generate graph languages consisting of all complete graphs with 2^n nodes for all $n \geq 1$. The emptiness problem for nonblocking eNCE graph grammars can again be solved in linear time.

In this paper we consider *confluent* and *boundary* eNCE graph grammars. In confluent eNCE graph grammars, the order in which the productions are applied is irrelevant for the resulting graph. In boundary eNCE graph grammars nonterminal nodes are never adjacent. Such eNCE graph grammars are always confluent. Boundary eNCE graph grammars are analyzed in [9]. From an algorithmic point of view, confluent and boundary eNCE graph grammars are the most interesting subclasses of eNCE graph grammars; see, for example, [17].

We show that the emptiness problem is DEXPTIME-complete[3] for confluent and boundary eNCE graph grammars. The exponential time algorithm for deciding emptiness of confluent eNCE graph grammars is based on a transformation of a confluent eNCE graph grammar into a nonblocking confluent eNCE graph grammar generating the same language. This transformation only takes exponential time.

The complexity of the emptiness problem always provides a lower bound for deciding non-trivial and monotone [4] properties on the graphs generated by a grammar. For example, the problem whether a C-eNCE or B-eNCE graph

---

[3] DEXPTIME is the class of problems solvable in exponential ($= 2^{\text{poly}(n)}$) time by deterministic Turing machines.

[4] A graph property Π is monotone if $\Pi(G)$ implies $\Pi(H)$ for each subgraph G of H.

grammar can generate a disconnected graph, a graph with a cycle, a non-planar graph, or a non-bipartite graph is also DEXPTIME-hard.

2 Preliminaries

We define eNCE graph grammars in a sequence of definitions; however, we try to be as informal and intuitive as possible.

Definition 1. (Graphs) Let Σ and Γ be two finite sets of symbols (node and edge symbols, respectively). A *node/edge labeled graph* over Σ, Γ is a system $G = (V, E, \Phi)$, where

1. V is a finite set of nodes,
2. $\Phi : V \to \Sigma$ is a *node labeling* that associates with each node u a node label $\Phi(u)$,
3. $E \subseteq \{\{u, v\} \mid u, v \in V, u \neq v\} \times \Gamma$ is a finite set of labeled edges, i.e., each edge $e = (\{u, v\}, l)$ consists of two distinct nodes u, v and an edge label l.

A node or an edge labeled by $a \in \Sigma$ or $l \in \Gamma$ is called an *a-node* or *l-edge*, respectively.

We deal with undirected node and edge labeled graphs over Σ, Γ which we simply call *graphs*.

Next we define the composition of two graphs G and H by replacing a node u from G by H. This composition mechanism is used in derivation steps of eNCE graph grammars.

Definition 2. (Substitutions) An *embedding relation* for a graph H over Σ, Γ is a set

$$D \subseteq \Sigma \times \Gamma \times \Gamma \times V_H,$$

i.e., each tuple (a, l_1, l_2, w) from an embedding relation consists of a node label a, two edge labels l_1, l_2, and a node w of H.

Let G and H be two graphs over Σ, Γ. Let D an embedding relation for H, and let u be a node from G. The graph $G[u/_D H]$ is obtained by replacing node u by H with respect to D as follows:

1. Let J be the disjoint union of G, H without node u and its incident edges.
2. For each edge $(\{v, u\}, l_1)$ from G, insert an edge $(\{v, w\}, l_2)$ to J if and only if $(\Phi(v), l_1, l_2, w) \in D$. The resulting graph is $G[u/_D H]$.

Intuitively speaking, the substitution of a node u in G by a graph H is controlled by the embedding relation as follows: Let $N(u) = \{v \mid (\{u, v\}, l) \in E_G\}$ be the *node neighborhood* of u in G. If (a, l_1, l_2, w) is a tuple in the embedding relation then node w from H will be connected with an a-node v from $N(u)$ by an l_2-edge if and only if the a-node v was previously connected with u by an l_1-edge. Figure 1 shows an example of such a substitution.

We continue with the definition of eNCE graph grammars.

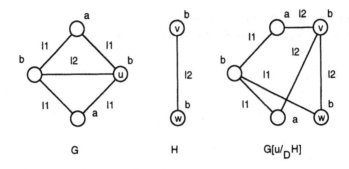

Fig. 1. The figure above shows three graphs G, H, and $G[u/_D H]$, where $D = \{(a, l_1, l_2, v), (b, l_2, l_1, w)\}$.

Definition 3. (eNCE graph grammars) An *eNCE (edge label neighborhood controlled embedding)* graph grammar is a system $\mathcal{G} = (\Sigma, \Delta, \Gamma, \Omega, S, P)$, where

1. Σ, Δ, $\Sigma - \Delta$, Γ, Ω, and $\Gamma - \Omega$ are finite sets of *node labels, terminal node labels, nonterminal node labels, edge labels, terminal edge labels,* and *nonterminal edge labels,* respectively,
2. S is a graph over Σ, Γ, the *axiom* of $\mathcal{G}$, and
3. P is a finite set of *productions*. Each *production* is a triple (A, G, C) where
   (a) A is a nonterminal node label from $\Sigma - \Delta$,
   (b) G is a graph over Σ, Γ,
   (c) C is an embedding relation for G.
   A is called the *left hand side* and (G, C) is called the *right hand side* of the production.

The definition of eNCE graph grammars divides the sets of node labels and edge labels into terminal and nonterminal labels. A node (edge) labeled by a terminal or nonterminal label is called a terminal or nonterminal node (edge), respectively. A graph is called terminal if all its nodes and edges are terminal.

The next definition shows how eNCE graph grammars define sets of graphs by derivations.

Definition 4. (Derivations and languages) Let $\mathcal{G}$ be an eNCE graph grammar and G, H be two graphs over Σ, Γ. We say that G *directly derives* H in $\mathcal{G}$, denoted by

$$G \Longrightarrow_{\mathcal{G}} H$$

if and only if G has some A-node u, $\mathcal{G}$ has a production (A, J, F), and $H = G[u/_F J]$.

We say that G *derives* H in $\mathcal{G}$, if

$$G \overset{*}{\Longrightarrow}_{\mathcal{G}} H,$$

where $\overset{*}{\Longrightarrow}_{\mathcal{G}}$ is the transitive and reflexive closure of $\Longrightarrow_{\mathcal{G}}$.

The *language* $L(\mathcal{G})$ of an eNCE graph grammar $\mathcal{G}$ is the set of all terminal graphs derivable from the axiom of $\mathcal{G}$.

We intend to analyze the complexity of problems with eNCE graph grammars as input. For our complexity analysis, the size of a grammar is just the size of the string that you get when writing down the grammar in the usual way.

3 Confluent eNCE graph grammars

An eNCE graph grammar $\mathcal{G}$ is called *confluent* (C-eNCE) if for each graph G derivable from the axiom of $\mathcal{G}$, all nonterminal nodes u, v in G, and all productions $(\Phi(u), H, D), (\Phi(v), J, F)$ in $\mathcal{G}$:

$$G[u/_D H][v/_F J] = G[v/_F J][u/_D H].$$

In confluent eNCE graph grammars, the order in which the productions are applied is irrelevant for the resulting graph.

Since only nonterminal nodes can be substituted, a nonterminal edge incident to two terminal nodes can never become terminal in any further derivation step. Thus, any further derivation can not produce a terminal graph. Such nonterminal edges that are incident to two terminal nodes are called *blocking edges*. An eNCE graph grammar is called *nonblocking* if all graphs derivable from its axiom have no blocking edges. In the following lines, we show that for each confluent eNCE graph grammar $\mathcal{G}$ it is possible to construct in time $O(2^{poly(size(\mathcal{G}))})$ a nonblocking confluent eNCE graph grammar $\mathcal{G}'$ such that $L(\mathcal{G}) = L(\mathcal{G}')$, where $poly()$ is some polynomial function. Up to now, such a construction was only known for boundary eNCE graph grammars; see [9].

The substitution mechanism used by eNCE graph grammars is unfortunately not associative; however, an associative substitution mechanism for eNCE graph grammars is obtained by a simple extension of the embedding mechanism as for example in [5, Definition 3.1] for directed eNCE graph grammars.

Definition 5. (Extended Substitutions) For a graph G and an embedding relation C for G, we call the pair $\mathbf{G} = (G, C)$ an *extended graph*.

Let $\mathbf{G} = (G, C)$ and $\mathbf{H} = (H, D)$ be two extended graphs and let u be a node of $\mathbf{G}$. The extended graph $\mathbf{J} = (J, F)$, denoted by $\mathbf{G}[u/\mathbf{H}]$, is defined by the graph

$$J := G[u/_D H]$$

and the embedding relation

$$F := \{(a, l_1, l_2, v) \in C \mid v \neq u\}$$
$$\cup \{(a, l_1, l_3, v) \mid (a, l_1, l_2, u) \in C \text{ and } (a, l_2, l_3, v) \in D \text{ for some } l_2\}.$$

Intuitively speaking, the embedding relation F is the embedding relation C in that each tuple (a, l_1, l_2, u) is replaced by all (a, l_1, l_3, v), where (a, l_2, l_3, v) is from D for some l_2. For example, if $(a, l_3, l_1, u), (b, l_1, l_2, u)$ are tuples in some embedding relation for graph G from Figure 1, then the embedding relation F for $G[u/_D H]$ as defined in Figure 1 contains the tuples $(a, l_3, l_2, v), (b, l_1, l_1, w)$.

The extended substitution is associative, i.e., for all extended graphs $\mathbf{G}, \mathbf{H}, \mathbf{J}$, all nodes u from $\mathbf{G}$, and all nodes v from $\mathbf{H}$:

$$\mathbf{G}[u/\mathbf{H}][v/\mathbf{J}] = \mathbf{G}[u/\mathbf{H}[v/\mathbf{J}]].$$

If only the underlying graphs are considered, then there is no fundamental difference between the extended substitution and the usual substitution. If in an eNCE graph grammar $\mathcal{G}$ the axiom is changed into an extended graph, then the new eNCE graph grammar generates the same graphs (with empty embedding relations) using the extended substitution as the original eNCE graph grammar using the usual substitution. Furthermore, the extended eNCE graph grammar is confluent if and only if the usual eNCE graph grammar is confluent.

Using the extended substitution, one can simply show that the confluence property is not restricted to those graphs derivable from the axiom, but carries over, for example, to all graphs on the right hand sides of the productions. (We assume that for each production (A, G, C) there is a graph derivable from the axiom which has a node labeled by A.) A more general statement is shown in the following lemma, and in [5, Lemma 3.2] for directed eNCE graph grammars.

Lemma 6. *Let* $\mathbf{K}$, $\mathbf{G}$, $\mathbf{H}$, $\mathbf{J}$ *be extended graphs,* u, v *be nodes of* $\mathbf{G}$, *and* w *be a node of* $\mathbf{K}$. *If*

$$\mathbf{K}[w/\mathbf{G}][u/\mathbf{H}][v/\mathbf{J}] = \mathbf{K}[w/\mathbf{G}][v/\mathbf{J}][u/\mathbf{H}]$$

then

$$\mathbf{G}[u/\mathbf{H}][v/\mathbf{J}] = \mathbf{G}[v/\mathbf{J}][u/\mathbf{H}].$$

Proof. The embedding relations of $\mathbf{G}[u/\mathbf{H}][v/\mathbf{J}]$ and $\mathbf{G}[v/\mathbf{J}][u/\mathbf{H}]$ are equal by the definition of the extended substitution. The graphs underlying $\mathbf{G}[u/\mathbf{H}][v/\mathbf{J}]$ and $\mathbf{G}[v/\mathbf{J}][u/\mathbf{H}]$ are equal, because they can be obtained by removing all nodes of $\mathbf{K}$ except w (and their incident edges) from $\mathbf{K}[w/\mathbf{G}[u/\mathbf{H}][v/\mathbf{J}]]$ and $\mathbf{K}[w/\mathbf{G}[v/\mathbf{J}][u/\mathbf{H}]]$, which are equal by the assumption and the associativity of the extended substitution. $\square$

The idea of our intended transformation is based on the definition of a finite congruence on extended graphs with respect to the extended substitution mechanism. Similar concepts are considered, for example, in [4, 16, 17] for the hyperedge replacement embedding and the NLC embedding. Here, we restrict the definition of the congruence to extended graphs whose nodes are all terminal.

In the following, we assume that each graph over Σ, Γ is also associated with the sets Δ, Ω of terminal labels. This allows us to talk about terminal nodes and edges, nonterminal nodes and edge, and blocking edges. We call a graph in which all nodes are terminal a *node-terminal graph*.

Definition 7. (Replaceability) Two extended node-terminal graphs $\mathbf{H} = (H, D)$ and $\mathbf{J} = (J, F)$ are called *replaceable*, denoted by $\mathbf{H} \sim \mathbf{J}$, if and only if

$$\mathbf{H} \text{ has a blocking edge } \Leftrightarrow \mathbf{J} \text{ has a blocking edge}$$

and

$$\{(a, l_1, l_2, \Phi_H(u)) \mid (a, l_1, l_2, u) \in D\} = \{(a, l_1, l_2, \Phi_J(u)) \mid (a, l_1, l_2, u) \in F\}.$$

Replaceability is an equivalence relation by definition. There are at most $O(|\Sigma| \cdot |\Gamma|^2 \cdot |\Delta|)$ equivalence classes for all node-terminal extended graphs.

Lemma 8. *Let* $\mathbf{H}, \mathbf{J}, \mathbf{H}', \mathbf{J}'$ *be four node-terminal extended graphs and* $\mathbf{G}$ *be an extended graph with two nonterminal nodes* u, v. *If* $\mathbf{G}[u/\mathbf{H}][v/\mathbf{J}] = \mathbf{G}[v/\mathbf{J}][u/\mathbf{H}]$, $\mathbf{H} \sim \mathbf{H}'$, *and* $\mathbf{J} \sim \mathbf{J}'$ *then*

1. $\mathbf{G}[u/\mathbf{H}'][v/\mathbf{J}'] = \mathbf{G}[v/\mathbf{J}'][u/\mathbf{H}']$ *and*
2. $\mathbf{G}[u/\mathbf{H}][v/\mathbf{J}] \sim \mathbf{G}[u/\mathbf{H}'][v/\mathbf{J}']$.

Proof. 1. The node sets and embedding relations of $\mathbf{G}[u/\mathbf{H}'][v/\mathbf{J}']$ and $\mathbf{G}[v/\mathbf{J}'][v/\mathbf{H}']$ are equal by the definition of the extended substitution. Furthermore, all edges between two nodes from $\mathbf{G}$, two nodes from $\mathbf{H}'$, or two nodes from $\mathbf{J}'$ exist in $\mathbf{G}[u/\mathbf{H}'][v/\mathbf{J}']$ if and only if they exist in $\mathbf{G}[v/\mathbf{J}'][v/\mathbf{H}']$. The same holds for all edges between nodes from $\mathbf{G}$ and $\mathbf{H}'$ and between nodes from $\mathbf{G}$ and $\mathbf{J}'$. Thus, it remains to show that $\mathbf{G}[u/\mathbf{H}'][v/\mathbf{J}']$ has an edge between some node from $\mathbf{H}'$ and some node from $\mathbf{J}'$ if and only if $\mathbf{G}[u/\mathbf{J}'][v/\mathbf{H}']$ has such an edge.

Let w_1' be a node from $\mathbf{H}'$ and w_2' be a node from $\mathbf{J}'$ such that the embedding relations of $\mathbf{H}'$ and $\mathbf{J}'$ have a tuple containing w_1' and w_2', respectively. Since $\mathbf{H} \sim \mathbf{H}'$ and $\mathbf{J} \sim \mathbf{J}'$, it follows that

1. $\mathbf{H}$ has a node w_1 labeled as node w_1',
2. $\mathbf{J}$ has a node w_2 labeled as node w_2'
3. the embedding relation of $\mathbf{H}$ has some tuple (a, l_1, l_2, w_1) if and only if the embedding relation of $\mathbf{H}'$ has some tuple (a, l_1, l_2, w_1') for some a, l_1, l_2, and
4. the embedding relation of $\mathbf{J}$ has some tuple (a, l_1, l_2, w_2) if and only if the embedding relation of $\mathbf{J}'$ has some tuple (a, l_1, l_2, w_2') for some a, l_1, l_2.

That is, $\mathbf{G}[u/\mathbf{H}][v/\mathbf{J}]$ has some l-edge between w_1 and w_2 for some edge label l if and only if $\mathbf{G}[u/\mathbf{H}'][v/\mathbf{J}']$ has some l-edge between w_1' and w_2', and $\mathbf{G}[v/\mathbf{J}][u/\mathbf{H}]$ has some l-edge between w_1 and w_2 for some edge label l if and only $\mathbf{G}[v/\mathbf{J}'][u/\mathbf{H}']$ has some l-edge between w_1' and w_2'. Since

$$\mathbf{G}[u/\mathbf{H}][v/\mathbf{J}] = \mathbf{G}[v/\mathbf{J}][u/\mathbf{H}]$$

the result follows.

2. The fact that $\mathbf{G}[u/\mathbf{H}][v/\mathbf{J}]$ has a blocking edge if and only if $\mathbf{G}[u/\mathbf{H}'][v/\mathbf{J}']$ has a blocking edge follows directly from the argumentation in the first part of this proof.

The fact that the embedding relations are equivalent in the sense of the definition of replaceability follows from the definition of the extended substitution and the replaceability of $\mathbf{H}, \mathbf{H}'$ and $\mathbf{J}, \mathbf{J}'$. $\square$

It is straight forward that Lemma 6 and 8 can be extended to the case in which more than two nonterminal nodes of G are substituted in a confluent way.

Now, we define a so-called reduced graph $red(\mathbf{H})$ for a node-terminal extended graph $\mathbf{H}$ such that, first, $\mathbf{H} \sim red(\mathbf{H})$ and, second, $red(\mathbf{H}) = red(\mathbf{J})$ if and only if $\mathbf{H} \sim \mathbf{J}$ for two node-terminal extended graphs $\mathbf{H}, \mathbf{J}$.

Definition 9. (Reduced graphs) For a node-terminal extended graph $\mathbf{H} = (H, D)$ the *reduced graph* $red(\mathbf{H})$ is the node-terminal extended graph (H', D') defined as follows:

1. Let $X := \{\Phi(u) \mid (a, l_1, l_2, u) \in D\}$.
2. H' has a node u_a labeled by a for each label a in X. H' has two additional nodes w_1 and w_2. If H has a blocking edge, then H' has a blocking edge between w_1 and w_2. (The labels of w_1, w_2, and the blocking edge are arbitrary but fixed.)
3. D' has a tuple $(a, l_1, l_2, u_{\Phi(v)})$ for each tuple $(a, l_1, l_2, v) \in D$.

The reduced graph $red(\mathbf{H})$ of $\mathbf{H}$ can be generated in linear time with respect to the size of $\mathbf{H}$. Each reduced graph has at most $|\Delta| + 2$ nodes, one edge, and an embedding relation with $|\Sigma| \cdot |\Gamma|^2 \cdot |\Delta|$ tuples. Clearly, $\mathbf{H}$ and $red(\mathbf{H})$ are always replaceable and $red(\mathbf{H}) = red(\mathbf{J})$ if and only if $\mathbf{H} \sim \mathbf{J}$.

The next lemma shows that we can determine in exponential time all equivalence classes of all node-terminal extended graphs derivable from the right hand sides of the productions of a confluent eNCE graph grammar.

Lemma 10. *Let $\mathcal{G}$ be a confluent eNCE graph grammar and $p = (A, G, C)$ be a production of $\mathcal{G}$. It is possible to generate in time $O(2^{poly(size(\mathcal{G}))})$ all $red(\mathbf{H})$, where $\mathbf{H}$ is some node-terminal extended graph derivable from (G, C).*

Proof. For production $p = (A, G, C)$, let L_p be the set of all $red(\mathbf{H})$, where $\mathbf{H}$ is some node-terminal extended graph derivable from (G, C). We describe an algorithm which successively generates all sets L_p in exponential time by a bottom-up processing.

1. For all productions $p = (A, G, C)$ do:
   (a) If G is node-terminal then initialize L_p by $\{red((G, C))\}$.
   (b) Otherwise, initialize L_p by the empty set.
2. For all productions $p = (A, G, C)$, where G has nonterminal nodes $u_1, \ldots, u_k$ do:
   (a) Generate all reduced graphs

   $$red((G, C)[u_1/\mathbf{H_1}] \ldots [u_k/\mathbf{H_k}]),$$

   where $\mathbf{H_i}$, $1 \le i \le k$, is from a set $L_{p'}$ and p' contains the label of u_i on its left hand side.
   (b) Insert those of the generated reduced graphs into L_p which are not existing in L_p yet.
3. Repeat step 2 until no set L_p is extended between two iterations of step 2.

The correctness basically follows from Lemma 8 which implies that

$$(G, C)[u_1/\mathbf{H_1}] \ldots [u_k/\mathbf{H_k}] \sim (G, C)[u_1/red(\mathbf{H_1})] \ldots [u_k/red(\mathbf{H_k})].$$

Then the definition of the reduced graphs implies

$$red((G, C)[u_1/\mathbf{H_1}] \ldots [u_k/\mathbf{H_k}]) = red((G, C)[u_1/red(\mathbf{H_1})] \ldots [u_k/red(\mathbf{H_k})]).$$

After the ith iteration of step 2, the set L_p for $p = (A, G, C)$ contains all node-terminal reduced graphs $red(\mathbf{H})$ where $\mathbf{H}$ is derivable from (G, C) with a derivation depth of at most i, i.e., the extended graph $\mathbf{H}$ has a parse tree of height at most i.

The number of reduced graphs in each L_p is bounded by the number of equivalence classes of $\sim$ which is at most $O(2^{|\Sigma| \cdot |\Gamma|^2 \cdot |\Delta|})$. That is, the algorithm halts after $O(2^{|P| \cdot (|\Sigma| \cdot |\Gamma|^2 \cdot |\Delta|)})$ iterations of step 2. In each iteration of step 2, $O(2^{k \cdot (|\Sigma| \cdot |\Gamma|^2 \cdot |\Delta|)})$ graphs of size $O(size(G) + k \cdot (|\Delta| + |\Sigma| \cdot |\Gamma|^2 \cdot |\Delta|))$ are considered. $\square$

Now, we can show the main theorem of this section.

Theorem 11. *For each confluent eNCE graph grammar $\mathcal{G}$, there is a nonblocking confluent eNCE graph grammar $\mathcal{G}'$ such that $L(\mathcal{G}) = L(\mathcal{G}')$ and $size(\mathcal{G}') \in O(2^{poly(size(\mathcal{G}))})$.*

Proof. We number the equivalence classes of $\sim$ by the integers $1, \ldots, N$. This allows us to talk about the rth equivalence class of $\sim$.

Let $A_1, \ldots, A_n$ be the nonterminal node labels of $\mathcal{G}$. Then $\overline{A}, A_{1,r}, \ldots, A_{n,r}$ for $r = 1, \ldots, N$ are the nonterminal node labels of $\mathcal{G}'$. The terminal node labels and all edge labels of $\mathcal{G}'$ are those from $\mathcal{G}$.

If G is a graph with k nonterminal nodes $u_1, \ldots, u_k$ labeled by $A_{i_1}, \ldots, A_{i_k}$, respectively, then $G(u_1/r_1, \ldots, u_k/r_k)$ denotes the graph G in which the k nonterminal nodes $u_1, \ldots, u_k$ are relabeled by $A_{i_1,r_1}, \ldots, A_{i_k,r_k}$, respectively. All other nodes and edges remain unchanged.

Let (A_i, G, C) be a production from $\mathcal{G}$, where G has k nonterminal nodes $u_1, \ldots, u_k$ labeled by $A_{i_1}, \ldots, A_{i_k}$, respectively. Then $\mathcal{G}'$ contains all productions

$$(A_{i,r}, G', C')$$

such that

$$
\begin{aligned}
G' &:= G(u_1/r_1, \ldots, u_k/r_k), \\
C' &:= \{(a, l_1, l_2, v) \in C \mid a \in \Delta\} \\
&\quad \cup \{(A_{i',r'}, l_1, l_2, v) \mid (A_{i'}, l_1, l_2, u) \in C \wedge 1 \leq r' \leq N\},
\end{aligned}
$$

and $(G, C)[u_1/\mathbf{H}_1] \ldots [u_k/\mathbf{H}_k]$ is from the rth equivalence class if $\mathbf{H}_l$, $1 \leq l \leq k$, is from the r_lth equivalence class and derivable in $\mathcal{G}$ from a pair (J, F), where (A_{i_l}, J, F) is a production of $\mathcal{G}$. It follows that a graph H is derivable in $\mathcal{G}$ from the right hand side of a production (A_i, G, C) if and only if H is derivable in $\mathcal{G}'$ from the right hand side of some production $(A_{i,r}, G', C')$ for some r. Additionally, the right hand side of a production $(A_{i,r}, G', C')$ can only derive into a node-terminal extended graph from the rth equivalence class.

Let S be the axiom of G with k nonterminal nodes $u_1, \ldots, u_k$ labeled by $A_{i_1}, \ldots, A_{i_k}$, respectively. Then $\mathcal{G}'$ contains all productions

$$(\overline{A}, S(u_1/r_1, \ldots, u_k/r_k), \emptyset)$$

such that $(S, \emptyset)[u_1/\mathbf{H_1}]\ldots[u_k/\mathbf{H_k}]$ has no blocking edges if $\mathbf{H_l}$, $1 \leq l \leq k$, is from the r_lth equivalence class and derivable in $\mathcal{G}$ from a pair (J, F), where (A_{i_l}, J, F) is a production of $\mathcal{G}$. These productions exactly select the derivation combinations which do not generate node-terminal graphs with blocking edges from the axiom. The axiom of $\mathcal{G}'$ consists of a single node labeled by $\overline{A}$.

The resulting eNCE graph grammar $\mathcal{G}'$ is confluent (by the confluence of $\mathcal{G}$), nonblocking (by the selection of the productions generated from the axiom of G), and generates exactly all terminal graphs which are generated by $\mathcal{G}$. The size of $\mathcal{G}'$ is in $O(2^{poly(size(\mathcal{G}))})$. $\square$

It follows the exponential time decidability of emptiness for confluent eNCE graph grammars.

Corollary 12. *The emptiness problem for C-eNCE graph grammars can be solved in exponential time.*

Proof. Modify the confluent eNCE graph grammar $\mathcal{G}$ into a nonblocking confluent eNCE graph grammar $\mathcal{G}'$ and then check whether the language of $\mathcal{G}'$ is empty. The first step can be done in exponential time, (see Lemma 10 and Theorem 11). The emptiness investigation of $\mathcal{G}'$ can then be done in linear time with respect to the size of $\mathcal{G}'$ which is at most exponential in the size of $\mathcal{G}$. By the construction of $\mathcal{G}'$, it is sufficient to test whether $\mathcal{G}$ has a production with $\overline{A}$ on its left hand side. $\square$

4 Boundary eNCE graph grammars

An eNCE graph grammar $\mathcal{G}$ is called *boundary* (B-eNCE) if nonterminal nodes are never adjacent in the axiom of $\mathcal{G}$ and in all graphs of the productions of $\mathcal{G}$. We show that the emptiness problem is DEXPTIME-hard even for boundary eNCE graph grammars. Boundary graph grammars are always confluent, because there is no interdependence between not adjacent nodes.

The class DEXPTIME is also characterized by alternating Turing machines using polynomially bounded space [2]. An *alternating Turing machine* is a system

$$M = (Q_E, Q_U, \{0, 1\}, \delta, q_S, q_F)$$

in which Q_E is the finite set of *existential states*, Q_U is the finite set of *universal states*, $Q := Q_E \cup Q_U$ is the set of all states,

$$\delta \subseteq Q \times \{0, 1\} \times Q \times \{0, 1\} \times \{left, right\}$$

is the *next move relation*, $q_S \in Q$ is the *start state*, and $q_F \in Q$ is the *final state*. In general, there are more than two tape symbols and more than one final state, but the restriction to two tape symbols $0, 1$ and one final state q_F has no influence to our complexity analysis.

A *composite symbol* is a pair (x, q) in which x is a tape symbol from $\{0, 1\}$ and q is either a state or the symbol $\#$. An *instantaneous description* (ID) for M

of length m is a sequence of m composite symbols. In each ID there is exactly one composite symbol with a state on the right side. The position of this composite symbol in the ID indicates the *head position* of M. The set of all composite symbols is denoted by **CS**. An ID is called *initial* or *final* if it contains the state q_S or q_F, respectively.

Since we consider only polynomially space bounded Turing machines, we assume that each ID has length $m = f(I)$, where $f(I)$ is the polynomial space bound of M for instance I. So the IDs need not to be extended step by step from the input size n up to the maximal length m.

An ID $\beta' = (X'_1, \ldots, X'_m)$ is a *successor ID* of some $\beta = (X_1, \ldots, X_m)$ if and only if β, β' differ in at most two positions $j, j+1$, and

$$
\begin{array}{ll}
(X_j, X_{j+1}) = ((x,q),(y,\#)) & (X_j, X_{j+1}) = ((y,\#),(x,q)) \\
(X'_j, X'_{j+1}) = ((z,\#),(y,r)) \quad \text{or} \quad & (X'_j, X'_{j+1}) = ((y,r),(z,\#)) \\
(q,x,r,z,right) \in \delta & (q,x,r,z,left) \in \delta
\end{array}
$$

An alternating Turing machine M *accepts* an ID β if and only if either β is a final ID, or β contains an existential state and some successor ID of β is accepted by M, or β contains a universal state and all successor IDs of β are accepted by M. M *accepts* an input $I = (I_1 I_2 \cdots I_n) \in \{0,1\}^*$ if and only if M accepts the initial ID

$$
\beta_I = ((I_1, q_S), (I_2, \#), \ldots, (I_n, \#), (0, \#), \ldots, (0, \#))
$$

of length m.

Theorem 13. *The emptiness problem for B-eNCE graph grammars is DEXPTIME-hard.*

Proof. We design a B-eNCE graph grammar $\mathcal{G}$ for an arbitrary alternating polynomially space bounded Turing machine M and an input $I = (I_1 I_2 \ldots I_n)$ for M, such that M accepts I if and only if the language of $\mathcal{G}$ is not empty.

Let m be the polynomial space bound of M. $\mathcal{G}$ has the node labels $a_1, \ldots, a_m, A, b$, where $a_1, \ldots, a_m, b$ are terminal and A is nonterminal. The edge labels are $\bar{l}$ and l_X, where X is a composite symbol. The labels $l_{(0,\#)}, l_{(1,\#)}, l_{(0,q_F)}, l_{(1,q_F)}$ are terminal edge labels. The remaining labels $\bar{l}, l_{(0,q)}, l_{(1,q)}$ for $q \in Q$ and $q \neq q_F$ are nonterminal edge labels.

The axiom of $\mathcal{G}$ has $m+1$ nodes $u_1, \ldots, u_m, v$ labeled by $a_1, \ldots, a_m, A$, respectively, and edges $(\{u_i, v\}, l_{X_i})$ for $i = 1, \ldots, m$, where $\beta_I = (X_1, \ldots, X_m)$ is the initial ID defined by the input I for M.

Let $G = (V, E, \Phi)$ be any graph without $\bar{l}$-edges derivable from the axiom of $\mathcal{G}$. The grammar $\mathcal{G}$ is designed such that each nonterminal A-node u of G has an environment

$$
\{(\Phi(v), l) \mid (\{u, v\}, l) \in E\}
$$

that represents an ID. For example, the environment of the nonterminal A-node v in the axiom of $\mathcal{G}$ represents the initial ID β_I.

We have three types of productions. Productions of the first type change the environment of an A-node with respect to an existential computation step. Productions of the second type change the environment of an A-node with respect to a universal computation step. Productions of the third type change nonterminal A-nodes into terminal b-nodes.

$\mathcal{G}$ has the following productions.

Type 1 productions for a move to the right

For each $j = 1, \ldots, m - 1$ and each $(q, x, r, z, right) \in \delta$, where q is an existential state, there is a production (A, G, C), where G is the graph consisting of a single A-node u and

$$C = \{(a_j, l_{(x,q)}, l_{(z,\#)}, u)\}$$
$$\cup \{(a_j, l_X, \bar{l}, u) \mid X \in \mathbf{CS}, \neq (x, q)\}$$
$$\cup \{(a_{j+1}, l_{(y,\#)}, l_{(y,r)}, u) \mid y \in \{0, 1\}\}$$
$$\cup \{(a_{j+1}, l_X, \bar{l}, u) \mid X \in \{0, 1\} \times Q\}$$
$$\cup \{(a_i, l_X, l_X, u) \mid X \in \mathbf{CS}, 1 \leq i \leq m, i \neq j, i \neq j + 1\}$$
$$\cup \{(a_i, \bar{l}, \bar{l}, u) \mid 1 \leq i \leq m\}.$$

If one of these productions is applied without generating a nonterminal $\bar{l}$-edge, then the environment of the new node u represents the old ID changed with respect to the move $\delta(q, x) \in (r, z, right)$.

Type 1 productions for a move to the left

According to the type 1 productions for a move to the right.

Type 2 productions

For each $j = 1, \ldots, m$, each tape symbol $x \in \{0, 1\}$, and each universal state q, there is a production (A, G, C) defined as follows. The graph G and the embedding relation C is the disjoint union of all graphs and embedding relations, respectively, of productions of type 1, which would exist for j, all $(q, x, r, y, left) \in \delta$, and all $(q, x, r, y, right) \in \delta$, if q would be considered as an existential state.

Type 3 productions for elimination of nonterminal nodes

There is a production (A, G, C), where G is the graph consisting of a single b-node u and
$$C = \{(a_j, l_X, l_X, u) \mid X \in \mathbf{CS}, 1 \leq j \leq m\}$$
$$\cup \{(a_j, \bar{l}, \bar{l}, u) \mid 1 \leq j \leq m\}.$$

A close inspection of the defined productions shows that the language of $\mathcal{G}$ is not empty if and only if M accepts its input I. It is also straightforward that the generation of $\mathcal{G}$ can be done in logarithmic space with respect to the the size of I. $\square$

We have shown by Corollary 12 and Theorem 13 the following result.

Corollary 14. *The emptiness problem for C-eNCE and B-eNCE graph grammars is log-space complete for DEXPTIME.*

5 Acknowledgments

We are grateful for many fruitful suggestions and comments by some anonymous referee of an earlier version of this paper.

References

1. F.J. Brandenburg. On partially ordered graph grammars. In [7], volume 291 of *LNCS*, pages 99–111, 1987.
2. A.K. Chandra, D.C. Kozen, and L.J. Stockmeyer. Alternation. *Journal of the ACM*, 28:114–133, 1981.
3. V. Claus, H. Ehrig, and G. Rozenberg. *Proceedings of Graph-Grammars and Their Application to Computer Science '78*, volume 73 of *LNCS*. Springer-Verlag, 1979.
4. B. Courcelle. The monadic second-order logic of graphs I: Recognizable sets of finite graphs. *Information and Computation*, 85:12–75, 1990.
5. B. Courcelle, J Engelfriet, and G. Rozenberg. Handle-rewriting hypergraph grammars. *Journal of Computer and System Sciences*, 46:218–270, 1993.
6. H. Ehrig, H.J. Kreowski, and G. Rozenberg. *Proceedings of Graph-Grammars and Their Application to Computer Science '90*, volume 532 of *LNCS*. Springer-Verlag, 1991.
7. H. Ehrig, M. Nagl, A. Rosenfeld, and G. Rozenberg. *Proceedings of Graph-Grammars and Their Application to Computer Science '86*, volume 291 of *LNCS*. Springer-Verlag, 1987.
8. H. Ehrig, M. Nagl, and G. Rozenberg. *Proceedings of Graph-Grammars and Their Application to Computer Science '82*, volume 153 of *LNCS*. Springer-Verlag, 1983.
9. J. Engelfriet, G. Leih, and E. Welzl. Boundary graph grammars with dynamic edge relabeling. *Journal of Computer and System Sciences*, 40:307–345, 1990.
10. A. Habel. *Hyperedge Replacement: Grammars and Languages*, volume 643 of *LNCS*. Springer-Verlag, 1992.
11. D. Janssens and G. Rozenberg. On the structure of node label controlled graph languages. *Information Science*, 20:191–216, 1980.
12. D. Janssens and G. Rozenberg. Restrictions, extensions, and variations of NLC grammars. *Information Science*, 20:217–244, 1980.
13. D. Janssens and G. Rozenberg. Graph grammars with neighbourhood-controlled embedding. *Theoretical Computer Science*, 21:55–74, 1982.
14. M. Kaul. Syntaxanalyse von Graphen bei Präzedenz-Graphgrammatiken. Dissertation, Universität Osnabrück, Osnabrück, Germany, 1985.
15. M. Kaul. Practical applications of precedence graph grammars. In [7], volume 291 of *LNCS*, pages 326–342, 1987.
16. T. Lengauer and E. Wanke. Efficient analysis of graph properties on context-free graph languages. *Journal of the ACM*, 40(2):368–393, 1993.
17. E. Wanke. Algorithms for graph problems on BNLC structured graphs. *Information and Computation*, 94(1):93–122, 1991.

Time-Optimal Tree Computations on Sparse Meshes [*]

D. Bhagavathi[1], V. Bokka[2], H. Gurla[2], S. Olariu[2], J. L. Schwing[2]

[1] Department of Computer Science, Southern Illinois University, Edwardsville,
IL 62026
[2] Department of Computer Science, Old Dominion University, Norfolk, VA 23529

Abstract. The main goal of this work is to fathom the suitability of the mesh with multiple broadcasting architecture (MMB) for some tree-related computations. We view our contribution at two levels: on the one hand we exhibit time lower bounds for a number of tree-related problems both on the CREW-PRAM and on the MMB. On the other hand, we show that these lower bounds are tight by exhibiting time-optimal tree algorithms on the MMB. Specifically, we show that the task of encoding and/or decoding n-node binary and ordered trees cannot be solved faster than $\Omega(\log n)$ time even if the MMB has an infinite number of processors. We then go on to show that this lower bound is tight. We also show that the task of reconstructing n-node binary trees from their traversals can be performed in O(1) time on the same architecture. Our algorithms rely on novel time-optimal algorithms on sequences of parentheses that we also develop.

Keywords: meshes with multiple broadcasting, binary trees, ordered trees, encoding, decoding, traversals, tree reconstruction, parentheses algorithms.

1 Introduction

Due to its simple and regular interconnection topology the mesh is particularly well suited for solving various problems in image processing, pattern recognition, graph theory, and computer graphics. At the same time, its large diameter renders the mesh less effective in computing with data spread over processing elements far apart. To overcome this problem, the mesh architecture has been enhanced by various types of bus systems. Early solutions involving the addition of one or more global buses, shared by all processors, have been implemented on a number of massively parallel machines. Recently, a more powerful architecture, referred to as mesh with multiple broadcasting (MMB, for short), has been obtained by adding one bus to every row and to every column of the mesh [7, 14]. The MMB has been implemented in VLSI and is used in the DAP family of computers [14].

An MMB of size $M \times N$ consists of MN synchronous processors positioned on a rectangular array overlaid with a bus system. The processors are connected to their nearest neighbors and are assumed to know their own coordinates within

---

[*] Work supported by NASA grant NCC1-99

the mesh. In addition, in every row of the mesh the processors are connected to a horizontal bus; similarly, in every column the processors are connected to a vertical bus as illustrated in Figure 1.

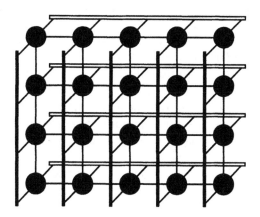

Fig. 1. A mesh with multiple broadcasting of size 4×5

Processor $P(i, j)$ is located in row i and column j $(1 \leq i \leq M; 1 \leq j \leq N)$, with $P(1, 1)$ in the north-west corner of the mesh. Every processor has a constant number of registers of size $O(\log MN)$; in unit time, the processors perform an arithmetic or boolean operation, communicate with one of their neighbors using a local connection, broadcast a value on a bus, or read a value from a specified bus. All these operations involve handling at most $O(\log MN)$ bits of information. For practical reasons, only one processor is allowed to broadcast on a given bus at any one time. By contrast, all the processors on the bus can simultaneously read the value being broadcast. In accord with other researchers [7, 10, 14], we assume that communications along buses take $O(1)$ time. Although inexact, recent experiments with the DAP and the YUPPIE multiprocessor array system, seem to indicate that this is a reasonable working hypothesis [10, 14]. Being of theoretical interest as well as commercially available, the MMB has attracted a great deal of attention [1, 2, 3, 7, 8, 12, 14].

A PRAM [5] consists of synchronous processors, all having unit-time access to a shared memory. In the CREW-PRAM, a memory location can be simultaneously accessed in reading but not in writing. From a theoretical point of view, an MMB can be perceived as a restricted version of the CREW-PRAM machine: the buses are nothing more than *oblivious* concurrent read-exclusive write registers with the access restricted to certain sets of processors. Indeed, in the presence of p CREW-PRAM processors, groups of $\sqrt{p}$ of these have concurrent read access to a register whose value is available for one time unit, after which it is lost. Given that the MMB is, in this sense, *weaker* than the CREW-PRAM, it is very often quite a challenge to design algorithms in this model that match the performance of their CREW-PRAM counterparts. Typically, for the same running time, the MMB uses more processors.

The main goal of this work is to fathom the suitability of the MMB architecture for some tree-related computations. Our contribution is to show tight time lower bounds and to provide time-optimal tree algorithms on the MMB architecture. Specifically, we show that the following tasks can be solved in $\Theta(\log n)$ time on an MMB of size $n \times n$:

- Encode an n-node binary tree into a $2n$-bitstring;
- Encode an n-node ordered tree into a $2n$-bitstring;
- Recover an n-node binary tree from its $2n$-bit encoding;
- Recover an n-node ordered tree from its $2n$-bit encoding.

We also show that the task of reconstructing an n-node binary tree from its preorder and inorder traversals can be performed in $O(1)$ time on the same platform. Our algorithms rely on novel time-optimal algorithms involving sequences of parentheses, that we also develop.

The remainder of the paper is organized as follows. Section 2 presents our lower bound arguments. Section 3 discusses a number of fundamental results that are instrumental in our algorithms. Section 4 discusses the details of our parentheses algorithms. Section 5 presents time-optimal algorithms for encoding and decoding binary and ordered trees. Section 6 addresses the problem or reconstructing binary trees from their traversals. Finally, Section 7 offers concluding remarks and open problems.

2 Lower Bounds

The purpose of this section is to provide lower bounds for a number of fundamental problems that establish the optimality of our algorithms. Our lower bounds will be stated first in the CREW-PRAM and then extended to the MMB using a recent result of Lin *et al.* [9]. All our arguments for the CREW-PRAM rely directly or indirectly on the following fundamental result of Cook *et al.* [4].

Proposition 2.1. [4] The task of computing the logical OR of n bits has a time lower bound of $\Omega(\log n)$ on the CREW-PRAM, regardless of the number of processors and memory cells used. □

LEFTMOST ONE: Given a sequence of n bits, find the position of the leftmost 1.

It is a trivial observation that OR reduces to LEFTMOST ONE in $O(1)$ time. Therefore, Proposition 2.1 implies the following result.

Corollary 2.2. LEFTMOST ONE has a time lower bound of $\Omega(\log n)$ on the CREW-PRAM, regardless of the number of processors and memory cells used. □

To obtain lower bounds for the problems of interest to us on the MMB, we rely on the following recent result of Lin *et al.* [9].

Proposition 2.3. [9] Any computation that takes $O(t(n))$ computational steps on an n-processor MMB can be performed in $O(t(n))$ computational steps on an n-processor CREW-PRAM with $O(n)$ extra memory. □

It is important to note that Proposition 2.3 guarantees that if $T_M(n)$ is the execution time of an algorithm for solving a given problem on an n-processor MMB, then there exists a CREW-PRAM algorithm to solve the same problem in $T_P(n) = T_M(n)$ time using n processors and $O(n)$ extra memory. In other words, "too fast" an algorithm on the MMB implies "too fast" an algorithm for the CREW-PRAM.

A sequence $\sigma = x_1 x_2 \ldots x_n$ of parentheses is said to be *well-formed* if it contains the same number of left and right parentheses and in every prefix of σ the number of right parentheses does not exceed the number of left parentheses. Next we define the classic parentheses matching problem.

MATCHING: Given a well-formed sequence $\sigma = x_1 x_2 \ldots x_n$ of parentheses, for each parenthesis in σ, find its match.

Lemma 2.4. MATCHING has a time lower bound of $\Omega(\log n)$ on the CREW-PRAM, regardless of the number of processors and memory cells used.

Proof. We reduce OR to MATCHING. For this purpose, let the input to OR consist of n bits $b_1, b_2, \ldots, b_n$. We convert this input to a sequence $c_0 c_1 c_2 \ldots c_{2n+1}$ of parentheses by writing $c_0 = $ '(' and $c_{2n+1} = $ ')', and by setting for all j ($1 \le j \le n$):

- $c_{2j-1} = $ '(' and $c_{2j} = $ ')', whenever $b_j = 0$;
- $c_{2j-1} = $ ')' and $c_{2j} = $ '(', whenever $b_j = 1$.

An easy inductive argument on the number of 1's in $b_1, b_2, \ldots, b_n$ shows that the sequence $c_0 c_1 c_2 \ldots c_{2n+1}$ is always well-formed. Furthermore, the matching pair of c_0 is c_{2n+1} if and only if the answer to the OR problem is 0. The conclusion follows by Proposition 2.1. $\square$

One is often interested in the solution of the following problem.

ENCLOSING PAIR: Given a well-formed sequence $\sigma = x_1 x_2 \ldots x_n$ of parentheses, for every matching pair of parentheses in σ, find the closest pair of parentheses that encloses it.

Lemma 2.5. ENCLOSING PAIR has a time lower bound of $\Omega(\log n)$ on the CREW-PRAM, regardless of the number of processors and memory cells used.

Proof. We reduce LEFTMOST ONE to ENCLOSING PAIR. Assume that the input to LEFTMOST ONE is $b_1, b_2, \ldots, b_n$. Construct a sequence $c_1 c_2 \ldots c_{4n+2}$ of parentheses as follows:

- $c_{2n+1} = $ '('; $c_{2n+2} = $ ')';

furthermore, for all j ($1 \le j \le n$) set

- $c_{2n-2j+1} = $ '('; $c_{2n-2j+2} = $ ')'; $c_{2n+2j+1} = $ '('; $c_{2n+2j+2} = $ ')', whenever $b_j = 0$;
- $c_{2n-2j+1} = $ '('; $c_{2n-2j+2} = $ '('; $c_{2n+2j+1} = $ ')'; $c_{2n+2j+2} = $ ')', whenever $b_j = 1$.

Our construction guarantees that the sequence is well-formed and that every parenthesis knows its match; in particular, c_{2n+1} and c_{2n+2} are a matching pair. Furthermore, the closest enclosing parentheses for the pair (c_{2n+1}, c_{2n+2}) is $(c_{2n-2k+2}, c_{2n+2k+1})$, if and only if k is the position of the leftmost 1 in $b_1, b_2, \ldots, b_n$. Now the conclusion follows by Corollary 2.2. $\square$

A binary tree T is either empty or consists of a root and two disjoint binary trees, called the left subtree, T_L and the right subtree, T_R. For later reference, we assume that every node in a binary tree maintains pointers to its left and right children. In many contexts, it is desirable to encode the shape of T as succinctly as possible. In this paper, we are interested in one such encoding scheme recursively defined as follows[3]:

$$\sigma(T) = \begin{cases} \epsilon & \text{if } T \text{ is empty;} \\ 1\sigma(T_L)0\sigma(T_R) & \text{otherwise.} \end{cases} \tag{1}$$

Note that under (1) an arbitrary n-node binary tree is encoded into $2n$ bits.

BINARY TREE ENCODING: Given an n-node binary tree, find its encoding.

Lemma 2.6. BINARY TREE ENCODING has a time lower bound of $\Omega(\log n)$ on the CREW-PRAM, regardless of the number of processors and memory cells used.

Proof. We reduce OR to BINARY TREE ENCODING. Assume that $b_1, \ldots, b_n$ is an arbitrary input to OR. We assume that b_n is 0 for otherwise the answer is trivial. Convert this bit sequence to an n-node binary tree T with nodes $1, 2, 3, \ldots, n$. Specifically, we associate with every bit b_j $(1 \leq j \leq n)$ the node j of T, such that:

- 1 is the root of T;
- for every i $(1 \leq i \leq n - 1)$, node $i + 1$ is the unique child of i. Moreover, $i + 1$ is the left child of i if $b_i = 1$ and the right child otherwise.

Clearly the construction of T takes O(1) time. Let $\sigma(T) = c_1 c_2 \ldots c_{2n}$ be the $2n$-bit encoding of T. We note that $c_{2n-1} = 1$ if and only if the answer to OR is 0. Thus, once the encoding is available, one can determine in O(1) time the answer to OR. Now the conclusion follows by Proposition 2.1. $\square$

The converse operation of recovering a binary tree from its encoding is of interest in a number of practical applications. We state the problem as follows.

BINARY TREE DECODING: Recover an n-node binary tree from its encoding.

Lemma 2.7. BINARY TREE DECODING has a time lower bound of $\Omega(\log n)$ on the CREW-PRAM, regardless of the number of processors and memory cells used.

Proof. We reduce OR to BINARY TREE DECODING. Let $b_1, b_2, \ldots, b_n$ be an arbitrary input to OR. First, if $b_1 = 1$, then the answer to OR is 1. We may, therefore, assume that $b_1 = 0$. Construct a well-formed sequence of parentheses, $c_0 c_1 \ldots c_{2n+1}$ as described below:

- $c_0 =$ '(' ; $c_{2n+1} =$ ')'.
- $c_{2i-1} =$ '('; $c_{2i} =$ ')', whenever $b_i = 0$;
- $c_{2i-1} =$ ')' and $c_{2i} =$ '(', whenever $b_i = 1$ and $b_{i-1} = 0$;
- $c_{2i-1} =$ '(' and $c_{2i} =$ ')', whenever $b_i = 1$ and $b_{i-1} = 1$.

It is easy to verify that the resulting sequence is well-formed and so, interpreting every '(' as a 1 and every ')' as a 0, $c_0 c_1 \ldots c_{2n+1}$ is the encoding of a binary tree T with $n + 1$ nodes. Notice that $\text{root}(T)$ has two children if and only if the OR of the input sequence is 1. Therefore, once the decoding is available,

---

[3] This scheme is similar to the one reported in [15].

one can solve the OR problem in $O(1)$ time. Now Proposition 2.1 implies that any algorithm that performs the decoding must take $\Omega(\log n)$ time. $\square$

An ordered tree T is either empty or it contains a root and disjoint ordered subtrees, $T_1, T_2, \ldots, T_k$. For later reference, we assume that ordered trees are specified by parent pointers. The encoding $\sigma(T)$ of T, is defined as follows:

$$\sigma(T) = \begin{cases} \epsilon & \text{if } T \text{ is empty} \\ 1\sigma(T_1)\sigma(T_2)\ldots\sigma(T_k)0 & \text{otherwise.} \end{cases} \tag{2}$$

Note that the encoding of an n-node ordered tree is a sequence of $2n$ bits.

ORDERED TREE ENCODING: Given an ordered tree, find its encoding.

Lemma 2.8. On the CREW-PRAM, ORDERED TREE ENCODING has a time lower bound of $\Omega(\log n)$, regardless of the number of processors used.

Proof. We reduce OR to ORDERED TREE ENCODING. Let $b_1, b_2, \ldots, b_n$ be an arbitrary input to OR. We add two bits $b_0 = 1$ and $b_{n+1} = 0$. The new sequence $b_0, b_1, \ldots, b_{n+1}$ is converted to an ordered tree T on nodes $\{0, 1, \ldots, n+1\}$ as follows:

• node 0 is the root;
• for all i $(1 \leq i \leq n+1)$, the parent of node i is 0 if $b_i = 0$ and $n+1$ otherwise.

Let $c_1 c_2 \ldots c_{2n+4}$ be the $2(n+2)$-bit encoding of T. Observe that $c_{2n+2} = 1$ if and only if the OR of the input sequence is 0, and so the answer to OR can be obtained in $O(1)$ time, once the encoding is available. Therefore, by Proposition 2.1, the encoding algorithm must take $\Omega(\log n)$ time. $\square$

The converse problem is stated as follows.

ORDERED TREE DECODING: Recover an ordered tree from its encoding.

Lemma 2.9. On the CREW-PRAM, ORDERED TREE DECODING has a time lower bound of $\Omega(\log n)$, regardless of the number of processors used.

Proof. We reduce ENCLOSING PAIR to ORDERED TREE DECODING. Let the input to ENCLOSING PAIR be $s_1 s_2 \ldots s_{2n}$. Augment this sequence with s_0 = '(' and s_{2n+1} = ')'. Thus, interpreting every '(' as a 1 and every ')' as a 0 we obtain the valid encoding of some ordered tree T under (2). Now, consider any algorithm that correctly recovers T from the encoding above. It is easy to see that the setting of parent pointers gives exactly the solution to the ENCLOSING PAIR problem for the augmented sequence. The conclusion follows from Lemma 2.5. $\square$

Now Proposition 2.3 together with Lemmas 2.4, 2.5, 2.6, 2.7, 2.8, and 2.9 imply the following result.

Theorem 2.10. MATCHING, ENCLOSING PAIR, BINARY TREE ENCODING, BINARY TREE DECODING, ORDERED TREE ENCODING, and ORDERED TREE DECODING have a lower bound of $\Omega(\log n)$ on an MMB of size $n \times n$. $\square$

3 Basics

The purpose of this section is to review a number of fundamental results for the MMB that will be instrumental in the design of our algorithms.

The problem of list ranking is to determine the rank of every element in a given list, stored as an unordered array, that is, the number of elements following it in the list. Recently, Olariu *et al.* [12] have proposed a time-optimal algorithm for list ranking on MMB's.

Proposition 3.1. [12] The task of ranking an n-element linked list stored in one row of an MMB of size $n \times n$ can be performed in $O(\log n)$ time. Furthermore, this is time-optimal. $\square$

The All Nearest Smaller Values problem (ANSV, for short) is formulated as follows: given a sequence of n real numbers $a_1, a_2, \ldots, a_n$, for each a_i $(1 \leq i \leq n)$, find the nearest element to its left and the nearest element to its right. Recently, Olariu *et al.* [12] have proposed a time-optimal algorithm for the ANSV problem.

Proposition 3.2. [11] Any instance of size n of the ANSV problem can be solved in $O(\log n)$ time on an MMB of size $n \times n$. Furthermore, this is time-optimal. $\square$

The *prefix sums* problem is a key ingredient in many parallel algorithms and is stated as follows: given a sequence $a_1, a_2, \ldots, a_n$ of items, compute all the sums of the form $a_1, a_1 + a_2, a_1 + a_2 + a_3, \ldots, a_1 + a_2 + \cdots + a_n$.

Proposition 3.3. [7, 12] The prefix sums (also maxima or minima) of a sequence of n real numbers stored in one row of an MMB of size $n \times n$ can be computed in $O(\log n)$ time. Furthermore, this is time-optimal. $\square$

Merging two sorted sequences is one of the fundamental operations in computer science. Recently, Olariu *et al.* [12] have proposed a constant time algorithm to merge two sorted sequences of total length n stored in one row of an MMB of size $n \times n$.

Proposition 3.4. [12] Let $S_1 = (a_1, a_2, \ldots, a_r)$ and $S_2 = (b_1, b_2, \ldots, b_s)$, with $r + s = n$, be sorted sequences stored in one row of an MMB of size $n \times n$, with $P(1, i)$ holding a_i $(1 \leq i \leq r)$ and $P(1, r + i)$ holding b_i $(1 \leq i \leq s)$. S_1 and S_2 can be merged in $O(1)$ time. $\square$

Recently, the simple merging algorithm of Proposition 3.4 was used to derive a time-optimal sorting algorithm for MMB's [12].

Proposition 3.5. [12] An n-element sequence of items from a totally ordered universe stored one item per processor in the first row of an MMB of size $n \times n$ can be sorted in $O(\log n)$ time. Furthermore, this is time-optimal. $\square$

4 Time-Optimal Parentheses Algorithms

The purpose of this section is to present two time-optimal algorithms involving sequences of parentheses on an MMB of size $n \times n$. In addition to being interesting in their own right, these algorithms are instrumental in our subsequent tree algorithms.

Consider a sequence $\sigma = x_1 x_2 \ldots x_n$ of parentheses stored one item per processor in the first row of an MMB of size $n \times n$, with x_k $(1 \leq k \leq n)$ stored by $P(1, k)$. Assuming that the sequence is well-formed, we present an algorithm to find all the matching pairs. The idea is as follows. First, we compute a sequence $w_1, w_2, \ldots, w_n$ obtained from σ by setting $w_1 = 0$ and by defining w_k $(2 \leq k \leq n)$

as follows:

$$w_k = \begin{cases} 1 & \text{if both } x_{k-1} \text{ and } x_k \text{ are left parentheses;} \\ -1 & \text{if both } x_{k-1} \text{ and } x_k \text{ are right parentheses;} \\ 0 & \text{otherwise.} \end{cases}$$

We now compute the prefix sums of $w_1, w_2, \ldots, w_n$ and let the result be $e_1, e_2, \ldots, e_n$. By Proposition 3.3, this operation is performed in $O(\log n)$ time. It is easy to see that left and right parentheses x_i and x_j are a matching pair if and only if x_j is the first right parenthesis to the right of x_i for which $e_i = e_j$.

Further, with each parenthesis x_k $(1 \le k \le n)$, we associate the tuple (e_k, k). On the set of these tuples we define a binary relation $\prec$ by setting

$$(e_i, i) \prec (e_j, j) \text{ whenever } (e_i < e_j) \text{ or } [(e_i = e_j) \text{ and } (i < j)].$$

It is an easy exercise to show that $\prec$ is a linear order. Now, sort the sequence $(e_1, 1), \ldots, (e_n, n)$ in increasing order of $\prec$. By Proposition 3.5, sorting the ordered pairs can be done in $O(\log n)$ time. The key observation is that, as a result of sorting, the matching pairs occur next to one another. Consequently, we have the following.

Theorem 4.1. Given a well-formed sequence of n parentheses as input, all matching pairs can be found in $O(\log n)$ time on an MMB of size $n \times n$. Furthermore, this is time-optimal. $\square$

Next, we are interested in a time-optimal solution to the ENCLOSING PAIR problem stated in Section 2. Consider a well-formed sequence $\sigma = x_1 x_2 \ldots x_n$ of parentheses, stored one item per processor in the first row of the mesh. The details of the algorithm follow.

Step 1. Find the match of every parenthesis in σ; every processor $P(1, i)$ stores in a local variable the position j of the match x_j of x_i.

Step 2. Solve the corresponding instance of the ANSV problem.

It is not hard to see that at the end of Step 2 every processor knows the identity of the closest enclosing pair. By Proposition 3.2 and Theorem 4.1, the running time of this simple algorithm is bounded by $O(\log n)$. By Theorem 2.10, this is the best possible on this architecture. Thus, we have proved the following result.

Theorem 4.2. Given a well-formed sequence of n parentheses stored one item per processor in the first row of an MMB of size $n \times n$, the ENCLOSING PAIR problem can be solved in $O(\log n)$ time. Furthermore, this is time-optimal. $\square$

5 Encoding and Decoding Trees

The purpose of this section is to show that the task of encoding n-node binary and ordered trees into a $2n$-bitstring can be carried out in $O(\log n)$ time on an MMB of size $n \times n$. By virtue of Theorem 2.10, this is time-optimal.

Consider an n-node binary tree T with left and right subtrees T_L and T_R, respectively. We assume that the nodes of T are stored, one item per processor, in the first row of an MMB of size $n \times n$. First, we show how to associate with T the unique encoding $\sigma(T)$ defined in (1). Our encoding algorithm can be seen as a variant of the classic Euler-tour technique [5]. We proceed as follows. Replace every node u of T by 3 copies, u^1, u^2, and u^3. If u has no left child, then set $link(u^1) \leftarrow u^2$, else if v is the left child of u, set $link(u^1) \leftarrow v^1$ and $link(v^3) \leftarrow u^2$. Similarly, if u has no right child, then set $link(u^2) \leftarrow u^3$ else if w is the right child of u then set $link(u^2) \leftarrow w^1$ and $link(w^3) \leftarrow u^3$. It is worth noting that the processor associated with node u can perform the pointer assignments in $O(1)$ time. What results is a linked list starting at $root(T)^1$ and ending at $root(T)^3$, with every edge of T traversed exactly once in each direction. It is easy to confirm that the total length of the linked list is $O(n)$. Finally, assign to every u^1 a 1, to every u^2 a 0 and delete all elements of the form u^3. It is now an easy matter to show that what remains represents the encoding of T specified in (1).

The correctness of this simple algorithm being easy to see, we turn to the complexity. Computing the Euler tour amounts to setting pointers. Since all the information is available locally, this step takes $O(1)$ time. The task of eliminating every node of the form u^3 can be reduced to list ranking, prefix computation, and compaction in the obvious way. By virtue of Propositions 3.1 and 3.3 these tasks can be performed in $O(\log n)$ time. By Theorem 2.10, this is the best possible. Consequently, we have the following result.

Theorem 5.1. The task of encoding an n-node binary tree can be performed in $O(\log n)$ time on an MMB of size $n \times n$. Furthermore, this is time-optimal. $\square$

It is worth noting here that the encoding algorithm described above is quite general and can be used for other purposes as well. For example, the *preorder-inorder* traversal of T is obtained by replacing for every node u of T, u^1 and u^2 by the label of u (see [13] for details). We will further discuss properties of the preorder-inorder traversal in the context of reconstructing binary trees from their preorder and inorder traversals in Section 6.

Our encoding algorithm for ordered trees is very similar to the one described for binary trees. Consider an n-node ordered tree T. It is well-known [11] that for the purpose of getting the encoding (2) of T we only need to convert T into a binary tree BT as in [6] and then to encode BT using (1). It is easy to confirm that the resulting encoding is exactly the one defined in (2). The conversion of T into BT can be performed in $O(1)$ time since it amounts to resetting pointers only. By Theorem 5.1, the encoding of BT takes $O(\log n)$ time. By Theorem 2.10 this is the best possible. Consequently, we have the following result.

Theorem 5.2. The task of encoding an n-node ordered tree can be performed in $O(\log n)$ time on an MMB of size $n \times n$. Furthermore, this is time-optimal. $\square$

Before addressing the task of recovering binary and ordered trees from their encodings, we introduce some notation and review a few technical results. Let T be a binary tree and let v be a node of T. We let T^v stand for the subtree of T rooted at v. A bitstring τ is termed *feasible* if it contains the same number of 0's and 1's and in every prefix the number of 0's does not exceed the number of 1's. Recently, Olariu *et al.* [11] have shown that every feasible bitstring is the encoding of some binary tree. For later reference, we state the following technical result [11].

Proposition 5.3. A nonempty bitstring τ is feasible if and only if for every 1 in τ there is a *unique* matching 0 such that τ can be written as $\tau_1 1 \tau_2 0 \tau_3$, with both τ_2 and $\tau_1 \tau_3$ feasible. $\square$

Proposition 5.3 motivates us to associate with every 1 and its matching 0, a node v in T. The following simple observation [11] will justify our decoding procedure.

Observation 5.4. The corresponding decomposition of τ as $\tau_1 1 \tau_2 0 \tau_3$ has the property that $\sigma(T_L{}^v) = \tau_2$, and $\sigma(T_R{}^v)$ is a prefix of τ_3. $\square$

Observation 5.4 motivates our algorithm for recovering a binary tree from its encoding. Let τ be a feasible bitstring. For every 1 in τ we find the unique matching 0 guaranteed by Proposition 5.3. The corresponding (1,0) pair is associated with a node v in the binary tree T corresponding to τ. We then compute the left and right children of v. The details of the algorithm are spelled out as follows. Begin by ranking the 1's of τ and use the ranks as indices in T. For every 1, find its unique matching 0. Let v_i be the node of T corresponding to the 1 of rank i and to its matching 0; let p_i and q_i denote the positions in τ of the 1 of rank i and that of its matching 0, respectively. The processor in charge of v_i sets pointers as follows:

- $\text{left}(v_i) \leftarrow \textbf{nil}$ in case $q_i = p_i + 1$, and $\text{left}(v_i) \leftarrow v_{i+1}$ otherwise;
- $\text{right}(v_i) \leftarrow v_j$ if $p_j = q_i + 1$, and $\textbf{nil}$ otherwise.

The correctness follows immediately from Proposition 5.3 and Observation 5.4. Therefore, we turn to the complexity. Note that to rank all the 1's we need to compute their prefix sum. By Proposition 3.3, this task can be performed in $O(\log n)$ time. By Theorem 4.2, the matching takes $O(\log n)$ time. Finally, the setting of pointers can be done in $O(1)$ time. Thus, we have the following result.

Theorem 5.5. The task of recovering an n-node binary tree from its encoding takes $O(\log n)$ time on an MMB of size $n \times n$. Furthermore, this is time-optimal. $\square$

The task of recovering an n-node ordered tree T from its $2n$-bit encoding is similar. We begin by perceiving the encoding of T as the encoding of a binary tree BT. Once, this tree has been recoved as we just described, we proceed to convert BT to T using the classic ordered-to-binary conversion [6]. As it turns out, this latter task can be carried out in $O(\log n)$ time using the sorting algorithm of Proposition 3.5. Due to space limitations the details are omitted.

Theorem 5.6. The task of recovering an n-node ordered tree from its $2n$-bit encoding can be performed in $O(\log n)$ time on an MMB of size $n \times n$. Furthermore, this is time-optimal. $\square$

6 Reconstructing Binary Trees from Traversals

It is well-known that a binary tree can be reconstructed from its inorder traversal along with either its preorder or its postorder traversal [6]. Our goal is to show that this task can be performed in $O(1)$ time on the MMB. The main idea of our algorithm is borrowed from Olariu *et al.* [13], where the reconstruction process was reduced to that of merging two sorted sequences.

Let T be an n-node binary tree. For simplicity, we assume that the nodes of T are $\{1, 2, \ldots, n\}$. Let c_1, c_2, ..., c_n and d_1, d_2, ..., d_n be the preorder and inorder traversals of T, respectively. We may think of $c_1, c_2, \ldots, c_n$ as $1, 2, \ldots, n$, the case where $c_1, c_2, \ldots, c_n$ is a permutation of $1, 2, \ldots, n$ reducing easily to this case [13]. In preparation for merging, we construct two sequences of triples. The first sequence is $(1, j_1, c_1), (1, j_2, c_2), \ldots, (1, j_n, c_n)$ such that $d_{j_i} = c_i$, $(i = 1, 2, \ldots, n)$. In other words, the second coordinate j_i of a generic triple represents the position of c_i in the inorder sequence $d_1, d_2, \ldots, d_n$. The second sequence consists of the triples $(2, 1, d_1), (2, 2, d_2), \ldots, (2, n, d_n)$. Denote by $\prod$ the set of triples

$$\{(1, j_1, c_1), (1, j_2, c_2), \ldots, (1, j_n, c_n), (2, 1, d_1), (2, 2, d_2), \ldots, (2, n, d_n)\},$$

and define a binary relation $\prec$ on $\prod$ as follows: for arbitrary triples (α, β, γ) and $(\alpha', \beta', \gamma')$ in $\prod$ we have:
Rule 1. $((\alpha = 1) \wedge (\alpha' = 1)) \rightarrow (((\alpha, \beta, \gamma) \prec (\alpha', \beta', \gamma')) \leftrightarrow (\gamma < \gamma'))$;
Rule 2. $((\alpha = 2) \wedge (\alpha' = 2)) \rightarrow (((\alpha, \beta, \gamma) \prec (\alpha', \beta', \gamma')) \leftrightarrow (\beta < \beta'))$;
Rule 3. $((\alpha = 1) \wedge (\alpha' = 2)) \rightarrow (((\alpha, \beta, \gamma) \prec (\alpha', \beta', \gamma')) \leftrightarrow$
$$((\beta \leq \beta') \vee (\gamma \leq \gamma'))).$$
In view of the rather forbidding aspect of Rules 1, 2 and 3, an explanation is in order. First, note that Rules 1 and 2 confirm that with respect to the relation $\prec$ both sequences $(1, j_1, c_1), (1, j_2, c_2), \ldots, (1, j_n, c_n)$ and $(2, 1, d_1), (2, 2, d_2), \ldots,$ $(2, n, d_n)$ are sorted. Intuitively, Rule 3 specifies that in the preorder-inorder traversal any pair of distinct labels u and v must occur in the order "$\ldots u \ldots v \ldots v \ldots u \ldots$" or "$\ldots u \ldots u \ldots v \ldots v \ldots$" [13].

Consider the sequence e_1, e_2, ..., e_{2n} obtained by extracting the third coordinate of the triples in the sequence resulting from merging the two sequences above. As argued in [13], the sequence e_1, e_2, ..., e_{2n} is the preorder-inorder traversal of T.

Let $c_1 = 1$, $c_2 = 2$, ..., $c_n = n$ and d_1, d_2, ..., d_n be the preorder and inorder traversals of a binary tree. We assume that these sequences are stored in the first row of an MMB of size $n \times 2n$ in left to right order, with the c_i's stored to the left of the d_i's. It is easy to modify the algorithm to work on a mesh of size $n \times n$. To construct the sets of triples discussed above, every processor storing c_i needs to determine the position of the second copy of c_i in the inorder traversal. Notice that every processor storing a d_j can construct the corresponding triple without needing any further information. The details follow.
Step 1. Begin by replicating the contents of the first row throughout the mesh. This is done by tasking every processor $P(1, i)$ to broadcasts the item it stores on the bus in its own column. Every processor reads the bus and stores the value broadcast.
Step 2. Every processor $P(i, i)$ $(1 \leq i \leq n)$ broadcasts c_i on the bus in row i. The unique processor storing the second copy of label c_i will inform $P(i, i)$ of its position in the inorder sequence. A simple data movement now sends this information to $P(1, i)$. Clearly, at the end of Step 2, every processor in the first row of the mesh can construct the corresponding triple.
Step 3. Merge the two sequences of triples using Proposition 3.4 and store the

result in the first row of the mesh. Finally, every processor retains the third coordinate of the triple it receives by merging.

The correctness of the algorithm is easy to see. Since all steps take $O(1)$ time, we have proved the following result.

Lemma 6.1. Given the preorder and inorder traversals of an n-node binary tree, the corresponding preorder-inorder traversal can be constructed in $O(1)$ time on an MMB of size $n \times n$. $\square$

Our next goal is to show that once the preorder-inorder traversal $e_1, \ldots, e_{2n}$ is available, the corresponding binary tree can be reconstructed in $O(1)$ time. Recall, that every label of a node in T occurs twice in the preorder-inorder traversal. Furthermore, by virtue of Step 2 above, the first copy of a label knows the position of its duplicate, and vice-versa.

We associate a node u with every pair of identical labels in $e_1, e_2, \ldots, e_{2n}$. Let e_i and e_j be the first and second copy of a given label. The processor holding e_i assigns children pointers as follows:

- if e_{i+1} is the first copy of a label v, then left(u) $\leftarrow v$; otherwise, left(u) $\leftarrow$ **nil**;
- if e_{j+1} is the first copy of a label w, then right(v_i) $\leftarrow w$; otherwise, right(v_i) $\leftarrow$ **nil**.

The setting of pointers takes $O(1)$ time. Therefore, Lemma 6.1 implies the following result.

Theorem 6.2. An n-node binary tree can be reconstructed from its preorder and inorder traversals in constant time on an MMB of size $n \times n$. $\square$

7 Concluding Remarks and Open Problems

In this paper, we have presented a number of time-optimal tree algorithms on meshes with multiple broadcasting. Specifically, we have shown that the tasks of encoding an n-node binary or ordered tree into a $2n$-bitstring as well as the task of recovering an n-node binary or ordered tree from its $2n$-bit encoding can be solved in $\Theta(\log n)$ time. We have also shown that the problem of reconstructing an n-node binary tree from its preorder and inorder traversal can be solved in $O(1)$ time on an MMB of size $n \times n$.

Our algorithms rely heavily on time-optimal algorithms for sequences of parentheses that we developed. Specifically, we have shown that the tasks of finding all the matching pairs in a well-formed sequence of parentheses and of determining the closest enclosing pair for every matching pair in a well-formed sequence can be solved in $\Theta(\log n)$ time.

A number of problems are open. In particular, it is not known whether reconstructing an ordered tree in parent-pointer format can be done in less than $O(\log n)$ time. It is clear that such an algorithm using the closest enclosing pair can be devised. A very hard and important problem is to determine the *smallest* size MMB on which instances of size n of the above tree-related computations run in $O(\log n)$ time, that is as fast as possible. To the best of our knowledge this question is still open.

205

References

1. D. Bhagavathi, P. J. Looges, S. Olariu, J. L. Schwing, and J. Zhang, A fast selection algorithm on meshes with multiple broadcasting, *IEEE Transactions on Parallel and Distributed Systems*, 5, (1994), 772–778.

2. D. Bhagavathi, S. Olariu, W. Shen, and L. Wilson, A time-optimal multiple search algorithm on enhanced meshes, with Applications, *Journal of Parallel and Distributed Computing*, 22, (1994), 113–120.

3. D. Bhagavathi, S. Olariu, J. L. Schwing, and J.Zhang, Convex polygon problems on meshes with multiple broadcasting, *Parallel Processing Letters*, 2 (1992) 249–256.

4. S. A. Cook, C. Dwork, and R. Reischuk, Upper and lower time bounds for parallel random access machines without simultaneous writes, *SIAM Journal on Computing*, 15 (1986) 87–97.

5. J. JáJá, *An introduction to parallel algorithms*, Addison-Wesley, Reading, MA, 1991.

6. D. Knuth, *The Art of Computer Programming: Fundamental Algorithms,* vol. 1, 2nd edition, Addison-Wesley, Reading, MA, 1973.

7. V. P. Kumar and C. S. Raghavendra, Array processor with multiple broadcasting, *Journal of Parallel and Distributed Computing*, 2, (1987) 173–190.

8. V. P. Kumar and D. I. Reisis, Image computations on meshes with multiple broadcast, *IEEE Trans. on Pattern Analysis and Machine Intelligence*, 11, (1989), 1194–1201.

9. R. Lin, S. Olariu, J. L. Schwing, and J. Zhang, Simulating enhanced meshes, with applications, *Parallel Processing Letters*, 3 (1993) 59–70.

10. M. Maresca and H. Li, Connection autonomy in SIMD computers: a VLSI implementation, *Journal of Parallel and Distributed Computing*, 7, (1989), 302–320.

11. S. Olariu, J. L. Schwing, and J. Zhang, Optimal parallel encoding and decoding algorithms for trees, *International Journal of Foundations of Computer Science*, 3 (1992), 1–10.

12. S. Olariu, J. L. Schwing, and J. Zhang, Time-optimal sorting and applications on $n \times n$ enhanced meshes, *Proc. IEEE Internat. Conf. on Computer Systems and Software Engineering*, The Hague, May 1992.

13. S. Olariu, C. M. Overstreet, and Z. Wen, Parallel reconstruction of binary trees, *Journal of Parallel and Distributed Computing*, to appear.

14. D. Parkinson, D. J. Hunt, and K. S. MacQueen, The AMT DAP 500, 33rd *IEEE Comp. Soc. International Conf.*, 1988, 196–199.

15. S. Zaks, Lexicographic generation of ordered trees, *Theoretical Computer Science*, 10 (1980), 63–82.

Prefix Graphs and Their Applications [*]

Shiva Chaudhuri and Torben Hagerup

Max-Planck-Institut für Informatik, Im Stadtwald, D–66123 Saarbrücken, Germany
shiva@mpi-sb.mpg.de   torben@mpi-sb.mpg.de

Abstract. The *range product problem* is, for a given set S equipped with an associative operator $\circ$, to preprocess a sequence $a_1, \ldots, a_n$ of elements from S so as to enable efficient subsequent processing of queries of the form: Given a pair (s, t) of integers with $1 \leq s \leq t \leq n$, return $a_s \circ a_{s+1} \circ \cdots \circ a_t$. The generic range product problem and special cases thereof, usually with $\circ$ computing the maximum of its arguments according to some linear order on S, have been extensively studied. We show that a large number of previous sequential and parallel algorithms for these problems can be unified and simplified by means of prefix graphs.

1 Introduction

In 1983 Chandra, Fortune and Lipton introduced a computational paradigm closely related to the Ackermann function and used it to study the computation of semigroup products on unbounded-fanin circuits [6, 7]. Since then the paradigm was rediscovered several times, under different names and in different guises, and exploited in the design of sequential and parallel algorithms. In particular, Berkman and Vishkin developed the "recursive star-tree data structure" [4] and used it in a series of papers. We unify much of the previous research by observing that it is concerned simply with solving special cases of the range product problem, and by giving a generic algorithm for the range product problem in terms of graphs known as *prefix graphs*. We prove few new results; however, the machinery developed here allows us, with a modest effort, to obtain simpler proofs of the results of several previous papers, and to exhibit the intimate connection between such seemingly disparate problems as binary addition on a circuit and linear-range merging (i.e., merging sequences of length n with elements in $\{1, \ldots, n\}$) on a PRAM. We hope that our effort at unification and simplification will contribute to a wider understanding and appreciation of the underlying paradigm.

2 Definition of Prefix Graphs

Informally, a prefix graph has n vertices arranged in a column on the left and n vertices arranged in a column on the right, some vertices in between, and enough

---

[*] Supported by the ESPRIT Basic Research Actions Program of the EU under contract No. 7141 (project ALCOM II).

edges to allow us to go from left to right, provided that we also go down by at least one level.

Definition 1. For all $n \in \mathbb{N}$, a *prefix graph* of *width* n is a directed acyclic graph $G = (V, E)$ with n distinguished *input vertices* $x_1, \ldots, x_n$ of indegree zero and n distinguished *output vertices* $y_1, \ldots, y_n$ of outdegree zero and with the following properties, where the *span* of a vertex $v \in V$, span(v), is defined as $\{i : 1 \leq i \leq n$ and G contains a path from x_i to $v\}$.
  (1) For $i = 1, \ldots, n$, span$(y_i) = \{1, \ldots, i-1\}$ (for $i = 1$ this is $\emptyset$);
  (2) For all $v \in V$, span(v) is either empty or an "interval" of the form $\{s, \ldots, t\}$, for some integers s and t with $1 \leq s \leq t \leq n$;
  (3) Any two vertices in V with a common successor have disjoint spans.
The *depth* of a vertex v in G is the length of a longest path in G from an input vertex to v, and the depth of G is the maximum depth of any of its vertices.

Note that there is a natural linear order on the set of edges entering a vertex v in a prefix graph. If $e = (u, v)$ and $e' = (u', v)$ are two such edges, e precedes e' if and only if the elements of the span of u are smaller than the elements of the span of u' (if either span is empty, the order is undefined, but irrelevant). We call this order the *canonical ordering* of the edges entering v.

3   Generic Prefix and Range Product Algorithms

In the context of a fixed semigroup $(S, \circ)$, the *composition problem* defined by n elements $a_1, \ldots, a_n$ of S is to compute $a_1 \circ \cdots \circ a_n$. The corresponding *prefix product problem* is to compute all the prefix products $a_1, a_1 \circ a_2, \ldots, a_1 \circ \cdots \circ a_n$, and the corresponding *range product problem* is to preprocess the sequence $a_1, \ldots, a_n$ so that in response to a *range query* $[s, t]$, where $1 \leq s \leq t \leq n$, one can quickly compute $a_s \circ \cdots \circ a_t$.

Prefix graphs suggest very natural and simple reductions of the prefix product problem to the composition problem, and of the range product problem to the prefix product problem. In order to solve the prefix product problem defined by n elements $a_1, \ldots, a_n$ of a semigroup $(S, \circ)$, take a prefix graph $G = (V, E)$ of width n, apply $a_1, \ldots, a_n$ to the inputs of G and let values of S "percolate" through G from the inputs to the outputs, each vertex composing the values reaching it over its incoming edges, in the order given by the canonical ordering of these edges, and sending the resulting product over all of its outgoing edges (we say that the vertex solves its *local composition problem*). It is easy to see by induction on the vertex depth that each vertex with span $\{s, \ldots, t\}$ computes $a_s \circ \cdots \circ a_t$; in particular, the solution to the prefix product problem can be read off the output vertices.

In order to solve the range product problem for $a_1, \ldots, a_n$, we begin by carrying out the same computation. Further, each vertex in V computes all suffix products of the sequence of values that reached it (it solves its *local suffix problem*) and saves these, which ends the preprocessing.

For any set J of the form $J = [s, t] = \{s, s + 1, \ldots, t\}$ with $1 \le s \le t \le n$, let $P(J) = a_s \circ a_{s+1} \circ \cdots \circ a_t$. We claim that if $v \in V$ is of depth $d \ge 1$ in G and $\text{span}(v) = [s, t]$, then for any integer m with $s \le m \le t$, $P([m, t])$ is the product of at most d values computed during the preprocessing. The proof is by induction on d, and the claim is obvious if $d = 1$. Otherwise choose (the unique) $(u, v) \in E$ such that $m \in \text{span}(u)$, apply the induction hypothesis to compute $P(\text{span}(u) \cap [m, t])$ and note that if $J = [m, t] \backslash (\text{span}(u) \cap [m, t]) \ne \emptyset$, then $P(J)$ is one of the suffix products stored at v.

Applying the claim above to the output vertices, we see that if G is of depth $d \ge 1$, then the answer to any range query can be obtained as the composition of at most d of the values computed in the preprocessing phase. The relevant values can be found in $O(d)$ sequential time during a backwards scan in G, as in the proof above; we omit the details.

The algorithms that we will describe always just simulate the generic prefix and range product algorithms. The only variable parameters will be the semigroup $(S, \circ)$ under consideration and the model of computation.

4   Existence of Prefix Graphs

Define $I_0 : \mathbb{N} = \{1, 2, \ldots\} \to \mathbb{N}$ by $I_0(n) = \lceil n/2 \rceil$, for all $n \in \mathbb{N}$. Inductively, for $k = 1, 2, \ldots$, define $I_k : \mathbb{N} \to \mathbb{N}$ by $I_k(n) = \min\{i \in \mathbb{N} : I_{k-1}^{(i)}(n) = 1\}$, for all $n \in \mathbb{N}$, where superscript (i) denotes i-fold repeated application. Finally, for all $n \in \mathbb{N}$, take $\alpha(n) = \min\{k \in \mathbb{N} : I_k(n) \le k\}$. It can be shown that for all $n, k \in \mathbb{N}$, $I_{k+1}(n) \le I_k(n) \le I_k(n + 1)$ and, if $n \ge 2$, $I_k(n) < n$.

An important fact about prefix graphs is that for all $n, k \in \mathbb{N}$, there is a prefix graph $G_{n,k}$ of width n, depth at most $2k$ and $O(nkI_k(n))$ edges. The simplest demonstration of this fact proceeds by simultaneous induction on n and k. We hence describe $G_{n,k}$ in terms of graphs $G_{n',k'}$, where (n', k') precedes (n, k) lexicographically. Let $m = I_{k-1}(n)$. We will assume that m divides n. The construction of $G_{n,k}$ is shown in Fig. 1. We take n input vertices and n output vertices and partition both input and output vertices into groups of size m. The vertices in corresponding groups are connected via copies of $G_{m,k}$. Furthermore we create a new vertex for each input or output group and add edges from each input to the new vertex associated with its group and from each vertex associated with an output group to all vertices in its group. Finally, if $k > 1$, we connect the n/m new vertices on the left with the n/m new vertices on the right via a copy of $G_{n/m,k-1}$. If $k = 1$, we have $n/m = 2$, and we instead identify the upper new vertex on the left with the lower new vertex on the right, i.e., we connect the new vertices via a prefix graph of depth zero and with no edges.

It is easy to see that the resulting graph is a prefix graph of depth at most $2k$. Assume that for all (n', k') that precede (n, k) lexicographically, $G_{n',k'}$ contains at most $2n'k'I_{k'}(n')$ edges. Then the copies of $G_{m,k}$ contribute a total of at most $2nkI_k(m) = 2nk(I_k(n) - 1)$ edges, the copy of $G_{n/m,k-1}$, if present, contributes at most $2(n/m)(k - 1)I_{k-1}(n/m) \le 2n(k - 1)$ edges, and with the remaining $2n$ edges this yields a grand total of at most $2nkI_k(n)$ edges.

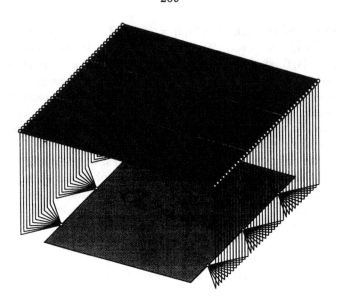

Fig. 1. A doubly-recursive construction of the prefix graph $G_{n,k}$.

The construction above, suitably modified if m does not divide n, suffices for all our applications, except those in Section 5.5. We now describe an alternative, but similar construction that suffices for all our applications, incorporates rounding and leads more directly to a fast parallel construction algorithm, one recursion of depth $\Theta(\log n)$ having been converted to iteration. The vertices in the graph will be classified as either *front* or *back vertices* as they are introduced; input vertices are always front vertices, and output vertices that are not also input vertices are back vertices.

The induction now is only on k. Without loss of generality we will assume that n is a power of 2. Take $l = I_k(n) - 1$ and let $m_0 \geq m_1 \geq \cdots \geq m_l \geq m_{l+1}$ be a sequence of powers of 2 with $m_0 = n$, $m_l = O(1)$ and $m_{l+1} = 1$; the exact values will be specified later. $G_{n,k}$ consists of $l+1$ "layers", illustrated in Fig. 2. Layer i, for $i = 0, \ldots, l$, partitions both input and output vertices of $G_{n,k}$ into groups of m_{i+1} consecutive vertices, joins all input vertices in each group to a new front vertex, joins one new back vertex for each group to all output vertices in the group, and finally connects every set of m_i/m_{i+1} consecutive front vertices to the corresponding back vertices via a copy of $G_{m_i/m_{i+1},k-1}$.

It is easy to see that the graph $G_{n,k}$ thus constructed is indeed a prefix graph (cf. Fig. 2): For every input vertex x and every higher-numbered output vertex y, exactly one of the layers contains a path from x to y (for this it is essential that the groups in layer l are trivial, i.e., of size 1). We still need to choose $m_1, \ldots, m_l$, which will depend on a constant parameter $q \in \mathbb{N}$. If $k = 1$, take $m_i = I_0^{(i)}(n) = n/2^i$, for $i = 1, \ldots, l$. The recursive construction then depends only on $G_{2,0}$, which we take to be the prefix graph of width 2 with no edges. If $k \geq 2$, we will assume for convenience that $((2q+1)I_1(n))^{2q} \leq n$; this

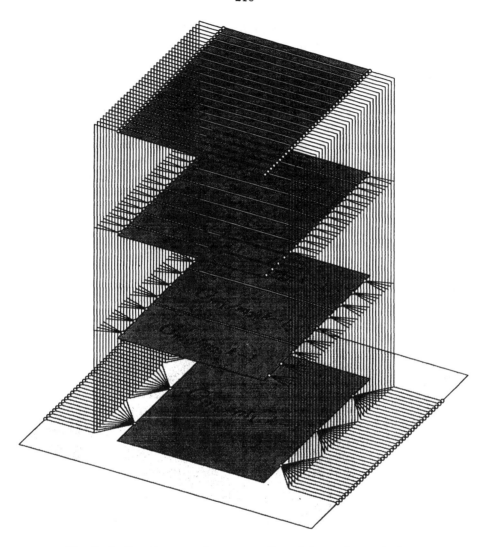

Fig. 2. An iterative-recursive construction of the prefix graph $G_{n,k}$.

excludes only a finite set of values of n. We then take m_i as the smallest power of 2 no smaller than $((2q + 1)I_{k-1}^{(i)}(n))^{2q}$, for $i = 1, \ldots, l$; one can show that $I_{k-1}^{(I_k(n)-1)}(n) = 2$ for all $n, k \geq 2$, so that indeed $m_l = O(1)$, as required.

Let V_F and V_B be the sets of front and back vertices, respectively, in $G_{n,k}$, and denote by $\deg(v)$ the indegree of each vertex v. We are interested in the quantity

$$R_q(n, k) = \sum_{v \in V_F} \sqrt{\deg(v)|\mathrm{span}(v)|} + \sum_{v \in V_B} (\deg(v))^q,$$

called the *order-q root-span* of $G_{n,k}$. We will show that $R_q(n, k) = O(nk(I_k(n))^q)$.

If $k \geq 2$, assume by induction that $R_q(m, k-1) \leq rm(k-1)(I_{k-1}(n))^q$ for some constant r and for all $m \in I\!N$, and reconsider the construction of $G_{n,k}$. Input vertices of $G_{n,k}$ contribute nothing to $R_q(n, k)$, whereas the output (back) vertices have indegree $l+1 = I_k(n)$ and thus contribute a total of $n(I_k(n))^q$. For $i = 0, \ldots, l$, layer i contains n/m_i copies of $G_{m_i/m_{i+1},k-1}$. The (front) input vertices of these copies contribute exactly n to $R_q(n, k)$. We will determine the remaining contribution of layer i to $R_q(n, k)$. Consider three cases. (1) $k = 1$: The remaining contribution is zero. (2) $k \geq 2$ and $i = l$: Since $m_l = O(1)$, the remaining contribution is $O(n)$. (3) $k \geq 2$ and $i < l$: Since the size of the span of every vertex in the copies of $G_{m_i/m_{i+1},k-1}$ increases by a factor of m_{i+1} when the copies are incorporated into $G_{n,k}$, the remaining contribution of layer i to $R_q(n, k)$ is at most

$$\frac{n}{m_i}\sqrt{m_{i+1}}R_q(m_i/m_{i+1}, k-1) \leq \frac{rn(k-1)(I_{k-1}(m_i/m_{i+1}))^q}{\sqrt{m_{i+1}}}.$$

We have $m_i/m_{i+1} \leq 2(I_{k-1}^{(i)}(n))^{2q}$ (the factor of 2 is due to the rounding to the nearest power of 2). One can show that for all $m \in I\!N$, $I_{k-1}(2m^{2q}) \leq (2q+1)I_{k-1}(m)$. Hence $I_{k-1}(m_i/m_{i+1}) \leq (2q+1)I_{k-1}(I_{k-1}^{(i)}(n)) = (2q+1)I_{k-1}^{(i+1)}(n)$. But then the contribution of layer i to $R_q(n, k)$ is at most

$$\frac{rn(k-1)((2q+1)I_{k-1}^{(i+1)}(n))^q}{\sqrt{((2q+1)I_{k-1}^{(i+1)}(n))^{2q}}} = rn(k-1).$$

Putting the cases together and observing that the number of layers is $l+1 = I_k(n)$, we altogether obtain $R_q(n, k) \leq n(I_k(n))^q + O(n) + I_k(n)(n + rn(k-1))$, which, for the constant r chosen sufficiently large, is bounded by $rnk(I_k(n))^q$.

We summarize the results of this section as follows.

Theorem 2. *For all $n, k \in I\!N$ and all fixed $q \in I\!N$, there is a prefix graph of width n, depth at most $2k$ and order-q root-span $O(nk(I_k(n))^q)$. In particular, for all $n \in I\!N$ and all fixed $q \in I\!N$, there is a prefix graph of width n, depth $O(\alpha(n))$ and order-q root-span $O(n(\alpha(n))^{q+1})$.*

The lesson to be learned from Theorem 2 is that if we allocate resources (such as time or processors) proportional to $\sqrt{\deg(v)|\text{span}(v)|}$ to each front vertex v of a prefix graph, and proportional to any fixed power of $\deg(v)$ to each back vertex v, then the total amount of resources used will be only slightly superlinear. In most cases it suffices to allocate resources proportional to $\deg(v)$ to each vertex v (if v is a front vertex, $\deg(v) \leq |\text{span}(v)|$ and hence $\deg(v) \leq \sqrt{\deg(v)|\text{span}(v)|}$); this can be viewed as placing a constant amount of resources (e.g., one processor) at each edge of the prefix graph.

Assuming constant-time access to a few tables, most notably one that gives the value of $I_k^{(i)}(m)$ for all positive integers m, k and i with $m \leq n$, $k \leq \alpha(n)$ and $i \leq \log n$, a suitable representation of a prefix graph with the properties described in Theorem 2 can be computed in constant time on a CREW PRAM with $nk(I_k(n))^q$ processors, and hence in $O(nk(I_k(n))^q)$ sequential time. Suitable tables can be constructed in constant time on an n-processor CRCW PRAM, and hence in linear sequential time, whereas the best construction known for the CREW PRAM needs $O(\log(\log^* n))$ time and $O(n)$ operations. We omit the details due to lack of space, referring the reader to [4] and [5] for some of the arguments.

5 Applications of Prefix Graphs

In this section we describe a number of concrete instances of the generic algorithms introduced in Section 3. In each case Theorem 2 is used to bound the resource requirements of the resulting algorithms. We give algorithms with slightly superlinear processor-time products; simple arguments presented in Section 6 reduce this quantity to $O(n)$ in each case. Our examples span sequential computation, unbounded-fanin circuits and PRAMs.

5.1 Sequential Range Products

Assuming that the operator $\circ$ of the semigroup $(S, \circ)$ can be evaluated in constant sequential time, it is trivial to verify that each vertex v of a prefix graph can solve its local composition and suffix problems in linear time $O(\deg(v))$. By the generic prefix and range product algorithms and Theorem 2, we obtain the following result of Alon and Schieber [2].

Theorem 3. *For all $n, k \in \mathbb{N}$, instances of size n of the range product problem can be solved with preprocessing time $O(nkI_k(n))$ and query time $O(k)$.*

5.2 Addition on Unbounded-Fanin Circuits

In this section we consider unbounded-fanin circuits with AND, OR and NOT gates. The size of a circuit is defined to be the total number of gates and wires in the circuit, and the depth is the longest path from an input to an output. The *find-first problem* of size n is, given n bits $u_1, \ldots, u_n$, to compute the n bits $v_1, \ldots, v_n$ with $v_i = 1$ iff $u_i = 1$ and $u_1 = u_2 = \cdots = u_{i-1} = 0$, for $i = 1, \ldots, n$. As shown in [6, Theorem 3.4], find-first problems of size n can be solved by a circuit of constant depth and size $O(n)$.

Consider the problem of computing the $(n+1)$-bit sum $z_{n+1} \cdots z_1$ of two n-bit numbers $x_n \cdots x_1$ and $y_n \cdots y_1$. There is a well-known reduction of this problem to that of computing prefix sums of elements of the semigroup with three elements, S, R and P, and the operator $\circ$ defined as follows: For all $u, v \in \{S, R, P\}$, $u \circ v = v$ if $v \in \{S, R\}$, and $u \circ P = u$. For $i = 1, \ldots, n$, let $a_i = S$ if $x_i = y_i = 1$, $a_i = R$ if $x_i = y_i = 0$ and $a_i = P$ otherwise. S, R and P are "set", "reset" and "propagate" indicators for the carry bit. It is easy to see that there is a carry into bit position $i + 1$ iff $c_{i+1} = S$, for $i = 1, \ldots, n$, where $c_{i+1} = a_1 \circ \ldots \circ a_i$, so that adding $x_n \cdots x_1$ and $y_n \cdots y_1$ essentially boils down to computing $c_2, \ldots, c_{n+1}$. For this, imagine that the edges of a prefix graph can carry the values S, R and P and that the vertices can compute the product (with operator $\circ$) of the values on their incoming edges. Then the outputs of the prefix graph on input $a_1, \ldots, a_n$ are $c_2, \ldots, c_{n+1}$. The computation at a vertex of the prefix graph is essentially a find-first problem, since the value obtained by composing the values in its input sequence is simply the rightmost non-P value in the sequence, or P if there is no such value. Using the circuit of [6] mentioned above, we can simulate the computation at v with a boolean circuit of constant depth and size $O(\deg(v))$. An edge in the prefix graph can be simulated with two parallel wires, so that we altogether obtain a circuit of size and depth within constant factors of the size and depth of the prefix graph. Applying Theorem 2, we get the following result of [7, Corollary 3.5].

Theorem 4. *For all $n, k \in \mathbb{N}$, two n-bit numbers can be added with a circuit of depth $O(k)$ and size $O(nkI_k(n))$.*

As an alternative to the theorem above, two n-bit numbers can also be added with a circuit of depth $O(\alpha(n))$ and size $O(n)$. This follows by choosing $k = \alpha(n)$, using prefix-product circuits of [10] of logarithmic depth and linear size to reduce the problem size by a factor of $\Theta((\alpha(n))^2)$ and using Theorem 4 to solve the reduced problem, much in the spirit of Section 6.

In the remainder of Section 5 we describe PRAM algorithms in which the local composition and suffix problems are solved in parallel at each vertex of a prefix graph.

5.3 Segmented Broadcasting

The *segmented broadcasting problem* of size n is, given an array A of n cells, some of which contain *significant* objects, while others do not, to replace each insignificant object by the nearest significant object to its left, if any. If we let S be the set of objects and denote an insignificant object by 0, the segmented broadcasting problem is the prefix product problem associated with the semigroup $(S, \circ)$, where for all $a, b \in S$, $a \circ 0 = a$ and $a \circ b = b$ if $b \neq 0$. Instances of size n of the corresponding composition problem can be solved in constant time on an n-processor CRCW PRAM by a straightforward simulation of the find-first circuits of [6] mentioned in the previous subsection. By the generic prefix product algorithm and Theorem 2, we therefore obtain the following result of Berkman and Vishkin [4] and Ragde [11, Theorem 4], who solved versions of segmented broadcasting called *all nearest zero bits* and *ordered chaining*, respectively.

Theorem 5. *For all $n, k \in \mathbb{N}$, segmented broadcasting problems of size n can be solved in time $O(k)$ on a CRCW PRAM with $nkI_k(n)$ processors.*

5.4 Linear-Range Merging

An instance of the *linear-range merging problem* of size n is to merge two sorted sequences, each containing n elements in the range $1 \ldots n$. A linear-range merging problem of size n reduces to a segmented broadcasting problem of size $2n$. To see this, multiply each input element by 2 and subtract 1 from each element in one input sequence only. This ensures that no value occurs in both input sequences. It now suffices to determine for each input element x its rank in the opposite input sequence, i.e., the number of elements $\leq x$ in that sequence, since the position of the element in the output sequence can be taken as its rank in the opposite input sequence plus its position in its own sequence.

The remaining task therefore is, given a sorted sequence $X = (x_1, \ldots, x_n)$ of n integers in the range $1 \ldots 2n$, to compute a table $Rank[1 \ldots 2n]$ such that for $j = 1, \ldots, 2n$, $Rank[j]$ is the rank of j in X (this task is carried out twice, once for each input sequence). Suppose that we begin by computing preliminary ranks as follows: Initialize $Rank[j]$ to 0, for $j = 1, \ldots, 2n$, and take $x_{n+1} = \infty$, where ∞ is an integer larger than $2n$. Then, for $i = 1, \ldots, n$, if $x_i \neq x_{i+1}$, then set $Rank[x_i] = i$. This computes the correct rank of every value that occurs in X. Furthermore, the correct value to be stored in each entry in the rank table is the maximum of the preliminary ranks stored in the entries to its left (including the entry under consideration), so that the problem at hand is the segmented broadcasting problem defined by the preliminary ranks.

By Theorem 5 and the discussion above, linear-range merging problems can be solved very efficiently on a CRCW PRAM. Exploiting additional information furnished by the reduction from merging to segmented broadcasting, however, we can solve linear-range merging problems equally efficiently on the CREW PRAM: For $j = 1, \ldots, 2n$, initialize $Rank[j]$ to $(j, 0, 0)$, rather than to 0. Then, for $i = 1, \ldots, n$, instead of setting $Rank[x_i] = i$ if $x_i \neq x_{i+1}$, set $Rank[x_i] = (x_i, x_{i+1}, i)$ if $x_i \neq x_{i+1}$. The third component of each triple is the actual rank information, the second component is a pointer to the next triple with a nonzero third (or second) component, and the first component is just the position of the triple itself. Let S be the set of triples of integers and define the semigroup $(S, \circ)$ as follows: For all $(a_1, b_1, c_1), (a_2, b_2, c_2) \in S$, $(a_1, b_1, c_1) \circ (a_2, b_2, c_2) = (a_1, b_1, c_1)$ if $c_2 = 0$, and $(a_1, b_1, c_1) \circ (a_2, b_2, c_2) = (a_2, b_2, c_2)$ otherwise.

Let $(a_1, b_1, c_1), \ldots, (a_n, b_n, c_n)$ be the set of triples generated for an instance of the linear-range merging problem. For such a sequence, it is easy to see that for any integers s and t with $1 \leq s \leq t \leq n$, $(a_s, b_s, c_s) \circ \cdots \circ (a_t, b_t, c_t)$ is (a_i, b_i, c_i) for the unique $i \in \{s, \ldots, t\}$ with $b_i > a_t$, or (a_s, b_s, c_s) if there is no such i. As a consequence, if we execute the associated instance of the generic prefix product algorithm on a prefix graph G, the local composition problem occurring at each vertex v in G can be solved in constant time even on a CREW PRAM with $\deg(v)$ processors.

The final rank table can be obtained from the prefix products, computed above, of the preliminary rank table by replacing each triple by its third component, which is the actual rank information. Using Theorem 2, we therefore obtain the following result, due to Berkman and Vishkin [5].

Theorem 6. *For all $n, k \in \mathbb{N}$, linear-range merging problems of size n can be solved in $O(k)$ time on a CRCW PRAM or a CREW PRAM with $nkI_k(n)$ processors. In the case of the CREW PRAM, an appropriate prefix graph must be supplied as part of the input.*

Using a slightly smaller prefix graph, we obtain the following new result.

Theorem 7. *For all $n, k \in I\!N$, two sorted sequences of length n each of integers drawn from a range of size $O(n/I_k(n))$ can be merged in $O(k)$ time using $O(n)$ operations on a CRCW PRAM or a CREW PRAM. In the case of the CREW PRAM, an appropriate prefix graph must be supplied as part of the input.*

5.5 Prefix and Range Maxima of c-Bounded Sequences

A sequence $a_1, \ldots, a_n$ of integers is called *c-bounded*, for $c \in I\!N$, if $|a_i - a_{i+1}| \le c$, for $i = 1, \ldots, n-1$. In this section we develop algorithms for the prefix maxima and range maxima problems for c-bounded input sequences. These results were first proved by Berkman and Vishkin [4]. The following simple fact is a consequence of [8, Theorem 4(c)].

Lemma 8. *For all $n, m \in I\!N$, the maximum of n integers $a_1, \ldots, a_n$ with $\max_{1 \le i \le j \le n} |a_i - a_j| < m$ can be computed in constant time on a CRCW PRAM with $O(n + \sqrt{m})$ processors.*

If a c-bounded sequence $a_1, \ldots, a_n$ is applied to the inputs of a prefix graph and each vertex computes the maximum of the values entering it, the maximum difference between two values entering a vertex v will be at most $c(|\mathrm{span}(v)| - 1)$. A front vertex v can therefore execute the algorithm of Lemma 8 with $O(\deg(v) + \sqrt{c|\mathrm{span}(v)|})$ processors; a back vertex v can use the trivial brute-force algorithm requiring $(\deg(v))^2$ processors. By Theorem 2, we obtain:

Theorem 9. *For all $n, k, c \in I\!N$, the prefix maxima of a c-bounded sequence of length n can be computed on a CRCW PRAM with $O(\sqrt{cn}k(I_k(n))^2)$ processors in $O(k)$ time.*

We will solve the range maxima problem for c-bounded sequences by executing the algorithm of Theorem 9 (or, rather, the corresponding suffix maxima algorithm) at each vertex of a prefix graph $G = (V, E)$. For this we must establish an upper bound $c(v)$ on the difference between successive values entering each vertex $v \in V$. Let $\Gamma^-(v) = \{u \in V : (u, v) \in E\}$ and note that $2\max\{c|\mathrm{span}(u)| : u \in \Gamma^-(v)\}$ is such an upper bound (in fact, the factor of 2 is not needed). By induction, one can show that $|\mathrm{span}(u)| = |\mathrm{span}(u')|$ for all $u, u' \in \Gamma^-(v)$, provided that v is a front vertex, i.e., in this case we can take $c(v) = 2c|\mathrm{span}(v)|/|\Gamma^-(v)| = 2c|\mathrm{span}(v)|/\deg(v)$.

Plugging the value for $c(v)$ determined above into the processor count of Theorem 9, we see that the number of processors needed at a front vertex v is $O(\sqrt{2c|\mathrm{span}(v)|/\deg(v)} \deg(v)k(I_k(n))^2) = O(\sqrt{c\deg(v)|\mathrm{span}(v)|}k(I_k(n))^2)$. By Theorem 2, this sums over all front vertices to $O(\sqrt{cn}k^2(I_k(n))^3)$. Since suffix maxima problems are simple to solve in constant time with a cubic number of processors, the same number of processors suffices for the back vertices. Noting that once the prefix product problem has been solved, the local suffix problem can be solved for all vertices in parallel, we have:

Theorem 10. *For all $n, k, c \in \mathbb{N}$, the preprocessing of a c-bounded sequence of length n for subsequent sequential range queries in $O(k)$ time can be done in $O(k)$ time on a CRCW PRAM with $O(\sqrt{c}nk^2(I_k(n))^3)$ processors.*

5.6 Randomized Prefix and Range Maxima

The algorithm in the previous subsection was able to deal only with c-bounded sequences because maxima of general sequences cannot be computed deterministically in constant time. The situation changes if we allow randomization. However, since the randomized algorithm is applied to many small inputs, namely once at each vertex, we have to cope with the failure of the execution at some vertices. The number of affected vertices being small, we can subsequently allocate enough resources to each such vertex, by means of an algorithm for so-called *interval allocation* [9], to let it run a deterministic algorithm. We now provide the details.

Lemma 11. *Let $f, g, h : \mathbb{N} \to \mathbb{N}$ satisfy $n \leq f(n) \leq g(n)$ for all $n \in \mathbb{N}$ and suppose that instances of size n of a composition problem can be solved in constant time both with $nh(n)$ processors and failure probability at most $1/g(n)$ and deterministically with $f(n)$ processors. Then, for every $k \in \mathbb{N}$, instances of size n of the corresponding prefix product problem can be solved with probability at least $1 - 1/g(f^{-1}(\sqrt{n})) - 2^{-\sqrt{n}}$ using $O(k)$ time and $nh(n)$ processors plus the resources needed for solving k successive interval allocation problems of size $nI_k(n)$. Here $f^{-1}(\sqrt{n})$ is to be understood as $\min\{m \in \mathbb{N} : f(m) \geq \sqrt{n}\}$.*

Proof. We allocate $h(n)$ processors to each edge of a prefix graph of width n and depth $2k$ with at most $rnI_k(n)$ edges, where $r \geq 1$ is a constant, and proceed level by level from the input vertices to the output vertices. At each level, each vertex first carries out five attempts to solve its local composition problem using the given randomized algorithm. Subsequently each vertex v at which all trials failed requests $f(\deg(v))$ processors and applies the given deterministic algorithm, after which we proceed to the next level.

The analysis must bound the probability that too many processors are requested. Assume that n is large enough to make $rI_k(n) \leq n$. Take $n_0 = f^{-1}(\sqrt{n})$ and say that a vertex v is of *high degree* if $\deg(v) \geq n_0$. Since $g(n_0) \geq f(n_0) \geq \sqrt{n}$, the probability that all five trials fail for some high-degree vertex is at most $rnI_k(n)(1/g(n_0))^5 \leq 1/g(n_0)$. The number of processors requested by the low-degree vertices is a weighted sum $X = \sum_v w_v X_v$ of independent Bernoulli variables, where $w_v = f(\deg(v)) \leq f(n_0 - 1) \leq \sqrt{n}$ and $E(X_v) = 1/g(\deg(v))$ for all v. We have $E(X) = \sum_v w_v E(X_v) \leq \sum_v f(\deg(v))/g(\deg(v)) \leq rnI_k(n)$. By the Chernoff bound of Raghavan for weighted sums [12, Theorem 1], $\Pr(X \geq 6rnI_k(n)) \leq 2^{-6rnI_k(n)/\sqrt{n}} \leq 2^{-\sqrt{n}}$.

Abstractly speaking, the argument above shows how to solve an instance of a certain problem at each vertex of a prefix graph, given a randomized and a (more wasteful) deterministic algorithm for the problem under consideration.

The analysis of Lemma 11 applies, in particular, when the problem at each vertex is its local suffix problem, and the randomized algorithm used is the prefix product algorithm of Lemma 11. Recalling that all the local suffix problems can be solved in parallel, we obtain an algorithm for the range product problem.

A CRCW PRAM with n processors can compute the maximum of n numbers in constant time with probability $1 - 2^{-n^{\Omega(1)}}$ [1, Theorem 3.9], and a deterministic algorithm for the problem that uses n^2 processors and constant time is obvious. Furthermore, interval allocation problems of size n can be solved in $O(\log^* n)$ time on an n-processor CRCW PRAM with probability $1 - 2^{-n^{\Omega(1)}}$ [9, Theorem 5.1], while processor allocation is free in the *parallel comparison-tree* (PCT) model of Valiant [13]. Using these facts in the general framework, we obtain the following results of [3].

Theorem 12. *For all $n, k \in \mathbb{N}$, range maxima preprocessing problems of size n can be solved (a) in $O(k \log^* n)$ time on a CRCW PRAM and (b) in $O(k)$ time on a PCT, so that in each case the number of processors needed is $nkI_k(n)$, the resulting sequential query time is $O(k)$, and the failure probability is $2^{-n^{\Omega(1)}}$.*

6 Making the algorithms work-optimal

In this section we show how to make all of the algorithms work-optimal without affecting the time performance. For any operator $\circ$ such that $a \circ b$ can be evaluated in constant sequential time (which is the case in all our applications), it is easy to see that range product problems of size n can be preprocessed in $O(\log n)$ time with $O(n)$ operations for a subsequent query time of $O(\log n)$ by means of a balanced binary tree.

Proposition 13. *If a range product problem of size n can be preprocessed in time $t(n)$ with $O(n2^{t(n)})$ operations, yielding query time $O(t(n))$, then it can be preprocessed in time $O(t(n))$ with $O(n)$ operations, yielding query time $O(t(n))$.*

Proof. Partition the input elements into $O(n/2^{t(n)})$ groups of size $O(2^{t(n)})$ each, preprocess each group using the optimal logarithmic-time algorithm above and preprocess the group products using the nonoptimal algorithm assumed in the proposition, each of which uses $O(t(n))$ time and $O(n)$ operations. Since each range is the disjoint union of two ranges within groups and one range spanning a number of whole groups, we can subsequently answer any range query in $O(t(n))$ time.

Proposition 14. *If a range product problem can be preprocessed in time $t(n)$ with $O(n)$ operations, yielding query time $O(t(n))$, then the corresponding prefix product problem can be solved in time $O(t(n))$ with $O(n)$ operations.*

Proof. First preprocess the input for range queries. Then partition the input elements into $O(n/t(n))$ groups of size $O(t(n))$ each, compute the prefix products for the first elements of all groups by means of range queries, which needs

$O(t(n))$ time and $O(n)$ operations, and then compute all remaining prefix products sequentially within each group, which also needs $O(t(n))$ time and $O(n)$ operations.

Theorems 3, 5, 6, 12(b) and, for constant c, Theorems 9 and 10 imply that each of the respective problems can be solved in $O(\alpha(n))$ time using $O(n(\alpha(n))^q)$ operations, for some fixed q. Theorem 12(a) implies that the problem can be solved in $O(\log^* n)$ time using $O(n \log^* n)$ processors. Applying Propositions 13 and 14, we obtain

Corollary 15. *The problems of Theorems 3, 5, 6, 12(b) and, for constant c, Theorems 9 and 10 can all be solved optimally in $O(\alpha(n))$ time; for the range product problems, both the preprocessing time and the sequential query time is $O(\alpha(n))$. The problem of Theorem 12(a) can be solved optimally in $O(\log^* n)$ time, yielding constant sequential query time.*

References

1. N. Alon and N. Megiddo, Parallel Linear Programming in Fixed Dimension Almost Surely in Constant Time, *J. Assoc. Comput. Mach.* **41** (1994), pp. 422–434.
2. N. Alon and B. Schieber, Optimal Preprocessing for Answering On-line Product Queries, Tech. Rep. No. 71/87, Tel-Aviv University, 1987.
3. O. Berkman, Y. Matias and U. Vishkin, Randomized Range-Maxima in Nearly-Constant Parallel Time, *Comput. Complexity* **2** (1992), pp. 350–373.
4. O. Berkman and U. Vishkin, Recursive Star-Tree Parallel Data Structure, *SIAM J. Comput.* **22** (1993), pp. 221–242.
5. O. Berkman and U. Vishkin, On Parallel Integer Merging, *Inform. and Computation* **106** (1993), pp. 266–285.
6. A. K. Chandra, S. Fortune and R. Lipton, Lower Bounds for Constant Depth Circuits for Prefix Problems, Proc. 10th International Colloquium on Automata, Languages and Programming (1983), Springer Lecture Notes in Computer Science, Vol. 154, pp. 109–117.
7. A. K. Chandra, S. Fortune and R. Lipton, Unbounded Fan-in Circuits and Associative Functions, *J. Comput. Syst. Sci.* **30** (1985), pp. 222–234.
8. D. Eppstein and Z. Galil, Parallel Algorithmic Techniques for Combinatorial Computation, *Ann. Rev. Comput. Sci.* **3** (1988), pp. 233–283.
9. T. Hagerup, The Log-Star Revolution, Proc., 9th Annual Symposium on Theoretical Aspects of Computer Science (1992), Springer Lecture Notes in Computer Science, Vol. 577, pp. 259–278.
10. R. E. Ladner and M. J. Fischer, Parallel Prefix Computation, *J. Assoc. Comput. Mach.* **27** (1980), pp. 831–838.
11. P. Ragde, The Parallel Simplicity of Compaction and Chaining. *J. Alg.* **14** (1993), pp. 371–380.
12. P. Raghavan, Probabilistic Construction of Deterministic Algorithms: Approximating Packing Integer Programs, *J. Comput. Syst. Sci.* **37** (1988), pp. 130–143.
13. L. G. Valiant, Parallelism in Comparison Problems, *SIAM J. Comput.* **4** (1975), pp. 348–355.

The Complexity of Broadcasting
in Planar and Decomposable Graphs

Andreas Jakoby* and Rüdiger Reischuk and Christian Schindelhauer

Med. Universität zu Lübeck***

Abstract. Broadcasting in processor networks means disseminating a
single piece of information, which is originally known only at some nodes,
to all members of the network. The goal is to inform everybody using as
few rounds as possible, that is to minimize the broadcasting time.

Given a graph and a subset of nodes, the sources, the problem to determine its specific broadcast time, or more general to find a broadcast
schedule of minimal length has been shown to be $\mathcal{NP}$-complete. In contrast to other optimization problems for graphs, like vertex cover or traveling salesman, little was known about restricted graph classes for which
polynomial time algorithms exist, for example for graphs of bounded
treewidth. The broadcasting problem is harder in this respect because
it does not have the finite state property. Here, we will investigate this
problem in detail and prove that it remains hard even if one restricts
to planar graphs of bounded degree or constant broadcasting time. A
simple consequence is that the minimal broadcasting time cannot even
be approximated with an error less than 1/8, unless $\mathcal{P} = \mathcal{NP}$.

On the other hand, we will investigate for which classes of graphs this
problem can be solved efficiently and show that broadcasting and even
a more general version of this problem becomes easy for graphs with
good decomposition properties. The solution strategy can efficiently be
parallelized, too. Combining the negative and the positive results reveals
the parameters that make broadcasting difficult. Depending on simple
graph properties the complexity jumps from $\mathcal{NC}$ or $\mathcal{P}$ to $\mathcal{NP}$.

Classification: graph algorithms, graph decomposition, computational
complexity.

1 Introduction

Broadcasting in processor networks means disseminating a single piece of information, which is originally known only at some nodes, called the *sources,* to
all members of the network. This is done in a sequence of rounds by pairwise
message exchange over the communication lines of the network. In one round
each processor can send a message to at most one of its neighbours. The goal is

---

* supported by Deutsche Forschungsgemeinschaft, Grant DFG Re672/2, and
  German-Israeli Foundation for Scientific Research & Development,
  Grant GIF I-2007-199.06/91

*** Institut für Theoretische Informatik, Wallstraße 40, 23560 Lübeck,Germany
  email: jakoby/reischuk/schindel@informatik.mu-luebeck.de

to inform everybody using as few rounds as possible. This number is called the *minimum broadcasting time* of the network.

Broadcasting is a basic task for multiprocessor systems that should be supported by the topology of the network. This problem has been studied extensively, mostly in the case of a single source – for an overview see [HHL88]. In several papers the broadcast capabilites of well known families of graphs like hypercubes, cube-connected-cycles, shuffle exchange graphs or de Bruijn graphs have been investigated and compared. In [HJM90] Hromkovič, Jeschke and Monien have studied the relation between the broadcasting time and the time for solving the related gossiping problem for special families of graphs.

On the other hand, one has tried to find optimal topologies for networks with a given number of nodes such that the broadcasting time is best possible. Here the worst case over all nodes as the single source should be minimized. The problem gets more complicated when restricting to graphs of bounded degree. In [LP88] Listman and Peters have studied several classes of bounded degree graphs in this respect, see also [BHLP92]. Balanced binary trees already achieve a broadcasting time of logarithmic order, therefore the question is the optimal constant factor in front of the logarithm.

In this paper we will investigate the optimization problem for arbitrary networks. That means, given a graph and a subset of nodes as sources, determine its specific broadcast time or more general find a broadcast schedule of minimal length. This problem in general is $\mathcal{NP}$-complete. We will show that this property remains even if one restricts to planar graphs of bounded degree or constant broadcast time. Furthermore, the problem cannot be solved approximately with an arbitrary precision unless $\mathcal{P} = \mathcal{NP}$.

On the other hand, we will investigate for which classes of graphs this problem can be solved efficiently. All what seems to be known was that broadcasting is easy for trees shown by Slater, Cockayne, and Hedetniemi in [SCH81]. Many combinatorial optimization problems for graphs have been shown to be solvable in polynomial sequential time and even in polylogarithmic parallel time for more general classes of graphs: graphs of bounded treewidth (see for example the paper by Arnborg, Lagergren and Seese [ALS91]) and graphs of small connectivity ([R91a]) – an overview can be found in [R91b].

The broadcasting problem seems to be more difficult in this respect since it does not have the finite-state-property or a bounded number of equivalence classes. Thus the methods of [ALS91] and [R91a] are not directly applicable. Still, modifying the framework developed in [R91a] we can show that broadcasting becomes easy for graphs with good decomposition properties. For this purpose we consider edge decompositions of a graph. This notion seems more appropriate for the broadcasting problem than the dual approach based on vertex decompositions. Furthermore, a careful inspection of the possibilities how information can flow within a component and between different components of a graph will be required. For the internal flow components that are connected behave most favourably, but in general connectivity cannot always be achieved by an edge decomposition that generates small components. Disconnected components complicate the analysis of the broadcasting problem significantly, whereas most other optimization problems become easier in such a case. The algorithm even works for a more general version of the broadcasting problem. Furthermore, it can be parallelized efficiently to yield $\mathcal{NC}$-solutions.

As a conclusion we can say that combining these new negative and positive results the parameters that make broadcasting difficult are determined quite precisely. The complexity of this problem jumps from $\mathcal{P}$ to $\mathcal{NP}$ depending on the internal structure of the networks.

2   Definitions and Previous Results

A formal definition of the broadcasting problem can be given as follows.

Definition 1 *Let $G = (V, E)$ be a (directed) graph with a distinguished subset of vertices $V_0 \subseteq V$, the **sources**, and $T^* \in \mathbb{N}$ be a deadline. The task is to decide whether there exists a **broadcast schedule**, that is a sequence of subsets of edges $E_1, E_2, \ldots, E_{T^*-1}, E_{T^*}$ with the property $V_{T^*} = V$, where for $i > 0$ we define $V_i := V_{i-1} \cup \{ v \mid (u, v) \in E_i \text{ and } u \in V_{i-1} \}$ and require $E_i \subseteq \{ (u, v) \in E \mid u \in V_{i-1} \}$ and $\forall u \in V_{i-1} : |E_i \cap (\{u\} \times V)| \leq 1$.*
*Let us distinguish between the general multiple source problem **MB** and the restricted version with only a single source **SB**.*   □

The meaning of the sets E_i and V_i is the following: V_i denotes the set of nodes that have received the broadcast information by round i. For $i = 0$ this is just the set of sources. By the deadline T^* the set V_{T^*} should include all nodes of the network. E_i is the set of edges that are used to send information at round i, where each processor $u \in V_{i-1}$ can use at most one of its outgoing edges.

MB (denoted by ND49 in [GJ79]) has shown to be $\mathcal{NP}$-complete.

Theorem A: *MB for graphs with unbounded degree is $\mathcal{NP}$-complete, even if restricted to a fixed deadline $T^* \geq 4$.*

For a fixed deadline the number of sources obviously has to grow linearly in the size of the whole graph. But even the single source problem is difficult, in this case the deadline has to grow at least logarithmically.

Theorem B: *SB for graphs with unbounded degree is $\mathcal{NP}$-complete.*

The proofs of both results were published by Slater, Cockayne, and Hedetniemi ([SCH81]). For the second result, their reduction of the 3-dimensional matching problem to SB requires a deadline of order $\sqrt[3]{|V|}$ for the broadcast problem. Furthermore, in the same paper it is shown

Theorem C: *SB can be solved in linear time for trees. This also holds for the constructive version of this problem finding an optimal broadcast schedule.*

3   New Results

All theorems above can be improved significantly. For the lower bounds it suffices to consider undirected graphs, the upper bounds given below also hold for the more general case of directed graphs.

3.1 Lower Bounds

Designing more complicated reductions of the 3-dimensional matching problem and the 3-SAT problem we can show

Theorem 1 *MB restricted to planar graphs with bounded degree at least 4 and a fixed deadline T^* at least 3 is $\mathcal{NP}$-complete.*

The reduction to prove this result uses graphs of a specific kind that are guaranteed to have a broadcast schedule of length 4. Now consider an approximation algorithm that for a network G gives an estimate of its minimum broadcast time $T(G)$. The estimate $\tilde{T}(G)$ may be an arbitrary real number, but is required to be within a precision γ of the correct value: $(1+\gamma)^{-1} \cdot T(G) \leq \tilde{T}(G) \leq (1+\gamma) \cdot T(G)$. In this case, any estimate with a precision $\gamma \leq \frac{1}{8}$ could be used to solve the decision problem. Thus we also get

Theorem 2 *There exists no polynomial-time approximating algorithm for MB with a precision $1/8$, unless $\mathcal{P} = \mathcal{NP}$.*

If for this minimization problem one restricts to integer values and approximations from above then the statement of the theorem can be improved from precision $1/8$ to any value less than $1/3$.

Broadcasting with a single source does not become substantially easier, even for bounded degree graphs with a logarithmic diameter.

Theorem 3 *SB restricted to graphs $G = (V, E)$ with bounded degree at least 3 is $\mathcal{NP}$-complete, even if the deadline grows at most logarithmically in the size of the graph.*

Also planarity does not make things much simpler as the following result shows.

Theorem 4 *SB restricted to planar graphs $G = (V, E)$ of degree 3 is $\mathcal{NP}$-complete (in this case the deadline grows like $\sqrt{|V|}$).*

3.2 Upper Bounds

On the positive side, we will extend the classes of graphs for which the broadcasting problem can be solved fast. The general approach is to partition a given graph into smaller components and then solve a generalized broadcast problem on each component separately and later combine the partial solutions. A split into smaller components can be done by removing edges (edge separators) or by removing nodes (node separators). In general, edge separation seems to be a weaker notion since graphs that contain nodes of very large degree cannot be split into small pieces by separators of small size, whereas small node separators may exist (consider a star). However, for graphs of bounded degree both notions are closely related. Here we will restrict ourself to edge separators because then the analysis of the broadcast problem is somewhat easier. But similar algorithmic techniques can also be used for graph decompositions based on node separators.

For the purpose of decomposing a graph G it suffices to consider only the case of undirected graphs. Thus, if G is directed in the following definition we simply mean the corresponding undirected graph.

Definition 2 *A graph $H = (V_H, E_H)$ is an **edge decomposition graph** of a graph $G = (V, E)$ if the following conditions hold:*
- *The nodes G_i of V_H represent induced subgraphs $G_i = (V_i, E_i)$ of G such that the V_i are pairwise disjoint and $V = \bigcup_{G_i \in V_H} V_i$.*

- *$\{G_i, G_j\} \in E_H$ iff there is an edge between a node of G_i and a node of G_j.*

*H is called an **edge decomposition tree** of G if H is a tree.*
*Define the **cut** of an edge $\{G_i, G_j\}$, the cut of a node G_i, and the cut of H as those edges of G that connect G_i and G_j, resp. connect G_i to other components, or connect any pair of components:*

$$\text{cut}(G_i, G_j) := \{\ \{u, v\} \in E \mid u \in V_i \text{ and } v \in V_j\} \quad \text{for } i \neq j\ ,$$

$$\text{cut}(G_i) := \bigcup_{\{G_i, G_j\} \in E_H} \text{cut}(G_i, G_j) \quad \text{and} \quad \text{cut}(H) := \bigcup_{G_i \in V_H} \text{cut}(G_i)\ .$$

*A graph $G = (V, E)$ is $(\boldsymbol{\kappa, \mu, c})$–**edge decomposable** if there exists an edge decomposition graph $H = (V_H, E_H)$ such that for all $G_i \in V_H$:*

$$|\text{cut}(G_i)| \leq \kappa\ , \qquad |V_i| \leq \mu \qquad \text{and} \qquad \text{cc}(G_i) \leq c\ ,$$

where $\text{cc}(G_i)$ denotes the number of connected component of G_i. $\qquad\square$

The decomposition process partitions a graph into different components. Each component G_i itself may be connected or fall into several connected components. When constructing an optimal broadcast schedule this issue turns out to be important. For example, a $\sqrt{n} \times \sqrt{n} - 2$-dimensional grid is $(O(\sqrt{n}), (O(\sqrt{n}), 1)$-edge decomposable into a tree. The parameters are $(4, 2, 2)$ for a circle of length n. Taking the number of connected components within each component into consideration will allow us to determine good upper bounds for the algorithmic effort to solve the broadcasting problem.

Other approaches have been proposed how to decompose a graph into smaller components, based on the notions of treewidth ([RS86]), see for example [ALS91, BK91, L90], and k-connected components using minimal node separators [H90, HR89]. In general, it is $\mathcal{NP}$-complete to construct optimal decompositions with respect to the relevant parameters. However, suboptimal solutions can be found in polynomial time.

In the following we assume that an arbitrary edge decomposition of the network is given and analyse only the complexity of constructing an optimal broadcast schedule with the help of that decomposition. Note that the broadcast time will always be best possible, only the complexity of finding the schedule will depend on the parameters of the decomposition.

Theorem 5 *For a graph $G = (V, E)$ of maximal degree d with a given (κ, μ, c)-edge decomposition tree the MB-problem can be solved in time*

$$O\Big(|V|^c \cdot (2(\kappa + \mu))^{4\kappa} \cdot (|V| \cdot (\mu + \kappa)^3 \cdot (d+1)^\mu + |E|)\Big)$$

$$\leq \exp O\Big(c \cdot \log |V| + \kappa \cdot \log(\kappa + \mu) + \mu \cdot \log d\Big).$$

The algorithm we have designed actually works for a more general version of the broadcasting problem, in which the sources may receive the broadcast information at different rounds and each node of the network may have its individual deadline. Let us call this the *general broadcasting problem* **GB** (see definition 3).

The time bound becomes polynomial for classes of graphs that can be decomposed into smaller components using not too large separators. Let **llog** and **lllog** denote the logarithm function iterated twice, resp. three times.

Corollary 1. *Restricted to graphs $G = (V, E)$ with*

- $(O(\frac{\log n}{\text{llog } n}), O(\frac{\log n}{\text{llog } n}), O(1))$–*edge decomposition trees or*

- *to graphs with bounded degree and $(O(\frac{\log n}{\text{lllog } n}), O(\log n), O(1))$–edge decomposition trees*

the MB–problem (and even the GB-problem) can be solved in polynomial time.

So far, we have only considered the decision version of the MB-problem, resp. the task to determine the minimal length of a broadcast schedule. But applying ideas similar to the one in [R91a] one can also design an algorithm for constructing an optimal broadcast schedule by using the same techniques as for the decision problem.

Theorem 6 *Constructing an optimal broadcast schedule can be done in the same time bounds as stated for the decision problem in Theorem 5.*

Using the machinery developed in [R91a] we can also derive a fast and processor efficient parallel algorithm. Even if the decomposition tree is not nicely balanced using path compression techniques the problem can be solved with a logarithmic number of iterations (with respect to the number of components) on a standard parallel machine model like a PRAM with concurrent write capabilities. The basic task one has to solve is how a chain of 2 components can be replaced by a single component that externally behaves identically with respect to broadcasting.

Theorem 7 *For a graph $G = (V, E)$ of maximal degree d with a given (κ, μ, c)–edge decomposition tree MB can be solved in parallel time*

$$O\Big(\log |V| \cdot (c \cdot \log |V| + \kappa \cdot \log(\kappa + \mu) + \mu \cdot \log d)\Big)$$

with a processor bound of $\exp O\Big(c \cdot \log |V| + \kappa \cdot \log(\kappa + \mu) + \mu \cdot \log d\Big).$

For nicely decomposable classes of graphs these bounds put MB into $\mathcal{NC}$.

Corollary 2. *Restricted to graphs $G = (V, E)$ with*

- $(O(\frac{\log n}{l \log n}), O(\frac{\log n}{l \log n}), O(1))$-*edge decomposition trees or*

- *to graphs with bounded degree and* $(O(\frac{\log n}{l \log n}), O(\log n), O(1))$-*edge decomposition trees*

MB is in $\mathcal{NC}^2$.

All these bounds apply to the GB–problem as well. Also the constructive variant can be solved with the same effort.

In the remaining part of this extended abstract we can only present some of the basic ideas to prove these results. For a complete version see [JRS93].

4   Proofs of the Lower Bounds

The $\mathcal{NP}$-hardness of·the minimum broadcasting time problem will be obtained by a reduction of the 3DM problem (3-Dimensional Matching, see [GJ79]). The graph $G' = G(A, B, C, M)$ of an instance (A, B, C, M) with $M \subseteq A \times B \times C$ of the 3DM problem is defined as follows: Each element of the sets A, B, and C and each triple of M is represented by a vertex. The membership relation between set elements and triples defines the edges between these vertices.

$$G' := (V', E') \text{ with } V_A := \{\, \alpha_x \mid x \in A \,\}, \ V_B := \{\, \beta_x \mid x \in B \,\},$$
$$V_C := \{\, \gamma_x \mid x \in C \,\}, V_M := \{\, \mu_y \mid y \in M \,\}, V' := V_A \cup V_B \cup V_C \cup V_M,$$
$$E' := \{\, (\mu_y, v_x) \mid y \in M, \ v_x \in V_A \cup V_B \cup V_C \text{ and } x \in y \,\}$$

The reduction will use a restricted planar version of the 3DM problem, which is still $\mathcal{NP}$-complete [DF86]. For an instance (A, B, C, M) the following properties are required:

- $G(A, B, C, M)$ is planar.

- For each element x of $A \cup B \cup C$ there are at most 3 triples in M containing x (thus, $|M|$ is bounded by $3q$ where $q := |A| = |B| = |C|$).

   Proof of Theorem 1 and 2: Planar 3DM will be reduced to the broadcasting problem as follows:

Let (A, B, C, M) be an instance of 3DM with $|A| = q$ and let $G' = G(A, B, C, M)$ be the matching graph. The corresponding broadcasting graph G is obtained by replacing each node $\alpha_i \in V_a$ of G' by a chain $\alpha_{i,1}$, $\alpha_{i,2}$, and $\alpha_{i,3}$ of length 3 (see Figure 1). The other nodes and edges remain unchanged. $V_{A,1}$ is chosen as the set of sources, and the deadline is set to 3.

Observe that G has degree 4 and is planar if G' is planar.

Lemma 1 *G has a broadcast schedule of length 3 iff the q sources in round 1 send the information to a subset of the nodes in V_M that defines a matching.*

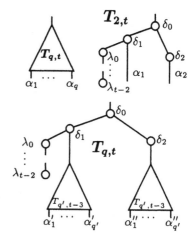

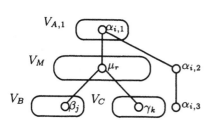

Fig. 1. *The broadcasting graph corresponding to an instance of the 3DM problem.*

Fig. 2. *The recursive construction of the trees $T_{q,t}$. The λ_i form a chain that forces the left son of δ_0 to be informed first in order to obey the deadline t. For $q > 2$ the tree $T_{q,t}$ is obtained from 2 copies of $T_{\lceil q/2 \rceil, t-3}$.*

Theorem 1 and 2 now follow easily. ∎

Proof of Theorem 3: Consider the tree $T_{q,t}$ in figure 2, where its root is the only source and $q \leq \exp t/3$ outgoing edges α_i. It has the following properties:

- Within $\delta := 3\lceil \log q \rceil - 1$ rounds $T_{q,t}$ can reach a state such that in the next round the information of the source can be propagated simultaneously over all outgoing edges α_i.

- If a broadcast schedule for $T_{q,t}$ finishes by round t then none of these edges can propagate the information before round $\delta + 1$.

Connect each leaf of the tree with a node of $V_{A,1}$ of the graph G defined above and call this new graph G'. Let $t := 3\lceil \log q \rceil + 3$ and let w be the the root of $T_{q,t}$. Then, G' with source w has a broadcast schedule of length at most $3 \log q + 3$ iff M contains a matching. The resulting graph has degree 5, but is not necessarily planar. By additional effort G' can be modified to decrease the node degree to 3. ∎

Proof of Theorem 4: To achieve planarity in the single source case we construct a direct reduction for the satisfiability problem 3SAT. Although a planar version of 3SAT remains $\mathcal{NP}$-complete it does not help much in this case because the connections to the source will destroy the planarity. A simple exchange of an edge crossing by a planar subgraph with 4 input/output-edges also does not seem to work for the broadcasting problem.

We have found a way to make such a replacement legal under special circumstances, namely if the direction of the information flow over the edges is known in advance and if at most one of the two input edges is used. The first property can easily be achieved for 3SAT, while the second requires special coding tricks.

The details are quite long and can be found in the full version [JRS93]. ∎

5 Efficient Algorithms for Decomposable Graphs

We start with a generalization of the broadcast problem. So far, each source node has got the broadcast information at round 0. In the more general case, a source may get the information with an arbitrary delay of $\sigma(v)$ rounds. Furthermore, we require that the information has to reach each node v by round $\rho(v)$, instead of a global deadline T^* identical for all nodes. This generalization may be of less interest with respect to practical applications. Nevertheless, it is necessary in order to apply an approach based on graph decompositions, as it has been for several other graph theoretic decision and optimization problems.

Definition 3 *General Broadcasting Problem* **GB**:
Given a graph $G = (V, E)$ and two partial functions $\sigma, \rho : V \rightarrow \mathbb{N}$, decide whether there exists a broadcast schedule $E_1, E_2, \ldots$, with

$$V_i = V_{i-1} \cup \{ v \mid (u,v) \in E_i \text{ and } u \in V_{i-1} \} \cup \{ v \in V \mid \sigma(v) = i \} ,$$

$$E_i \subseteq \{ (u,v) \in E \mid u \in V_{i-1} \} \quad \text{and} \quad \forall u \in V_{i-1} : \mid E_i \cap (\{u\} \times V) \mid \leq 1$$

such that $\forall v \in V : v \in V_{\rho(v)}$. □

The following exponential upper bounds can be achieved by standard enumeration methods.

Lemma 2 *GB can be solved in time $O(|V|^3 (d+1)^{|V|})$, where d denotes the degree of the graph $G = (V, E)$.* ∎

Lemma 3 *GB restricted to graphs $G = (V, E)$ with maximum degree d can be solved by a CRCW-PRAM with $O(|V| \cdot (d+1)^{|V|})$ processors in time $O(|V|^2)$.* ∎

Definition 4 *Let a graph $G = (V, E)$ and a broadcast schedule $\mathcal{E} = E_1, E_2, \ldots$ for G be given. The round in which a node receives the information to be broadcasted first is called its **starting round**. A node u is called **active** at a round t if sends the information to another node v ($\{u,v\} \in E_t$) at this round. A node u is called **busy** if it is active after its starting round until all its neighbors have received the information. $\mathcal{E}$ is **busy** if all nodes are busy and no node receives the broadcast information several times.* □

Observe that each broadcast schedule can easily be tranformed into a busy broadcast schedule of the same or smaller length.

Proof of Theorem 5: Let $H = (V_H, E_H)$ be a (κ, μ, c)–edge decomposition tree of the graph $G = (V, E)$ with $V_H = \{G_1, \ldots, G_k\}$. Figure 3 shows a component G_2 which is connected to three other components G_1, G_3 and G_4.

Let $\mathcal{G}_i := \{ G_i^1, \ldots, G_i^{c_i} \}$ with $G_i^a = (V_i^a, E_i^a)$ be the set of connected components of the G_i and define $cut(G_i^a)$ as the set of edges of $cut(G_i)$ with one endpoint in G_i^a.

$$cut(G_i^a, G_j) := cut(G_i, G_j) \cap cut(G_i^a), \quad cut(G_i^a, G_j^b) := cut(G_i^a, G_j) \cap cut(G_i^b) .$$

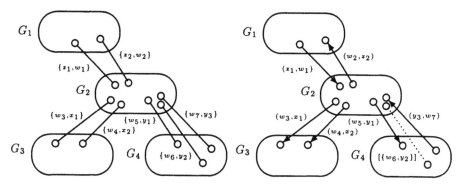

Fig. 3. A node G_2 of an edge decomposition tree and its neighbors.

Fig. 4. A possible information flow within a broadcast schedule: from G_2 to other components in one or in both directions, some edges may not be used.

To describe a broadcasting schedule of G, each edge $\{u, v\}$ of G is labelled by a tupel $(\tau(u, v), r(u, v))$. The first value $\tau(u, v)$ denotes the round when this edge is used and the second $r(u, v)$ the direction (from u to v or form v to u if u and v are connected in both directions). If this edge is not used we set $\tau(u, v) := -1$.

If we restrict $\mathcal{E}$ to $\text{cut}(G_i^a, G_j^b)$ it suffices to denote the round $\tau(G_i^a, G_j^b)$ when an edge of $\text{cut}(G_i^a, G_j^b)$ is used for the first time and for each other edge $\{u, v\}$ with $\tau(u, v) \geq 0$ the **relative round** $\hat{\tau}(u, v) := \tau(u, v) - \tau(G_i^a, G_j^b)$. If no edge in $\text{cut}(G_i^a, G_j^b)$ is used let $\tau(G_i^a, G_j^b) := -1$. For G_i^a let $\tau(G_i^a)$ be the first round when a node of G_i^a receives the broadcast information. Similarly define $\hat{\tau}(G_i^a, G_j^b) := \tau(G_i^a, G_j^b) - \tau(G_i^a)$ if $\tau(G_i^a, G_j^b) \geq 0$, else let $\hat{\tau}(G_i^a, G_j^b) := -1$.

The following two lemmata show that with the help of the relative rounds $\hat{\tau}$ the number of possible protocolls of information exchange between two components can be bounded quite substantially. This property will be basic for the time efficiency of the algorithm described below. It serves as a replacement for the "finite state property" shared by graph theoretic problems like independent set or vertex cover, for which efficient solutions based on the decompositional approach have been given before.

Lemma 4 *If $\mathcal{E}$ is a busy broadcast schedule then for all $\{u, v\} \in \text{cut}(G_i^a, G_j^b)$ holds: the numbers $\hat{\tau}(G_i^a, G_j^b) + \hat{\tau}(u, v)$ and $\hat{\tau}(G_j^b, G_i^a) + \hat{\tau}(u, v)$ are smaller than $|\text{cut}(G_i)| + |V_i| \leq \kappa + \mu$.* ∎

Let G_i, G_j be neighbouring components and G_i^a and G_j^b connected components of G_i, resp. G_j with nonempty $\text{cut}(G_i^a, G_j^b)$. Define a **state** between the two neighbours as a tupel $\gamma_{i,j} := \left(\left[\hat{\tau}(G_i^a, G_j^b), \hat{\tau}(G_j^b, G_i^a) \mid \text{cut}(G_i^a, G_j^b) \neq \emptyset \right], \left[(\hat{\tau}(e), r(e)) \mid e \in \text{cut}(G_i, G_j) \right] \right)$.

Figure 4 illustrates a complex information flow between a component and its neighbours. The **surface** $\Gamma_{i,j}$ is the set of all possible states $\gamma_{i,j}$ of busy broadcast schedules. A **state** S_i of a component G_i is a vector consisting of the starting round $\tau(G_i^a)$ for all connected components of G_i and tupels $\gamma_{i,j}$ for all neighbouring components of G_i. As above, let Γ_i be the set of all possible states S_i that may appear in busy schedules.

Lemma 5 *For a component G_i with cutsize $|\text{cut}(G_i)| \le \kappa_i$, size $|V_i| \le \mu_i$ and $c_i = \text{cc}(G_i)$ connected components, the size of Γ_i is bounded by $\gamma(\kappa_i, \mu_i, c_i) :=$ $|V|^{c_i} \cdot (2(\kappa_i + \mu_i))^{3\kappa_i}$.*

Proof:
$$|\Gamma_i| \le \prod_{G_i^a \in \mathcal{G}_i} |V| \cdot \prod_{\text{cut}(G_i,G_j) \ne \emptyset} |\Gamma_{i,j}|$$
$$\le |V|^{c_i} \cdot \prod_{\text{cut}(G_i,G_j) \ne \emptyset} (\kappa_i + \mu_i)^{2|\text{cut}(G_i,G_j)|} \cdot (2(\kappa_i + \mu_i))^{|\text{cut}(G_i,G_j)|}$$
$$\le |V|^{c_i} \cdot (2(\kappa_i + \mu_i))^{3\kappa_i} . \qquad \blacksquare$$

The following strategy solves the minimum broadcasting time problem for graphs $G = (V, E)$ with a given (κ, μ, c)–edge decomposition tree $H = (V_H, E_H)$. Let $\Delta(G_i, S_i)$ denote the minimal schedule length of the local broadcast problem for the graph G_i and external information exchange as specified by state S_i ($= \infty$ if there is no schedule for state S_i). Observe that this value is independent of the structure of G outside of G_i.

Step 1: For each component G_i and each state $S_i \in \Gamma_i$ determine $\Delta(G_i, S_i)$.

Step 2: Choose an arbitrary component G_r and declare G_r as the root of H. Let $G_{i,0}$ be the father of G_i in H according to the orientation with respect to G_r. Let G_i^* denote the subgraph of G containing G_i and all its descendents. Evaluate the function $\Delta(G_i^*, S_{i,0})$ for all G_i and $S_{i,0}$ starting with the leaf components of H.

Lemma 6 *Let $G_{i,1}, \ldots, G_{i,\ell_i}$ denote the sons of G_i and let $S_{i,j}$ be a state connecting G_i and $G_{i,j}$. For all G_i the minimal deadline for the general broadcast problem for G_i^* with respect to external information exchange $S_{i,0}$ can be computed as*

$$\Delta(G_i^*, S_{i,0}) = \min_{\substack{S_i = (\tau_1, \ldots, \tau_{cc(G_i)}), \\ S_{i,0}, \ldots, S_{i,\ell_i}) \in \Gamma_i}} \max_{j \in [1 \ldots l_i]} \{\Delta(G_{i,j}^*, S_{j,i})\} \cup \{\Delta(G_i, S_i)\} .$$

Proof: This property can be shown by induction on the depth of the subgraphs G_i. $\qquad \blacksquare$

Therefore, $\Delta(G_r^*) = \Delta(G_r^*, \lambda)$ denotes the minimal schedule length for the graph G itself.

The correctness of step 1 follows directly from the definition of a surface and the definition of the general broadcast problem. The correctness of step 2 follows directly from Lemma 6.

According to lemma 2 for each G_i and S_i the computation of $\Delta(G_i, S_i)$ requires at most $O((\mu_i + \kappa_i)^3 \cdot (d+1)^{\mu_i})$ many steps. Since lemma 5 gives a bound on the number of states of each component step 1 can be executed in time

$$\sum_{G_i} \gamma(\kappa_i, \mu_i, c_i) \cdot O\left((\mu_i + \kappa_i)^3 (d+1)^{\mu_i}\right) \leq O\left(|V|^{c+1} (2(\kappa+\mu))^{3\kappa} \cdot (\mu+\kappa)^3 \cdot (d+1)^{\mu}\right).$$

The computation of $\Delta(G_i^*, S_{i,0})$ is independent of the remaining structure of G. Thus for each G_i, given all values $\Delta(G_i, S_i)$ and $\Delta(G_{i,j}^*, S_{j,i})$ the computation of all $\Delta(G_i^*, S_{i,0})$ can be executed in time $O(\gamma(\kappa_i, \mu_i, c_i) \cdot \ell_i)$. Summing up over all G_i gives the bound

$$\sum_{G_i} O(\gamma(\kappa_i, \mu_i, c_i) \cdot \ell_i) \leq O\left(\gamma(\kappa, \mu, c) \cdot \sum_{G_i} \ell_i\right) \leq O(|E| \cdot |V|^c \cdot (2(\kappa+\mu))^{3\kappa}) .$$

$\blacksquare$

By using tree contraction methods the evaluation of the Δ-function can also be done in parallel requiring only a logarithmic number of iterations. The details are described in [R91a].

Conclusions

We have shown that the single source broadcasting problem remains hard for planar graphs with high internal connectivity. After we have presented the lower bound in Theorem 1 Middendorf was able to improve our construction to degree 3 and deadline 2 [M93].

On the other hand, even a much more general version with many sinks and individual deadlines can be solved efficiently on graphs with moderate internal connections.

The algorithms described are based on edge decomposition of graphs. The same technique with a slightly worse time bound due to a larger number of states also works for other graph decomposition methods, for example based on treewidth. These issues are included in the full paper [JRS93].

References

[ALS91]   S. Arnborg, J. Lagergren, and D. Seese, *Easy Problems for Tree-Decomposable Graphs* J. Algorithms 12, 1991, 308-340.

[BHLP92] J.-C. Bermond, P. Hell, A. Liestman, and J. Peters,   *Broadcasting in Bounded Degree Graphs*, SIAM J. Disc. Math. 5, 1992, 10-24.

[BK91]   H. Bodlaender, T. Kloks, *Better Algorithms for the Pathwidth and Tree-width of Graphs*, Proc. 18 ICALP, 1991, 544-555.

[DF86]   M. Dyer and A. Frieze, *Planar 3DM Is $\mathcal{NP}$-Complete*, J. Algorithms 7, 1986, 174-184.

[GJ79]   M. Garey and D. Johnson, *Computers and Intractability, A Guide To the Theory of $\mathcal{NP}$-Completeness*, Freeman 1979.

[H90]   W. Hohberg, *The Decomposition of Graphs into k-Connected Components for Arbitrary k,* Technical Report, TH Darmstadt, 1990.

[HHL88]   S. Hedetniemi, S. Hedetniemi, and A. Liestman, *A Survey of Gossiping and Broadcasting in Communication Networks,* Networks 18, 1988, 319-349.

[HJM90]   J. Hromkovič, C.-D. Jeschke, and B. Moinien, *Optimal Algorithms for Dissemination of Information in Some Interconnection Networks,* Proc. 15. MFCS, 1990, 337-346.

[HR89]   W. Hohberg, R. Reischuk, *Decomposition of Graphs – A Uniform Approach for the Design of Fast Sequential and Parallel Algorithms on Graphs,* Technical Report, TH Darmstadt, 1989.

[JRS93]   A. Jakoby, R. Reischuk, C. Schindelhauer, *The Complexity of Broadcasting in Planar and Decomposable Graphs,* Technical Report, TH Darmstadt, 1993.

[L90]   J. Lagergren, *Efficient Parallel Algorithms for Tree-Decomposition and Related Problems,* Proc. 31. FoCS, 1990, 173-182.

[LP88]   A. Liestman and J. Peters, *Broadcast Networks of Bounded Degree,* SIAM J. Disc. Math. 4, 1988, 531-540.

[M93]   M. Middendorf, *Minimum Broadcast Time is $\mathcal{NP}$-complete for 3-regular planar graphs and deadline 2,* IPL 46, 1993, 281-287.

[R91a]   R. Reischuk, *An Algebraic Divide-and-Conquer Approach to Design Highly Parallel Solution Strategies for Optimization Problems on Graphs,* Technical Report, TH Darmstadt, 1991.

[R91b]   R. Reischuk, *Graph Theoretical Methods for the Design of Parallel Algorithms,* Proc. 8. FCT, 1991, 61-67.

[RS86]   N. Robertson, P. Seymour, *Graph Minors II. Algorithmic Aspects of Tree-Width,* J. Alg. 7, 1986, 309-322.

[SCH81]   P. Slater, E. Cockayne, and S. Hedetniemi, *Information Dissemination in Trees,* SIAM J. Comput. 10, 1981, 692-701.

The Maximal f-Dependent Set Problem for Planar Graphs is in NC

Zhi-Zhong Chen

Department of Mathematical Sciences
Tokyo Denki University
Hatoyama, Saitama 350-03, Japan
Email: chen@r.dendai.ac.jp

Abstract. The maximal f-dependent set (Max-f-DS) problem is the following problem: Given a graph $G = (V, E)$ and a nonnegative integer-valued function f defined on V, find a maximal subset U of V such that no vertex $u \in U$ has degree $> f(u)$ in the subgraph induced by U. Whether the problem is in NC (or RNC) or not is an open question. Concerning this question, only a rather trivial result due to Diks, Garrido, and Lingas is known up to now, which says that the problem can be solved in NC if the maximum value of f is poly-logarithmic in the input size [*Proceedings of the 2nd International Symposium on Algorithms*, LNCS **557** (1991) 385-395]. In this paper, we show a nontrivial interesting result that the Max-f-DS problem for planar graphs can be solved in $O(\log^5 n)$ time with $O(n)$ processors on a CRCW PRAM, where n is the input size.

1 Introduction

The maximal independent set (MIS) problem is perhaps one of the most fundamental and familiar problem in parallel computation. This problem was conjectured to be not in NC by Valiant [Val82]. However, Karp and Wigderson disproved Valiant's conjecture in 1985 by giving an NC algorithm for the MIS problem [KW85]. Subsequently, more efficient NC or RNC algorithms for the problem were proposed in [ABI86, Lub86, GS89a, GS89b]. These algorithms have been proved to be very useful in designing NC or RNC algorithms for many other interesting problems [CH93, CY86, DGL91, GPS87, PV93, SM91].

In [DGL91], Diks *et al.* considered a natural generalization of the MIS problem, namely, the *maximal f-dependent set* (Max-f-DS) *problem*. The Max-f-DS problem is defined as follows: Given a graph $G = (V, E)$ and a nonnegative integer-valued function f defined on V, find a maximal subset U of V such that no vertex $u \in U$ has degree $> f(u)$ in the subgraph induced by U. Note that if $f(v) = 0$ for all vertices v of G, then the Max-f-DS problem is just the MIS problem. We here give a brief review of Diks *et al.*'s motivation of considering the Max-f-DS problem. Recall that there is a well-known notion of f-*matching* in the literature [LP86]. Let f be a positive integer-valued function defined on the set of vertices of a graph G. A subset F of the set of edges of G forms an f-matching if no vertex u of G has degree $> f(u)$ in the subgraph induced by F. A *maximal f-matching* of G is an f-matching of G that is not properly contained in any other f-matching of G. Since f-matching is a naturally generalized notion of matching [LP86], maximal f-matching is a naturally generalized notion of maximal matching. On the other hand, like the MIS problem, the maximal matching problem is also a well-known important problem in parallel computation [II86, IS86, GR88]. Thus, it is very natural to consider the *maximal f-matching problem*, namely, the problem of finding a maximal f-matching of a given graph. Now, to see the naturalness of the Max-f-DS problem, observe that the Max-f-DS problem is just the *vertex* counterpart of the maximal f-matching problem.

In [DGL91], Diks *et al.* proved that the maximal f-matching problem is in NC by extending the technique of Israeli and Shiloach used to design an NC algorithm for the maximal matching problem [IS86]. However, unlike the maximal f-matching problem, it seems impossible to obtain an NC algorithm for the Max-f-DS problem just by extending the known techniques used to design NC algorithms for the MIS problem. In fact, whether the Max-f-DS problem is in NC or not is left as an open question in Diks *et al.*'s paper [DGL91]. Concerning this question, only a rather trivial result due to Diks *et al.* is known up to now, which says that the Max-f-DS problem can be solved in NC if the maximum value of f is poly-logarithmic in the input size [DGL91]. It is worth mentioning that a similar result was independently obtained by Shoudai and Miyano [SM91].

In this paper, we show a nontrivial and interesting result that the Max-f-DS problem for planar graphs is in NC. The parallel algorithm proposed in this paper takes $O(\log^5 n)$ time using a linear number of processors on a CRCW PRAM. Thus, it is optimal within a poly-logarithmic factor. Our algorithm fully makes use of the planarity of the input graph. We here give a simple outline of our algorithm. First, we give a parallel algorithm for solving a special case of the Max-f-DS problem for planar graphs, where the input planar graph is a bipartite one in which one part contains only vertices of degree ≤ 6. This algorithm is an NC reduction to the maximal f-matching problem. Since the maximal f-matching problem is in NC [DGL91], we get an NC algorithm for solving the special case. Second, we show an NC reduction from the Max-f-DS problem for (general) planar graphs to the special case. The two results together give us an NC algorithm for solving the Max-f-DS problem for planar graphs.

2 Basic definitions

Unless stated otherwise, all graphs in this paper are simple ones, i.e., do not have parallel edges or self-loops. Let $G = (V, E)$ be a graph. The degree of a vertex v of G is denoted by $deg_G(v)$. For a subset U of V, $G[U]$ denotes the subgraph of G induced by U. Let f be a function from V to nonnegative integers. An *f-dependent set* (f-DS) of G is a subset U of V such that no vertex $u \in U$ has degree $> f(u)$ in the graph $G[U]$. A *maximal f-dependent set* (Max-f-DS) of G is an f-DS of G that is not properly contained in any other f-DS of G. The *maximal f-dependent set* (Max-f-DS) *problem* is the following: Given a graph $G = (V, E)$ and a nonnegative integer-valued function f defined on V, find a Max-f-DS of G. Without loss of generality, we may assume, throughout this paper, that any instance $\langle G, f \rangle$ of the Max-f-DS problem satisfies that for every v of G, $f(v)$ is not larger than the number of edges of G. If $f(v) = 0$ for all vertices v of G, then an f-DS of G is an *independent set* of G and similarly a Max-f-DS of G is a *maximal independent set* (MIS) of G.

Let $G = (V, E)$ be a multigraph without any self-loop (that is, G may have some parallel edges), and let f be a positive integer-valued function defined on V. An *f-matching* of G is a subset F of E such that no vertex v of G has degree $> f(v)$ in the subgraph induced by F. A *maximal f-matching* of G is an f-matching of G that is not properly contained in any other f-matching of G.

The model of parallel computation we use is the *concurrent read concurrent write parallel random access machine* (CRCW PRAM). The model consists of a number of identical processors and a common memory. Both concurrent reads and concurrent writes of the same memory location by different processors are allowed. In the latter case, we do not care which processor actually writes. (See [KR90] for a discussion of the PRAM models.)

3   The algorithm for special planar graphs

In this section, we present a parallel algorithm for solving a special case of the Max-f-DS problem for planar graphs. Let us first give a definition. A pair $\langle G, f \rangle$ is said to be *proper* if (1) G is a bipartite planar graph (X, Y, E) with $deg_G(x) \leq 6$ for all $x \in X$ and (2) f is a function from $X \cup Y$ to positive integers with $deg_G(x) \leq f(x)$ for all $x \in X$. The algorithm given in this section is for finding a Max-f-DS of G given a proper pair $\langle G, f \rangle$. The following lemma has been shown in [CH93].

Lemma 1. [CH93]. Let $G = (X, Y, E)$ be a bipartite planar graph in which $deg_G(x) \leq 6$ for all $x \in X$ and $deg_G(y) \geq 16$ for all $y \in Y$. Then, at least $|X|/6$ vertices of X have degree at most 2 in G.

Proof. See Appendix. □

The next lemma will play a key role in proving our main result of this paper.

Lemma 2. Let $\langle G, f \rangle$ be a proper pair with $G = (X, Y, E)$. Then, there are two disjoint subsets X' and X'' of X such that (i) $|X' \cup X''| \geq c|X|$ for some constant $c > 0$, (ii) $Y \cup X'$ is an f-DS of G, and (iii) for each $x \in X''$, $Y \cup X' \cup \{x\}$ is not an f-DS of G. Moreover, X' and X'' can be found in $O(\log^3 n)$ time using $O(n)$ processors, where $n = |X| + |Y|$.

Proof. We will design a parallel algorithm for finding X' and X''. Let us first give an explanation of the algorithm. The algorithm starts by partitioning Y into two subsets $Y_{\leq 15}$ and $Y_{\geq 16}$, where $Y_{\leq 15} = \{y \in Y : deg_G(y) \leq 15\}$ and $Y_{\geq 16} = \{y \in Y : deg_G(y) \geq 16\}$. Next, it finds a subset I of X such that (a) each vertex $x \in I$ is adjacent to at most two vertices of $Y_{\geq 16}$ in the graph G, (b) each vertex of $Y_{\leq 15}$ is adjacent to at most one vertex of I in the graph G, and (c) I contains a constant fraction of the vertices of X. We will later use Lemma 1 to prove that such an I exists and is easy to find. The remaining task of the algorithm is to partition I into X' and X''. By the condition (b) for I and the fact that $\langle G, f \rangle$ is a proper pair, the partition can be done without paying attention to the vertices of $Y_{\leq 15}$. To obtain X' and X'' from I, firstly the algorithm partitions I into three subsets I_0, I_1, and I_2, where for $0 \leq j \leq 2$, I_j consists of all vertices $x \in I$ such that x is adjacent to exactly j vertices of $Y_{\geq 16}$ in the graph G. Secondly, it decides which vertices of I_2 can simultaneously be put in X'. To do this, it first constructs a multigraph G'' (without any self-loop) from the graph $G[I_2 \cup Y_{\geq 16}]$ by replacing the two edges incident to each vertex $x \in I_2$ (say, $\{x, y_1\}$ and $\{x, y_2\}$) with a single edge $\{y_1, y_2\}$, and then finds a maximal f-matching M of G''. Note that there is a one-to-one correspondence between the vertices of I_2 and the edges of G''. This correspondence enables the algorithm to find out those vertices of I_2 that can simultaneously be put in X'. Thirdly, the algorithm decides which vertices of I_1 can simultaneously be added to X'. (Note: At the beginning of this step, X' is a subset of I_2.) This is an easy task, however, because each vertex of I_1 is adjacent to exactly one vertex of $Y_{\geq 16}$ in the graph G. Finally, the algorithm unconditionally adds all vertices of I_0 to X' and then set $X'' = I - X'$.

The precise specification of the algorithm follows:

Algorithm 1:
Input: A proper pair $\langle G, f \rangle$ with $G = (X, Y, E)$.
Output: Two subsets X' and X'' of X satisfying the conditions of Lemma 2.

1. Compute $Y_{\geq 16} = \{y \in Y \ : \ deg_G(y) \geq 16\}$ and $Y_{\leq 15} = \{y \in Y \ : \ deg_G(y) \leq 15\}$.
2. Let $G' = G[X \cup Y_{\geq 16}]$. Compute $X_{\leq 2} = \{x \in X \ : \ deg_{G'}(x) \leq 2\}$.
3. Construct a graph $K = (X_{\leq 2}, E_K)$, where E_K consists of all $\{x_1, x_2\}$ such that x_1 and x_2 are adjacent to a common $y \in Y_{\leq 15}$ in G.
4. Find an MIS I of the graph K.
5. Partition I into three subsets I_0, I_1, and I_2, where for $0 \leq j \leq 2$, I_j consists of all vertices $x \in I$ such that in the graph $G[I \cup Y_{\geq 16}]$, x has degree j.
6. From the graph $G[Y_{\geq 16} \cup I_2]$, construct a multigraph $G'' = (Y_{\geq 16}, E'')$ without any self-loop as follows: For each $x \in I_2$, replace the two edges incident to x (say, $\{x, y_1\}$ and $\{x, y_2\}$) with a single edge $\{y_1, y_2\}$ (this edge is called x's *brother*).
7. Compute a maximal f-matching M of G''.
8. Compute $J = \{x \in I_2 : x$'s brother is in $M\}$.
9. In parallel, for each vertex $y \in Y_{\geq 16}$, first compute $I_{1,y} = \{x \in I_1 : \{x, y\} \in E\}$, next set $J_y = I_{1,y}$ if $|I_{1,y}| \leq f(y) - deg_{G[Y_{\geq 16} \cup J]}(y)$ and set J_y to be the set of the first $f(y) - deg_{G[Y_{\geq 16} \cup J]}(y)$ vertices of $I_{1,y}$ otherwise. (Note: We here assume that the vertices of the input graph G are linearly ordered.)
10. Set $X' = J \cup (\cup_{y \in Y_{\geq 16}} J_y) \cup I_0$ and $X'' = I - X'$.

We will show that X' and X'' indeed satisfy the conditions stated in the lemma. We first estimate $|X' \cup X''|$ $(= |I|)$. Since $G' = G[X \cup Y_{\geq 16}]$ satisfies the conditions of Lemma 1, we know that $|X_{\leq 2}| \geq |X|/6$. The maximum degree of K is obviously no more than $6 \times 15 = 90$. This fact and the maximality of I imply that $|I| \geq |X_{\leq 2}|/91$. Thus, $|I| \geq |X|/546$ and the condition (i) is satisfied.

We next show that the condition (ii) is satisfied. Since $X' \subseteq X$ and $deg_G(x) \leq f(x)$ for all $x \in X$, no vertex $x \in X'$ has degree $> f(x)$ in the graph $G[Y \cup X']$. Thus, it suffices to show that $deg_{G[Y \cup X']}(y) \leq f(y)$ for all $y \in Y$. Fix an arbitrary $y \in Y$ to consider. Let us first suppose that $y \in Y_{\leq 15}$. Then, by Algorithm 1, y is adjacent to at most one vertex of X' in the graph $G[Y \cup X']$. This fact together with $f(y) \geq 1$ implies that $deg_{G[Y \cup X']}(y) \leq f(y)$. Thus, we are done in case $y \in Y_{\leq 15}$. Let us now assume that $y \in Y_{\geq 16}$. Note that for any $S \subseteq X$, y has the same degree in the graph $G[S \cup Y]$ as in the graph $G[S \cup Y_{\geq 16}]$. Thus, it suffices to show that $deg_{G[Y_{\geq 16} \cup X']}(y) \leq f(y)$. By steps $6 \sim 8$, it is easy to see that y has the same degree in the subgraph of the multigraph G'' induced by M as in the graph $G[Y_{\geq 16} \cup J]$. Since M is an f-matching, y has degree at most $f(y)$ in the subgraph of G'' induced by M and hence has degree at most $f(y)$ in the graph $G[Y_{\geq 16} \cup J]$. Since $I_{1,y} \cap I_{1,y'} = \emptyset$ for every two vertices y and y' in $Y_{\geq 16}$, step 9 guarantees that y has degree at most $f(y)$ in the graph $G[Y_{\geq 16} \cup (X - I_0)]$. Now, noting that y is adjacent to no vertex of I_0, we see that y has degree at most $f(y)$ in the graph $G[Y_{\geq 16} \cup X']$. Therefore, the condition (ii) is really satisfied.

To prove that the condition (iii) is satisfied, consider an arbitrary vertex $x \in X''$. Then, $x \in I_2 - J$ or $x \in I_1 - (\cup_{y \in Y_{\geq 16}} J_y)$. Let us first suppose that $x \in I_2 - J$. Let e be x's brother. By the definition of J and the maximality of M, the subgraph of the multigraph G'' induced by $M \cup \{e\}$ must contain a vertex $y \in Y_{\geq 16}$ with degree $> f(y)$. Since this vertex y has the same degree in the graph $G[Y_{\geq 16} \cup J \cup \{x\}]$ as in the subgraph of G'' induced by $M \cup \{e\}$, y must have degree $> f(y)$ in the graph $G[Y_{\geq 16} \cup J \cup \{x\}]$. Thus, we are done in case $x \in I_2 - J$. Let us now assume that $x \in I_1 - (\cup_{y \in Y_{\geq 16}} J_y)$. Then, by step 9, the vertex $y \in Y_{\geq 16}$ adjacent to x in the graph $G[Y_{\geq 16} \cup X' \cup \{x\}]$ must have degree $> f(y)$ in the graph $G[Y_{\geq 16} \cup X' \cup \{x\}]$. Hence, the condition (iii) is also satisfied in this case.

We finally analyze the complexity of Algorithm 1. As mentioned in [CH93], steps $1 \sim 4$ run in $O(\log n)$ time with $O(n)$ processors. Steps 5 and 6 trivially run in $O(\log n)$ time with $O(n)$ processors. In step 7, Diks et $al.$'s algorithm can be applied [DGL91]. Thus, step 7 can be implemented in $O(\log^3 n)$ time with $O(n)$ processors [DGL91]. It is easy to see that steps $8 \sim 10$ take $O(\log n)$ time using $O(n)$ processors. Therefore, Algorithm 1 takes $O(\log^3 n)$ time using $O(n)$ processors. $\square$

We now use Lemma 2 to design a parallel algorithm for finding, given a proper pair $\langle G = (X, Y, E), f \rangle$, a Max-$f$-DS U of G with $Y \subseteq U$. Let us first give an explanation of the algorithm. It starts by computing $X_0 = \{x \in X : deg_G(x) = 0\}$. Note that every Max-$f$-DS of G must contain the vertices of X_0. If $X_0 = X$, then clearly $X \cup Y$ is a Max-f-DS of G and so the algorithm outputs $X \cup Y$ and stops. Otherwise, using Algorithm 1, the algorithm finds two subsets

X' and X'' of $X - X_0$ from the graph $G[(X - X_0) \cup Y]$ and the function f. Note that $U' = Y \cup X_0 \cup X'$ is an f-DS of G. Next, using U', the algorithm modifies G and f to obtain a new proper pair $\langle G', f' \rangle$ such that any Max-f'-DS of G' extended with the vertices of U' is a Max-f-DS of G. So, the algorithm is a recursive one. Since G' is constructed in such a way that no vertex of $X' \cup X''$ is contained in G', the recursion depth of the algorithm is $O(\log n)$ by Lemma 2.

The following is the precise specification of the algorithm.

Algorithm 2:
Input: A proper pair $\langle G, f \rangle$ with $G = (X, Y, E)$.
Output: A Max-f-DS U of G with $Y \subseteq U$.
1. Compute $X_0 = \{x \in X : deg_G(x) = 0\}$.
2. If $X_0 = X$, then output $X \cup Y$ and halt.
3. Let $H = G[(X - X_0) \cup Y]$. Find two subsets X' and X'' of $X - X_0$ from H such that X' and X'' satisfy the conditions of Lemma 2 with the graph G (resp., X) in Lemma 2 replaced by H (resp., $X - X_0$).
4. Let $U' = Y \cup X_0 \cup X'$. Compute $X_1 = \{x \in X - (X_0 \cup X' \cup X'') : U' \cup \{x\}$ is not an f-DS of $G\}$.
5. Let $X_2 = X - (X_0 \cup X' \cup X'' \cup X_1)$ and $Y_1 = \{y \in Y : f(y) > deg_{G[U']}(y)\}$. Construct a function f' from $X_2 \cup Y_1$ to positive integers as follows: For each $y \in Y_1$, $f'(y) = f(y) - deg_{G[U']}(y)$ and for each $x \in X_2$, $f'(x) = f(x)$.
6. Let $G' = G[X_2 \cup Y_1]$. Recursively call the algorithm to find a Max-f'-DS W of the graph G' with $Y_1 \subseteq W$. (Note: $\langle G', f' \rangle$ is a proper pair.)
7. Output $U = U' \cup (W \cap X_2)$.

Lemma 3. Algorithm 2 is correct.

Proof. We show the lemma by induction on the recursion depth d of Algorithm 2. In case $d = 0$, Algorithm 2 is clearly correct. Assume that $d > 0$ and Algorithm 2 correctly works given any proper pair on which it makes at most $d - 1$ recursive calls. Let $\langle G, f \rangle$ be a proper pair on which Algorithm 2 makes exactly d recursive calls. Consider the behavior of Algorithm 2 on input $\langle G, f \rangle$. First, we show that the output U is an f-DS of G. Noting that $\langle G, f \rangle$ is a proper pair, we have that no vertex $x \in X_0 \cup X' \cup (W \cap X_2)$ has degree $> f(x)$ in the graph $G[U]$. Thus, we need only to show that no vertex $y \in Y$ has degree $> f(y)$ in the graph $G[U]$. Let $y \in Y$. Then, by Lemma 2, $deg_{G[U']}(y) \leq f(y)$. In case $y \in Y - Y_1$, $deg_{G[U']}(y) = f(y)$ by step 5 and so no vertex of X_2 can be adjacent to y in the graph G. Thus, if $y \in Y - Y_1$, y must have degree at most $f(y)$ in the graph $G[U]$. On the other hand, in case $y \in Y_1$, y must be adjacent to at most $f'(y) = f(y) - deg_{G[U']}(y)$ vertices of $W \cap X_2$ in the graph G' by the inductive hypothesis and so y has degree at most $f(y)$ in the graph $G[U' \cup (X_2 \cap W)]$ by the construction of G'. Therefore, U is an f-DS of G.

Next, we show the maximality of U. Let x be an arbitrary vertex in $V - U$. Then, $x \in X''$ or $x \in X_1$ or $x \in X_2 - W$. In case $x \in X''$, $U \cup \{x\}$ is not an f-DS of G by Lemma 2. In case $x \in X_1$, $U \cup \{x\}$ is obviously not an f-DS of G. In case $x \in X_2 - W$, some vertex $y \in Y_1$ must have degree $> f'(y)$ in the graph

$G[W \cup \{x\}]$ by the inductive hypothesis and hence must have degree $> f(y)$ in the graph $G[U \cup \{x\}]$ by the construction of f'. This completes the proof. $\square$

Lemma 4. Let $\langle G, f \rangle$ be a proper pair with $G = (X, Y, E)$. Then, a Max-f-DS U of G with $Y \subseteq U$ can be found in $O(\log^4 n)$ time using $O(n)$ processors, where $n = |X| + |Y|$.

Proof. By Lemma 3, Algorithm 2 correctly finds a Max-f-DS of G. It suffices to show that Algorithm 2 takes $O(\log^4 n)$ time using $O(n)$ processors. Obviously, all steps except steps 3 and 6 can be done in $O(\log n)$ time with $O(n)$ processors. By Lemma 2, step 3 can be done in $O(\log^3 n)$ time with $O(n)$ processors. Since Lemma 2 implies that the recursion depth of Algorithm 2 is $O(\log n)$, Algorithm 2 takes $O(\log^4 n)$ time with $O(n)$ processors. $\square$

4 The algorithm for general planar graphs

In this section, we use Algorithm 2 to design a parallel algorithm for solving the Max-f-DS problem for planar graphs. Let us first give an explanation of the algorithm. Given a pair $\langle G = (V, E), f \rangle$ of a planar graph and a nonnegative integer-valued function defined on V, the algorithm starts by computing an independent set X of G such that $|X| \geq |V|/42$ and $deg_G(x) \leq 6$ for all $x \in X$. It is known that such an X exists and can be found efficiently in parallel [GPS87]. If $X = V$, then clearly X is a Max-f-DS of G and so the algorithm outputs X and stops. Otherwise, the algorithm recursively finds a Max-f-DS U' of the graph $G[V - X]$. Note that U' must be also an f-DS of G. Next, using U', the algorithm modifies G and f to obtain a proper pair $\langle H, f' \rangle$ such that any Max-f'-DS W of H together with U' can be simply used to find a Max-f-DS of G. To find W, Algorithm 2 is used. Since the size of X is sufficiently large, the algorithm has recursion depth $O(\log n)$.

The following is the precise specification of the algorithm:

Algorithm 3:
Input: A pair $\langle G, f \rangle$, where $G = (V, E)$ is a planar graph and f is a function from V to nonnegative integers.
Output: A Max-f-DS of G.
1. Find an independent set X of G such that $|X| \geq |V|/42$ and $deg_G(x) \leq 6$ for each $x \in X$.
2. If $V = X$, then output X and halt.
3. Recursively call the algorithm to find a Max-f-DS U' of the graph $G[V - X]$.
4. Compute $X_1 = \{x \in X : U' \cup \{x\}$ is not an f-DS of $G\}$.
5. Compute $X_2 = \{x \in X - X_1 : f(x) = 0\}$.
6. Let $X' = X - (X_1 \cup X_2)$ and $Y = \{u \in U' : f(u) > deg_{G[U']}(u)\}$. Construct a function f' from $X' \cup Y$ to positive integers as follows: For each $y \in Y$, $f'(y) = f(y) - deg_{G[U']}(y)$ and for each $x \in X'$, $f'(x) = f(x)$.
7. Construct a bipartite planar graph $H = (X', Y, E_H)$ as follows: For each $x \in X'$ and each $y \in Y$, $\{x, y\} \in E_H$ if and only if $\{x, y\} \in E$.

8. Compute a Max-f'-DS W of H with $Y \subseteq W$ using Algorithm 2. (Note: $\langle H, f' \rangle$ is a proper pair.)
9. Output $U = U' \cup X_2 \cup (W \cap X')$.

Lemma 5. Algorithm 3 is correct.

Proof. We show the lemma by induction on the recursion depth d of Algorithm 3. In case $d = 0$, Algorithm 3 is clearly correct. Assume that $d > 0$ and Algorithm 3 correctly works given any input on which it makes at most $d-1$ recursive calls. Let $\langle G, f \rangle$ be an input on which Algorithm 3 makes exactly d recursive calls. Consider the behavior of Algorithm 3 on input $\langle G, f \rangle$. First, we show that the output U is an f-DS of G. By the inductive hypothesis, U' is an f-DS of G. Moreover, the definition of X_1 and the fact that X is an independent set of G imply that no vertex x of $X_2 \cup (X' \cap W)$ has degree $> f(x)$ in the graph $G[U]$. Thus, we need only to show that no vertex $u \in U'$ has degree $> f(u)$ in the graph $G[U]$. Let $u \in U'$. Then, by the inductive hypothesis, $deg_{G[U']}(u) \le f(u)$. In case $u \in U' - Y$, $deg_{G[U']}(u) = f(u)$ by step 6 and so no vertex of $X - X_1$ can be adjacent to u in the graph G. Thus, u has degree at most $f(u)$ in the graph $G[U]$ if $u \in U' - Y$. On the other hand, if $u \in Y$, then u is adjacent to at most $f'(u) = f(u) - deg_{G[U']}(u)$ vertices of $W \cap X'$ in the graph H and so u has degree at most $f(u)$ in the graph $G[U' \cup (X' \cap W)]$. Since no vertex of X_2 can be adjacent to a vertex of U' in the graph G, u still has degree at most $f(u)$ in the graph $G[U]$ if $u \in Y$. Therefore, the output U of Algorithm 3 is an f-DS of G.

Next, we show the maximality of U. Let v be an arbitrary vertex in $V - U$. Then, $v \in (V - X) - U'$ or $v \in X_1$ or $v \in X' - W$. In case $v \in (V - X) - U'$, $U \cup \{v\}$ is not an f-DS of G by the inductive hypothesis. In case $v \in X_1$, $U \cup \{v\}$ is obviously not an f-DS of G. In case $v \in X' - W$, $W \cup \{v\}$ is not an f'-DS of H and hence $U' \cup (W \cap X') \cup \{v\}$ is not an f-DS of G by the constructions of f' and H. $\square$

Theorem 6. A Max-f-DS of a planar graph G with n vertices can be found in $O(\log^5 n)$ time with $O(n)$ processors.

Proof. By Lemma 5, we only need to show that Algorithm 3 takes $O(\log^5 n)$ time using $O(n)$ processors. Clearly, all steps of Algorithm 3 except steps 1, 3 and 8 run in $O(\log n)$ time with $O(n)$ processors. Step 1 runs in $O(\log^* n)$ time with $O(n)$ processors [GPS87]. By Lemma 4, step 8 runs in $O(\log^4 n)$ time with $O(n)$ processors. Thus, all steps except step 3 of Algorithm 3 run in $O(\log^4 n)$ time with $O(n)$ processors. By step 1, the recursion depth of Algorithm 3 is $O(\log n)$ and hence Algorithm 3 runs in $O(\log^5 n)$ time with $O(n)$ processors. $\square$

5   Conclusion

We have shown that the Max-f-DS problem for planar graphs is in NC. The time-processor product of the algorithm presented in this paper is small. It would be very nice to design an NC or RNC algorithm for the Max-f-DS problem for general graphs. However, this seems to be a very difficult task. In fact, we even do not know an NC or RNC algorithm for the Max-f-DS problem for planar graphs with parallel edges. Our algorithm does not work for planar graphs with parallel edges because Lemma 1 does not hold for such graphs.

References

[ABI86]   N. Alon, L. Babai, and A. Itai, A Fast and Simple Randomized Parallel Algorithm for the Maximal Independent Set Problem, *J. Algorithms* **7** (1986) 567-583.

[BM80]   J. A. Bondy and U. S. R. Murty, *Graph Theory with Applications* (North-Holland, New York, 1980).

[CH93]   Z.-Z. Chen and X. He, Parallel Algorithms for Maximal Cycle-Free Sets, *Submitted for publication.*

[CY86]   M. Chrobak and M. Yung, Fast Algorithms for Edge-Coloring Planar Graphs, *J. Algorithms* **10** (1986) 35-51.

[DGL91]   K. Diks, O. Garrido and A. Lingas, Parallel Algorithm for Finding maximal k-Dependent Sets and Maximal f-Matchings, in: *Proc. 2nd International Symp. on Algorithms*, Lecture Notes in Computer Science, Vol. 557 (Springer, Berlin, 1991) 385-395.

[GR88]   A. Gibbons and W. Rytter, *Efficient Parallel Algorithms* (Cambridge University Press, Cambridge, 1988).

[GPS87]   A.V. Goldberg, S.A. Plotkin and G.E. Shannon, Parallel Symmetry-Breaking in Sparse Graphs, in: *Proc. 19th ACM Symp. on Theory of Computing* (ACM, 1987) 315-324.

[GS89a]   M. Goldberg and T. Spencer, A New Parallel Algorithm for the Maximal Independent Set Problem, *SIAM J. Comput.* **18** (1989) 419-427.

[GS89b]   M. Goldberg and T. Spencer, Constructing a Maximal Independent Set in Parallel, *SIAM J. Disc. Math.* **2** (1989) 322-328.

[II86]   A. Israeli and A. Itai, A Fast and Simple Randomized Parallel Algorithm for Maximal Matching, *Inform. Process. Lett.* **22** (1986) 77-80.

[IS86]   A. Israeli and Y. Shiloach, An Improved Maximal Matching Parallel Algorithm, *Inform. Process. Lett.* **22** (1986) 57-60.

[KR90]   R.M. Karp and V. Ramachandran, *Parallel Algorithms for Shared Memory Machines*, in: J. van Leeuwen ed., *Handbook of Theoretical Computer Science Vol. A* (Elsevier, Amsterdam, 1990) 868-941.

[KW85]   R.M. Karp and A. Wigderson, A Fast Parallel Algorithm for the Maximal Independent Set Problem, *J. ACM* **32** (1985) 762-773.

[LP86]   L. Lovász and M.D. Plummer, *Matching Theory, Annals of Discrete Mathematics* (29), North-Holland Mathematics Studies 121 (Elsevier Science Publisher B.V., 1986).

[Lub86]   M. Luby, A Simple Parallel Algorithm for the Maximal Independent Set Problem, *SIAM J. Comput.* **15** (1986) 1036-1053.

[PV93]   D. Pearson and V. V. Vazirani, Efficient Sequential and Parallel Algorithms for Maximal Bipartite Sets, *J. Algorithms* **14** (1993), 171-179.

[SM91]   T. Shoudai and S. Miyano, Using Maximal Independent Sets to Solve Problems in Parallel, in: *Proc. 17th International Workshop on Graph-Theoretic Concepts in Computer Science*, Lecture Notes in Computer Science, Vol. 570 (Springer, Berlin, 1991) 126-134.

[Val82]   L.G. Valiant, Parallel Computation, in: *Proc. 7th IBM Symposium on Mathematical Foundations of Computer Science* (1982) 173-189.

Appendix

Proof of Lemma 1: Let us assume first that X contains no vertex x with $deg_G(x) = 0$. Partition X into two subsets $X_{\leq 2}$ and $X_{\geq 3}$ as follows:

$$X_{\leq 2} = \{x \in X : 1 \leq deg_G(x) \leq 2\} \quad \text{and} \quad X_{\geq 3} = \{x \in X : 3 \leq deg_G(x) \leq 6\}.$$

Then, we have:

$$|E| = \sum_{x \in X_{\leq 2}} deg_G(x) + \sum_{x \in X_{\geq 3}} deg_G(x) \leq 2|X_{\leq 2}| + 6|X_{\geq 3}|.$$

We also know that $|E| = \sum_{y \in Y} deg_G(y) \geq 16|Y|$. Thus, $16|Y| \leq 2|X_{\leq 2}| + 6|X_{\geq 3}|$ and so $|Y| \leq (|X_{\leq 2}| + 3|X_{\geq 3}|)/8$. On the other hand, since G is a bipartite planar graph, $|E| \leq 2(|X| + |Y|) - 4$ (see, e.g., [BM80]) and so we have:

$$|X_{\leq 2}| + 3|X_{\geq 3}| \leq \sum_{x \in X_{\leq 2}} deg_G(x) + \sum_{x \in X_{\geq 3}} deg_G(x) = |E| < 2(|X_{\leq 2}| + |X_{\geq 3}| + |Y|).$$

Therefore,

$$|X_{\leq 2}| > |X_{\geq 3}| - 2|Y| \geq |X_{\geq 3}| - (|X_{\leq 2}| + 3|X_{\geq 3}|)/4.$$

This implies that $|X_{\geq 3}| \leq 5|X_{\leq 2}|$. Since $|X_{\leq 2}| + |X_{\geq 3}| = |X|$ by our assumption, it follows that $|X_{\leq 2}| \geq |X|/6$ as to be shown.

Now suppose that X contains some vertices x with $deg_G(x) = 0$. Let n_i be the number of vertices $x \in X$ with $deg_G(x) = i$ for $0 \leq i \leq 2$, and let $n_{\geq 1}$ be the number of vertices $x \in X$ with $deg_G(x) \geq 1$. By the arguments above, we know that $n_1 + n_2 \geq n_{\geq 1}/6$. Thus, there are at least $n_0 + n_1 + n_2 \geq |X|/6$ vertices in X with degree at most 2 in G. $\square$

On-Line Convex Planarity Testing [*]
(extended abstract)

Giuseppe Di Battista[1], Roberto Tamassia[2], and Luca Vismara[1]

[1] Dipartimento di Informatica e Sistemistica, Università di Roma "La Sapienza"
dibattista/vismara@iasi.rm.cnr.it
[2] Department of Computer Science, Brown University
rt@cs.brown.edu

Abstract. An important class of planar straight-line drawings of graphs are convex drawings, where all faces are drawn as convex polygons. A graph is said to be convex planar if it admits a convex drawing. We consider the problem of testing convex planarity in a dynamic environment, where a graph is subject to on-line insertions of vertices and edges. We present on-line algorithms for convex planarity testing with the following performance, where n denotes the number of vertices of the graph: convex planarity testing and insertion of vertices take worst-case time $O(1)$, insertion of edges takes amortized time $O(\log n)$, and the space requirement of the data structure is $O(n)$. Furthermore, we give a new combinatorial characterization of convex planar graphs.

1 Introduction

Graph drawing is receiving increasing attention in the graph theoretic and algorithmic communities (see, e.g., [3, 23, 15, 19, 24, 10, 7, 27, 4]). Surveys on graph drawing algorithms can be found in [33, 8].

Planar straight-line drawings of graphs are especially interesting for their combinatorial and geometric properties. A classical result independently established by Steinitz and Rademacher [32], Wagner [39], Fary [18], and Stein [31] shows that every planar graph has a planar straight-line drawing. A grid drawing is a drawing in which the vertices have integer coordinates. Independently, de Fraysseix, Pach, and Pollak [5], and Schnyder [29] have shown that every n-vertex planar graph has a planar straight-line grid drawing with $O(n^2)$ area.

An important class of planar straight-line drawings are convex drawings, where all faces are drawn as convex polygons (see Fig. 1.a, Fig. 1.b and Fig. 2). Convex drawings of graphs have been extensively studied in graph theory and in

[*] Research supported in part by the National Science Foundation under grant CCR-9007851, by the U.S. Army Research Office under grant DAAL03-91-G-0035 and DAAH04-93-0134, by the Advanced Research Projects Agency under contract N00014-91-J-4052, ARPA order 8225, by the NATO Scientific Affairs Division under collaborative research grant 911016, by the Progetto Finalizzato Sistemi Informatici e Calcolo Parallelo and the grant 94.00023.CT07 of the Italian National Research Council, and by the Esprit II BRA of the European Community (project ALCOM).

graph algorithms. A graph is said to be convex planar if it admits a convex drawing. Tutte [37, 38] considered strictly convex drawings, where faces are strictly convex polygons (i.e. π angles are not allowed). He showed that every triconnected planar graph is strictly convex planar and that a strictly convex drawing can be constructed by solving a system of linear equations. Tutte [37, 38], Thomassen [35, 36], and Chiba, Yamanouchi, and Nishizeki [1] gave combinatorial characterizations of convex and strictly convex planar graphs. In [1] linear time algorithms are given for convex planarity testing and for constructing convex drawings with real coordinates. Chiba, Onoguchi, and Nishizeki [2] extended the results of [1] to construct "quasi convex" drawings for graphs that are not convex planar. Kant [23] provided a linear time algorithm to construct convex drawings with integer coordinates and quadratic area. Finally, parallel algorithms for convex planarity testing and for constructing convex drawings were presented by Dehne, Djidjev and Sack [6] and by He and Kao [20].

In this paper we present new results on convex planarity:

- We consider the problem of testing convex planarity in a dynamic environment, where a graph is subject to on-line insertions of vertices and edges. Note that the convex planarity property is not monotone, namely, there exists a sequence of insertions of vertices and edges on a planar graph G such that G alternatively gains or loses convex planarity after each insertion operation. We present on-line algorithms for convex and strictly convex planarity testing with the following performance, where n denotes the number of vertices of the graph: (strictly) convex planarity testing and insertion of vertices take worst-case time $O(1)$, insertion of edges takes amortized time $O(\log n)$, and the space requirement of the data structure is $O(n)$.
- We give a new combinatorial characterization of convex planar and strictly convex planar graphs. Our characterization appears to be simpler and more intuitive than the ones that have been presented in the literature [37, 38, 35, 36, 1].

Previous work on dynamic planarity testing and dynamic graph drawing is in [11, 12, 13, 3, 14, 40, 16, 17, 25, 4].

Besides their theoretical significance, our results are motivated by the development of advanced graph drawing systems. A variety of visualization applications require to automatically draw graphs. Examples include software engineering tools (displaying, e.g., entity-relationship diagrams and subroutine-call graphs), algorithms animation systems (representing, e.g., data structures), and project planning systems (displaying, e.g., pert diagrams and organization charts).

In order to satisfy the increasing demand of graph drawing algorithms, several graph drawing systems have been recently devised (see, for example, [9, 21]). Such systems usually have a library of graph drawing algorithms, each devised to take into account a specific set of aesthetic requirements. Thus, in graph drawing systems the problem of selecting among a set of algorithms the one that provides the "best" visualization is of crucial importance.

Since graph drawing systems are used interactively, the above selection problem has to be solved under tight performance requirements, especially for large graphs. The problem becomes harder when the graphs to represent are subject to frequent updates. For example, given a graph, the system may need to quickly determine whether a convex drawing algorithm can be applied.

Open problems left by this work include:

- Reduce the amortized time complexity of the various operations to $O(\alpha(n))$. La Poutré has recently shown that on-line planarity testing can be done within this bound.
- Develop a fully dynamic convex planarity testing algorithm. The best result for fully dynamic planarity testing is $O(n^{\frac{1}{2}})$ [17].
- Characterize the area required by a strictly convex drawing. Kant [23] has shown that convex drawings with integer coordinates can be constructed with quadratic area. It is also known that drawing a cycle as a strictly convex polygon with integer vertex coordinates requires cubic area [26].

The rest of this paper is organized as follows. Preliminary definitions are given in Section 2. Section 3 contains the combinatorial characterization. The on-line algorithm for convex planarity testing is presented in Section 4.

2 Preliminaries

We assume familiarity with planar graphs and with graph connectivity [28].

A *drawing* of a graph maps each vertex to a distinct point of the plane and each edge (u, v) to a simple Jordan curve with end-points u and v. A drawing is *planar* if no two edges intersect, except, possibly, at common end-points. A graph is planar if it has a planar drawing. A *straight-line* drawing is a drawing such that each edge is mapped to a straight-line segment.

Two planar drawings of a planar graph G are *equivalent* if, for each vertex v, they have the same circular clockwise sequence of edges incident on v. Hence, the planar drawings of G are partitioned into equivalence classes. Each such class is called an *embedding* of G. An *embedded* planar graph (also *plane* graph) is a planar graph with a prescribed embedding. A triconnected planar graph has a unique embedding, up to a reflection. A planar drawing divides the plane into topologically connected regions delimited by cycles; such cycles are called *faces*. The *external* face is the boundary of the unbounded region. Two drawings with the same embedding have the same faces. Hence, one can refer to the faces of an embedding.

A *convex* drawing of a planar graph G is a planar straight-line drawing of G in which all the faces are convex polygons. A *strictly convex* drawing of a planar graph G is a planar straight-line drawing of G in which all the faces are strictly convex polygons (i.e., no π angle is allowed). A graph is said *(strictly) convex planar* if it admits a (strictly) convex drawing.

A *separating k-set* of a graph is a set of k vertices whose removal disconnects the graph. Separating 1-sets and 2-sets are called cut-vertices and separation

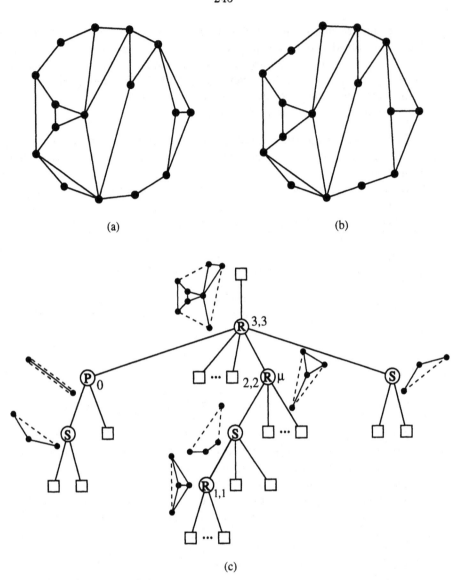

(a)

(b)

(c)

Fig. 1. (a) A strictly convex drawing of a biconnected planar graph G. (b) A convex drawing of G. (c) The SPQR-tree of G and the skeletons of its nodes.

pairs, respectively. A graph is biconnected if it is connected and has no cut-vertices, it is triconnected if it is biconnected and has no separation pairs.

Lemma 1. *A graph is (strictly) convex planar only if it is biconnected.*

Our data structure makes use of rooted trees. The dynamic trees of Sleator and Tarjan [30] support link/cut operations and various queries (such as finding the lowest-common ancestor of two nodes) in logarithmic time. As shown in [16],

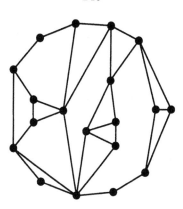

Fig. 2. A non-convex drawing of a biconnected planar graph.

they can be modified to support ordered trees and expand/contract operations. In the description of time bounds we use standard concepts of amortized complexity [34].

In the rest of this section, the SPQR-tree presented in [11, 12, 13, 14] is described.

Let G be a biconnected graph. A *split pair* of G is either a pair of adjacent vertices or a separation pair. A *split component* of a split pair $\{u, v\}$ is either an edge (u, v) or a maximal subgraph C of G such that $\{u, v\}$ is not a split pair of C. In the former case the split component is said *trivial*, in the latter *non-trivial*. Let $\{s, t\}$ be a split pair of G. A *maximal split pair* $\{u, v\}$ of G with respect to $\{s, t\}$ is such that for any other split pair $\{u', v'\}$, vertices u, v, s, and t are in the same split component.

Let $e = (s, t)$ be an edge of G, called *reference edge*. The *SPQR-tree* T of G with respect to e describes a recursive decomposition of G induced by its split pairs. Tree T is a rooted ordered tree whose nodes are of four types: S, P, Q, and R. Each node μ of T has an associated biconnected multigraph, called the *skeleton* of μ, and denoted by *skeleton*(μ). Also, it is associated with an edge of the skeleton of the parent ν of μ, called the *virtual edge* of μ in *skeleton*(ν). Tree T is recursively defined as follows.

Trivial Case: If G consists of exactly two parallel edges between s and t, then T consists of a single Q-node whose skeleton is G itself.

Parallel Case: If the split pair $\{s, t\}$ has at least three split components $G_1, \cdots,$ G_k ($k \geq 3$), the root of T is a P-node μ. Graph *skeleton*(μ) consists of k parallel edges between s and t, denoted $e_1, \cdots, e_k$, with $e_1 = e$.

Series Case: If the split pair $\{s, t\}$ has exactly two split components, one of them is the reference edge e, and we denote with G' the other split component. If G' has cut-vertices $c_1, \cdots, c_{k-1}$ ($k \geq 2$) that partition G into its blocks $G_1, \cdots, G_k$, in this order from s to t, the root of T is an S-node μ. Graph *skeleton*(μ) is the cycle $e_0, e_1, \cdots, e_k$, where $e_0 = e$, $c_0 = s$, $c_k = t$, and e_i connects c_{i-1} with c_i ($i = 1, \cdots, k$).

Rigid Case: If none of the cases above applies, let $\{s_1, t_1\}, \cdots, \{s_k, t_k\}$ be the maximal split pairs of G with respect to $\{s, t\}$ ($k \geq 1$), and for $i = 1, \cdots, k$, let G_i be the union of all the split components of $\{s_i, t_i\}$ but the one containing the reference edge e. The root of T is an R-node μ. Graph $skeleton(\mu)$ is obtained from G by replacing each subgraph G_i with the edge e_i between s_i and t_i.

Except for the trivial case, μ has children $\mu_1, \cdots, \mu_k$ in this order, such that μ_i is the root of the SPQR-tree of graph $G_i \cup e_i$ with respect to reference edge e_i ($i = 1, \cdots, k$). The tree so obtained has a Q-node associated with each edge of G, except the reference edge e. We complete the SPQR-tree by adding another Q-node, representing the reference edge e, and making it the parent of μ so that it becomes the root. An example of SPQR-tree is shown in Fig. 1.c.

The *virtual edge* of node μ_i is edge e_i of $skeleton(\mu)$. A virtual edge is said *trivial* if the corresponding node μ_i is a Q-node, *non-trivial* otherwise. The endpoints of e_i are called the *poles* of μ_i. Graph G_i is called the *pertinent graph* of node μ_i, and the *expansion graph* of edge e_i.

Let μ be a node of T. We have:

- if μ is an R-node, then $skeleton(\mu)$ is a triconnected graph;
- if μ is an S-node, then $skeleton(\mu)$ is a cycle;
- if μ is a P-node, then $skeleton(\mu)$ is a triconnected multigraph consisting of a bundle of multiple edges;
- if μ is a Q-node, then $skeleton(\mu)$ is a biconnected multigraph consisting of two multiple edges.

The skeletons of the nodes of T are homeomorphic to subgraphs of G. Also, the union of the sets of split pairs of the skeletons of the nodes of T is equal to the set of split pairs of G. It is possible to show that SPQR-trees of the same graph with respect to different reference edges are isomorphic and are obtained one from the other by selecting a different Q-node as the root. SPQR-trees are closely related to the classical decomposition of biconnected graphs into triconnected components [22]. Namely, the triconnected components of a biconnected graph G are in one-to-one correspondence with the internal nodes of the SPQR-tree: the R-nodes correspond to triconnected graphs, the S-nodes to polygons, and the P-nodes to bonds. SPQR-trees of planar graphs were introduced in [11] and applied to the problem of on-line planarity testing.

The SPQR-tree T of a planar graph with n vertices and m edges has m Q-nodes and $O(n)$ S-, P-, and R-nodes. Also, the total number of vertices of the skeletons stored at the nodes of T is $O(n)$.

3 A New Characterization of (Strictly) Convex Planar Graphs

Let Γ be a planar straight-line drawing of a biconnected planar graph G. A vertex of G is said *external* (resp., *internal*) in Γ if it is (resp., it is not) a vertex

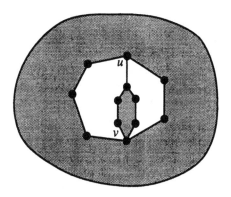

Fig. 3. A split component drawn inside Γ.

of the external face of Γ. A subgraph G' of G is *drawn outside* (resp., *inside*) Γ if G' has (resp., does not have) edges on the external face of Γ. The *external cycle* of a split component C in Γ is the cycle of G bounding the region of the plane in which C is drawn.

Lemma 2. *Let G be a biconnected planar graph and let Γ be a strictly convex drawing of G. The non-trivial split components of G are drawn outside Γ.*

Sketch of Proof. Suppose, for a contradiction, that a non-trivial split component C of a split pair $\{u, v\}$ is drawn inside Γ. Vertices u and v divide the external cycle of C into two paths (possibly having edges in common). Each path is part of a distinct internal face of G. By easy geometric considerations, it follows that only one of those two faces can be drawn as a strictly convex polygon in Γ (see Fig. 3). Thus Γ is not strictly convex: a contradiction. □

A corollary follows from the previous lemma.

Corollary 3. *Let G be a biconnected planar graph and let Γ be a strictly convex drawing of G. For each separation pair $\{u, v\}$, vertices u and v must be external in Γ.*

Sketch of Proof. Suppose, for a contradiction, that at least one vertex of separation pair $\{u, v\}$ is internal in Γ. All the split components of $\{u, v\}$ but one are drawn inside Γ, thus, by Lemma 2, Γ is not strictly convex: a contradiction. □

We are now ready to state the main result of this section.

Theorem 4. *Let G be a biconnected planar graph and let T be the SPQR-tree of G. Graph G is strictly convex planar if and only if, for each node μ of T, all the non-trivial virtual edges in skeleton(μ) are on the same face.*

Sketch of Proof. Only if. If μ is an S-node, then *skeleton*(μ) is a cycle and the claim is trivially true.

If μ is a P-node, then suppose, for a contradiction, that $skeleton(\mu)$ contains at least three (parallel) non-trivial virtual edges with common endpoints u and v. Even if u and v are external vertices in Γ, one of the expansion graphs of the virtual edges has to be drawn between the other two, that is, drawn inside Γ. Thus, by Lemma 2, Γ is not strictly convex: a contradiction.

If μ is an R-node, consider the expansion graphs of the non-trivial virtual edges in $skeleton(\mu)$. By Lemma 2, such expansion graphs must be drawn outside Γ. It follows that if we replace them in Γ with straight-line segments (representing their virtual edges), we obtain a planar straight-line drawing Γ_μ of $skeleton(\mu)$, in which all the non-trivial virtual edges are on the external face. The claim is then proved if we consider that $skeleton(\mu)$ is a triconnected graph and thus it has a unique embedding.

If. We show how to draw Γ in a circle c while visiting T. For each node μ of T, we choose as external the face of $skeleton(\mu)$ containing the virtual edges and we draw $skeleton(\mu)$ in a circular segment of c.

At the beginning of the visit, the circular segment coincides with c and we draw the skeleton of the root of T (two parallel edges, one of which virtual) as a chord of c. At each following step, let μ be the node currently visited and let ν be its parent; the virtual edge e_μ in $skeleton(\nu)$ is represented by a chord of c which identifies a circular segment s_μ (see Fig. 4).

If μ is a P-node, $skeleton(\mu)$ is drawn by placing the poles of μ (i.e. the endpoints of e_ν and of the non-virtual edge in $skeleton(\mu)$) at the endpoints of the chord identifying s_μ.

If μ is an S-node, $skeleton(\mu)$ is drawn by placing the poles of μ (i.e. the endpoints of e_ν in $skeleton(\mu)$) at the endpoints of the chord identifying s_μ, and the other vertices at distinct points of the circular arc of s_μ.

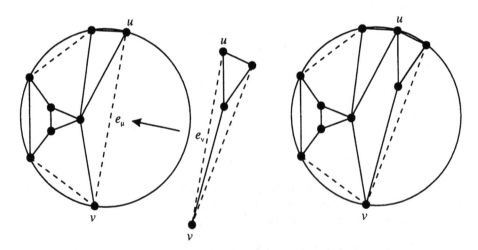

Fig. 4. Illustration of the construction in the proof of Theorem 4.

If μ is an R-node, a strictly convex drawing of $skeleton(\mu)$ is obtained by using the algorithm of Tutte [38] or the algorithm of Chiba et al. [1, 2], with the poles of μ (i.e. the endpoints of e_ν in $skeleton(\mu)$) at the endpoints of the chord identifying s_μ and the other external vertices at distinct points of the circular arc of s_μ.

Then e_μ and e_ν are removed from the drawing. If μ is a P-node the whole process consists of replacing a virtual edge of the drawing with two parallel edges (one of which virtual). If μ is an S-node or an R-node, it consists of appending a strictly convex polygon to the drawing along a virtual (external) edge, which is then removed (see Fig. 4).

Notice that, at each step, the following invariants hold for the drawing that is being constructed:

- the virtual edges are on the external face, thus represented by chords of c;
- the internal face generated by the removal of e_μ and e_ν is a strictly convex polygon, being the endpoints of e_μ and e_ν on c;
- the external face is a strictly convex polygon, being all its vertices on c.

The planarity of Γ follows from the planarity of the drawing of $skeleton(\mu)$ and from the invariants above. □

In the rest of this section we extend the characterization of Theorem 4 to non-strictly convex drawings.

A split component with split-pair $\{u, v\}$ is said to be a (u, v)-*chain* if it is a path. A chain is *maximal* if it is not a subpath of another chain.

Lemma 5. *Let Γ be a convex drawing of a biconnected planar graph G, and let C be a maximal (u, v)-chain of G drawn inside Γ. We have:*

- *there are at most three (u, v)-chains;*
- *the (u, v)-chains distinct from C are drawn outside Γ;*
- *u and v are not adjacent.*

Sketch of Proof. The proof follows from Lemma 2 considering that, if edge (u, v) does not exist, (only) one (u, v)-chain can be drawn placing vertices and edges on a straight-line segment. □

The *reduced* graph of a biconnected graph G is the biconnected graph G' homeomorphic to G obtained from G by replacing each maximal (u, v)-chain such that u and v are not adjacent with edge (u, v), called *short-cut*.

Theorem 6. *A biconnected graph is convex planar if and only if its reduced graph is strictly convex planar.*

Sketch of Proof. Let G be a biconnected graph and let G' be its reduced graph.

Only if. If G is a cycle the claim is trivially proved. If G is not a cycle, let $\{u, v\}$ be a nonadjacent split pair of G with at least one maximal (u, v)-chain. Let Γ be a convex drawing of G. Two cases are possible (see Lemma 5).

- One of the maximal (u, v)-chains of $\{u, v\}$, denoted C, is drawn inside Γ. Because of the convexity of Γ, the vertices of C are placed on a straight-line segment.
- All the maximal (u, v)-chains of $\{u, v\}$ are drawn outside Γ. In this case, not being G a cycle, there exists only one maximal (u, v)-chain C. The vertices of C may or may not be placed on a straight-line segment in Γ.

In both cases, we replace C with short-cut (u, v), drawn as a straight-line segment; notice that the resulting drawing is still convex. By repeating this process for all the split pairs with the property above, we obtain a convex drawing Γ_c of the reduced graph G'. Notice that in Γ_c there may still be π angles, but only around vertices of degree at least three, while all the other angles are less than π. It follows that a strictly convex drawing Γ' of G' can always be obtained from Γ_c by local adjustment of those vertices.

If. Let Γ' be a strictly convex drawing of G'. A convex drawing of G can be obtained from Γ' by replacing each short-cut (u, v) with the corresponding maximal (u, v)-chain, drawn placing its vertices on a straight-line segment. $\square$

4   On-Line Convex Planarity Testing

We consider a dynamic environment where a planar graph G is updated by the insertion of vertices and edges that preserve planarity. The repertory of query and update operations extends the one given in [13]:

Convex: Determine whether G is convex planar.

StrictlyConvex: Determine whether G is strictly convex planar.

Test(v_1, v_2): Determine whether edge (v_1, v_2) can be added to G while preserving planarity.

InsertEdge(e, v_1, v_2): Add edge e between vertices v_1 and v_2 to graph G. The operation is allowed only if the resulting graph is itself planar.

InsertVertex(e, v, e_1, e_2): Split edge e into two edges e_1 and e_2 by inserting vertex v.

AttachVertex(e, v, u): Add vertex v and connect it to vertex u by means of edge e.

MakeVertex(v): Add an isolated vertex v.

Note that graph G may be non-biconnected (and even non-connected).

The data structure extends the one for on-line planarity testing given in [13]. Namely, we use the following additional structures:

- For each R-node μ:
  - For each face f of *skeleton*(μ) (recall that the embedding of the skeleton of an R-node is unique), a balanced binary tree $B(f)$ associated with f, where each leaf of $B(f)$ is associated with an edge e of f, and stores value 1 or 0 according to whether e is a non-trivial virtual edge or not, and each internal node stores the sum of the values of the leaves in its subtree (see

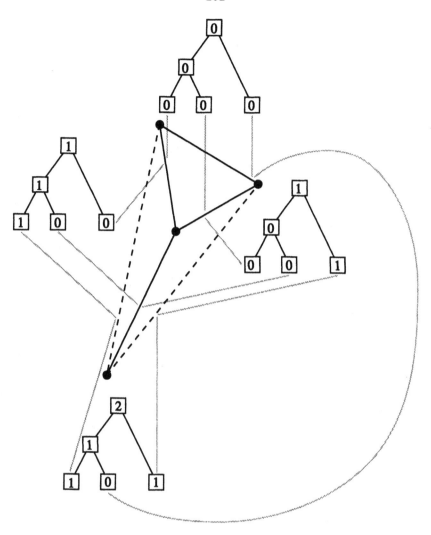

Fig. 5. The skeleton of R-node μ in Fig. 1.c and the balanced binary trees for its faces.

Fig. 5). Hence, the root of $B(f)$ stores the number of non-trivial virtual edges of f, denoted $Rvirtual(f)$.

- A variable $totalRvirtual(\mu) = \frac{1}{2}\sum_f Rvirtual(f)$, storing the total number of non-trivial virtual edges in $skeleton(\mu)$;
- A variable $maxRvirtual(\mu) = \max_f\{Rvirtual(f)\}$, storing the maximum value of $virtual(f)$ over all faces f of $skeleton(\mu)$.

- For each P-node μ:

  - An indicator $Pthree(\mu)$, where $Pthree(\mu) = 1$ if $skeleton(\mu)$ has three or more non-trivial virtual edges, and $Pthree(\mu) = 0$ otherwise.

- Variables storing the following values:

- The total number of non-trivial virtual edges in the skeletons of the R-nodes of T, denoted

$$sumtotalRvirtual(G) = \sum_{R-node\mu \in T} totalRvirtual(\mu);$$

- The sum of the values $maxRvirtual(\mu)$ over all the R-node of T, denoted

$$summaxRvirtual(G) = \sum_{R-node\mu \in T} maxRvirtual(\mu);$$

- The number of P-nodes with at least three non-trivial virtual edges, denoted

$$sumPthree(G) = \sum_{P-node\mu \in T} Pthree(\mu);$$

- The total number of biconnected components of G, denoted $totalbico(G)$.

For each R-node μ of the SPQR-tree in Fig. 1c, the values of $totalRvirtual(\mu)$ and $maxRvirtual(\mu)$ are indicated (separated by a comma); likewise, for each P-node μ the value of $Pthree(\mu)$ is indicated. For graph G, $sumtotalRvirtual(G) = 6$, $summaxRvirtual(G) = 6$, and $sumPthree(G) = 0$.

Theorem 7. *Operation StrictlyConvex returns true if and only if the following conditions hold:*

1. $totalbico(G) = 1$;
2. $sumPthree(G) = 0$;
3. $sumtotalRvirtual(G) = summaxRvirtual(G)$.

Sketch of Proof. Condition 1 expresses the fact that G is biconnected (see Lemma 1). Condition 2 expresses the fact that every P-node has no more than two non-trivial virtual edges, i.e., every split-pair has no more than two non-trivial split-components. Condition 3 is equivalent to $maxRvirtual(\mu) = totalRvirtual(\mu)$ for each R-node μ, which expresses the fact that all non-trivial virtual edges of the skeleton of an R-node are on the same face. Thus, Conditions 1, 2, and 3 are equivalent to Theorem 4. □

Theorem 8. *Let G be a planar graph that is updated on-line by adding vertices and edges, and let n be the current number of vertices of G. There exists a data structure for on-line convex planarity testing of G with the following performance: the space requirement is $O(n)$; operations MakeVertex, Convex and StrictlyConvex take worst-case time $O(1)$; operations Test, AttachVertex and InsertVertex take worst-case time $O(\log n)$; and operation InsertEdge takes amortized time $O(\log n)$.*

Sketch of Proof. In an update operation, the auxiliary data structure can be updated in time proportional to $\log n$ times the number of elementary updates in the skeletons of the R-nodes. Hence, by extending the amortized analysis of [13], updating the auxiliary data structure takes $O(\log n)$ time amortized. □

References

1. N. Chiba, T. Yamanouchi, and T. Nishizeki, "Linear Algorithms for Convex Drawings of Planar Graphs," in Progress in Graph Theory, J.A. Bondy and U.S.R. Murty (eds.), pp. 153-173, Academic Press, 1984.

2. N. Chiba, K. Onoguchi, and T. Nishizeki, "Drawing Planar Graphs Nicely," Acta Informatica, vol. 22, pp. 187-201, 1985.

3. R.F. Cohen, G. Di Battista, R. Tamassia, I.G. Tollis, and P. Bertolazzi, "A Framework for Dynamic Graph Drawing," Proc. 8th Symp. on Computational Geometry, pp. 261-270, 1992.

4. R.F. Cohen, G. Di Battista, R. Tamassia, and I.G. Tollis, "A Framework for Dynamic Graph Drawing," to appear in SIAM Journal of Computing.

5. H. de Fraysseix, J. Pach, and R. Pollack, "How to Draw a Planar Graph on a Grid," Combinatorica, vol. 10, pp. 41-51, 1990.

6. F. Dehne, H. Djidjev, J.-R. Sack, "An Optimal PRAM Algorithm for Planar Convex Embedding," Proc. ALCOM Workshop on Graph Drawing, pp. 75-77, 1993.

7. G. Di Battista, P. Eades, H. de Fraysseix, P. Rosenstiehl, and R. Tamassia (eds.), Proc. ALCOM Workshop on Graph Drawing, 1993.

8. G. Di Battista, P. Eades, R. Tamassia, and I.G. Tollis, "Algorithms for Drawing Graphs: an Annotated Bibliography," to appear in Computational Geometry Theory and Applications.

9. G. Di Battista, A. Giammarco, G. Santucci, and R. Tamassia, "The Architecture of Diagram Server," Proc. IEEE Workshop on Visual Languages, pp. 60-65, 1990.

10. G. Di Battista, G. Liotta, and F. Vargiu, "Spirality of Orthogonal Representations and Optimal Drawings of Series-Parallel Graphs and 3-Planar Graphs," Proc. 3rd Workshop on Algorithms and Data Structures, LNCS, vol. 709, pp. 151-162, 1993.

11. G. Di Battista and R. Tamassia, "Incremental Planarity Testing," Proc. 30th IEEE Symp. on Foundations of Computer Science, pp. 436-441, 1989.

12. G. Di Battista and R. Tamassia, "On-line Graph Algorithms with SPQR-trees," Proc. 17th Int. Colloquium on Automata, Languages and Programming, Lecture Notes in Computer Science, vol. 443, pp. 598-611, 1990.

13. G. Di Battista and R. Tamassia, "On-line Planarity Testing," Dept. Computer Science, Brown Univ., Technical Report CS-92-39, 1992.

14. G. Di Battista and R. Tamassia, "On-line Maintenance of Triconnected Components with SPQR-Trees," Dept. Computer Science, Brown Univ., Technical Report CS-92-40, 1992.

15. G. Di Battista and L. Vismara, "Angles of Planar Triangular Graphs," Proc. 25th ACM Symp. on the Theory of Computing, pp. 431-437, 1993.

16. D. Eppstein, G.F. Italiano, R. Tamassia, R.E. Tarjan, J. Westbrook, and M. Yung, "Maintenance of a minimum spanning forest in a dynamic planar graph," Journal of Algorithms, vol. 13, pp. 33-54, 1992.

17. D. Eppstein, Z. Galil, G.F. Italiano, and T.H. Spencer, "Separator Based Sparsification for Dynamic Planar Graph Algorithms," Proc. 25th ACM Symp. on the Theory of Computing, pp. 208-217, 1993.

18. I. Fary, "On Straight Lines Representation of Planar Graphs," Acta Sci. Math. Szeged, vol. 11, pp. 229-233, 1948.

19. A. Garg, M. Goodrich, and R. Tamassia, "Area-Efficient Upward Tree Drawings," Proc. 9th Symp. on Computational Geometry, pp. 359-368, 1993.

20. X. He and M.-Y. Kao, "Parallel Construction of Canonical Ordering and Convex Drawing of Triconnected Planar Graphs," Proc. 4th Int. Symp. on Algorithms and Computation, Lecture Notes in Computer Science, vol. 762, pp. 303-312, 1993.

21. M. Himsolt, "A View to Graph Drawing through Graph[Ed]," Proc. ALCOM Workshop on Graph Drawing, pp. 117-118, 1993.

22. J. Hopcroft and R.E. Tarjan, "Dividing a Graph into Triconnected Components," SIAM Journal of Computing, vol. 2, pp. 135-158, 1973.

23. G. Kant, "Drawing Planar Graphs using the lmc-ordering," Proc. 33rd IEEE Symp. on Foundations of Computer Science, pp. 101-110, 1992.

24. G. Kant, "A More Compact Visibility Representation," to appear in Proc. 19th Workshop on Graph-Theoretic Concepts in Computer Science, Lecture Notes in Computer Science, vol. 790, pp. 411-424, 1993.

25. J. La Poutré, "Alpha-Algorithms for Incremental Planarity Testing," Proc. 26th ACM Symp. on the Theory of Computing, pp. 706-715, 1994.

26. Y.-L. Lin and S.S. Skiena, "Complexity Aspects of Visibility Graphs," Dept. Computer Science, State Univ. of New York, Stony Brook, Technical Report 92-08, 1992.

27. S. Malitz and A. Papakostas, "On the Angular Resolution of Planar Graphs," SIAM Journal on Discrete Mathematics, vol. 7, pp. 172-183, 1994.

28. T. Nishizeki and N. Chiba, Planar Graphs: Theory and Algorithms, Annals of Discrete Mathematics, North Holland, 1988.

29. W. Schnyder, "Embedding Planar Graphs on the Grid," Proc. ACM- SIAM Symp. on Discrete Algorithms, pp. 138-148, 1990.

30. D.D. Sleator and R.E. Tarjan, "A Data Structure for Dynamic Trees," Journal of Computer System Sciences, vol. 24, pp. 362-381, 1983.

31. S.K. Stein, "Convex Maps," Proc. Amer. Math. Soc., vol. 2, pp. 464-466, 1951.

32. E. Steinitz and H. Rademacher, Vorlesung uber die Theorie der Polyeder, Springer, Berlin, 1934.

33. R. Tamassia, G. Di Battista, and C. Batini, "Automatic Graph Drawing and Readability of Diagrams," IEEE Transactions on Systems, Man and Cybernetics, vol. SMC-18, no. 1, pp. 61-79, 1988.

34. R.E. Tarjan, "Amortized Computational Complexity," SIAM Journal on Algebraic Discrete Methods, vol. 6, n. 2, pp. 306-318, 1985.

35. C. Thomassen, "Planarity and Duality of Finite and Infinite Planar Graphs", J. Combinatorial Theory, Series B, vol. 29, pp. 244-271, 1980.

36. C. Thomassen, "Plane Representations of Graphs," in Progress in Graph Theory, ed. J.A. Bondy and U.S.R. Murty, pp. 43-69, Academic Press, 1984.

37. W.T. Tutte, "Convex Representations of Graphs," Proc. London Math. Soc., vol. 10, pp. 304-320, 1960.

38. W.T. Tutte, "How to Draw a Graph," Proc. London Math. Soc., vol. 13, pp. 743-768, 1963.

39. K. Wagner, "Bemerkungen zum Vierfarbenproblem," Jber. Deutsch. Math.-Verein, vol. 46, pp. 26-32, 1936.

40. J. Westbrook, "Fast Incremental Planarity Testing," Proc. 19th Int. Colloquium on Automata, Languages and Programming, Lecture Notes in Computer Science, vol. 623, pp. 342-353, 1992.

Book Embeddings and Crossing Numbers

Farhad Shahrokhi[1], Ondrej Sýkora[2,*], László A. Székely[3], Imrich Vrt'o[2,**]

[1] Department of Computer Science, University of North Texas
P.O.Box 13886, Denton, TX, USA
[2] Institute for Informatics, Slovak Academy of Sciences
Dúbravská 9, 842 35 Bratislava, Slovak Republic
[3] Department of Computer Science, Eőtvős University
1088 Budapest, Múzeum krt. 6-8, Hungary

Abstract. The paper introduces the book crossing number problem which can be viewed as a variant of the well-known plane and surface crossing number problem or as a generalization of the book embedding problem. The book crossing number of a graph G is defined as the minimum number of edge crossings when the vertices of G are placed on the spine of a k-page book and edges are drawn on pages, so that each edge is contained by one page. We present polynomial time algorithms for drawing graphs in books with small number of crossings. One algorithm is suitable for sparse graphs and gives a drawing in which the number of crossings is within a multiplicative factor of $O(\log^2 n)$ from the optimal one under certain conditions. Using these drawings we improve the best known upper bound on the rectilinear crossing number, provided that $m \geq 4n$. We also derive a general lower bound on the book crossing number of any graph and present a second polynomial time algorithm to generate a drawing of any graph with $O(m^2/k^2)$ many edge crossings. This number of crossings is within a constant multiplicative factor from our general lower bound of $\Omega(m^3/n^2 k^2)$, provided that $m = \Theta(n^2)$. For several classes of well-known graphs, we also sharpen our algorithmic upper bounds by giving specific drawings.

1 Introduction

Several linear layout problems have been a subject of study recently. They are motivated as simplified mathematical models of many different tasks in computer science. Given a set of modules, the linear layout consists of placing the modules on a line and then wiring together the modules according to the given specifications. Typical representatives of the linear layout problems are: the bandwidth problem [CCDG82, Ch88], the pagenumber problem [BK79, CLR87, Kn90] and the boundary VLSI layout problem [Ls80, Ul84]. Surveys on graph layout problems can be found in [Dz92, Yn86a].

* This research was partially supported by EC Cooperative Action IC1000 "Project ALTEC" and the Slovak Academy of Sciences Grant No. 2/1138/94
** This research was supported by the European Community programme "COST" and done at Laboratory for Computer Science, University of Paris XI

In this paper we introduce a new linear layout problem, the *book crossing number problem*. Informally, given a graph G and a book of k pages, we look for an embedding of the vertices of G in a line along the spine of the book and its edges on the pages, such that each edge is contained by one page, no three edges cross in one point, and the sum of crossings on all pages is to be minimized. Motivated by VLSI design, the cases $k = 1, 2$ were investigated in [MKNF87, MKNF90] where the corresponding decision problems were proven to be NP-complete. The case $k = 1$ was mentioned in [Kn90] as the outerplanar crossing number but no results were shown. The case $k = 2$ was studied in several papers (e.g. [MKT71, Nc68]), in connection with the design of printed circuit boards, as a restricted version of the plane crossing number problem. In these papers, the vertices of graphs are placed on a line in the linear order determined by a hamiltonian cycle. Edges are drawn as semicircles above and below the line. Minimizing procedures were proposed, which permute edges between two sides until a satisfactory drawing is found. The book crossing number problem is closely related to the pagenumber problem which have recently attracted considerable attention [MWW88, Yn86b]. The pagenumber of a graph G is the minimum number k such that vertices of G can be embedded in the spine of the k-page book and edges on pages (each edge on exactly one page) *without* crossings. The problem can also be viewed as a variant of the plane and surface crossing number problem which is extensively studied in mathematical literature, see [EG73, SSSV] for references.

We study basic properties of the book crossing number problem. We derive general lower bounds and establish several optimal bounds on the book crossing numbers for specific families like complete graphs, complete bipartite graphs and hypercubes. We propose a polynomial time algorithm that produces a few book crossing drawing of any graph by means of drawing of the complete graph. The algorithm finds asymptotically optimal crossing numbers at least for dense graphs.

In this paper we have also investigated the *rectilinear crossing number*, that is, the minimum of crosssings when each edge is drawn using one straight line segment. This parameter is closely related to the 1-page crossing number. Provided that $m \geq 4n$, we substantially improve the best known result of [BD92] on relationship between the planar and rectilinear crossing numbers.

2 The Problem Definition and Basic Notions

A *book* consists of a *spine* and k *pages*, $k \geq 1$. The spine of the book is a line. Each page is a half-plane that has the spine as its boundary. The embedding of an undirected graph $G = (V, E)$ in the book consists of two steps. The first step places the vertices on the spine in some order. The second step draws each edge of the graph in one page of the book with a curve such that any curve has only its two end-points on the spine and no three curves intersect in one point unless it is an end-point in common. Let us denote by $\nu_k(G)$ the minimum number of crossings among all k-page book embeddings of G. An optimal drawing does not contain an edge which crosses itself, nor two edges with more than one point in common. Let $p(G)$ denote the pagenumber of G, i.e. the smallest k such that $\nu_k(G) = 0$. Let $cr_0(G)$ denote the minimum number of edge crossings when G is drawn in the plane. Let $\overline{cr_0}$ denote the minimum number of crossings when each edge is drawn in the plane using one straight line segment. Let Δ and δ denote the maximum and minimum degree of vertices of a graph.

3 Lower Bounds

In this section we shall study general techniques for obtaining lower bounds for $\nu_k(G)$. It is useful to consider the following equivalent problem to the book crossing problem. Find a bijection of vertices of an n-vertex graph G into vertices of the n-vertex regular polygon $P(n)$ and colour the edges of G (chords and sides of $P(n)$) by k colours so that the number of crossings of edges of the same colour is minimized. Clearly, the edges coloured by the same colour would correspond to edges drawn on the same page.

Before proving our lower bound theorem we need two lemmas:

Lemma 3.1 For any graph $G = (V, E)$ with $|V| \geq 2$ holds

$$\nu_1(G) \geq |E| - 2|V| + 3.$$

Proof. Assume G is connected, otherwise we may add edges. Consider a drawing of G in one page with $\nu_1(G)$ crossings. Replace crossings by new artificial vertices. We obtain a planar graph with $|V| + \nu_1(G)$ vertices, $|E| + 2\nu_1(G)$ edges and hence with $|E| - |V| + \nu_1(G) + 2$ faces by applying the Euler's polyhedral formula. One face has size at least $|V|$ and others have at least 3-edge boundaries. Thus

$$|V| + 3(|E| - |V| + \nu_1(G) + 2 - 1) \leq 2(|E| + 2\nu_1(G)),$$

which yields

$$\nu_1(G) \geq |E| - 2|V| + 3.$$

$\square$

The following Lemma is an analogue of Ajtai's et al. [ACNS82] and Leighton's [Lg83] lower bound for the plane crossing number.

Lemma 3.2 For any graph $G = (V, E)$ with $|V| = n, n \geq 4, |E| = m, m \geq 3n$

$$\nu_1(G) \geq \frac{|E|^3}{37|V|^2}.$$

Proof. Let D be a drawing of G in one page with $\nu_1(G)$ crossings. Let $r, r \geq 4$ be an integer. Let $G' = (V', E')$ be an induced subgraph of G with $|V'| = r$ and D' be a subdrawing in D associated with G'. Let $\nu_1(D')$ denote the number of crossings in D'. According to the previous Lemma, we have

$$\nu_1(D') \geq |E'| - 2r + 3.$$

Summing up these inequalities over all drawings associated with induced subgraphs G' with r vertices, we obtain

$$\sum_{\substack{G'=(V',E') \\ |V'|=r}} \nu_1(D') \geq m\binom{n-2}{r-2} - (2r-3)\binom{n}{r}.$$

Now consider a particular crossing in D. Then this crossing will appear in the drawing of exactly $\binom{n-4}{r-4}$ many induced subgraphs G' on r vertices, since any crossing in D is identified by 4 vertices and any 4 vertices of G are contained in $\binom{n-4}{r-4}$ many subsets of V of cardinality r. It follows that

$$\binom{n-4}{r-4}\nu_1(G) = \sum_{\substack{G'=(V',E') \\ |V'|=r}} \nu_1(D')$$

which implies

$$\nu_1(G) \geq \frac{(n-2)(n-3)}{(r-2)(r-3)}\left(m - (2r-3)\frac{n(n-1)}{r(r-1)}\right).$$

Setting $r = \lceil 3n^2/m \rceil$, we get

$$\nu_1(G) \geq \frac{n^2}{r^2}\left(m - \frac{2n^2}{r}\right) > \frac{m^3}{37n^2}.$$

$\square$

Theorem 3.1 Let $G = (V, E)$ be a graph satisfying $|E| \geq 3|V|$ and $k \leq \frac{|E|}{3|V|}$, then

$$\nu_k(G) \geq \frac{|E|^3}{37k^2|V|^2} - \frac{27k|V|}{37}.$$

Proof. Consider a drawing of G in a k-page book with $\nu_k(G)$ crossings. Let $G_i = (V, E_i)$ be the graph drawn on the ith page, $i = 1, 2, ..., k$. Define a function $f(x)$ with

$$f(x) = 0, \text{ for } x \leq 3|V|,$$
$$f(x) = \frac{x^3}{37|V|^2} - \frac{27|V|}{37}, \text{ for } x \geq 3|V|.$$

Clearly, f is a convex function and the Jensen's inequality applies to it. Hence

$$\nu_k(G) \geq \sum_{i=1}^{k} \nu_1(G_i) \geq \sum_{i=1}^{k} f(|E_i|) \geq kf\left(\frac{\sum|E_i|}{k}\right) = kf(|E|/k).$$

$\square$

In the next section we prove that the lower bound is achievable e.g. by complete graphs.

For $k = 2$ we may observe that $\nu_2(G) \geq cr_0(G)$. However, for $k > 2$ we do not know any relation between $\nu_k(G)$ and $cr_0(G)$. For some k it might not exist in general.

Proposition 3.1 *There exist graphs G_1 and G_2 such that*

$$\nu_3(G_1) < cr_0(G_1) \text{ and } \nu_3(G_2) > cr_0(G_2).$$

Proof. Set $G_1 = K_6$, then $\nu_3(K_6) = 0$ as $p(K_6) = 3$ and $cr_0(K_6) > 0$. Let G_2 be the planar graph proposed by Yannakakis [Yn86b] such that $p(G_2) = 4$. Then $\nu_3(G_2) > 0$ and $cr_0(G_2) = 0$. □

4 Upper Bounds

Let $G = (V, E)$. Assume that $E = \{e_1, e_2, ..., e_m\}$ and define $E_t = \{e_1, e_2, ..., e_t\}$, $1 \leq t \leq m$ and $E_0 = \emptyset$. Let D be a drawing of G in a 1-page book. Let D^i be a copy of D which is constructed on page i, $1 \leq i \leq k$ of the k-page book. We can draw an edge set $E' \subseteq E$ using D^i's as follows: for any $e \in E'$, we select a particular page j and then draw e in that page using the drawing of e in D^j in this page. This way we obtain a drawing for the subgraph of G which is spanned by the edges in E' in k pages.

Theorem 4.1 *Let $\dot{D}$ be the drawing of a graph $G = (V, E)$ in a 1-page book. Then, we can obtain a drawing $\bar{D}$ of G in a k-page book, in polynomial time, so that*

$$\nu_k(\bar{D}) \leq \frac{1}{k}\nu_1(D).$$

Proof. Our algorithm to construct $\bar{D}$ has m iterations. Initially, we generate a copy D^l of D on any page, $1 \leq l \leq k$. At iteration i, $1 \leq i \leq m$, we construct a drawing of E_i on k pages. At iteration 1, we draw e_1 on page 1 using the drawing of e_1 in D^1.

Next, we describe a simple greedy strategy which shows how to construct a drawing of E_i from the drawing of E_{i-1} for any $2 \leq i \leq m$. Let c_i denote the number of edges in E_{i-1} which cross e_i in the original drawing D of G. Then, since we have k pages, there should be at least one page j, which contains at most c_i/k many of the edges which cross e_i, $2 \leq i \leq m$. We will select one such page j and draw e_i using the drawing of e_i in D^j. Note that, then drawing of e_i will create at most c_i/k many crossings. This way, at iteration i of the algorithm the total number of crossings for drawing E_i is at most

$$\sum_{j=1}^{i} \frac{c_j}{k}$$

and therefore when the algorithm terminates the total number of crossings is at most

$$\sum_{j=1}^{m} \frac{c_j}{k} = \frac{\nu_1(D)}{k}.$$

Clearly, our algorithm runs in polynomial time. □

In view of Theorem 4.1 it is useful to have drawings of any graph with small number of crossings in one page. This is our next result.

Let D be a drawing of G in one page. Let l_x denote a curve that connects vertex x to infinity with the smallest possible number of crossings with edges of the drawing, $|l_x|$. Set $l^D = \max_{x \in V(G)} |l_x|$. Let $b(G)$ denote the bisection width of G and set $\bar{b}(G) = \max_{H \subseteq G} b(H)$, where the maximum is taken over all subgraphs of G.

Theorem 4.2 *Assume that we have an approximation algorithm for bisecting any graph G so that the number of edges in the cut is at most $R(n)b(G)$, where $R(n)$ is some nondecreasing functional measure of error. Then we can produce a drawing of G in a 1-page book with at most*

$$O(\log^2 n R^2(n)(cr_0(G) + \sum_{i \in V(G)} d_i^2(G)))$$

crossings, where $d_i(G), i = 1, 2, ..., n$ denotes the degree sequence of the graph G.

Proof. Using our approximation algorithm, we recursively partition G, to G_1 and G_2, draw G_1 and G_2 in the 1-page book so that all vertices of G_1 precede all vertices of G_2 in the spine, and then insert and draw the edges in the partition. The process of recursively partitioning G gives rise to a binary tree T, the partition tree of G, whose vertices are subgraphs of G. Let H be a node of T; we denote by H_1 and H_2, the left and right children of H, respectively. Let D_{H_1}, and D_{H_2} be the drawings of H_1 and H_2, which are obtained by the algorithm. Let edge $e = xy$ be an edge in the partition. We draw e using pieces of the curves that connect x and y to infinity realizing $l^{D_{H_1}}$ and $l^{D_{H_2}}$ to leave the two drawings and connecting the two curves with a third curve. Note that then,

$$l^{D_H} \leq \max\{l^{D_{H_1}}, l^{D_{H_2}}\} + R(n)b(H).$$

Solving the above recurrence relation over a subtree of T rooted at H, we get,

$$l^{D_H} = O(\log n R(n)\bar{b}(H)). \tag{1}$$

Moreover, for any vertex H in T, it is easy to verify that

$$\nu_1(D_H) \leq \nu_1(D_{H_1}) + \nu_1(D_{H_2}) + R^2(n)b^2(H) + 2l^{D_H}(R(n)b(H))$$

and hence by (1) we get

$$\nu_1(D_H) \leq \nu_1(D_{H_1}) + \nu_1(D_{H_2}) + O(R^2(n)\bar{b}^2(H)\log n).$$

Now according to [PSS94], use the fact that

$$\bar{b}(H) \leq 7 \sqrt{cr_0(H) + \sum_{i \in V(H)} d_i^2(H)}.$$

Then, using the above inequality we have,

$$\nu_1(D_H) \le \nu_1(D_{H_1}) + \nu_1(D_{H_2}) + O(R^2(n)(cr_0(H) + \sum_{i \in V(H)} d_i^2(H)) \log n). \quad (2)$$

Next, observe that for any vertex H of T,

$$cr_0(H_1) + \sum_{i \in V(H_1)} d_i^2(H_1) + cr_0(H_2) + \sum_{i \in V(H_2)} d_i^2(H_2) \le cr_0(H) + \sum_{i \in V(H)} d_i^2(H),$$

and use these two inequalities to solve recurrence relation (2) over a subtree of T rooted at H. We get,

$$\nu_1(D_H) = O(\log^2 n R^2(n)(cr_0(H) + \sum_{i \in V(H)} d_i^2(H))).$$

This finishes the proof, since we can take H to be G. □

Very recently [CY94] polynomially time bounded algorithm for approximately computing the bisection within a constant multiplicative factor from the optimal were discovered. Hence, we have the following:

Theorem 4.3 *We can draw any graph G in a k-page book in polynomial time with at most*

$$O\left(\frac{\log^2 n}{k}(cr_0(G) + \sum_{i \in V} d_i^2(G))\right)$$

crossings. Moreover, assume that G has $m \ge 4n$, and $\Delta = O(\delta^{1.5})$, then for $k = 1, 2$, the number of crossings in our drawing is within a factor of $O(\log^2 n)$ from the optimal value.

The significance of drawing G on one page in Theorem 4.2 is best understood by its relevance to near optimal drawing of graphs with straight lines.

Corollary 4.1 *We can draw any graph G with straight lines on the plane in polynomial time with at most*

$$O\left(\log^2 n(cr_0(G) + \sum_{i \in V} d_i^2(G))\right)$$

crossings. Moreover, the number of crossings in our drawing is within a multiplicative factor of $O(\log^2 n)$ from the optimal value for any G with $m \ge 4n$, and $\Delta = O(\delta^{1.5})$.

Proof sketch. Note that a 1-page drawing of G is equivalent to a drawing in which the vertices of G are placed at the corners of a convex polygon on the plane, in the same order as they are ordered on the book spine, and the edges of G are drawn using sides or chords of the polygon. To verify the suboptimality, one can show that for $m \ge 4n$ and $\Delta = O(\delta^{1.5})$, $cr_0(G)$ assymtotically dominates the sum of the square of degrees. □

Bienstock and Dean [BD92] have extensively investigated the relationship between $cr_0(G)$ and $\overline{cr_0}(G)$ and have obtained the following.

Theorem 4.4 *For any graph G, we have,*

$$\overline{cr_0}(G) = O(\Delta cr_0^2(G))$$

The result in [BD92] is the first general result in the literature, regarding the relationship between $cr_0(G)$ and $\overline{cr_0}(G)$. However, this result does not provide for a provably near optimal solution in polynomial time.

For $m \geq 4n$ Corollary 4.1 implies,

$$\overline{cr_0}(G) = O(\Delta cr_0(G) \log^2(cr_0(G))),$$

which improves the result in [BD92].

Our next result in this section is to provide good drawing of dense graphs in a k-page book. To do this we first obtain an upper bound on $\nu_k(K_n)$ by providing a suitable drawing of K_n. Then we use this drawing to draw any graph.

Theorem 4.5 *For $k < \lceil n/2 \rceil$*

$$\nu_k(K_n) < \frac{2}{k^2}\left(1 - \frac{1}{2k}\right)\binom{n}{4} + \frac{n^3}{2k}.$$

Proof. Assume n is even. For n odd the proof is similar. Let $k > 1$ as the case $k = 1$ is trivial. It is known that $p(K_n) = \lceil n/2 \rceil$, [BK79]. Let $n/2 = kl+r$, where $0 \leq r < k$. We consider K_n as the regular n-polygon with vertices labeled by $0, 1, 2, ..., n-1$ in the counter-clockwise direction and edges (straight lines) uv, for $0 \leq u < v \leq n-1$. Let C_n denote the cycle $012...(n-1)$. Let G_0 denote the subgraph of K_n defined as follows:

$$V(G_0) = V(K_n),$$

$$E(G_0) = \{uv| u + v = \frac{n}{2} \text{ mod } n\} \cup \{uv| u + v = \frac{n}{2} + 1 \text{ mod } n\} - E(C_n)$$

Note that G_0 is an outerplanar graph. Let G_i denote the graph obtained from G_0 by rotating it by the angle $2\pi i/n$ around the center of the polygon, for $i = 1, 2, ..., \frac{n}{2} - 1$. Hence

$$V(G_i) = V(K_n)$$

$$E(G_i) = \{uv| u + v = \frac{n}{2} + 2i \text{ mod } n\} \cup \{uv| u + v = \frac{n}{2} + 2i + 1 \text{ mod } n\} - E(C_n).$$

This easily implies that $E(G_i) \cap E(G_j) = \emptyset$ for $i, j = 0, 1, 2, ..., n/2-1, i \neq j$ and

$$E(K_n) = E(C_n) \cup \bigcup_{i=0}^{\frac{n}{2}-1} E(G_i).$$

Now we describe a drawing of K_n on k pages, i.e. we colour the edges of K_n by k colours $1, 2, ..., k$ in the following way:
We colour $E(C_n)$ by the first colour.
Further, we successively divide the graphs $G_0, G_1, ..., G_{\frac{n}{2}-1}$ into r groups, each containing $l + 1$ graphs and $k - r$ groups, each containing l graphs and colour

the groups by k different colours. More precisely:

If $1 \leq m \leq r$, then colour $G_{(m-1)(l+1)+j}$, for $0 \leq j \leq l$ by the mth colour.

If $r+1 \leq m \leq k$, then colour $G_{(m-1)l+r+j}$, for $0 \leq j \leq l-1$ by the mth colour.

Note that the graphs coloured by the colours $1, 2, ..., r$ are isomorphic to $\cup_{i=0}^{l} G_i$ and graphs coloured by colours $r+1, r+2, ..., k$ are isomorphic to $\cup_{i=0}^{l-1} G_i$.

If H is a one-colour subgraph of K_n let $c(H)$ denote the number of crossings of its edges.

A detailed counting analysis shows that

$$c(G_0 \cup G_t) = 4nt - 8t^2 - 2n + 1,$$

for $1 \leq t \leq n/2 - 1$. Because $c(G_i \cup G_j) = c(G_0 \cup G_{j-i})$, for $0 \leq i < j \leq l$, we get

$$\nu_k(K_n) \leq r \sum_{0 \leq i < j \leq l} c(G_i \cup G_j) + (k-r) \sum_{0 \leq i < j \leq l-1} c(G_i \cup G_j)$$

$$< \frac{2kl}{3} l^2 (n-l) + \frac{ln}{k}(n - 2r + 2k)r.$$

Because the function $l^2(n-l)$ is nondecreasing for $0 \leq l \leq n/2k$ and similarly the function $(n - 2r + 2k)r$ is a nondecreasing function for $0 \leq r \leq k$, setting $l = n/2l$ and $r = k$ we get the claimed upper bound. □

In what follows we propose a general drawing algorithm which draws any graph G by means of drawing of K_n.

Corollary 4.2 *For any graph $G = (V, E)), |V| = n, |E| = m$, we can construct a drawing in a k-page book in polynomial time, with at most*

$$\nu_k(G) < \left(\frac{1}{3k^2}\left(1 - \frac{1}{2k}\right) + O\left(\frac{1}{kn}\right) \right) m^2$$

many crossings.

Proof. First, we show the existence of a suitable drawing. Consider our drawing of complete graph $K_n = (V', E')$ in a k-page book. Thus any vertex of V' is placed on the spine of the book and any edge is drawn in some page of the book. For any bijection $h : V \rightarrow V'$ we can obtain a drawing of G in the following way: draw any edge $e = ij \in E$ using the drawing the edge $h(i)h(j)$ of K_n. Now let $\bar{h}$ be a random bijection from V to V', thus we pick up one bijection randomly and uniformly. Then, it is easy to verify that the expected number of crossings in a drawing associated with this random bijection is at most

$$\frac{8\binom{m}{2}}{n(n-1)(n-2)(n-3)} \left(\frac{2}{k^2}\left(1 - \frac{1}{2k}\right)\binom{n}{4} + \frac{n^3}{2k}\right).$$

To actully obtain such a drawing of G, one can remove the randomness from our construction in polynomial time by verifying that the expected values of the conditional probabilities can be computed in polynomial time [AES92]. The

details are similar to proofs of Lemma 3.2 and Theorems 3.3 and 3.4 in [SSSV94].
□

The upper bound from Corollary 4.2 is asymptotically optimal at least for dense graphs i.e. $m = \Theta(n^2)$ and $k \leq (1 - \varepsilon)m/3n$, for $0 < \varepsilon < 1$, as follows from Theorem 3.1.

5 Specific Drawings

In this section we sharpen our general upper bounds for some specific cases.
For $k = 2$ we slightly modify the embedding of K_n in Theorem 4.4 in the following way.
If $n = 0 \bmod 4$ then place edges of G_i, for $0 \leq i \leq n/4 - 1$ on the first page and the remaining pages on the second page.
If $n = 1 \bmod 4$ then place edges of G_i, for $0 \leq i \leq (n-1)/4 - 1$ on the first page and the remaining edges on the second page.
If $n = 2 \bmod 4$ then place edges of G_i, for $0 \leq i \leq (n-6)/4$, and edges ij, $i+j = n/2 \bmod n$ on the first page and the remaining edges on the second page.
If $n = 3 \bmod 4$ then place edges of G_i, for $0 \leq i \leq (n-3)/4$ on the first page and the remaining edges on the second page.
This is actually the drawing of K_n in the plane, having a hamiltonian cycle free of crossings, obtained by Guy et al. [GJS68]. The number of crossings is

$$\frac{1}{4} \left\lfloor \frac{n}{2} \right\rfloor \left\lfloor \frac{n-1}{2} \right\rfloor \left\lfloor \frac{n-2}{2} \right\rfloor \left\lfloor \frac{n-3}{2} \right\rfloor .$$

For $k = 2$ and $K_{m,n}$, Corollary 4.2 implies

Theorem 5.1

$$\nu_2(K_{m,n}) < \frac{m^2 n^2}{16} + O\left(\frac{m^2 n^2}{m+n} \right) .$$

This upper bound is very close to the best known upper bound for $cr_0(K_{m,n})$ proposed by Zarankiewicz [Zr54] (see [BCL76] for the history of the problem)

$$cr_0(K_{m,n}) \leq \frac{1}{4} \left\lfloor \frac{m}{2} \right\rfloor \left\lfloor \frac{m-1}{2} \right\rfloor \left\lfloor \frac{m}{2} \right\rfloor \left\lfloor \frac{m-1}{2} \right\rfloor .$$

We conjecture that there exists a drawing of $K_{m,n}$ in two pages with the same number of crossings. Such a construction would differ from the Zarankiewicz's drawing because it would contain a hamiltonian cycle free of crossings.

Another construction of $K_{m,n}$ in two pages could be obtained from the $(m+2n)/2$ page embedding of $K_{m,n}$ of Muder et al. [MWW88], by identifying suitable pages.

Let Q_n denote the n-dimensional hypercube graph usually defined by means of the cartesian product of graphs: $Q_1 = K_2, Q_n = Q_{n-1} \times K_2$, for $n \geq 2$. The algorithm presented in Theorem 4.3 provides drawings of Q_n with $O(4^n n^2/k)$ crossings in a k-page book, since it is known that $cr_0(Q_n) = \Theta(4^n)$. In this section we give better drawings of Q_n with smaller number of edge crossings.

Theorem 5.2 *For $k < n - 1$*

$$\nu_k(Q_n) < \frac{4^n}{2(2^k - 1)}.$$

Proof. First we construct a recursive one page drawing of Q_n. To obtain the drawing of Q_n from that of Q_{n-1} we take two copies of the drawing of Q_{n-1} and place one next to other on the horizontal line (spine). We then add the edges for dimension n. Figure 1. illustrates the drawings produced for Q_1, Q_2 and Q_3.

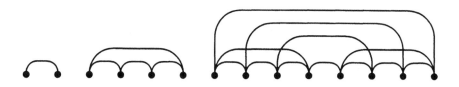

Figure 1. One-page drawings of Q_1, Q_2 and Q_3.

Now we describe a drawing of Q_n on k pages. It is known that $p(Q_n) \leq n-1$, [BK79]. Assume $n - 1 = kl + r$, where $0 \leq r < k$. We add $k - 1$ pages to the 1-page book considered above and move the edges of dimensions $ik + m + 1$ to the m-th page, for $i = 0, 1, 2, ..., l$ if $2 \leq m \leq r$ and for $i = 0, 1, 2, ..., l - 1$ if $r + 1 \leq k$. We count the total number of crossings on all pages. It is easy to see that for given dimensions d_1 and $d_2, 2 \leq d_1 < d_2$, on a page, there are less than 2^{n-2+d_1} crossings of edges of dimensions d_1 and d_2 on that page. Hence

$$\nu_k(Q_n) < \sum_{m=1}^{r} \sum_{i=0}^{l-1} 2^{n-1+ik+m}(l - i) + \sum_{m=r+1}^{k} \sum_{i=0}^{l-2} 2^{n-1+ik+m}(l - 1 - i)$$

$$= 2^{n-1} \sum_{m=1}^{r} 2^m \sum_{i=0}^{l-1} 2^{ik}(l - i) + 2^{n-1} \sum_{m=r+1}^{k} 2^m \sum_{i=0}^{l-2} 2^{ik}(l - 1 - i)$$

$$< 2^{n-1}(2^{r+1} - 2)\frac{2^{k(l+1)}}{(2^k - 1)^2} + 2^{n+r}(2^{k-r+1} - 1)\frac{2^{kl}}{(2^k - 1)^2} = \frac{4^n}{2(2^k - 1)}.$$

$\square$

Theorem 5.2 implies $\nu_1(Q_n) < 4^n/2$. Using the theory of embedding graphs into graphs first introduced by Leighton [Lg83] and generalized and refined by Shahrokhi and Székely [SS92] and Shahrokhi, Sýkora, Székely and Vrt'o [SSSV], one can obtain lower bounds on $\nu_1(G)$ for any graph G. This way it can be shown that

$$\nu_1(Q_n) > \frac{1}{6}4^n - n(n + 4)2^{n-3},$$

and hence the upper bound of Theorem 5.2 is within a constant factor from the optimal value.

For $k = 2$ Theorem 5.2 implies $\nu_2(Q_n) < 4^n/6$. Asymptotically optimal lower bounds follow from $\nu_2(Q_n) \geq cr_0(Q_n) > 4^n/20 - (n^2 + 1)2^{n-1}$ [SV93].

6 Conclusions

Our paper leaves several open problems:

- Are the upper bounds for $k > 2$ in Theorem 5.2 optimal up to constant factors?
- Does there exist a drawing of $K_{m,n}$ in two pages with the same number of crossings as in the Zarankiewicz' construction?
- It would be interesting to find exact crossing numbers at least up to the first order term for some specific graphs and choices of k. E.g. is it true that $\nu_2(K_n) = n^4/64 + O(n^3)$?
- Find new lower bound arguments for $\nu_k(G)$ based on structural properties.
- Determine a relation between $cr_0(G)$ and $\nu_k(G)$, for $k \geq 4$, e.g. is it true that $\nu(G) \geq \nu_4(G)$?

References

[ACNS82] Ajtai, M., Chvátal, V., Newborn, M., M., Szemerédy, E., "Crossing-free subgraphs", *Annals of Discrete Mathematics* **12** (1982), 9-12.

[AES92] Alon, N., Spencer, J.H., Erdős, P., "The Probabilistic Method", Wiley and Sons, New York, 1992.

[BCL76] Behzad, M., Chartrand, G., Lesniak-Foster, L., "Graphs and Digraphs", Wadsworth International Group, Belmont, 1976.

[BK79] Bernhart, F., Kainen, P. C., "The book thickness of a graph", *J. Combinatorial Theory, Series B* **27** (1979), 320–331.

[BD92] Bienstock, D., Dean, N., "New results on rectilinear crossing Numbers and Plane Embeddings", *J. Graph Theorey*, **16** (1992), 389–398.

[CCDG82] Chinn, P. Z., Chvátalová, L., Dewdney, A. K., Gibbs, N. E., "The bandwidth problem for graphs and matrices—a survey", *J. Graph Theory* **6** (1982), 223–253.

[Ch88] Chung, F. R. K., "Labeling of graphs", in: "Selected Topics in Graph Theory 3", (L. Beineke and R. Wilson, eds.), Academic Press, New York, 1988, 151–168.

[CLR87] Chung, F. R. K., Leighton, F. T., Rosenberg, A. L., "Embeddings graphs in books: A layout problem with applications to VLSI design", *SIAM J. Algebraic and Discrete Methods* **8** (1987), 33–58.

[CY94] Chung, F. R. K., Yau, S.T., "A near optimal algorithm for edge separators", in: Proc. *28th ACM Annual Symposium on Theory of Computing*, ACM Press, 1994, 1–8.

[Dz92] Díaz, J., "Graph layout problems", in: Proc. *17th Intl. Symposium on Mathematical Foundations of Computer Science*, LNCS **629**, Springer Verlag, Berlin, 1992, 15–23.

[EG73] Erdős, P., Guy, R. P, "Crossing number problems", *American Mathematical Monthly* **80** (1973), 52–58.

[GJS68] Guy, R. P., Jenkyns, T., Schaer, J., "The toroidal crossing number of the complete graph", *J. Combinatorial Theory* **4** (1968), 376–390.

[Kn90] Kainen, P. C., "The book thickness of a graph, II", *Congressus Numerantium* **71** (1990), 127–132.

[Lg83]   Leighton, F. T., "Complexity Issues in VLSI", M.I.T. Press, Cambridge, 1983.

[Ls80]   Leiserson, C. E., "Area efficient graph layouts (for VLSI)", in: Proc. *21st Annual IEEE Symposium on Foundations of Computer Science*, IEEE Computer Society Press, Los Alamitos, 1980, 270–281.

[MKNF87] Masuda, S., Kashiwabara, T., Nakajima, K., Fujisawa, T., "On the NP-completeness of a computer network layout problem", in: Proc. *1987 IEEE Intl. Symposium on Circuits and Systems*, IEEE Computer Society Press, Los Alamitos, 1987, 292–295.

[MKNF90] Masuda, S., Kashiwabara, T., Nakajima, K., Fujisawa, T., "Crossing minimization in linear embeddings of graphs", *IEEE Transactions on Computers* **39** (1990), 124–127.

[MKT71]   Melikhov, A. N., Koreichik, V. M., Tishchenko, V. A., "Minimization of the number of intersections of edges of a graph", (in Russian), *Vichislityelnie Sistemi Vip.* **41** (1971), 32–40.

[MWW88]   Muder, J. D., Weawer, M. L., West, D. B., "Pagenumber of complete bipartite graphs", *J. Graph Theory* **12** (1988), 469–489.

[Nc68]   Nicolson, T. A. J., "Permutation procedure for minimizing the number of crossings in a network", *Proc. Inst. Elec. Engnrs.* **115** (1968), 21–26.

[PSS94]   Pach, J., Shahrokhi, F., Szegedy, M., "Applications of crossing numbers", in: Proc. *10th Annual ACM Symposium on Computational Geometry*, ACM Press, New York, 1994.

[SS92]   Shahrokhi, F., Székely, L. A., "Effective lower bounds for crossing number, bisection width and balanced vertex separators in terms of symmetry", in: Proc. *Integer Programming and Combinatorial Optimization*, CMU Press, Pittsburgh, 1992, 102–113.

[SSSV]   Shahrokhi, F., Sýkora, O., Székely, L. A., Vrt'o, I., "The crossing number of a graph on a compact 2-manifold", *Advances in Mathematics*, to appear.

[SSSV94] Shahrokhi, F., Sýkora, O., Székely, L. A., Vrt'o, I., "Improved bounds for the crossing numbers on surfaces of genus g", in: Proc. *19-th Intl. Workshop on Graph-Theoretic Concepts in Computer Science WG'93*, LNCS **790**, Springer Verlag, Berlin, 1994, 388–397.

[SV93]   Sýkora, O., Vrt'o, I., "On crossing numbers of hypercubes and cube connected cycles", *BIT* **3** (1993), 232–237.

[Ul84]   Ullman, J. D., "Computational Aspects of VLSI", Computer Science Press, Rockville, 1984.

[Yn86a]   Yannakakis, M., "Linear and book embeddings of graphs", in: Proc. *Aegean Workshop on Computing*, LNCS **227**, Springer Verlag, Berlin, 1986, 229–240.

[Yn86b]   Yannakakis, M., "Four pages are necessary and sufficient for planar graphs", in: Proc. *18th ACM Annual Symposium on Theory of Computing*, ACM Press, New York, 1986, 104–108.

[Zr54]   Zarankiewicz, K., "On a problem of P. Turán concerning graphs, *Fundamenta Mathematica* **41** (1954), 137–145.

Measuring the Distance to Series–Parallelity by Path Expressions

Valeska Naumann[*]

TU Berlin

Abstract. Many graph and network problems are easily solved in the special case of series-parallel networks, but are highly intractable in the general case. This paper considers two complexity measures of *two-terminal directed acyclic graphs* (st-dags) describing the "distance" of an st-dag from series-parallelity. The two complexity measures are the *factoring complexity* $\psi(G)$ and *the reduction complexity* $\mu(G)$. Bein, Kamburowski, and Stallmann [3] have shown that $\psi(G) \leq \mu(G) \leq n - 3$, where G is an st-dag with n nodes. They conjectured that $\psi(G) = \mu(G)$. This paper gives a proof for this conjecture.

1 Introduction

Many combinatorial problems defined on graphs are NP-complete, and hence there is probably no polynomial-time algorithm for any of them. But it is shown that there exist efficient algorithms for many combinatorial problems if an input graph is restricted to the class of *series-parallel graphs* [9], [2], [6], [7]. Two examples are the *generalized matching problem* and the *decision problem*. In a generalized matching problem we would like to find a maximum number of vertex-disjoint copies of a fixed graph contained in an input graph [6]. In a decision (i. e. yes–no) problem we would like to decide whether an input graph satisfies a property which is often characterized by a finite number of forbidden (induced or homeomorphic) subgraphs [4].

Complexity measures describing the distance from series-parallelity are crucial for the design of polynomial-time algorithms, since the results of problems on series-parallel graphs can often be extended to algorithms which are exponential only in the complexity of the underlying st-dag, rather than in its size (see Section 3 for an example).

Any st-dag G can be reduced to a single edge by means of three kinds of reduction, the *parallel reduction*, the *series reduction* and the *node reduction* (see Section 3). If and only if series reductions and parallel reductions are sufficient to reduce G to a single edge, then G is series-parallel. If G is not series parallel, we can use node reductions together with series and parallel reductions, to reduce G to a single edge. The *reduction complexity* $\mu(G)$ is the minimum number of

---
[*] TU Berlin, Fachbereich Mathematik, Sekr. MA 6-1, Strasse des 17.Juni 136, D-10623 Berlin, e-mail: naumann@math.tu-berlin.de

node reductions sufficient (along with series and parallel reductions) to reduce G to a single edge.

The second complexity measure, the *factoring complexity* of G, is based on a special way of describing all paths from the *source* of G to the *target* of G.

The descriptions of all source-target paths (so called *factorings*) are algebraic expressions and consist of names of edges, parentheses and the operators $+$ (disjoint union) and $\cdot$ (concatenation). The *cost* of a factoring $\mathcal{F}$ of an st-dag G is the number of subexpressions of $\mathcal{F}$ that appear duplicated in a certain way (see Section 4 for details). The minimum cost of any factoring $\mathcal{F}$ of G is called the *factoring complexity of G*, denoted by $\psi(G)$.

In this paper we will show that $\psi(G) = \mu(G)$. The paper is organized as follows. Section 2 reviews standard graphtheoretic definitions and notations that will be used throughout the paper. Sections 3 and 4 introduce the reduction complexity and the factoring complexity, respectively. Section 5 gives an algorithm which reduces a given st-dag G with any corresponding factoring $\mathcal{F}$ to a single edge and returns a *node reduction sequence* with a length less than or equal to the *cost* of the factoring $\mathcal{F}$. This implies $\psi(G) \geq \mu(G)$, and together with the result of Bein, Kamburowski and Stallmann [3] that $\psi(G) \leq \mu(G)$, we obtain $\psi(G) = \mu(G)$.

2 Preliminaries

A directed graph $G = (V, E)$ consists of a finite set of vertices V and a finite set of edges E (multiple edges between the same two vertices are permitted). Each edge e is an ordered pair (v, w) of vertices, where v is the *source*, $s(e) = v$, and w is the *target*, $t(e) = w$, of e. The *in-degree* (*out-degree*) of the vertex v in G is the number of distinct edges with target (source) v and is denoted by $in(v, G)$ $(out(v, G))$.

A *path P from vertex v to vertex w*, is a sequence of edges $e_1, e_2, \ldots, e_k$ with $e_i \in E$ $(i = 1, ..., k)$ and $s(e_1) = v$, $t(e_k) = w$. A path from vertex v to v is a *cycle*.

A *two-terminal directed acyclic-graph* (st-dag) G is a directed graph without any cycle, having a unique source s and a unique target t. This implies that an st-dag is weakly connected, namely, there is a path from s to any vertex and from any vertex to t.

When we say an st-dag is *series-parallel* we mean that it is two-terminal edge series-parallel (see [10] for a discussion of the relationship between vertex and edge series-parallelity). The series-parallel st-dags are recursively defined as follows:

- An st-dag having a single edge e is two-terminal series-parallel (with $s(e) = s$ and $t(e) = t$)
- If $G_1 = (V_1, E_1)$ and $G_2 = (V_2, E_2)$ are two-terminal series-parallel, so are the graphs obtained by the following operations:

1. Parallel composition: identifying the sources and the targets, respectively, with each other; $G_p = (V_p = V_1 \cup V_2, E_p = E_1 \cup E_2)$ with $|V_p| = |V_1| + |V_2| - 2$ and $|E_p| = |E_1| + |E_2|$

2. Series composition: identifying the target of G_1 with the source of G_2; $G_p = (V_p = V_1 \cup V_2, E_p = E_1 \cup E_2)$ with $|V_p| = |V_1| + |V_2| - 1$ and $|E_p| = |E_1| + |E_2|$

In other words, an st-dag G is series-parallel, if series reductions and parallel reductions are sufficient to reduce G to a single edge.

An st-dag $G_1 = (V_1, E_1)$ is a *subdag* of another st-dag $G_2 = (V_2, E_2)$ if $V_1 \subseteq V_2$ and $E_1 \subseteq E_2$.

The notation $v < w$ means that there is a path P from v to w in G. For every pair of nodes $v, w \in G$ with $v < w$ there is a unique *subdag H* of G, *induced by* the nodes v, w . The *subdag H induced* by v, w is formed by all paths from v to w in G. This subdag H has source $v = s(H)$ and target $w = t(H)$.

An st-dag $\tilde{G}$ with source v and target w is an *autonomous subdag* of G if it is a subdag and satisfies the following additional property: For every path P from s to t the set of edges $P \cap \tilde{G}$ is either empty or forms a path from v to w. In other words, v is the only entry point, and w is the only exit of $\tilde{G}$. An autonomous subdag which contains only one edge, is a *trivial autonomous subdag*.

3 Reduction Complexity

We are interested here in three kinds of reductions: *parallel reduction, series reduction* and *node reduction*. A *parallel reduction* at v, w replaces two or more parallel edges $e_1, \ldots, e_k$ joining v to w by a single edge $g = (v, w)$. A *series reduction* at v is possible if $in(v, G) = out(v, G) = 1$; then $e = (u, v)$ and $f = (v, w)$ are replaced by $g = (u, w)$. A *node reduction* is a generalization of a series reduction. It is necessary that v has $in(v, G) = 1$ or $out(v, G) = 1$. Suppose v has in-degree one, and let $e = (u, v)$ be the edge into v. Let $f_1 = (v, w_1), \ldots, f_k = (v, w_k)$ be the edges out of v. Replace $e, f_1, \ldots, f_k$ by $g_1, \ldots, g_k$, where $g_i = (u, w_i)$. If $out(v, G) = 1$, the set of edges $f_1 = (u_1, v), \ldots, f_k = (u_k, v)$, $e = (v, w)$ will be replaced by $g_1 = (u, w_1), \ldots, g_k = (u, w_k)$.

For convenience, let $G \circ v$ denote the result of a node reduction at v, and let $[G]$ denote the graph that results when all possible series and parallel reductions have been applied to G. An st-dag G is said to be *irreducible* if $[G]=G$.

The following complexity measure of an st-dag describes how nearly is an st-dag to series-parallelity.

Definition 1. The minimal number of node reductions sufficient to reduce G to a single edge is called the *reduction complexity* of G, denoted by $\mu(G)$. In other words, $\mu(G)$ is the smallest number c for which there is a sequence $v_1, \ldots, v_c$ so that $[\ldots [[[G] \circ v_1] \circ v_2] \ldots \circ v_c]$ is a single edge. Such a sequence is called a *node reduction sequence*.

The following example taken from [3] shows that the cost of computation of the *two-terminal reliability* may grow exponentially with increasing distance from series-parallelity. Let G be an st-dag wherein each edge e is assigned a failure probability $p(e)$. Then $R(G)$, the *two-terminal reliability* of G, is the probability that there exists at least one source-target path with no failed edges in G. If G is series-parallel, $R(G)$ can be computed in linear time using series and parallel reductions [1]. For each reduction the failure probability of the new edge can be computed so that the graph has the same reliability before and after the reduction. For a series reduction, which replaces the edges e and f by g, one obtains

$$p(g) = 1 - (1 - p(e))(1 - p(f)).$$

For a parallel reduction, which replaces $e_1, \ldots, e_k$ by g, one obtains

$$p(g) = p(e_1) \ldots p(e_k).$$

Let G be the st-dag before the node reduction; let $\tilde{G}$ be the st-dag after the node reduction with $p(g_i) = p(f_i)$, $i = 1, \ldots, k$; let $\dot{G} = G - \{e, f_1, \ldots, f_k\}$. That is, $\tilde{G}$ is derived from G under the condition that e does not fail, while $\dot{G}$ is derived from G in case e fails. To make $\dot{G}$ an st-dag, we also remove from it all vertices and edges that are not on any path from s to t. Then we apply the recurrence

$$R(G) = (1 - p(e))R(\tilde{G}) + p(e)R(\dot{G})$$

Therefore the cost of computing $R(G)$ grows exponentially with $\mu(G)$.

4   Factoring Complexity

Another complexity measure is the *factoring complexity*. This complexity measure is based on descriptions of all source-target paths of G. If G is an st-dag, it is possible to define an algebraic expression α, a so-called *factoring of G*, for the set of all source-target paths in G. The expression α consists of names of edges, parentheses and the operators $+$ (disjoint union) and $\cdot$ (concatenation). If G is series-parallel, $+$ corresponds to the parallel composition and $\cdot$ to the series composition.

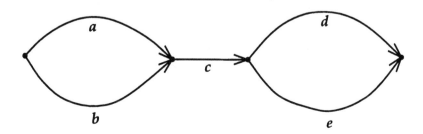

Fig. 1. a series-parallel st-dag

Figure 2 shows a series-parallel st-dag G. Some possible factorings of G are $(a+b)c(d+e)$, $(ac+bc)(d+e)$, $(a+b)(cd+ce)$ and $acd+ace+bc(d+e)$. This example shows that already a small st-dag can have many different factorings.

A formal definition of a factoring is given now. A *path expression* α is an algebraic expression for a set of paths between two specific vertices of an st-dag. This means all paths represented by α have the same start vertex, denoted by $s(\alpha)$, and the same target vertex, denoted by $t(\alpha)$. Let $\mathcal{P}(\alpha)$ denote the set of all paths represented by α. Any algebraic expression is a valid path expression if it can be obtained iteratively in the following way:

- The name of a single edge e from v to w is a path expression with $P(e) = \{e\}$, $s(e) = v$ and $t(e) = w$.
- If α_1 and α_2 are path expressions with $s(\alpha_1) = s(\alpha_2)$, $t(\alpha_1) = t(\alpha_2)$, and $\mathcal{P}(\alpha_1) \cap \mathcal{P}(\alpha_2) = \emptyset$, then $\alpha_1 + \alpha_2$ is a path expression with $s(\alpha_1 + \alpha_2) = s(\alpha_1) = s(\alpha_2)$, $t(\alpha_1 + \alpha_2) = t(\alpha_1) = t(\alpha_2)$, and $\mathcal{P}(\alpha_1 + \alpha_2) = \mathcal{P}(\alpha_1) \cup \mathcal{P}(\alpha_2)$.
- If α_1 and α_2 are path expressions with $t(\alpha_1) = s(\alpha_2)$, then $\alpha_1 \cdot \alpha_2$ is a path expression with $s(\alpha_1 \cdot \alpha_2) = s(\alpha_1)$, $t(\alpha_1 \cdot \alpha_2) = t(\alpha_2)$, and $\mathcal{P}(\alpha_1 \cdot \alpha_2) = \{P_1 P_2 \mid P_1 \in \mathcal{P}(\alpha_1),\ P_2 \in \mathcal{P}(\alpha_2)\}$.

If G is series-parallel, there are factorings of G which have no *duplicated expressions*. In other words, all names of edges of G appear exactly once in $\mathcal{F}$. If G is not series-parallel, all factorings have duplicate occurrences of subexpressions. In the following Example 3, the subexpressions $(a+b)c$, a, b, c, $(f+gh)$, dh, i and h are duplicated subexpressions.

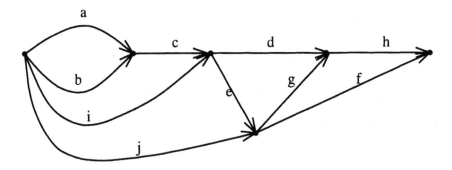

Fig. 2. An st-dag which is not series-parallel

Example 3: $(a+b)cdh + ((a+b)c+i)e(f+gh) + j(f+gh) + idh$
is one possible factoring for the st-dag shown in Figure 3

Now we are interested in the number of maximal duplicated subexpressions. In Example 3 the names of the edges a, b, c occur twice, like the subexpression $(a+b)c$. But a, b, c occur only in the subexpression $(a+b)c$. This means that a, b, c are not duplicated subexpressions themselves. In contrast to this, h occurs in the duplicated subexpression dh and in the duplicated subexpression $(f+gh)$

and there is no larger duplicated subexpression containing $(f + gh)$ and dh. Therefore, h is a maximal duplicated subexpression.

Definition 2. A subexpression α of a path expression $\mathcal{F}$ is a *duplicated expression* (i.e. α is a maximal duplicated subexpression) if the following holds:

1. α occurs k times ($k > 1$) in $\mathcal{F}$.
2. Any larger expression α' containing α appears less than k times in $\mathcal{F}$.

Definition 3. The *cost* of a factoring $\mathcal{F}$ is the number of distinct duplicated expressions in $\mathcal{F}$. The minimum cost of any factoring $\mathcal{F}$ of an st-dag G is called the *factoring complexity of G*, denoted by $\psi(G)$.

The factoring $\mathcal{F}$ from Example 3 which corresponds to the graph shown in Fig. 3, has cost 5. The corresponding duplicated expressions are $(a+b)c$, $(f+gh)$, dh, h, i. Another factoring $\mathcal{F}$ of G from Fig. 3 is $((a+b)c+i)(e(f+gh)+dh)+j(f+gh)$, which has optimal cost 2 ($\psi(G) = 2$). The corresponding duplicated expressions are $(f + gh)$ and h.

Bein, Kamburowski and Stallmann [3] have shown that $\psi(G) \leq \mu(G) \leq n - 3$ where G is an st-dag with n nodes. They conjectured that $\psi(G) = \mu(G)$.

This paper gives a proof for $\psi(G) \geq \mu(G)$. The proof will be done in the following way. Consider an st-dag G and a corresponding factoring $\mathcal{F}$. Then the algorithm given in Section 5 reduces G to a single edge, so that the length of the node reduction sequence generated by the algorithm is less than or equal to the number of duplicated expressions of $\mathcal{F}$. This holds for any st-dag and any corresponding factoring $\mathcal{F}$, from which $\psi(G) \geq \mu(G)$ follows. For the algorithm we need some more definitions and properties related to factorings of G.

The the commutative law relating to disjoint union, the associative law and the distributive laws relating to both operators can be applied to any factoring. This has the favorable effect that any factoring $\mathcal{F}$ of G can be transformed into a *standard form*

$$\mathcal{F} = P^1 + P^2 + \ldots + P^r,$$

where P^i, $i \leq i \leq r$, is a path from s to t. Therefore a factoring of an st-dag G can be transformed into any other factoring of G.

Among the duplicated expressions there are some special ones, the *minimal duplicated expressions*. If α is a duplicated expression and α itself contains no duplicated expression, then we say α is a *minimal duplicated expression* .

For the autonomous subdags there are, in general, no closed representations inside a factoring. This means that, for a given st-dag G and corresponding factoring $\mathcal{F}$, it is possible to have an autonomous subdag H of G, so that H has no corresponding subexpression in $\mathcal{F}$. But it is possible to construct a path expression h representing H with $h = T_1 + T_2 + \ldots + T_k$, where every T_i, $i = 1 \ldots k$, is a subexpression in $\mathcal{F}$.

Lemma 4. *Consider an st-dag G with source s, target t and a corresponding factoring $\mathcal{F}$. Then for every autonomous subdag H of G, there are subexpressions T_i in $\mathcal{F}$, $i = 1 \ldots k$, so that $T_1 + T_2 + \ldots + T_k$ is a factoring of H.*

Proof. Let the source of H be v ($s(H) = v$) and the target of H be w ($t(H) = w$). Since any st-dag is connected, there exist a path $\tilde{P}$ from s to v and a path $\grave{P}$ from w to t. Denote all paths from v to w lying in H by P^j, $j = 1 \ldots k$. We look now for all k paths P_j from s to t of the form $P_j = \tilde{P} P^j \grave{P}$, $j = 1 \ldots k$. When we have all paths P_j, we have all paths P^j of H, too.

We look for the minimal subexpressions T_i with $s(T_i) = v$, $t(T_i) = w$. From these subexpressions we choose the subexpressions that describe paths P_j from v to w lying in H.

In $\mathcal{F}$ the descriptions of any of the paths $P_j = \tilde{P} P^j \grave{P}$ can be transformed by deleting names of edges, parentheses, and operators, into the form:

$$\mathcal{F}_{\tilde{P}} \ \cdot \ T_i \ \cdot \ \mathcal{F}_{\grave{P}},$$

where T_i is a minimal subexpression, and $\mathcal{F}_{\tilde{P}}$ and $\mathcal{F}_{\grave{P}}$ are factorings of $\tilde{P}$ and $\grave{P}$, respectively. Since H is an autonomous subdag, for all paths P from v to w either $P \cap H = \emptyset$ or P is a path in H. By the choice of the T_i, every T_i contains at least one path from v to w contained in H. If for any T_i there is additionally a path P from v to w with $P \cap H = \emptyset$, this contradicts the minimality of T_i (because this holds for all paths from v to w and therefore T_i is the union of two disjoint subexpressions with source v and target w).

Every T_i is a subexpression of $\mathcal{F}$ with the same start and end vertex, and the paths $P_j = \tilde{P} P^j \grave{P}$ are pairwise different. Therefore the disjoint union $+$ is applicable. This means that $T_1 + T_2 + \ldots + T_k$ is a path expression. Since any path from v to w, lying in H, is described by one of these T_i, $i = 1 \ldots k$, $P(T_1) \cup P(T_2) \cup \ldots \cup P(T_k)$ contains all paths P^j of H. This means that $T_1 + T_2 + \ldots + T_k$ is a factoring of H.

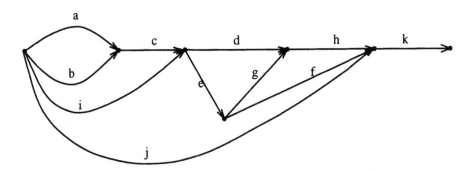

Fig. 3. An st-dag with a nontrivial autonomous subdag

Example 4: $(a + b)c(d + eg)hk + jk + (bc + i)efk + idhk + ieghk + acefk$
one possible factoring for the st-dag shown in Figure 4

Example 4 illustrates this Lemma. Let H be the autonomous subdag induced by the nodes $s(d)$ and $t(h)$. We choose $\tilde{P} = (a, c)$, $\grave{P} = (k)$. Then all paths of the form described above are: (a, c, d, h, k), (a, c, e, g, h, k) and (a, c, e, f, k). This means, H contains exactly the three paths $P_1 = (d, h)$, $P_2 = (e, g, h)$, $P_3 = (e, f)$ from $s(d)$ to $t(h)$. The minimal subexpressions T_i with $s(T_i) = s(d)$, $t(T_i) = t(h)$ are $T_1 = (d + eg)h$, $T_2 = ef$, $T_3 = dh$, $T_4 = egh$, $T_5 = ef$.

Now we look for the T_i which are used for the description of the three paths and obtain $T_1 = (d + eg)h$ and $T_2 = ef$, so that $(d + eg)h + ef$ is a factoring of H.

5 The Algorithm

Consider an st-dag G and a corresponding factoring $\mathcal{F}$. Then the following algorithm reduces G to a single edge, so that the length of the node reduction sequence generated by the algorithm is less than or equal to the number of duplicated expressions of $\mathcal{F}$. Since this holds for any st-dag and any corresponding factoring $\mathcal{F}$, $\psi(G) \geq \mu(G)$ follows.

Reduction-algorithm

1. $i := 0$;
2. $G := [G]$, $\mathcal{F} := \mathcal{F}_{[]}$;
3. $G_i := G$; $\mathcal{F}_i := \mathcal{F}$;
4. **While** G_i has more than one edge **do**
   (a) $G_{i+1} := G_i$; $\mathcal{F}_{i+1} := \mathcal{F}_i$; $i := i + 1$
   (b) Search for an edge $e = (v, w) \in E_{G_i}$ so that the name e is a minimal duplicated expression in $\mathcal{F}_i$ and $in(w, G_i) = 1$ or $out(v, G_i) = 1$.
   (c) The current reduction node is w for $in(w, G_i) = 1$ or v for $out(v, G_i) = 1$, respectively. Realize a node reduction at the current reduction node in G_i and reduce $\mathcal{F}_i$ accordingly (see below).
   (d) Store the current reduction node w as node u_i in the node reduction sequence.
   (e) Realize all possible series and parallel reductions in G_i and reduce $\mathcal{F}_i$ accordingly (see below)
5. For $i = 0$, G is series-parallel, and for $i > 1$, the algorithm terminates with a node reduction sequence of length i: $u_1, u_2, \ldots, u_i$.

One way to realize the reductions in the Reduction-algorithm follows now. Consider $G = (V, E)$ and a factoring $\mathcal{F}$ of G.

Series reduction in $\mathcal{F}$ at node v (steps 2 and 4(d)) The edges $e_1 = (u, v)$ and $e_2 = (v, w)$ are replaced by $g = (u, v)$. Every path P from s to t containing e_2, contains e_1 as a neighbor of e_2, too.

- Delete in $\mathcal{F}$ all appearances of e_1.
- Replace in $\mathcal{F}$ every appearance of the name e_2 by the name g.

Parallel reduction in $\mathcal{F}$ at v, w (steps 2 and 4(d)) The edges $e_1 = (v, w)$ and $e_2 = (v, w)$ are replaced by $g = (v, w)$. All paths containing e_2 can be obtained from all paths P containing e_1 by replacing e_1 with e_2.

- Replace in $\mathcal{F}$ every appearance of e_1 by g.
- Delete in $\mathcal{F}$ all appearances of e_2 and all parts of $\mathcal{F}$ (i.e. $+$, $\cdot$, $($, $)$, names of edges $d \in E$) which are used only for the representation of st-paths containing e_2.

The last item of parallel reduction in $\mathcal{F}$ can be realized for every appearance of e_2 as follows:

Delete all parentheses which are not necessary. Then go from e_2 to the right until there is a $)$ with the corresponding $($ standing on the left of e_2 or until there is a name of an edge e, so that e does not lie on a way from $t(e_2)$ to t. Finally delete everything between e_2 and $)$ or e, respectively.

The same will be done with the left side of e_2. This means, go from e_2 to the left until there is a $($ with the corresponding $)$ standing on the right of e_2 or until there is a name of an edge e, so that e does not lie on a way from s to $s(e_2)$. Delete everything between e_2 and $($ or e, respectively.

After this the name of e_2 will be deleted from $\mathcal{F}$.

Node reduction in $\mathcal{F}$ at node v (step 4(a)) A node reduction in $\mathcal{F}$ at node v can be realized analogously to a series reduction.

Assume that $in(v, G)=1$. Then the edges $e=(u, v)$, $f_1 = (v, w_1)$, $\ldots, f_k = (v, w_k)$ are replaced by the edges $g_1 = (u, w_1)$, $\ldots$, $g_k = (u, w_k)$. Every path P containing f_i, contains e as a neighbor of f_i, too.

- Delete in $\mathcal{F}$ all appearances of e.
- Replace in $\mathcal{F}$ every appearance of the name f_i by g_i, for $i = 1 \ldots k$.

For $out(v, G) = 1$ the edges $f_1 = (u_1, v), \ldots, f_k = (u_k, v)$, $e = (v, w)$ are replaced by the edges $g_1 = (u_1, w)$, $\ldots$, $g_k = (u_k, w)$. Every path P containing f_i contains e as a neighbor of f_i, too.

6   Correctness

For correctness, we have to show that all steps of the Reduction-algorithm can be carried out (see Lemma 5), that the algorithm terminates (see Lemma 6), and that the length of the node reduction sequence returned by the Reduction-algorithm is less than or equal to the cost of the factoring we started with (see Lemma 7). This implies that $\psi(G) \geq \mu(G)$, and together with the result of Bein, Kamburowski, and Stallmann [3], we obtain $\psi(G) = \mu(G)$.

Lemma 5. *Consider an st-dag G and any corresponding factoring $\mathcal{F}$, then all steps of the Reduction-algorithm can be realized.*

Proof. We have already shown in Section 5 that it is possible to realize series, parallel and node reductions in $\mathcal{F}$ in the same way as in G. Therefore we have only to show how to realize step 4(b). This means that we have to show:

If G is irreducible and $|E| > 1$, then there is an edge $e = (v, w) \in E_G$ with the properties:

$$\otimes \begin{cases} \text{The name } e \text{ is a minimal duplicated expression and} \\ in(w, G) = 1 \text{ or } out(v, G) = 1. \end{cases}$$

The proof is organized as follows:

1. We take an autonomous subgraph $\tilde{G} = (\tilde{V}, \tilde{E})$ of G with $|E| > 1$, which contains only trivial autonomous subdags. Then we construct a special corresponding factoring $\tilde{\mathcal{F}}$. We show that the search for an edge with the properties $\otimes$ can be restricted to $\tilde{G}$.
2. Since any st-dag can be topologically sorted, we can assume that $\tilde{G}$ is already topologically sorted with $s = 1$. If the edge $(1, 2)$ is not an edge with the properties $\otimes$, we construct a subdag H, which has only one exit $(=t(H))$ but in contrast to autonomous subdags, possibly more than one entry point. Then one of the edges e with $t(e) = t(H)$ fulfills the properties $\otimes$.
3. To find e, we construct a sequence of subdags $H = H_0, H_1, \ldots, H_k$ with the same unique exit, so that H_{j+1} is a subdag of H_j and H_k is an autonomous subdag.
4. By definition of $\tilde{G}$ (see item 1), H_k consists of only one edge. We show that this edge has the properties $\otimes$.

We now consider all four steps separately in detail.

1. Because of Lemma 4, $\tilde{G}$ has a corresponding factoring $\tilde{\mathcal{F}} = T_1 + T_2 + \ldots + T_l$ with T_i, $i = 1 \ldots l$, being a subexpression of $\mathcal{F}$. All subexpressions T_i, $i = 1 \ldots l$, have the same source and a common target and no two subexpressions T_i, T_j describe the same source–target path of $\tilde{G}$. Therefore, any duplicated expression of the factoring $\tilde{\mathcal{F}}$ is a duplicated expression of $\mathcal{F}$, too. Since, additionally, $\tilde{G}$ is an autonomous subgraph of G, any edge e which satisfies the properties $\otimes$ with respect to $\tilde{G}$ and $\tilde{\mathcal{F}}$ satisfies the properties $\otimes$ with respect to G and $\mathcal{F}$ as well. Therefore it is sufficient to show that $\tilde{G}$ has an edge e with the properties $\otimes$.
2. Let $\tilde{G} = (\tilde{V}, \tilde{E})$ with $\tilde{V} = \{1, \ldots, k\}$ be topologically sorted, where the source of $\tilde{G}$ is 1 and the target of $\tilde{G}$ is k. As $\tilde{G}$ is irreducible and has, nonetheless, more than one edge, $\tilde{\mathcal{F}}$ must have duplicated expressions. Therefore, $\tilde{\mathcal{F}}$ must have minimal duplicated expressions. In addition, there is an edge a with $s(a) = 1$ and $t(a) = 2$. Since $\tilde{G}$ is topologically sorted, $in(2, \tilde{G}) = 1$.

   If a is a duplicated expression in $\tilde{\mathcal{F}}$, a is an edge we were looking for. If a is not a duplicated expression, there is a subdag H induced by the nodes 2

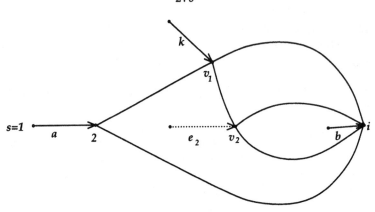

Fig. 4. The st-dag $\tilde{G}$

and i, say, so that all st–paths containing the edge a contain a path from the node 2 to the node i, too. This means that a occurs only in concatenation with a subexpression $\mathcal{F}_H$ in $\tilde{\mathcal{F}}$, so that $a \bullet (\ \mathcal{F}_H \ )$ describes the subpath from 1 to i for every st–path containing a. Therefore, the factoring $\tilde{\mathcal{F}}$ looks as follows:

$$\ldots a \bullet (\ \mathcal{F}_H \ ) \ \ldots$$

with either a $+$ or nothing directly before a. After $)$ there may stand $+$, $\bullet$, or nothing. In contrast to autonomous subdags, the node 2 is not the only entry point of H, in general.

3. At first we set $H_0 = H$. The sequence terminates with the first autonomous subdag. Since $\tilde{G}$ is irreducible, $out(2, \tilde{G}) > 1$. Therefore, H is no autonomous subdag, and thus $k > 0$. It remains to show how to construct H_j from H_{j-1}. For the construction of H_1 we set $H_{-1} = \tilde{G}$.

Look for a node $v_j \in H_{j-1}$ having an incoming edge e_j of $H_{j-2} \setminus H_{j-1}$ and without an incoming edge of H_{j-2} into the subdag induced by the nodes v_j and i. Then the subdag induced by the nodes v_j and i is H_j.

At every iteration in the loop the current H_j is either an autonomous subdag and then $H_j = H_k$ or has incoming edges not only in v_j or i, respectively. Outgoing edges of H_j are only possible at node i, because H_j is a subdag of H (which has outgoing edges only at i) and H_j contains all paths from v_j to i. Every H_j has less edges than H_{j-1}. Therefore the sequence of subdags is finite.

4. Recall that H_k consists of a single edge b. Then b occurs once inside of $(\ \mathcal{F}_H \ )$, describing the subdag H, and once outside of $(\ \mathcal{F}_H \ )$, describing a path P from s to t that contains the edges e_1 and b. Then there is a subexpression $\mathcal{F}_A$ in $\mathcal{F}$ describing a subdag A, so that $\mathcal{F}_A \cdot b$ is a duplicated expression in $\mathcal{F}$. By construction there is a H_j, so that A is a subdag of H_{j-1} but not a subdag of H_j. The st–paths P containing the edges a and e_j are not described by means of the subexpression $\mathcal{F}_A \cdot b$. Therefore there

is yet another subexpression, say $\mathcal{F}_C \cdot b$, used for describing P. This means that b is a duplicated expression, and since b is a single edge, it is a minimal duplicated expression.

Lemma 6. *The Reduction-algorithm terminates.*

Proof. It was already shown in Lemma 5 that all steps can be realized. It is clear that the number of edges will be reduced by every iteration. Since E is finite, the Reduction-algorithm terminates with a finite node reduction sequence.

Lemma 7. *For a given st-dag $G = (V, E)$, and any corresponding factoring $\mathcal{F}$, the parallel and series reductions realized in the Reduction-algorithm (see Section 5) do not increase the number of duplicated expressions of $\mathcal{F}$. In addition, for every realized node reduction the cost of the factoring $\mathcal{F}$ decreases by at least one.*

Proof. The number of duplicated expressions after a series or a parallel reduction in $\mathcal{F}$ is not larger than before, because g is a duplicated expression if and only if e_2 or $e_1 \cdot e_2$ or $e_1 + e_2$ was a duplicated expression, and these three subexpressions do not appear in $\mathcal{F}$ after the parallel or series reduction, respectively. By deleting parts of $\mathcal{F}$ no newly duplicated expressions can be created. After a parallel or series reduction the number of duplicated expressions is even less than before, if e_1 is a duplicated expression.

By a node reduction the number of duplicated expressions does not increase either, because g_i is a duplicated expression if and only if e or $e \cdot f_i$ (resp. $f_i \cdot e$) was a duplicated expression in $\mathcal{F}$ before. The cost of $\mathcal{F}$ is less than before, because e is a duplicated expression. Since e is an edge, the name e was a minimal duplicated expression in $\mathcal{F}$.

This means that the number of duplicated expressions (i.e. the cost) of $\mathcal{F}$ decreases by every node reduction.

We are now able to prove our main result.

Theorem 8. *For every realized node reduction the cost of the factoring $\mathcal{F}$ decreases by at least one. In particular we have $\psi(G) = \mu(G)$.*

Proof. By every realized node reduction the number of duplicated expressions (i.e. the cost) of $\mathcal{F}$ decreases. Because of Lemma 7, the length of the node reduction sequence the algorithm returns, is less than or equal to the cost of $\mathcal{F}$. This holds for any st-dag G and any factoring $\mathcal{F}$ corresponding to G. Hence $\psi(G) \geq \mu(G)$. Bein, Kamburowski and Stallmann [3] have shown that $\psi(G) \leq \mu(G)$. Therefore $\psi(G) = \mu(G)$ holds.

7 Acknowledgments

I thank Prof. Rolf H. Möhring for introducing me to this interesting problem and the referees for several careful readings. A special thank goes to Karsten Weihe for many helpful comments on earlier drafts of this paper.

References

1. A. Agrawal and A. Satyanarayana, An $O(|E|)$ time algorithm for computing the reliability of a class of directed networks, Oper. Res., 32 (1984), pp. 493–515.

2. W. Bein, P. Brucker, and A. Tamir, Minimum cost flow algorithms for series parallel networks, Discrete Appl. Math., 10 (1985), pp. 117–124.

3. W. Bein, J. Kamburowski, and F. M. Stallmann, Optimal reductions of two-terminal directed acyclic graphs, SIAM J. Comput., 6 (1992), pp. 1112–1129.

4. G. Chartrand, D. Geller, and S. Hedetniemi, Graphs with forbidden subgraphs, J. Combinatorial Theory, 10 (1971), pp. 12-41.

5. R. Duffin, Topology of series-parallel networks, J. Math. Anal. Appl., 10 (1965), pp. 303–318.

6. D. G. Kirkpatrick, and P. Hell, On the completeness of a generalized matching problem, Proc. 10th ACM Symp. on Theory of Computing, San Diego, Calif., 1978, pp. 265–274

7. R. H. Möhring, Computationally tractable classes of ordered sets, In: I. Rival, ed., Algorithms and Order, Kluwer Acad. Publ.; Dordrecht, 1989, pp. 105–194

8. J. S. Provan, The complexity of reliability computations in planar and acyclic graphs, SIAM J.Comput., 15 (1986), pp. 694–702.

9. K. Takamizawa, T. Nishizeki, and N. Saito, Linear-time computability of combinatorial problems on series-parallel graphs, J. Assoc. Comput. Mach., 29 (1982), pp. 623–641.

10. J. Valdes, R. Tarjan, and E. Lawler, The recognition of series parallel digraphs, SIAM J. Comput., 11 (1982), pp. 298–313.

Labelled Trees and Pairs of Input-Output Permutations in Priority Queues

M. Golin[1] and S. Zaks[2]

[1] Department of Computer Science, Hong Kong University of Science
& Technology, Clear Water Bay, Kowloon, Hong Kong, golin@cs.ust.hk
[2] Department of Computer Science, Technion, Haifa, Israel,
zaks@cs.technion.ac.il

Abstract. A priority queue can transform a permutation π of a set of size n to some but not necessarily all permutations σ. A recent result of Atkinson and Thiyagarajah [1] states that the number of distinct pairs (π, σ) is $(n+1)^{n-1}$. Recall that Cayley's Theorem ([2]) states that the number of labelled trees on $n + 1$ nodes is also equal to $(n+1)^{n-1}$. We present a direct correspondence between these labelled trees and these pairs of permutations and discuss related problems.

1 Introduction

In this paper we study the combinatorics of how a priority queue, a data structures that supports the INSERT and DELETE-MIN operations, can transform one permutation into another.

To see how a priority queue can transform one permutation into another start with some permution[3] π of $1, 2, \ldots, n$. Now scan through π from left-to-right. At each step of the process arbitrarily either insert the next element of π into the priority queue or – if the priority queue is not empty – delete the minimum item from the prority queue and write it out. When the process is complete the elements of π will have been written out in a (possibly) different order.

For example, for $n = 3$, if the elements are input into the priority queue in the order specified by the permutation (2 3 1) (that is, 2 enters first, followed by 3 and 1), then they can be output in two ways: (2 3 1) (by the sequence of operations <INSERT,DELETE-MIN,INSERT,DELETE-MIN, INSERT,DELETE-MIN> or <INSERT,INSERT,DELETE-MIN,DELETE-MIN,INSERT,DELETE-MIN>, or (1 2 3) (by the sequence of operations <INSERT,INSERT, INSERT,DELETE-MIN,DELETE-MIN, DELETE-MIN>, while the input sequence (1 2 3) can be output only as (1 2 3), regardless of which of the five possible orderings for the INSERT and DELETE-MIN operations is used. Each such pair of $\left(\begin{array}{c} input\ permutation \\ output\ permutation \end{array}\right)$ is termed *allowable* (following [1]). Thus the pair $\left(\begin{array}{ccc} 2 & 3 & 1 \\ 1 & 2 & 3 \end{array}\right)$ is allowable, while the

---

[3] A permutation of a set S is denoted as a sequence of all the elements of S.

pair $\begin{pmatrix} 2\,3\,1 \\ 3\,1\,2 \end{pmatrix}$ is not. The set of all allowable pairs of permutations of size n is denoted by $\mathcal{P}_n$.

For a given input permutation of size n, the possible sequences of priority queue operations of n INSERT and n DELETE-MIN operations correspond to legal parenthetic expressions with n '('s and n ')'s, which are in one-to-one correspondence with binary trees with n internal nodes (see, e.g., [4]). The number of such legal expressions is given by the Catalan number $\frac{1}{n+1}\begin{pmatrix} 2n \\ n \end{pmatrix}$.

It was recently shown that

Theorem 1 (Atkinson & Thiyagarajah [1]): The number of allowable pairs of permutations of n elements is $(n+1)^{n-1}$.

Note that the set of n elements that are permuted may be arbitrary. For ease of presentation, in the rest of the paper, except where otherwise stated, we will implicitly assume that the n elements being permuted are precisely $1, \ldots, n$.

Recall that Cayley's Theorem ([2]) states that the number of labelled trees on n nodes is also equal to n^{n-2}. We denote this set of trees by $\mathcal{T}_n$. The proof of Theorem 1 given in [1] is a purely counting one that makes no reference to Cayley's theorem. A combinatorial proof of Theorem 1 would exhibit a bijection between the set $\mathcal{P}_n$ of allowable pairs and the set $\mathcal{T}_{n+1}$ of labelled trees. This, for example, has recently been done for another problem related to the combinatorics of permutations. It was known since 1959 that the number of $(n-1)$ tuples of transpositions in S_n (the set of permutations on n items) whose ordered product is a given cycle of length n is n^{n-2} which is the number of labelled trees with n nodes. It was only in 1989 that a bijection between the two sets was demonstrated (see [3] for a short history and yet another bijection for that correspondence).

In this paper we present (Section 2) a new proof of Theorem 1 by demonstrating a bijection between the set $\mathcal{P}_n$ and the set $\mathcal{T}_{n+1}$. In addition, we present a linear time algorithm for constructing the tree corresponding to an allowable pair (Section 4) and a lower bound of $\Omega(n \log n)$ on finding the allowable pair corresponding to a tree, along with an optimal time $O(n \log n)$ algorithm to do so (Section 5). We also improve the $O(n \log n)$ expected time algorithm given in [1], for computing the number $s(\sigma)$ of allowable pairs $\begin{pmatrix} \pi \\ \sigma \end{pmatrix}$ for a given output permutation σ, to $O(n)$ (Section 3). Most proofs are only sketched in this Extended Abstract.

2 The correspondence

In this section we demonstrate our correspondence between the set $\mathcal{P}_n$ of allowable pairs of permutations of n elements and the set $\mathcal{T}_{n+1}$ of labelled trees on $n+1$ nodes. For a given pair $\begin{pmatrix} \pi \\ \sigma \end{pmatrix} \in \mathcal{P}_n$, we construct a tree $T_{\pi \to \sigma} \in \mathcal{T}_{n+1}$, such that all trees in $\mathcal{T}_{n+1}$ are covered by this correspondence, and such that different pairs in $\mathcal{P}_n$ correspond to different trees. We demonstrate our construction on the pair $\begin{pmatrix} \pi \\ \sigma \end{pmatrix} = \begin{pmatrix} 879146532 \\ 781493562 \end{pmatrix} \in \mathcal{P}_9$.

We denote the largest element of a permutation π by $max(\pi)$. If we examine the process of inputting π and outputting σ from the priority queue, we notice that when $max(\pi)$ (9 in the example) is output, the queue must be empty. When the largest element of the remaining elements of π (6 in the example) is output, the queue must again be empty, and so on. Thus, the pair of allowable permutations can be partitioned into $\begin{pmatrix} \pi \\ \sigma \end{pmatrix} = \begin{pmatrix} \pi_1 \ \pi_2 \ \cdots \ \pi_k \\ \sigma_1 \ \sigma_2 \ \cdots \ \sigma_k \end{pmatrix}$, where each of the pairs $\begin{pmatrix} \pi_i \\ \sigma_i \end{pmatrix}$ is allowable.

In the example, we have

$$\begin{pmatrix} \pi \\ \sigma \end{pmatrix} = \begin{pmatrix} 879146532 \\ 781493562 \end{pmatrix} = \begin{pmatrix} 87914 \ \ 653 \ \ 2 \\ 78149 \ \ 356 \ \ 2 \end{pmatrix}.$$

The above factorization of $\begin{pmatrix} \pi \\ \sigma \end{pmatrix}$ is unique. Moreover, this partition can be found by scanning the output permutation σ, and identifying the right-to-left maxima. That is, the first such element m_1 is the rightmost element of σ. The next one is the rightmost element m_2 that is to the left of m_1 and larger than it, the next one is the rightmost element m_3 that is to the left of m_2 and larger than it, and so on. In the example these right-to-left maxima are 2, 6 and 9.

Given $\begin{pmatrix} \pi \\ \sigma \end{pmatrix} \in \mathcal{P}_n$, we now construct the tree $T_{\pi \to \sigma} \in \mathcal{T}_{n+1}$, labelled with $\{1, \cdots, n\} \cup \{*\}$. We first label one node with $*$. This node has k neighbors, corresponding to the k elements in the partition $\begin{pmatrix} \pi \\ \sigma \end{pmatrix} = \begin{pmatrix} \pi_1 \ \pi_2 \ \cdots \ \pi_k \\ \sigma_1 \ \sigma_2 \ \cdots \ \sigma_k \end{pmatrix}$. In the example, this node will have 3 neighbors. The subtrees T_i will correspond to the pairs $\begin{pmatrix} \pi_i \\ \sigma_i \end{pmatrix}$ in a recursive manner. The first step in building the tree for our example is depicted in Figure 1 (left figure).

In order to build the trees T_i, we note that for each of $\begin{pmatrix} \pi_i \\ \sigma_i \end{pmatrix}$, the last element of σ_i is the largest. For a permutation α, let α' be the permutation obtained by deleting the largest element from α, and consider the pairs of permutations $\begin{pmatrix} \pi_i' \\ \sigma_i' \end{pmatrix}$. Clearly each such pair is also allowable. In our example, the pairs are $\begin{pmatrix} 8714 \\ 7814 \end{pmatrix}$, $\begin{pmatrix} 53 \\ 35 \end{pmatrix}$, and $(\ )$.

Given $\begin{pmatrix} \pi_i' \\ \sigma_i' \end{pmatrix}$, we need the following information in order to uniquely reconstruct $\begin{pmatrix} \pi_i \\ \sigma_i \end{pmatrix}$: (i) the value of $max(\pi_i)$, and (ii) the position of $max(\pi_i)$ in π_i (we know that $max(\pi_i)$ is the last element in σ_i).

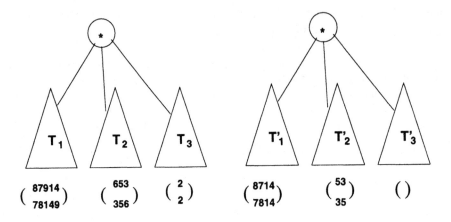

Fig. 1. The recursive construction - step 1 (left hand figure) and 2 (right hand one).

Thus, we will recursively build the trees $T_{i'} = T_{\pi_i' \to \sigma_i'}$, as depicted in Figure 1 (right figure), where the node that is marked $*$ in $T_{i'}$ gets the label $max(\pi_i)$. In addition, the edge connecting $*$ in $T_{\pi \to \sigma}$ to the tree T_i will be connected to the successor of $max(\pi_i)$ in π_i.

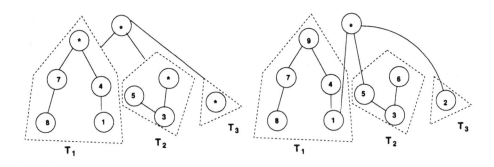

Fig. 2. The recursive construction - step 3 and the final output.

Suppose we have built the trees T_i', corresponding to $\begin{pmatrix} 8714 \\ 7814 \end{pmatrix}$, $\begin{pmatrix} 53 \\ 35 \end{pmatrix}$, and ( ), as depicted in Figure 2 (left figure). The tree T is then built by replacing each $*$-labelled node in T_i' with the element $max(\pi_i)$ (9,6 and 2 in our example), and then connecting the $*$ node to the successors of $max(\pi_i)$ in π_i, or to $max(\pi_i)$ in case it is the rightmost element in π_i (1 , 5 and 2, in our example). The resulting tree is shown in Figure 2 (right figure). (The reader can verify that by applying this construction recursively we do get the trees of Figure 2.)

In summary, using our transformation, we have the following:

Theorem 2: The above correspondence establishes a one-to-one correspondence between $\mathcal{P}_n$ and $\mathcal{T}_{n+1}$.

3 Calculating $s(\sigma)$ in $O(n)$ time

Suppose σ is a permutation on n items. Let

$$s(\sigma) = \left| \left\{ \pi : \begin{pmatrix} \pi \\ \sigma \end{pmatrix} \in \mathcal{P}_n \right\} \right|$$

be the number of permutations that can be transformed into σ by the priority queue operations. In this section we describe how the insights into the structure of allowable pairs developed in the previous section yield an $O(n)$ time algorithm for calculating $s(\sigma)$. This improves upon the algorithm described in [1] which runs in expected $O(n \log n)$ time on random permutations. Our main reason for presenting this algorithm here is that it introduces the basic tools that will be needed in the next section to develop an $O(n)$ time algorithm for constructing the tree $T_{\pi \to \sigma}$ corresponding to an allowable pair $\begin{pmatrix} \pi \\ \sigma \end{pmatrix}$.

For a permutation α, $\alpha[i]$ is the i'th element of α, and $\alpha^{-1}[j]$ is the location of j in α. In what follows we assume that both σ and its inverse σ^{-1} have already been processed so that, for any index i, the values of both $\sigma[i]$ and $\sigma_{-1}[i]$ can be accessed in constant time.

In the previous section we described how to uniquely decompose an allowable pair $\begin{pmatrix} \pi \\ \sigma \end{pmatrix}$ into k allowable pairs $\begin{pmatrix} \pi_i \\ \sigma_i \end{pmatrix}$, $i = 1, \ldots k$ where k is the number of right-to-left maxima in σ. The decomposition is effected by finding the k right-to-left maxima which are the rightmost elements of σ_i, $i = 1, \ldots k$. The structure of the decomposition therefore depends solely upon σ and not upon π. It is not difficult to see that if π_i, $i = 1, \ldots k$, are any sequences such that $\begin{pmatrix} \pi_i \\ \sigma_i \end{pmatrix}$, $i = 1, \ldots k$ are all allowable then $\begin{pmatrix} \pi \\ \sigma \end{pmatrix}$ is also allowable where $\pi = \pi_1 \pi_2 \cdots \pi_k$. This proves that

$$s(\sigma) = \prod_{1 \leq i \leq k} s(\sigma_i) \tag{1}$$

and suggests a simple algorithm for calculating $s(\sigma)$: scan through σ from right to left, identify the σ_i's, calculate all of the $s(\sigma_i)$'s, and then multiply them all together. The catch is that this requires being able to calculate the $s(\sigma_i)$ in an online fashion. To do this we recall that each σ_i is in a very special form; its largest element is its rightmost element.

Let $m_i = \max(\sigma_i)$ which is the rightmost item in σ_i. Recall that σ_i' is the sequence obtained by deleting m_i from σ_i. Suppose π_i is such that $\binom{\pi_i}{\sigma_i}$ is allowable and let π_i' be the sequence obtained by deleting m_i from π_i. Then $\binom{\pi_i'}{\sigma_i'}$ is allowable. Now suppose that π_i' is any sequence such that $\binom{\pi_i'}{\sigma_i'}$ is allowable. Insert m_i into any of the $|\pi_i'| + 1 = |\sigma_i|$ positions either at the front, back, or between two elements in π_i' and call the resulting permutation π_i. It is not difficult to see that $\binom{\pi_i}{\sigma_i}$ is allowable. This implies that

$$s(\sigma_i) = |\sigma_i| \cdot s(\sigma_i'). \tag{2}$$

In our our canonical example $\sigma_1 = 78149$ so $\sigma_1' = 7814$, and $m_1 = 9$. Let $\pi_1' = 8714$; $\binom{\pi_1'}{\sigma_1'}$ is allowable. Then each of the five possibilities $\pi_1 = 98714$, $\pi_1 = 89714$, $\pi_1 = 87914$, $\pi_1 = 87194$, and $\pi_1 = 87149$ yield allowable pairs $\binom{\pi_1}{\sigma_1}$. Also $s(78149) = 5 \cdot s(7814)$.

We can now develop the algorithm. It sweeps over σ from right to left, calculating, for each i, the length of σ_i and, recursively, the value of $s(\sigma_i')$. It then uses equation (2) to calculate $s(\sigma)$.

Calculating the length of the σ_i is easy: the scan starts at m_i, the last value in σ_i. It then scans to the left until it hits m_{i-i}, the first value larger than m_i. The distance covered between m_i and m_{i-1} is $|\sigma_i|$. (If the scan reaches the left end of σ without hitting a larger item then it stops and pretends that it has hit a larger item because it has reached the end of σ_1.) When the algorithm finishes sweeping through σ_i it has swept through σ_i' and has recursively calculated $s(\sigma_i')$. Let

$$b[i] := \max\{0, \max\{j < i : \sigma[j] > \sigma[i]\}\}$$

be the index of the first element in σ to the left of location i which contains a value greater than $\sigma[i]$. Equations (1) and (2) along with the discussion of the previous paragraph imply that

$$s(\sigma) = \prod_{1 \le i \le n} (i - b(i))$$

(this equation was also derived in [1] in a different fashion). This leads us to the following algorithm:

```
s := 1; PUSH(n + 1, S);
FOR j := n DOWNTO 1
    WHILE (σ[j] > TOPOFSTACK) DO
        m = POP(S);
        s := s * (σ⁻¹[m] − j);
    PUSH(σ[j], S);
```

—Pop remaining items off S. —
WHILE $(TOPOFSTACK \neq n + 1)$ DO
    $m = POP(S)$;
    $s := s * (\sigma^{-1}[m])$;

The function $Push(\alpha, S)$ pushes α onto stack S and $POP(S)$ pops the top value from S and returns it. The value $n + 1$ serves as a marker for the bottom of the stack.

Formally the algorithm works as follows: it starts with $s = 1$ and concludes with $s = s(\sigma)$. It uses a stack which originally contains only the value $n + 1$. As it sweeps from right to left stopping at location j it compares $\sigma[j]$ to the items in the stack, pops off all that are less than $\sigma[j]$ and then pushes $\sigma[j]$ onto the stack. An element m is popped off of the stack when the first element to its left with greater value is reached. For example, element m_i is popped off the stack when m_{i-1} is scanned. If element m is popped off the stack when j is scanned then $j = b(\sigma^{-1}[m])$. Since every element is popped off the stack exactly once the algorithm therefore calculates

$$\prod_{1 \leq m \leq n} \left(\sigma^{-1}[m] - b\left(\sigma^{-1}[m]\right)\right) = \prod_{1 \leq i \leq n} (i - b(i)) = s(\sigma).$$

Also, since every element is pushed onto and popped off of the stack exactly once, the algorithm runs in $O(n)$ time. We have just sketched the proof of

Theorem 3: Given a permutation σ on the set $\{1, \ldots, n\}$ the algorithm described above correctly calculates $s(\sigma)$ in $O(n)$ time.

4 Building the tree for a given pair of permutations

In Section 2 we described a bijection between labelled trees on $n + 1$ nodes and the set of allowable pairs of permutations of size n. The description of the bijection could be used to construct a recursive algorithm for constructing the tree $T_{\pi \to \sigma}$ corresponding to an allowable pair $\begin{pmatrix} \pi \\ \sigma \end{pmatrix}$, but that algorithm would not be particularly efficient. In this section we describe an $O(n)$ algorithm for building the tree $T_{\pi \to \sigma}$. The algorithm uses ideas similar to those utilized in the algorithm of the previous section for calculating $s(\sigma)$; it scans σ from right to left pushing and popping elements from the stack in exactly the same order as that one did.

In the remainder of this section we briefly sketch the algorithm for construction $T_{\pi \to \sigma}$ and the intuition behind it. We leave a formal proof of correctness for the full paper.

The algorithm starts with $n + 1$ isolated labelled nodes $\{1, \ldots n, *\}$. The algorithm uses a doubly linked list LL containing the permutation π in order; elements will be deleted from this list as the algorithm progresses. Finally, the algorithm also stores a list of pointers to the elements in LL. This will enable constant time location of the successor of element j in the list and constant time deletion of element j from the list. The algorithm follows:

```
—List LL stores π; S is a stack —
PUSH(*, S);
FOR j := n DOWNTO 1
        WHILE (σ[j] > TOPOFSTACK) DO
                m = POP(S);
        — Connect tree that has σ[j] as max value to its father —
        s := successor of σ[j] in LL.
        IF σ[j] is last item on list THEN s := σ[j].
        Make s a child of TOPOFSTACK
        Delete σ[j] from LL
        PUSH(σ[j], S)
```

As a convention we assume that $*$ is larger than any number.

The algorithm is best understood through reference to the recursive definition of the construction described in Section 2. Recall that the recursive definition builds the trees $T_{\pi'_i \to \sigma'_i}$ corresponding to the pairs $\binom{\pi'_i}{\sigma'_i}$, $i = 1, \ldots k$, substitutes the value m_i for the roots $(*)$ of $T_{\pi'_i \to \sigma'_i}$ and then attaches the root of the main tree to the node in $T_{\pi'_i \to \sigma'_i}$ which contains the successor of m_i in π_i (if m_i is the last element in π_i then m_i is considered its own successor). This is implemented by the right to left sweep of the algorithm.

The algorithm stores the $*$ node, the root of the tree, in the stack. It then sweeps from right to left over σ identifying the right-to-left maxima m_i, $i = k, k-1, \ldots, 1$ and, from them, the σ_i, $i = k, \ldots, 1$. When the algorithm reaches m_i it knows that it will now start sweeping σ_i. Because m_i is larger than anything that has been seen before, it pops everything except for the $*$ node off of the stack. It then references LL to find the successor of m_i in π and connects this to the root of the tree whose label is stored at the top of the stack. This properly connects the root to the ith tree.

Next it deletes $\sigma[j]$ from LL. This ensures that the last $|\pi'_i|$ items in LL are exactly π'_i in order. Then it pushes m_i onto the stack where it will stay until something larger is encountered, i.e., m_{i-1}. While it is sweeping from m_i to m_{i-1} it is building the tree $T_{\pi'_i \to \sigma'_i}$ in the same manner that it builds $T_{\pi \to \sigma}$. The node that it uses as the root for this subtree is the node which is on the top of the stack when this phase starts, i.e., m_i.

Since each node is pushed onto the stack only once and popped at most once, this algorithm uses only $O(n)$ time. We can prove:

Theorem 4: The algorithm described above uses $O(n)$ time to correctly construct the tree $T_{\pi \to \sigma}$ corresponding to allowable pair $\binom{\pi}{\sigma}$.

Figure 3 provides an example of the algorithm.

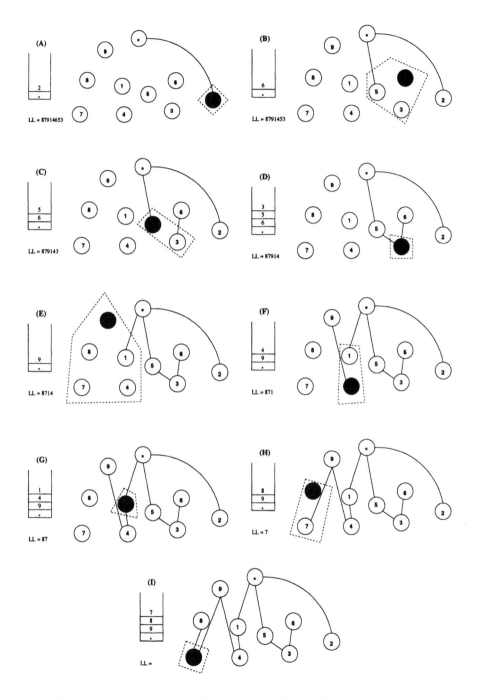

Fig. 3. The algorithm run on $\pi = 879146532$, $\sigma = 781493562$. The diagram illustrates the state of the data structures after the element in the shaded node is scanned (which will be the element on the top of the stack). The dotted lines enclose the current tree being built.

5 Building the pair of permutations for a given tree

In this section we sketch the construction of the pair of permutations $\begin{pmatrix} \pi \\ \sigma \end{pmatrix}$ for a given tree T, such that $T = T_{\pi \to \sigma}$. We should emphasize that in this section the tree is labelled with a '*' and any n labels, not necessarily $1, \ldots, n$. The extension of the correspondence of Section 2 to arbitrary labels is clear. We first note the following:

Theorem 5: Given a labelled tree T on $n + 1$ nodes, one of which labelled with '*', any comparison-based algorithm to construct the pair of allowable permutations $\begin{pmatrix} \pi \\ \sigma \end{pmatrix}$, satisfying $T = T_{\pi \to \sigma}$, requires at least $\Omega(n \log n)$ time.

Proof: Given any $x_1, x_2, \ldots, x_n$, let T be the tree with a node labelled '*' connected to n nodes, labelled $x_1, x_2, \ldots, x_n$. the allowable pair of permutations $\begin{pmatrix} \pi \\ \sigma \end{pmatrix}$ corresponding to this tree has $\pi = \sigma$ being the permutation of $x_1, x_2, \ldots, x_n$ in decreasing value, and the $\Omega(n \log n)$ thus follows from the standard lower bound for sorting.

$\square$

Note that by reversing the construction in our correspondence, the construction of the tree is easy, but is not evidently $O(n \log n)$. However, by using the standard dynamic (link-cut) trees (see [6]), the following can be proven:

Theorem 5: Given tree T, it is possible to construct the pair of permutations $\begin{pmatrix} \pi \\ \sigma \end{pmatrix}$ such that $T = T_{\pi \to \sigma}$ in $O(n \log n)$ time.

References

1. M. D. Atkinson and M. Thiyagarajah, *The permutational power of a priority queue*, BIT, 33 (1993), 2-6.
2. A. Cayley, *A theorem on trees*, Quart. J. Math., Vol. 23 (1889), 376-378.
3. I.P. Goulden and S. Pepper *Labelled trees and factorizations of a cycle into transpositions*, Discrete Mathematics, Vol. 113 (1993) 263-268.
4. D.E. Knuth, *The Art of Computer Programming, Vol. 1: Fundamental Algorithms*, Addison-Wesley, Reading, MA, 1968.
5. L. Lovász, *Combinatorial Problems and Exercises*, North-Holland, 1979.
6. D. D. Sleator and R. E. Tarjan, *A data structure for dynamic trees*, J. Comput. and System Sci., vol. 26 (1983) 362-391.

Rankings of graphs

H.L. Bodlaender[1]*, J.S. Deogun[2]**, K. Jansen[3], T. Kloks[4], D. Kratsch[5]***,
H. Müller[5], and Zs. Tuza[6]†

[1] Department of Computer Science
Utrecht University, P.O. Box 80.089
3508 TB Utrecht, The Netherlands
[2] Department of Computer Science and Engineering
University of Nebraska – Lincoln
Lincoln, NE 68588-0115, U.S.A.
[3] Fachbereich IV, Mathematik und Informatik
Postfach 3825, Universität Trier
D-54286 Trier, Germany
[4] Department of Mathematics and Computing Science
Eindhoven University of Technology
P.O.Box 513, 5600 MB Eindhoven
The Netherlands
[5] Fakultät für Mathematik und Informatik
Friedrich-Schiller-Universität
Universitätshochhaus
07740 Jena, Germany
[6] Computer and Automation Institute
Hungarian Academy of Sciences
H-1111 Budapest, Kende u. 13–17, Hungary

Abstract. A vertex (edge) coloring $c : V \rightarrow \{1, 2, \ldots, t\}$ ($c' : E \rightarrow \{1, 2, \ldots, t\}$) of a graph $G = (V, E)$ is a vertex (edge) *t-ranking* if for any two vertices (edges) of the same color every path between them contains a vertex (edge) of larger color. The *vertex ranking number* $\chi_r(G)$ (*edge ranking number* $\chi'_r(G)$) is the smallest value of t such that G has a vertex (edge) t-ranking. In this paper we study the algorithmic complexity of the VERTEX RANKING and EDGE RANKING problems. Among others it is shown that $\chi_r(G)$ can be computed in polynomial time when restricted to graphs with treewidth at most k for any fixed k. We characterize those graphs where the vertex ranking number χ_r and the chromatic number χ coincide on all induced subgraphs, show that $\chi_r(G) = \chi(G)$ implies $\chi(G) = \omega(G)$ (largest clique size) and give a formula for $\chi'_r(K_n)$.

---

* Email: hansb@cs.ruu.nl. This author was partially supported by the ESPRIT Basic Research Actions of the EC under contract 7141 (project ALCOM II)

** This author was partially supported by the Office of Naval Research under Grant No. N0014-91-J-1693

*** Research of this author was done while he visited IRISA, Rennes Cedex, France. This author was partially supported by Deutsche Forschungsgemeinschaft under Kr 1371/1-1

† This author was partially supported by the "OTKA" Research Fund of the Hungarian Academy of Sciences, Grant No. 2569

1 Introduction

In this paper we consider vertex rankings and edge rankings of graphs. The vertex ranking problem, also called the *ordered coloring problem* [15], has received much attention lately because of the growing number of applications. There are applications in scheduling problems of assembly steps in manufacturing systems [19], e.g., edge ranking of trees can be used to model the parallel assembly of a product from its components in a quite natural manner [6, 8, 12, 13, 14]. Furthermore the problem of finding an optimal vertex ranking is equivalent to the problem of finding a minimum-height elimination tree of a graph [6, 7]. This measure is of importance for the parallel Cholesky factorization of matrices [3, 9, 18]. Yet other applications lie in the field of VLSI-layout [17, 26].

The VERTEX RANKING problem 'Given a graph G and a positive integer t, decide whether $\chi_r(G) \leq t$' is NP-complete even when restricted to cobipartite graphs following a result of Pothen [20]. For trees there is a linear-time algorithm finding an optimal vertex ranking [24]. For the closely related edge ranking problem on trees an $O(n^3)$ algorithm was given in [8]. Efficient vertex ranking.algorithms for permutation, trapezoid, interval, circular-arc, circular permutation graphs, and cocomparability graphs of bounded dimension are presented in [7]. The vertex ranking problem is trivial on split graphs and solvable in linear time on cographs [25].

In [15], typical graph theoretical questions, as they are known from the coloring theory of graphs, are investigated. This also leads to a $O(\sqrt{n})$ bound for the vertex ranking number of a planar graph and the authors describe a polynomial-time algorithm which finds a vertex ranking of a planar graph using only $O(\sqrt{n})$ colors. For graphs in general there is an approximation algorithm of performance ratio $O(\log^2 n)$ for the vertex ranking number [3, 16]. In [3] it is also shown that one plus the pathwidth of a graph is a lower bound for the vertex ranking number of the graph (hence a planar graph has pathwidth $O(\sqrt{n})$, which is also shown in [16] using different methods).

Our goal is to extend the known results in both the algorithmic and graph theoretic directions. The paper is organized as follows. In Section 2, necessary notions and preliminary results are given. We study the algorithmic complexity of determining whether a graph G fulfils $\chi_r(G) \leq t$ and $\chi'_r(G) \leq t$, respectively, in Sections 3, and 4. The problem of computing vertex ranking number of graphs with bounded width is investigated in Section5. In Section 6 we characterize those graphs for which the vertex ranking number and the chromatic number coincide on every induced subgraph. These graphs are precisely the graphs containing no paths or cycles on four vertices as an induced subgraph; hence, we obtain a characterization of the *trivially perfect graphs* [11] in terms of rankings. Moreover we show that $\chi(G) = \chi_r(G)$ implies that the chromatic number of G is equal to its largest clique size. In Section 7 we give a recurrence relation for computing the edge ranking number of a complete graph.

2 Preliminaries

We consider only finite, undirected and simple graphs $G = (V, E)$. Throughout the paper n denotes the cardinality of the vertex set V and m denotes that of the edge set E of the graph $G = (V, E)$. For graph-theoretic concepts, definitions and properties of graph classes not given here we refer to [4, 5, 11].

Let $G = (V, E)$ be a graph. A subset $U \subseteq V$ is *independent* if each pair of vertices $u, v \in U$ is nonadjacent. A graph $G = (V, E)$ is *bipartite* if there is a partition of V into two independent sets A and B. The *complement* of the graph $G = (V, E)$ is the graph $\overline{G}$ having vertex set V and edge set $\{\{v, w\} \mid v \neq w, \{v, w\} \notin E\}$. For $W \subseteq V$ we denote by $G[W]$ the subgraph of $G = (V, E)$ induced by the vertices of W, and for $X \subseteq E$ we write $G[X]$ for the graph (V, X) with vertex set V and edge set X.

Definition 1. Let $G = (V, E)$ be a graph and let t be a positive integer. A *(vertex) t-ranking*, called *ranking* for short if there is no ambiguity, is a coloring $c : V \rightarrow \{1, \ldots, t\}$ such that for every pair of vertices x and y with $c(x) = c(y)$ and for every path between x and y there is a vertex z on this path with $c(z) > c(x)$. The *vertex ranking number* of G, $\chi_r(G)$, is the smallest t for which the graph G admits a t-ranking.

By definition adjacent vertices have different colors in any t-ranking, thus any t-ranking is a proper t-coloring. Hence $\chi_r(G)$ is bounded below by the *chromatic number* $\chi(G)$. A vertex $\chi_r(G)$-ranking of G is said to be an *optimal (vertex) ranking* of G.

The edge ranking problem is closely related to the vertex ranking problem.

Definition 2. Let $G = (V, E)$ be a graph and let t be a positive integer. An *edge t-ranking* is an edge coloring $c' : E \rightarrow \{1, \ldots, t\}$ such that for every pair of edges e and f with $c'(e) = c'(f)$, there is an edge g on every path between e and f with $c'(g) > c'(e)$. The *edge ranking number* $\chi'_r(G)$ is the smallest value of t such that G has an edge t-ranking.

Remark 3. There is a one-to-one correspondence between the edge t-rankings of a graph G and the vertex t-rankings of its line graph $L(G)$. Hence $\chi'_r(G) = \chi_r(L(G))$.

An edge t-ranking of a graph G is a particular proper edge coloring of G. Hence $\chi'_r(G)$ is bounded below by the *chromatic index* $\chi'(G)$. An edge $\chi'_r(G)$-ranking of G is said to be an *optimal edge ranking* of G.

As shown in [7], the vertex ranking number of a connected graph is equal to its minimum elimination tree height plus one. Thus (vertex) separators and edge separators are a convenient tool for investigating rankings of graphs. A subset $S \subseteq V$ of a graph $G = (V, E)$ is said to be a *separator* if $G[V \setminus S]$ is disconnected. A subset $R \subseteq E$ of a graph $G = (V, E)$ is said to be an *edge separator* (or *edge cut*) if $G[E \setminus R]$ is disconnected.

In this paper we use the *separator tree* for studying vertex rankings. This concept is closely related to elimination trees (cf.[3, 7, 18]).

Definition 4. Given a vertex t-ranking $c : V \rightarrow \{1, 2, \ldots, t\}$ of a connected graph $G = (V, E)$, we assign a rooted tree $T(c)$ to it by an inductive construction, such that a separator of a certain induced subgraph of G is assigned to each internal node of $T(c)$ and the vertices of each set assigned to a leaf of $T(c)$ have pairwise different colors:

1. If no color occurs more than once in G, then $T(c)$ consists of a single vertex r (called root), assigned to the vertex set of G.
2. Otherwise, let i be the largest color assigned to more than one vertex by c. Then the set S of vertices that are assigned colors $\{i + 1, i + 2, \ldots, t\}$ is a separator of G. We create a root r of $T(c)$ and assign S to r. (The induced subgraph of G corresponding to the subtree of T rooted at r will be G itself.) Assuming that a separator tree $T_i(c)$ with root r_i has already been defined for each connected component G_i of the graph $G[V \setminus S]$, the children of r in $T(c)$ will be the vertices r_i and the subtree of $T(c)$ rooted at r_i will be $T_i(c)$.

The rooted tree $T(c)$ is said to be a *separator tree* of G.

Notice that all vertices of G assigned to nodes of $T(c)$ on a path from a leaf to the root have different colors.

3   Unbounded ranking

It is still unknown whether the EDGE RANKING problem 'Given a graph G and a positive integer t, decide whether $\chi'_r(G) \leq t$' is NP-complete. Clearly, by Remark 3 this problem is equivalent to the VERTEX RANKING problem 'Given a graph G and a positive integer t, decide whether $\chi_r(G) \leq t$' when restricted to line graphs.

On the other hand, it is a consequence of the NP-completeness of the minimum elimination tree height problem shown by Pothen in [20] and the equivalence of this problem with the VERTEX RANKING problem [6, 7] that the latter is NP-complete even when restricted to cobipartite graphs.

We show that the analogous result holds for bipartite graphs as well.

Theorem 5. VERTEX RANKING *remains NP-complete for bipartite graphs.*

Proof. The transformation is from VERTEX RANKING for arbitrary graphs without isolated vertices. Given the graph G, we construct a graph $G' = (V', E')$. We take

$$V' = V \cup \{(e, i) \mid e \in E, 1 \leq i \leq t + 1\}$$

and

$$E' = \{\{v, (e, i)\} \mid v \in V, e \in E, 1 \leq i \leq t + 1 \text{ where } v \in e\}.$$

Clearly, the constructed graph G' is a bipartite graph. Now we show that G has a t-ranking if and only if G' has a $(t + 1)$-ranking.

Suppose G has a t-ranking $c : V \to \{1,\dots,t\}$. We construct a coloring $\hat{c}$ for G' in the following way. For the vertices $v \in V$ we set $\hat{c}(v) = c(v) + 1$ and for the vertices $(e,i) \in V' \setminus V$ we set $\hat{c}((e,i)) = 1$. Clearly $\hat{c}$ is a $(t+1)$-ranking of G'.

On the other hand, let $\hat{c} : V' \to \{1,\dots,t+1\}$ be a $(t+1)$-ranking of G'. We show that $\hat{c}(v) > 1$ for every vertex $v \in V$. Suppose not and let v be a vertex of V with $\hat{c}(v) = 1$. Let $e = \{v,w\}$ be an edge incident to v in G. Hence v is adjacent to $(e,1),(e,2),\dots,(e,t+1)$ in G'. Then $\hat{c}(v) = 1$ implies $\hat{c}((e,i)) > 1$ for $i = 1,2,\dots,t+1$. Since $\hat{c}$ is a $(t+1)$-ranking, there are l,l' with $l \neq l'$ such that $\hat{c}((e,l)) = \hat{c}((e,l'))$, implying a path $(e,l) - v - (e,l')$ which contradicts the assumption that $\hat{c}$ is a ranking. This proves that $\hat{c}(v) > 1$ holds for every vertex $v \in V$. As a consequence, for each edge $e = \{u,v\} \in E$, there is a vertex $(e,i) \in V'$ with $\hat{c}((e,i)) < \min(\hat{c}(u),\hat{c}(v))$. Thus, changing $\hat{c}$ on $V' \setminus V$ to $\hat{c}((e,i)) = 1$ for all $(e,i) \in V'$, we obtain another $(t+1)$-ranking of G'. Now we define $c(v) = \hat{c}(v) - 1$ for every $v \in V$. The coloring c is a t-ranking of G since the existence of a path between two vertices v and w of G such that $c(v) = c(w)$ and all inner vertices have smaller colors implies the existence of a path from v to w in G' with $\hat{c}(v) = \hat{c}(w)$ and all inner vertices having smaller colors, contradicting the fact that $\hat{c}$ is a $(t+1)$-ranking of G'. □

4 Bounded ranking

We show that the 'bounded' ranking problems 'Given a graph G, decide whether $\chi_r(G) \leq t$ $(\chi'_r(G) \leq t)$' are solvable in linear time for any fixed t. This will be done by verifying that the corresponding graph classes are closed under certain operations.

Definition 6. An *edge contraction* is an operation on a graph G replacing two adjacent vertices u and v of G by a vertex adjacent to all vertices that were adjacent to u or v. An *edge lift* is an operation on a graph G replacing two adjacent edges $\{v,w\}$ and $\{u,w\}$ of G by one edge $\{u,v\}$.

Definition 7. A graph H is a *minor* of the graph G if H can be obtained from G by a series of the following operations: vertex deletion, edge deletion, and edge contraction. A graph class $\mathcal{G}$ is *minor closed* if every minor H of every graph $G \in \mathcal{G}$ also belongs to $\mathcal{G}$.

Lemma 8. *The class of graphs satisfying $\chi_r(G) \leq t$ is minor closed for any fixed t.*

Proof. It is clear that the vertex ranking number of a graph cannot increase by vertex or edge deletions. If $G = (V,E)$ is a graph with c a t-ranking of G, and $H = (V',E')$ is obtained from G by contracting the edge $\{v,w\}$ to a vertex x, then the function $\hat{c}$, obtained from c by mapping x to $\max(c(v),c(w))$, and all other vertices y to $c(y)$, is a t-ranking of H. □

Corollary 9. *For each fixed t, the class of graphs satisfying $\chi_r(G) \leq t$ can be recognized in linear time.*

Proof. In [1], using results from Robertson and Seymour [22, 23], it is shown that every minor closed class of graphs that does not contain all planar graphs, has a linear time recognition algorithm. Now use Lemma 8. □

Theorem 10. *For each fixed t, the class of connected graphs satisfying $\chi'_r(G) \leq t$ can be recognized in constant time.*

Proof. For any fixed t, there are only a finite number of connected graphs G with $\chi'_r(G) \leq t$, as necessary conditions are that the maximum degree of G is at most t, and the diameter of G is bounded by $2^t - 1$. □

Certainly, the above theorem immediately implies that the graphs G satisfying $\chi'_r(G) \leq t$ can be recognized in linear time, by inspecting the connected components separately. This result might have also been obtained via more involved methods, by using results of Robertson and Seymour on graph immersions [21]. Similarly, one can show that for fixed t and d, the class of connected graphs with $\chi_r(G) \leq t$ and maximum vertex degree d can be recognized in constant time.

Definition 11. A graph H is an *immersion* of the graph G if H can be obtained from G by a series of the following operations: vertex deletion, edge deletion and edge lift. A graph class $\mathcal{G}$ is *immersion closed* if every immersion H of a graph $G \in \mathcal{G}$ also belongs to $\mathcal{G}$.

Lemma 12. *The class of graphs satisfying $\chi'_r(G) \leq t$ is immersion closed for any fixed t.*

Linear-time recognizability of the class of graphs satisfying $\chi'_r(G) \leq t$ now also follows from Lemma 12, the results of Robertson and Seymour, and the fact that graphs with $\chi'_r(G) \leq t$ have treewidth at most $2t + 2$.

5 Computing the vertex ranking number on graphs with bounded treewidth

In this section, we show that one can compute $\chi_r(G)$ of a graph G with treewidth at most k in polynomial time, for any fixed k. Such a graph is also called a partial k-tree. This result implies polynomial time computability of any class of graphs with a uniform upper bound on the treewidth, e.g., outerplanar graphs, series-parallel graphs, Halin graphs.

The notion of treewidth has been introduced by Robertson and Seymour (see e.g., [22]).

Definition 13. A *tree-decomposition* of a graph $G = (V, E)$ is a pair $(\{X_i \mid i \in I\}, T = (I, F))$ with $\{X_i \mid i \in I\}$ a collection of subsets of V, and $T = (I, F)$ a tree, such that

- $\bigcup_{i \in I} X_i = V$
- for all edges $\{v, w\} \in E$ there is an $i \in I$ with $v, w \in X_i$
- for all $i, j, k \in I$: if j is on the path from i to k in T, then $X_i \cap X_k \subseteq X_j$.

The width of a tree-decomposition $(\{X_i \mid i \in I\}, T = (I, F))$ is $\max_{i \in I} |X_i| - 1$. The treewidth of a graph $G = (V, E)$ is the minimum width over all tree-decompositions of G.

We often abbreviate $(\{X_i \mid i \in I\}, T = (I, F))$ as (X, T). A tree-decomposition (X, T) is rooted, if a node $i \in I$ is designated as root of T.

Definition 14. A rooted tree-decomposition $D = (S, T)$ with $S = \{X_i \mid i \in I\}$ and $T = (I, F)$ is a *nice* tree-decomposition, if the following conditions are satisfied:

1. every node of T has at most two children,
2. if a node i has two children j and k, then $X_i = X_j = X_k$. (i is called a **join** node.)
3. if a node i has one child j, then either $|X_i| = |X_j| + 1$ and $X_j \subset X_i$ (i is called an **introduce** node) or $|X_i| = |X_j| - 1$ and $X_i \subset X_j$ (i is called a **forget** node).
4. if a node i is a leaf of T, then $|X_i| = 1$. (i is called a **start** node.)

For every fixed k, there exists an algorithm, that given a graph $G = (V, E)$ determines whether the treewidth of G is at most k, and if so, finds a tree-decomposition of G of width at most k, and that uses $O(|V|)$ time [1]. It is not very hard to transform in $O(|V|)$ time this tree-decomposition into a nice one, with $|I| = O(|V|)$ (see [16]).

Definition 15. A *terminal graph* is a triple (V, E, X), with (V, E) an undirected graph, and $X \subseteq V$ a subset of the vertices, called the *terminals*.

To each node i of a rooted tree-decomposition (X, T) of graph $G = (V, E)$, we associate the terminal graph $G_i = (V_i, E_i, X_i)$, where $V_i = \bigcup \{X_j \mid j = i \vee j$ is a descendant of $i\}$, and $E_i = \{\{v, w\} \in E \mid v, w \in V_i\}$. As shorthand notation we write $p(v, w, G, c)$, if there is a path in G from v to w with all internal vertices having colors, smaller than $c(w)$.

Definition 16. Let $G = (V, E, X)$ be a terminal graph, and let $c : V \to \{1, \ldots, t\}$ be a vertex t-ranking of (V, E). The *characteristic* of c, $Y(c)$, is the quadruple $(c|_X, f_1, f_2, f_3)$, where

- $c|_X$ is the function c, restricted to domain X.
- $f_1 : X \times \{1, \ldots, t\} \to \{\text{true,false}\}$, is defined by: $f_1(v, i) = $ true if and only if there is a vertex $x \in V$ with $c(x) = i$ and $p(v, x, G, c)$.
- $f_2 : X \times X \times \{1, \ldots, t\} \to \{\text{true,false}\}$, is defined by: $f_2(v, w, i) = $ true, if and only if there is a vertex $x \in V$ with $c(x) = i$ and $p(v, x, G, c)$ and $p(w, x, G, c)$.

- $f_3 : X \times X \rightarrow \{1, \ldots, t, \infty\}$ is defined by: $f_3(v, w)$ is the smallest integer t' such that there exists a path from v to w in G with no internal vertices of color larger than t'. If there is no path from v to w in G, then $f_3(v, w) = \infty$.

A set S of characteristics of vertex t-rankings of G is a *full set* if and only if for every vertex t-ranking c of G, $Y(c) \in S$.

If $t = O(\log |V|)$, then a full set of characteristics of vertex t-rankings of $G = (V, E, X)$ (with $|X| \leq k+1$, k constant) has size polynomial in V: the numbers of possible values for $c|_X$, f_1, f_2, and f_3 are bounded by $O(log^{(k+1)+1}|V|)$, $2^{O(k \log |V|)}$, $2^{O((k+1)^2 \log |V|)}$, and $O(log^{\frac{1}{2}k(k+1)}|V|)$. The following lemma, given in [3], shows that we can ensure this property for graphs with treewidth at most k for fixed k.

Lemma 17. *If the treewidth of G is at most k, then $\chi_r(G) = O(k \cdot \log |V|)$.*

Let k be a fixed constant. Suppose we want to determine whether $\chi_r(G) \leq t$ for a graph $G = (V, E)$ with treewidth at most k. The first step is to compute a nice tree-decomposition $(\{X_i \mid i \in I\}, T = (I, F))$ of G of width at most k. Then we compute in a bottom-up order for every node i in the tree-decomposition a full set of characteristics for G_i. When we finally have a full set for the root node r we can easily determine whether $\chi_r(G) \leq t$.

Lemma 18. *Let i be a start (introduce, forget, join) node of nice tree-decomposition (X, T) of graph $G = (V, E)$ (with child j, children j_1 and j_2). A full set of characteristics of G_i can be computed in polynomial time (given the full sets of characteristiscs of G_j, of G_{j_1} and G_{j_2}).*

Lemma 19. *Let r be the root node of nice tree-decomposition (X, T) of $G = (V, E)$. Let S be a full set of characteristics of G_r. $\chi_r(G) \leq t$, if and only if $S \neq \emptyset$.*

Proof. Any element of S corresponds to a vertex t-ranking of G_r, or, equivalently, of G (as $G_r = (V, E, X_r)$), and vice versa. $\square$

Using the lemmas above, it follows directly that we can compute in polynomial time a full set of G_r, and hence determine whether $\chi_r(G) \leq t$.

Theorem 20. *For any fixed k, there exists a polynomial time algorithm, that determines the vertex ranking number of graphs with treewidth at most k.*

It is also possible to modify the algorithm, such that it also outputs a vertex t-ranking of the input graph, if existing, without increasing its running time by more than a constant factor.

6 The equality $\chi_r = \chi$

In this section we consider questions related to the equality of the chromatic number and the vertex ranking number of graphs.

Theorem 21. *If $\chi_r(G) = \chi(G)$ holds for a graph G, then G also satisfies $\chi(G) = \omega(G)$.*

Proof. Suppose that $G = (V, E)$ has a vertex t-ranking $c : V \to \{1, 2, \ldots, t\}$ with $t = \chi(G)$. We are going to consider the separator tree $T(c)$ of this t-ranking. Recall that $T(c)$ is a rooted tree and that every internal node of $T(c)$ is assigned to a subset of the vertex set of G which is a separator of the corresponding subgraph of G, namely more than one component arises when all subsets on the path from the node to the root are deleted from the graph. Furthermore, all vertices assigned to the nodes of a path from a leaf to the root of $T(c)$ have pairwise different colors.

The goal of the following recoloring procedure is to show that either $\chi(G) = \omega(G)$ or we can recolor G to obtain a proper coloring with a smaller number of colors. However, the latter contradicts the choice of the $\chi(G)$-ranking c.

We label the nodes of the tree $T(c)$ according to the following marking rules:

1. Mark a node s of $T(c)$ if the union $U(s)$ of all vertex sets assigned to all nodes on the path from s to the root is *not* a clique in G.
2. Also, mark a leaf l of $T(c)$ if the union $U(l)$ of all vertex sets assigned to all nodes on the path from l to the root is a clique in G, but $|U(l)| < t$.

Case 1: There is an unmarked leaf l. We have $|U(l)| = t$ and $U(l)$ is a clique. Hence, $\omega(G) = \chi(G)$.

Case 2: There is no unmarked leaf. We will show that this would enable us to recolor G saving one color, contradicting the choice of c.

Since every leaf of $T(c)$ is marked, every path from a leaf to the root consists of marked nodes eventually followed by unmarked nodes. Consequently, there is a collection of marked branches of $T(c)$, i.e., subtrees of $T(c)$ induced by one node and all its descendants consisting of all marked nodes. Further the father of the highest node of each branch is unmarked or the highest node is the root of $T(c)$ itself.

If the root of $T(c)$ is marked then we have exactly one marked branch, namely $T(c)$ itself. Then, by definition, the separator S assigned to the root is *not* a clique. However, none of its colors is used by the ranking for vertices in $V \setminus S$. Simply, any coloring of the separator S with fewer than $|S|$ colors will produce a coloring of G with fewer than $\chi(G)$ colors; contradiction.

If the root is unmarked, then we have to work with a collection of b marked branches, $b > 1$. Notice that all color-1 vertices of G are assigned to leaves of $T(c)$ and that any leaf of $T(c)$ belongs to some marked branch B. We are going to recolor the graph G by recoloring the marked branches one by one such that the new coloring of G does not use color 1. Let us consider a marked branch

B. Let h be its highest node in $T(c)$, and $S(h)$ the set assigned to h. Since h is marked but the root is unmarked, there must exist a vertex x of $S(h)$ and a vertex y belonging to $U(h)$ which are nonadjacent. Then $c(x) \neq c(y)$ since all vertices of $U(h)$ have pairwise different colors.

Assume $c(x) = 1$ or $c(y) = 1$. Then h is a leaf of $T(c)$. Hence, x and y, respectively, is the only color-1 vertex of G assigned to a node of B. We simply recolor x and y with $\max(c(x), c(y))$.

Finally consider the case $c(x) \neq 1$ and $c(y) \neq 1$. All color-1 vertices in the subgraph of G corresponding to B are recolored with $c(x)$ and x is recolored with $c(y)$. By the construction of $T(c)$, this does not influence other parts of the graph, since they are separated by vertex sets with higher colors.

Having done this operation in every marked branch, eventually we get a new color assignment of G which is still a proper coloring (though usually not a ranking). Since all leaves of $T(c)$ are marked, and no internal node of $T(c)$ contains color-1 vertices, color 1 is eliminated from G, contradicting the assumption $\chi_r(G) = \chi(G)$. Consequently, Case 2 cannot occur, implying $\chi(G) = \omega(G)$. This completes the proof. □

Clearly, $\chi_r(G) = \chi(G)$ does not imply that G is a perfect graph. (Trivial counterexamples are of the form $G = G' \cup K_{\chi_r(G')}$ where G' is an arbitrary imperfect graph.) On the other hand, if we require the equality on all induced subgraphs, then we remain with a relatively small class of graphs that is also called 'trivially perfect' in the literature (cf. [11]).

Theorem 22. *A graph $G = (V, E)$ satisfies $\chi_r(G[A]) = \chi(G[A])$ for every $A \subseteq V$ if and only if neither P_4 nor C_4 is an induced subgraph of G.*

Proof. The condition is necessary as $\chi_r(P_4) = \chi_r(C_4) = 3$ and $\chi(P_4) = \chi(C_4) = 2$.

Now let G be a P_4-free and C_4-free graph. The graphs with no induced P_4 and C_4 are precisely those in which every connected induced subgraph H contains a vertex w adjacent to all vertices of H [27]. Hence, the following efficient algorithm produces an optimal ranking in such graphs: If $H = (V', E')$ is connected, then we assign the color $\omega(H)$ to the dominating vertex w. Clearly, $\chi(H[V' \setminus \{w\}]) = \omega(H[V' \setminus \{w\}]) = \omega(H) - 1$, and it is easily seen that $\chi_r(H[V' \setminus \{w\}]) = \chi_r(H) - 1$ also holds; thus, induction can be applied. On the other hand, if H is disconnected, then its components can be found (in linear time), and an optimal ranking can be generated in each of them separately. □

7 Edge rankings of complete graphs

While obviously $\chi_r(K_n) = n$, it is not easy to give a closed formula for the edge ranking number of the complete graph. The most convenient way to determine $\chi'_r(K_n)$ seems to introduce a function $g(n)$ by the rules

$$g(1) = -1,$$
$$g(2n) = g(n),$$
$$g(2n + 1) = g(n + 1) + n.$$

In terms of the function $g(n)$, the following statement can be proved.

Theorem 23. *For every positive integer n,*

$$\chi'_r(K_n) = \frac{n^2 + g(n)}{3}.$$

Proof. The assertion is obviously true for $n = 1, 2, 3$. For larger values of n we are going to apply induction.

Similarly to vertex t-rankings, the following property holds for every edge t-ranking of a graph $G = (V, E)$: if i is the largest color occurring more than once, then the edges with colors $i + 1, i + 2, \ldots, t$ form an edge separator of G. Moreover, doing an appropriate relabeling of these colors $i + 1, i + 2, \ldots, t$ we get a new edge t-ranking of G with the property that there is a color $j > i$ such that all edges with colors $j, j + 1, \ldots, t$ form an edge separator of G which is minimal under inclusion.

We have to show that the best way to choose this edge separator R with respect to an edge ranking in a complete graph is by making the two components of $G[E \setminus R]$ as equal-sized as possible. Let us consider a K_n, $n \geq 4$. Let n_1 and n_2 be the numbers of vertices in the components, hence $n_1 + n_2 = n$ and the corresponding edge separator has size $n_1 n_2$. Every edge ranking starting with this separator has at least

$$n_1 n_2 + \max\{\chi'_r(K_{n_1}), \chi'_r(K_{n_2})\} = n_1 n_2 + \chi'_r(K_{\max\{n_1, n_2\}})$$

colors, and there is indeed one using exactly that many colors. Defining $a_1 := \min(n_1, n_2)$ and repeating the same argument for $n' := n - a_1$, and so on, we eventually get a sequence of positive integers $a_1, \ldots, a_s$, for some s, such that $\sum_{i=1}^{s} a_i = n$ and

$$a_i \leq \sum_{i < j \leq s} a_j \quad \text{for all } i, \quad 1 \leq i < s. \tag{1}$$

Notice that at least the last two terms of any such sequence are equal to 1. It is easy to see that the number of colors of any edge ranking represented by $a_1, \ldots, a_s$ is equal to $\sum_{1 \leq i < j \leq s} a_i a_j$, consequently

$$\chi'_r(K_n) = \min \sum_{1 \leq i < j \leq s} a_i a_j = \binom{n}{2} - \max \sum_{i=1}^{s} \binom{a_i}{2},$$

subject to the condition (1). Since a decreasing sort of the sequence maintains (1) we may assume $a_1 \geq a_2 \geq \ldots \geq a_s$. Thus, for each value of n, $\min \sum_{1 \leq i < j \leq s} a_i a_j$ is attained precisely by the unique sequence satisfying $a_i = \lfloor \frac{1}{2} \sum_{i \leq j \leq s} a_j \rfloor$ for all i, $1 \leq i < s$. In particular, we obtain

$$\chi'_r(K_n) = \chi'_r(K_{\lceil n/2 \rceil}) + \lfloor n/2 \rfloor \lceil n/2 \rceil.$$

Applying this recursion, it is not difficult to verify that, indeed, $\chi'_r(K_n)$ can be written in the form $\frac{1}{3}(n^2 + g(n))$, where $g(n)$ is the function defined above. $\square$

Observing that $g(2^n) = -1$ for all $n \geq 1$, we obtain the following result.

Corollary 24. $\chi'_r(K_{2^n}) = \frac{4^n - 1}{3}$.

8 Conclusions

We studied algorithmic and graph-theoretic properties of rankings of graphs. For many special classes of graphs, the algorithmic complexity of VERTEX RANKING is now known. However the algorithmic complexity of VERTEX RANKING when restricted to chordal graphs or circle graphs is still unknown. Furthermore it is not known whether the EDGE RANKING problem is NP-complete.

We started a graph-theoretic study of vertex ranking and edge ranking as a particular kind of proper (vertex) coloring and proper edge coloring, respectively. Much research has to be done in this direction. It is of particular interest which of the well-known problems in the theory of vertex colorings and edge colorings are also worth studying for vertex rankings and edge rankings.

References

1. H.L. BODLAENDER. A linear time algorithm for finding tree-decompositions of small treewidth. *Proceedings of the 25th Annual ACM Symposium on Theory of Computing*, ACM Press, New York, 1993, pp. 226–234.
2. H.L. BODLAENDER. A tourist guide through treewidth. *Acta Cybernetica* **11** (1993), 1–23.
3. H.L. BODLAENDER, J.R. GILBERT, H. HAFSTEINSSON, T. KLOKS. Approximating treewidth, pathwidth and minimum elimination tree height. *Proceedings of the 17th International Workshop on Graph-Theoretic Concepts in Computer Science WG'91*, Springer-Verlag, Lecture Notes in Computer Science 570, 1992, pp. 1–12. To appear in *Journal of Algorithms*.
4. J.A. BONDY, U.S.R. MURTY. *Graph Theory with Applications*. American Elsevier, New York, 1976.
5. A. BRANDSTÄDT. Special graph classes – a survey. *Schriftenreihe des FB Mathematik*, Universität Duisburg, SM–DU–199, 1991.
6. P. DE LA TORRE, R. GREENLAW, A.A. SCHÄFFER. Optimal edge ranking of trees in polynomial time. *Proceedings of the 4th Annual ACM-SIAM Symposium on Discrete Algorithms*, Austin, Texas, 1993, pp. 138–144.
7. J.S. DEOGUN, T. KLOKS, D. KRATSCH, H. MÜLLER. On vertex ranking for permutation and other graphs. *Proceedings of the 11th Annual Symposium on Theoretical Aspects of Computer Science*, P. Enjalbert, E.W. Mayr, K.W. Wagner, (eds.), Lecture Notes in Computer Science 775, Springer-Verlag, Berlin, 1994, 747–758.
8. J.S. DEOGUN, Y. PENG. Edge ranking of trees. *Congressus Numerantium* **79** (1990), 19–28.

9. I.S. DUFF, J.K. REID. The multifrontal solution of indefinite sparse symmetric linear equations. *ACM Transactions on Mathematical Software* **9** (1983), 302–325.
10. M.R. GAREY, D.S. JOHNSON. *Computers and Intractability: A Guide to the Theory of NP-completeness.* W.H. Freeman and Company, New York, 1979.
11. M.C. GOLUMBIC. *Algorithmic Graph Theory and Perfect Graphs.* Academic Press, New York, 1980.
12. A.V. IYER, H.D. RATLIFF, G. VIJAYAN. Optimal node ranking of trees. *Information Processing Letters* **28** (1988), 225–229.
13. A.V. IYER, H.D. RATLIFF, G. VIJAYAN. Parallel assembly of modular products—an analysis. Technical Report 88-06, Georgia Institute of Technology, 1988.
14. A.V. IYER, H.D. RATLIFF, G. VIJAYAN. On an edge ranking problem of trees and graphs. *Discrete Applied Mathematics* **30** (1991), 43–52.
15. M. KATCHALSKI, W. MC CUAIG, S. SEAGER. Ordered colourings. Manuscript, University of Waterloo, 1988.
16. T. KLOKS. *Treewidth.* Ph. D. Thesis, Utrecht University, The Netherlands, 1993.
17. C.E. LEISERSON. Area efficient graph layouts for VLSI. *Proceedings of the 21st Annual IEEE Symposium on Foundations of Computer Science*, 1980, pp. 270–281.
18. J.W.H. LIU. The role of elimination trees in sparse factorization. *SIAM Journal of Matrix Analysis and Applications* **11** (1990), 134–172.
19. J. NEVINS, D. WHITNEY. (eds.) *Concurrent Design of Products and Processes.* McGraw-Hill, New York, 1989.
20. A. POTHEN. The complexity of optimal elimination trees. Technical Report CS-88-13, Pennsylvania State University, U.S.A., 1988.
21. N. ROBERTSON, P. D. SEYMOUR. Graph minors. IV. Tree-width and well-quasi-ordering. *Journal on Combinatorial Theory* Series B **48** (1990), 227–254.
22. N. ROBERTSON, P. D. SEYMOUR. Graph minors. V. Excluding a planar graph. *Journal on Combinatorial Theory* Series B **41** (1986), 92–114.
23. N. ROBERTSON, P. D. SEYMOUR. Graph minors. XIII. The disjoint paths problem. Manuscript, 1986.
24. A.A. SCHÄFFER. Optimal node ranking of trees in linear time. *Information Processing Letters* **33** (1989/90), 91–96.
25. P. SCHEFFLER. Node ranking and searching on graphs (Abstract). in: U. Faigle and C. Hoede, (eds.), *3rd Twente Workshop on Graphs and Combinatorial Optimization*, Memorandum No.1132, Faculty of Applied Mathematics, University of Twente, The Netherlands, 1993.
26. A. SEN, H. DENG, S. GUHA. On a graph partition problem with application to VLSI layout. *Information Processing Letters* **43** (1992), 87–94.
27. E.S. WOLK. The comparability graph of a tree. *Proceedings of the American Mathematical Society* **3** (1962), 789–795.

Bypass Strong V-Structures and Find an Isomorphic Labelled Subgraph in Linear Time

Heiko Dörr

Institut für Informatik, Freie Universität, Berlin
doerr@inf.fu-berlin.de

Abstract. This paper identifies a condition for which the existence of an isomorphic subgraph can be decided in linear time. The condition is evaluated in two steps. First the host graph is analysed to determine its strong V-structures. Then the guest graph must be appropriately represented. If this representation exists, the given algorithm constructively decides the subgraph isomorphism problem for the guest and the host graph in linear time.

The result applies especially to the implementation of graph rewriting systems. An isomorphic subgraph must be determined automatically in each rewriting step. Thus the efficient solution presented in this paper is an important advancement for any implementation project.

1 Introduction

A main effort in any implementation of rule-based graph rewriting systems is directed to the solution of the subgraph isomorphism problem for labelled graphs. A graph rewriting rule is applicable to a host graph only if there is a subgraph isomorphic to the left-hand side of the rule. Hence the labelled subgraph problem must be solved in each rewriting step. Current implementations follow two strategies to deal with the problem. They either keep the problem open to the user who determines an isomorphic subgraph interactively [LoBe93], [Him89] or in advance [KlMa92], or they perform a full search [Zün92]. Neither strategy is satisfying with respect to a fast execution, which is a key condition for a broader use of graph rewriting.

In general the subgraph isomorphism problem lies in NP for undirected and directed labelled graphs [GaJo79]. When considering labelled graphs, more information on the problem instances is available. Vertices and edges become distinguishable by their labels. Our approach to an efficient algorithm for the labelled subgraph isomorphism makes heavy use of labels.

Let Σ_V, Σ_E be finite alphabets. A finite non-empty set V is the set of *vertices*, and $E \subseteq V \times \Sigma_V \times V$ is the set of *edges*. A total function $l : V \to \Sigma_V$ is a *vertex labelling*. The triple $G = (V, E, l)$ is a *directed labelled graph* over Σ_V, Σ_E, or just a *graph*. The set of unique vertex labels of G is defined as $uv(G) = \{ul \in \Sigma_V | \exists! v \in V \text{ such that } l(v) = ul\}$.

Let $G = (V, E, l)$ and $G' = (V', E', l')$ be graphs. The graph G' is a *subgraph* of G, $G' \subseteq G$ iff $V' \subseteq V$, $E' \subseteq E$ and $l' = l|_{V'}$. Let $h_V : V' \to V$ be a bijective vertex map such that for the corresponding edge map $h_E : E' \to E$, defined as $h_E(E') = \{(h_V(s), el, h_V(t)) | (s, el, t) \in E'\}$, holds $h_E(E') = E$ and $l = l' \circ h^{-1}$. Based on the vertex bijection h_V the *graph isomorphism* $\hat{h}$ maps the graph G' to $\hat{h}(G') = (h_V(V'), h_E(E'), l' \circ h^{-1})$. When clear from context we omit the indices of the vertex and edge maps h_E and h_V respectively and write h for both. The *host graph* G contains a *subgraph isomorphic* to the *guest graph* G' iff there is a graph isomorphism $\hat{h}$ such that $\hat{h}(G') \subseteq G$. The *subgraph isomorphism problem* for labelled graph is stated as follows: Given two graphs G and G', is there a graph isomorphism $\hat{h}$ such that $\hat{h}(G') \subseteq G$?

2 The Basic Algorithm

The labelled subgraph isomorphism can be constructed by traversing the guest graph starting at an arbitrary initial vertex. The first step of the procedure determines the set of initial partial handles according to the initial vertex. In each subsequent step the algorithm tests whether an edge can be mapped to an edge incident to a current partial handle. If possible, current partial handles and the induced vertex maps are extended and comprise the set of handles current to the next step. Thus all handles current to an iteration have an identical inverse image.

A *connected enumeration* represents a traversal of the guest graph. The elements of the enumeration are successively drawn by the algorithm. The enumeration must ensure that for all edges there is one adjacent edge preceding it in the enumeration because the main principle of our algorithm is the extension of partial handles. Otherwise an extension would be impossible. The initial edge deserves special treatment. The root of a connected enumeration is one vertex incident to the initial edge.

Definition 1. partial handles, connected enumeration

1. Let G and G' be two graphs. Any graph $G_p \subseteq G$ for which there exists a partial graph isomorphism $\hat{h}$ such that $\hat{h}(G') = G$ is a *partial handle*. If $\hat{h}$ is total, then G_p is a *full handle*. A partial handle $G_p = (\{v\}, \emptyset, l)$ with $v \in V$ is an *initial handle*.
2. Let $G = (V, E, l)$ be a graph with size q. A sequence (e_i) of edges $e_i \in E$, $i = 1 \ldots q$ is an *enumeration* of E iff $E = \{e_i | i = 1 \ldots q\}$. An enumeration $(e_i)_{i=1 \ldots q}$ of E is *connected* iff for all $i = 2 \ldots q$ there is a $j = 1 \ldots i$ such that e_j is adjacent with e_i. The *root vertex* of $(e_i)_{i=1 \ldots q}$ is a vertex incident to e_1.

For the specification of the algorithm, we assume that the graph is given in a *frame-based data structure*. Each frame stores the direct neighbourhood of a vertex. Each slot of a frame contains a list of isomorphic edges incident to the frame vertex. Isomorphic edges have the same edge label, same direction, and the labels of the incident vertices are equal too. With the frame data structure we can directly address incident edges by their labels and direction.

Figure 1 shows two frames. They have one segment for incoming and one for outgoing edges. Each segment again provides segments for any vertex label, and finally the slots for the edges of a certain edge label. In our example there are only *single entry slots* but they may also have *multiple entries*.

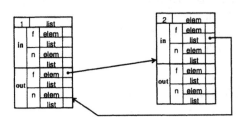

Fig. 1. Frame representation of a vertex and its incident edges

The algorithm maintains a set of morphisms. The final set should contain all morphisms which map the left-hand side into the host graph. The algorithm performs one of two main operations depending on the edge currently drawn from the input enumeration. The operation is either a simple *check or an extension*. The machine executes a simple check when the images of both endpoints of an edge are already found. In that case it selects those morphisms from the current set which are defined for that edge, i.e. the corresponding partial handles must contain an image of the edge. When the image of one endpoint is not determined already the machine tries to extend the current handles. Therefore it looks up the corresponding slots determined by the edge's label and direction.

Definition 2. basic operations

1. Let G_1 and G_2 be two graphs. Let $\hat{h}$ be a partial graph isomorphism from G' to G. The graph isomorphism $\hat{h}'$ *extends* $\hat{h}$ *along an edge* $e \in E_1$ iff for $e = (s, el, t)$
   (a) either $s \notin dom(h)$ or $t \notin dom(h)$,
   (b) there exists a vertex $v \in V_2$ such that
      if $s \notin dom(h) : (v, el, h(t)) \in E_2$ and $h'(s) = v$
      or if $t \notin dom(h) : (h(s), el, t) \in E_2$,and $h'(t) = v$
   (c) and $h'_V|_{dom(h_V)} = h_V$.
   Hence $\hat{h}(G_1) \subseteq \hat{h}'(G_1)$.
2. Let G and G' be two graphs. Let A be a set of partial graph isomorphisms from G' to G. The operations *check* and *extend* compute another set of graph isomorphisms with respect to an edge $e' \in E'$ as follows:

$$\text{check } A \, e' = \{\hat{h} \in A | h(e) \in E\}$$
$$\text{extend } A \, e' = \{\hat{h}' | \hat{h}' \text{ extends } \hat{h} \in A \text{ by } e'\}.$$

The connected enumeration of a guest graph G' and the host graph G are input to the algorithm for labelled subgraph matching. The algorithm performs a breadth first search for a graph isomorphism and decides the existence of a graph isomorphism by construction.

Definition 3. algorithm for a set of labelled isomorphic subgraphs
Let G and G' be graphs.
INPUT: G and a connected enumeration $(e_i)_{i=1...q}$ of E' with root vertex $v' \in V'$.
OUTPUT: a set of graph isomorphisms $\hat{h}$ with $\hat{h}(G') \subseteq G$.

1. **INITIALIZE** $A_0 := \{\hat{h}|\exists v \in V, l'(v') = l(v), h(v') = v\}$
2. **SET** $W_0 := \{v'\}$
3. **FOR** $i = 1...q$ **DO**
       LET $(s, el, t) = e_i$
       IF $s, t \in W_{i-1}$ **THEN** $A_i = $ check $A_{i-1}\ e_i$
                               ELSE $A_i = $ extend $A_{i-1}\ e_i$
       ENDIF
       SET $W_i = W_{i-1} \cup \{s, t\}$
4. **OUTPUT** A_q

We now prove the correctness of the algorithm.

Lemma 4. *Let G and G' be two graphs. Let $(e_i)_{i=1...q}$ be a connected enumeration of E' with root vertex $v' \in V'$. Let A be the output of the labelled subgraph algorithm for the input G, $(e_i)_{i=1...q}$, and v'.*

$$\text{For all } \hat{h} \in A \text{ it holds that } \hat{h}(G') \subseteq G.$$

Proof. We adopt the terminology of the labelled subgraph algorithm and prove $\hat{h}(G') \subseteq G$ by induction over i. By definition of A_0 it holds that $\hat{h}_0(G_0) \subseteq G$ for all $\hat{h}_0 \in A_0$ and initial handles $G_0 = (\{v'\}, \emptyset, l')$. Let $i \in \{0, \ldots, q-1\}$, $\hat{h}_i \in A_i$, and $(s, el, t) = e_{i+1}$ be the current edge.
Assume now that $\hat{h}_i(G_i) \subseteq G$ for the partial handle $G_i = (W_i, \{e_j | j = 1 \ldots i\}, l')$ and let $\hat{h}_{i+1} \in A_{i+1}$ be an extension of $\hat{h}_i$.
If both endpoints are already found, i.e. $s, t \in W_i$, then it holds for all $\hat{h}_{i+1} \in A_{i+1}$ that $h_{i+1}(e_{i+1}) \in E$. Since $\hat{h}_{i+1}$ extends a preceding $\hat{h}_i$ by the mapping of e_{i+1} it follows for the induced vertex maps that $dom(h_i) = dom(h_{i+1}) = W_i$. The same argument holds for the induced edges maps $dom(h_i) = dom(h_{i+1}) \backslash \{e_{i+1}\}$. With $\hat{h}_i(G_i) \subseteq G$ it follows that $\hat{h}_{i+1}(G_{i+1}) \subseteq G$.
Without loss of generality let us now assume that $s \notin W_i$. Since $\hat{h}_{i+1}$ extends $\hat{h}_i$ by s it holds that $dom(h_i) = dom(h_{i+1}) \backslash \{s\} = W_i$ and $dom(h_i) = dom(h_{i+1}) \backslash \{e_{i+1}\}$. The extended graph morphism maps s into V and e_{i+1} into E. Hence $\hat{h}_{i+1}(G)_{i+1} \subseteq G$ and the proof is complete.

The complexity of the algorithm depends strongly on the number of slot entries. In the case of multiple entries the algorithm must scan the list of entries to find the possible images of the given edge. Consequently the check takes time dependent on the number of entries. If we have to extend a match and the respective slot has multiple entries, the extension is performed for each entry and each current handle. Thus multiple extensions must be constructed.

The run time analysis of step 3 depends on the edge which is being processed. Let us determine the run time for the i-th iteration, i.e. for an edge $e_i = (s, el, t)$ with $i \in \{1, \ldots, q\}$. If both endpoints are included in the current match a *check* is performed for each handle. Let $n_{h(t),i}$ and $n_{h(s),i}$ be the number of entries in the respective slot of vertices $h(t)$ and $h(s)$ for morphisms $\hat{h} \in A_{i-1}$. Hence the number of tests is

$$\sum_{\hat{h} \in A_{i-1}} n_{h(t),i} \text{ or } \sum_{\hat{h} \in A_{i-1}} n_{h(s),i}$$

depending on the vertex at which the existence of the edge is checked.

When an *extension* is performed, assume that we extend the current handles by possible images of the source s, i.e. $s \notin W_{i-1}$. Let again $n_{h(t),i}$ be the number of entries in the respective slot of vertices $\{h(t) | h \in A_{i-1}\}$. When the slot in the image of t is not empty, i.e. $n_{h(t),i} > 0$, the algorithm extends all handles of A_{i-1}. For each item of a slot an extension must be defined. The number of extensions in the i-th step is then $\sum_{\hat{h} \in A_{i-1}} n_{h(t),i}$. To determine the overall run time, let p and q be the order and the size of the guest graph G'. In step 1, at most $|V|$ comparisons are necessary to find all initial handles. Then the algorithm must map $p-1$ vertices onto images in the graph G. The mappings are constructed by extensions of current handles. As a consequence, the existence of $q - (p-1)$ remaining edges must be checked. Without loss of generality, assume a connected enumeration which drives the isomorphism algorithm such that all extensions are performed before any edge existence is checked. Assume further that only source vertices must be found. Let $n_{h(t),i}$ be defined as above. The overall run time of the algorithm then is

$$|V| + \sum_{i=1}^{p-1} \left(\sum_{\hat{h} \in A_{i-1}} n_{h(t),i} \right) \text{ extensions } + \sum_{i=p}^{q} \left(\sum_{\hat{h} \in A_{i-1}} n_{h(t),i} \right) \text{ checks}.$$

An execution of the algorithm does not branch if the number of handles does not increase in any step. This is a restriction particular to the extension steps since checks by definition do not increase the number of morphisms. In case the set of initial morphisms has only one element, the non-branching execution of the algorithm computes a unique full handle if there is a match at all.

More formally let G and G' be two graphs and $(e_i)_{i=1 \ldots q}$ be a connected enumeration of E' with root vertex $v' \in V'$. The labelled subgraph isomorphism algorithm *executes without branches* iff for all $i \in \{1, \ldots, q-1\}$ and any $\hat{h} \in A_i$ there exists at most one extension $\hat{h}_{i+1} \in A_{i+1}$.

For an execution without branches we can derive the number of occurring extensions.

Lemma 5. *Let G and G' be two graphs with $p = |V'|$ and $q = |E'|$. Let $(e_i)_{i=1...q}$ be a connected enumeration of E' with root vertex $v' \in V'$. Let $m = |A_0|$ be the number of initial handles. If the labelled subgraph isomorphism algorithm executes non-branching it performs $m \times (p-1)$ extensions.*

Proof. From the premises it follows for all $i = 1...q$ and any $\hat{h} \in A_i$ that there exists at most one extension $\hat{h}_{i+1} \in A_{i+1}$. Hence the algorithm performs extensions only on single entry slots and either $n_{h(s),i} \leq 1$ or $n_{h(t),i} \leq 1$ depending on the direction of the extension. Furthermore it holds that $|A_{i+1}| \leq |A_i|$ for all $i = 1...q-1$; hence $|A_i| \leq |A_0|$ for $i = 1...q-1$. Let us assume that the elements of the enumeration are such that the mapping algorithm extends by the source vertex. For the number of extensions it follows that $\sum_{i=1}^{p-1} (\sum_{\hat{h} \in A_{i-1}} n_{h(t),i}) \leq \sum_{i=1}^{p-1} (|A_i|) \leq \sum_{i=1}^{p-1} (|A_0|) \leq m \times (p-1)$.

Theorem 6. *The algorithm takes time linear to the size of the guest graph if the extensions and checks are performed on single entry slots and there is exactly one initial handle.*

Proof. Lemma 5 states that if the algorithm executes without branches the number of extensions is bound by $m \times (p-1)$. If the root vertex of the input enumeration has a unique image in the host graph, the number of extensions equals $(p-1)$. The run time of the whole algorithm now still depends on the checks for the existence of the remaining $q - (p-1)$ edges. If these checks are performed on vertices which have at most one entry in the slot for the respective edge, the algorithm performs $q - (p-1)$ checks.

As a consequence, we must determine single entry slots of the host graph. Furthermore if we can give an appropriate enumeration of the guest graph we have proven the statement given as the title of the paper.

3 Strong V-Structures

There are two main factors for the combinatorial explosion of the subgraph algorithm. Firstly there may be several initial handles for the root vertex; secondly the connected enumeration of the guest graph may admit multiple extensions of partial handles. In that case the algorithm must process multiple partial handles in the next iteration.

There are in general several enumerations for a graph. Why should it not be possible to determine connected enumerations which do not lead to multiple extensions? The major question would then be:

How can we determine that enumeration, if it exists?

Figure 2 gives an example for a host graph for which a multiplying and a non-multiplying connected enumeration exist. The study of that graph provides a first impression of the problem. We have to perform the labelled subgraph matching for a guest graph G' and a host graph G. Assume that the algorithm has proceeded to the subgraph G_1. Only edges incident with G_1 are candidates for the next extension step. At this stage we have two alternatives. We could test the existence of an edge labelled with b joining the image of vertex 2 with a vertex labelled by C. The other alternative is a match for edge (1,e,3). The selection of the first alternative causes three extensions and thus three partial handles which have to be analysed further. Two of them cannot be completed. But this fact will not be observed before the next extension step. If we try to extend G_1 following the second alternative, we are lucky because we find a singular extension. Furthermore we can complete the handle in one additional step.

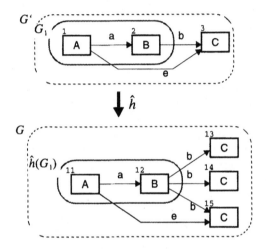

Fig. 2. Multiplying and non-multiplying connected enumerations

Obviously we do not want to rely on chance when we select a connected enumeration for a given rule. Thus we look for a procedure to distinguish "good" from "bad" enumerations. The closer analysis of the example given in Figure 3 provides a hint for this distinction. The graph G has a characteristic property leading to a multiplication of partial handles when extending the edge (2,b, 3). The graph contains three edges on which the extension of G_1 is possible. These edges, all incident to vertex 12, extend the partial handle $\hat{h}(G_1)$. In graph theoretical terms, there exists a number of non-trivial automorphisms for these edges. Pairs of these edges form *automorphic semipaths of length 2*. Such a path consists of two isomorphic edges which are incident in one of their endpoints. Whenever a graph contains such an automorphism, the matching algorithm for a rewriting rule may take non-linear time.

The example also includes the *key property for the solution* of the "multiplication problem". We have seen that there is one enumeration initiated in vertex 1 which performs a non-multiplying search although there are local automorphisms in G. When we choose enumeration $((1,e,3), (2,b,3), (1,a,2))$ and initiate the labelled subgraph matching at vertex 1, we do not fall into the automorphism trap: we traverse the host graph G in a deterministic manner without multiplication of partial handles. Thus the important information for the selection of a non-multiplying connected enumeration is the non-trivial automorphisms on semipaths of length 2 in the host graph.

Strong V-structures characterize automorphic semipaths of length 2 by their labels. Since only the labels of vertices and not their identity is of interest for the algorithm, this characterization is sufficient. We have chosen the adjective "strong" because the labels of the automorphic outer vertices are also significant. Weak V-structures, on the contrary, ignore the labels of the outer vertices.

Definition 7. strong V-structures
The set of all *strong V-structures* over alphabets Σ_V, Σ_E is defined as $SVS = \Sigma_V \times \Sigma_V \times \Sigma_E \times \{in, out\}$.
Let $G = (V, E, l)$ be a graph. A pair of edges $(e, e') \in E^2$ is an *instance* of a strong V-structure $vs = (vl_1, vl_2, el, d)$ iff there are vertices $x, y, z \in V$ with $y \neq z$, such that

1. $l(x) = vl_1, l(y) = l(z) = vl_2$ and
2. if $d = out$: $e = (x, el, z), e' = (x, el, y)$ and otherwise: $e = (z, el, x), e' = (y, el, x)$.

The set of *strong V-structures of a graph G* is given by

$$svs(G) = \{vs \in SVS | \exists (e, e') \in E^2 \text{ instance of } vs\}$$

Obviously the component vl_2 of a strong V-structure (vl_1, vl_2, el, d) must not be a unique label. The edges $((12,b,13)(12,b,14))$ of graph G in Figure 1 form an instance of the strong V-structure (B,C,b,out).

We can determine $svs(G)$ by sorting the incident edges and checking the occurrence of duplicates.

The strong V-structures of a graph are closely related to the frame data structure. We distinguished single and multiple entry slots depending on the number of entries. For each vertex of a graph G in the frame representation and from the knowledge of $svs(G)$ it follows whether a slot has definitely at most one entry or not. For all instances (e, e') of a strong V-structure (vl_1, vl_2, el, d) it holds that e and e' are isomorphic with respect to their center vertex. Thus the slot of the center vertex, which is determined by vl_2, el, and d contains at least e and e', and thus it is a multiple entry slot.

As a consequence of Lemma 5 on the number of extensions, and from the additional knowledge of $svs(G)$, we can decide whether the labelled subgraph isomorphism executes without branching. We therefore consider connected enumerations which *bypass* a set of strong V-structures.

Definition 8. bypassing a set of strong V-structures
Let G be a graph and $q = |E|$. Let $(e_i)_{i=1...q}$ be a connected enumeration of

E with root vertex v. Let W_i be the set of vertices incident to an edge which is part of the prefix of length $i = 1 \ldots q$ and $W_0 = v$. Let svs be a set of strong V-structures. The *connected enumeration* $(e_i)_{i=1\ldots q}$ *bypasses* svs iff for all $e_i = (s, el, t)$

$$\begin{array}{lll} \text{if } s \notin W_{i-1} : & (l(t), l(s), el, in) \notin svs \\ \text{or if } t \notin W_{i-1} : & (l(s), l(t), el, out) \notin svs \\ \text{or if } s, t \in W_{i-1} : \text{either } & (l(t), l(s), el, in) \notin svs \\ & \text{or } (l(s), l(t), el, out) \notin svs. \end{array}$$

Lemma 9. *Let G and G' be two graphs and let $(e_i)_{i=1\ldots q}$ be a connected enumeration of E' with root vertex $v' \in V'$. If (e_i), $i = 1 \ldots q$ bypasses $svs(G)$ then the algorithm for a labelled subgraph isomorphism executes without branching.*

Proof. We adopt the terminology of the algorithm. Let $i \in \{1, \ldots, q\}$ and $e_i = (s, el, t)$ be the current edge. When both endpoints are already found, i.e. $s, t \in W_{i-1}$, the algorithm does not branch by definition. Otherwise let $s \notin W_{i-1}$ and $\hat{h}', \hat{h}'' \in A_i$ be two extensions of $\hat{h}_{i-1} \in A_{i-1}$. They must be identical to their predecessor on its domain, i.e. $\hat{h}_{i-1} = \hat{h}'|_{dom(h_{i-1})} = \hat{h}''|_{dom(h_{i-1})}$, and $h'(s), h''(s) \in V \backslash h_i(V')$ with $h'(e_i), h''(e_i) \in E$. Since the connected enumeration bypasses $svs(G)$ it follows that $(l(t), l(s), el, in) \notin svs(G)$. Hence $h'(s) = h''(s)$ and the extended graph morphisms are identical: $\hat{h}' = \hat{h}''$.

Lemma 9 gives a sufficient condition for the non-branching execution of the labelled subgraph isomorphism. The remaining question is, how can we perform the existence checks in constant time? There is a small gap between the definition of bypassing and the algorithm in the case where both endpoints are found and an existence check is to be performed. The algorithm has no means to determine the endpoint on which the check is performed. For all edges $e_i = (s, el, t)$ of a bypassing enumeration with $s, t \in W_{i-1}$ either s or t has a single entry slot. Thus we refine the elements of the connected enumeration by this additional information. Consequently we modify the isomorphism algorithm to take that additional information into account. The modified isomorphism algorithm takes linear time for a refined enumeration if it bypasses the set of strong V-structures of the host graph.

Theorem 10. *Let G be a graph. If for G' there exists a refined connected enumeration bypassing $svs(G)$ and rooted in a uniquely labelled vertex v, then the algorithm for a labelled subgraph isomorphism takes $O(E')$ time.*

Proof. Let G and G' be graphs with $p = |V'|$. Let $(e_i)_{i=1\ldots q}$ be a connected enumeration of E' bypassing $svs(G)$ and rooted in a uniquely labelled vertex. Thus from Lemma 9 it follows that the algorithm executes without branching. This property also holds for the modified algorithm. Since the enumeration is rooted in a uniquely labelled vertex the algorithm builds $p - 1$ extensions, each extension consuming constant time. By the refinement of $(e_i)_{i=1\ldots q}$ we can drive the modified isomorphism such that each check is performed on a single entry slot. Thus each of the $q - (p - 1)$ checks takes constant time. Hence the whole isomorphism needs time linear to q.

4 Determination of Bypassing Connected Enumerations

One problem must still be solved: how can we construct an appropriate connected enumeration? We give a transformation to the rooted spanning tree problem for a guest graph and a set of strong V-structures. The transformation inspects first the *symmetric guest graph*. The shift to a symmetric graph reflects the assumed ability to search for an adjacent vertex independent of the direction of the joining edge. The algorithm decides whether it should extend the current partial handles by the source or the target of the current edge. All edges of the symmetric graph have a common interpretation: try to extend a current handle by the target vertex! As a consequence, each connected enumeration must contain one edge for each pair of symmetric edges. This holds except for symmetric edges already included in the left-hand side. In this case both edges must be part of the connected enumeration by definition.

Secondly, after modelling all possible enumerations, the transformation implements the information on strong V-structures in the symmetric graph. Some *edges of the symmetric graph* may be part of an instance of a strong V-structure. They must not be included in a bypassing connected enumeration. Thus they *are removed* from the symmetric graph. All remaining edges can be traversed without trapping into an instance of a strong V-structure during the algorithm's execution. In this transformation bypassing is equivalent to the existence of a directed spanning tree rooted in a uniquely labelled vertex. The edges of that tree form the first part of the bypassing connected enumeration. The remaining edges are put in the second part, still with respect to the set of strong V-structures. Hence the search algorithm maps firstly the vertices of the guest graph. Afterwards, it checks the existence of the remaining edges. The overall transformation costs time linear to the size of the guest graph.

Theorem 11. *Let G and G' be graphs. Let $p = |V'|$ and $q = |E'|$. Let*

$$\bar{E}' = \{(t, el, s)|(s, el, t) \in E'\}\backslash E'.$$

The elements of $\bar{E}'$ complete G' to a symmetric graph. Let

$$F' = \{(s, el, t) \in E'|(l'(s), l'(t), el, out) \notin svs(G)\} and$$

$$\bar{F}' = \{(t, el, s) \in \bar{E}'|(l'(t), l'(s), el, in) \notin svs(G)\}$$

be the edge sets of the symmetric graph cleared with respect to $svs(G)$. If there are

1. *a uniquely labelled vertex $u' \in V'$ with $l'(u') \in uv(G) \cap uv(G')$,*
2. *a directed spanning tree $S(G^\star, u')$ of $G^\star = (V', F' \cup \bar{F}', l')$ rooted in u',*
3. *and for all $\{(s, el, t) \in E'|(s, el, t) \notin S(G^\star, u')\}$ it holds that*
   either $(l'(s), l'(t), el, out) \notin svs(G)$
   or $(l'(t), l'(s), el, in) \notin svs(G)$,

then there exists a connected enumeration of E' bypassing $svs(G)$.

Proof. Let $(t_i)_{i=1...p-1}$ be a connected enumeration of the spanning tree of $G^\star$ rooted in u'. Let W_i be the set of vertices incident to the edges (t_j), $j = 1...i$ for $i = 1...q$ and $W_0 = \{u'\}$. The enumeration $(t_i)_{i=1...p-1}$ is defined such that for $t_i = (s, el, t)$ it holds that $t \notin W_{i-1}$. Let $(e_i)_{i=1...p-1}$ be a connected enumeration with $e_i = t_i$ if $t_i \in E'$ and if otherwise $t_i = (t, el, s) \notin E'$ then $e_i = (s, el, t)$. The enumeration $(e_i)_{i=1...p-1}$ bypasses $svs(G)$: let $i \in \{1, \ldots, p-1\}$ be fixed. Since the endpoints of e_i and t_i are identical it follows that $W_i = \bigcup_{j=1...i} inc_{G'}(e_j)$. Let further be $e_i = (s, el, t)$. If $e_i = t_i$ then it holds that $t \notin W_{i-1}$ and it follows by definition that $t_i \in F'$. Hence $(l'(s), l'(t), el, out) \notin svs(G)$. If $e_i \neq t_i$ it follows that $s \notin W_{i-1}$. By definition it holds that $t_i = (t, el, s) \in \bar{F}'$ and $(l'(t), l'(s), el, in) \notin svs(G)$. As a consequence it follows that $(e_i)_{i=1...p-1}$ is bypassing $svs(G)$.

For all remaining edges $e \in E' \backslash \{e_i | i = 1 \ldots p-1\}$ it holds that all vertices incident with e are member of W_{p-1}. From the premises it follows that any of these edges can be checked bypassing $svs(G)$. Hence we can add these edges to $(e_i)_{i=1...p-1}$ and receive a connected enumeration of E' bypassing $svs(G)$.

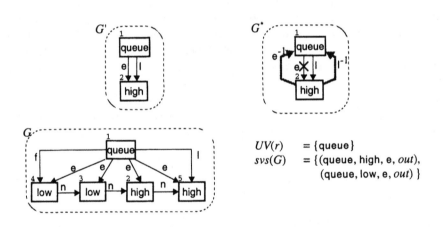

Fig. 3. Construction of a bypassing connected enumeration

We give an example for the construction of the theorem above in Figure 3. The upper half shows the guest graph G' and its transformed graph $G^\star$. The grey edges, i.e. $\bar{E}'$, are inserted in G' to give a symmetric graph. They are indicated by an "inverse" label. The edge $(1,e,2)$ is removed from $G^\star$. If the labelled subgraph matching traverses the host graph along the edge $(1, e, 2)$ in its original direction the search will branch because the corresponding slot has possibly multiple entries. Hence $\bar{F}' = \{(1, e, 2)\}$. Note that we can still traverse the edge $(1,e,2)$ in the opposite direction since the V-structure (high, queue, e, out) is not contained in $svs(G)$. We can find a spanning tree of the transformed graph rooted in the uniquely labelled vertex 1. It consists in our example just of the edge (1,l,2). This edge is the first part of the enumeration. The edge (1,e,2) completes the enumeration. It must be augmented with the information that the check for the existence of the edge must be performed in vertex 2. Then the modified algorithm for the labelled subgraph matching can execute the check by a constant time look-up.

5 Related Work

Our approach to reduce the complexity of the labelled subgraph isomorphism analyses the input to the algorithm. Corneil and Gotlieb take a similar approach for an efficient solution of the graph isomorphism problem [CoGo70]. In a preprocessing step a representative and a reordered representation of both input graphs are computed, and the problem is solved based on the transformed graphs. In contrast to our algorithm their procedure only gives an incomplete answer to the isomorphism problem. For some inputs it cannot decide whether the two input graphs are isomorphic.

The RETE-algorithm proposed by Bunke et al. [BGT91] addresses the labelled subgraph isomorphism problem in the context of graph rewriting systems. The algorithm is based on the observation that each rewriting step performs only local changes on the host graph. In a preprocessing step the rewriting system is analysed and the RETE-network created. Its topology represents the left-hand sides of the rewriting rules. It is supposed to carry all full handles. The network is initialized by input of the initial graph. In each rewriting step an appropriate isomorphic subgraph can be selected by inspection of the network. After execution of the rewriting step the information in the network is updated. This approach is not static as at run time the network must be updated. We, on the contrary, precompute the bypassing enumeration only once and apply the rewriting rule without auxiliary updates.

We share our interest in V-structures with Witt who studied locally unique graphs [Wit81]. He shows, by extension of the edge label alphabet, that it is possible to create a homomorphic and locally unique image for any graph with bounded degree. Furthermore he proves the existence of a linearizable hull for each locally unique graph. In his context a graph is linearizable iff each vertex of the graph has a unique address given as a list of edge labels. A connected enumeration which bypasses the set of strong V-structures can serve as a linear addressing scheme for a subset of its vertices. But that enumeration contains more information to solve the isomorphic subgraph problem efficiently. The notion of local uniqueness is not sufficient for that purpose because it cannot distinguish as many edges as strong V-structures can.

Unique vertex labels are a property already mentioned by Nagl and Göttler [Nag79], [Gött88]. Nagl introduces the "statische Verankerungsstruktur" (static anchor), Göttler uses a "Fixknoten" (fixed vertex) to define an application area of a rewriting rule and to force the application of subsequent rules to that area. The static anchor is used to program graph rewriting systems by means of graph rewriting systems only. None of the authors and even none of the recent publications on graph rewriting systems [Schü91], [KlMa92] have given a formal definition, not to mention a criterion, for that static anchor.

6 Conclusions

We can apply our results to graph rewriting systems if we infer information for all sentential forms. Then we can decide whether a connected enumeration drives a non-multiplying subgraph isomorphism for an arbitrary host graph. Thus we must analyse all graphs generated by a given graph rewriting system. An appropriate analysis technique is abstract interpretation. The major result of our analysis is an upper bound for a set of strong V-structures present in any sentential form. These results are of major importance for an implementation of graph rewriting systems [Dö94].

The strong V-structures of a host graph determine single entry slots. This information on the host graph enables us to select a connected enumeration as input for the labelled subgraph isomorphism algorithm. We have proven that the algorithm takes time linear in the size of the guest graph if two conditions hold for a connected enumeration:

1. its root must be a uniquely labelled vertex and,
2. it must bypass the set of strong V-structures of the host graph.

We have given a construction for bypassing enumerations. Based on our approach we can implement graph rewriting systems which perform the application test in linear time.

References

[BGT91] Bunke, H.; Glauser, T.; Tran, T.-H.: 'An efficient implementation of graph grammars based on the RETE matching algorithm', [EKR91], pp.174-189.

[CoGo70] Corneil, D.G.; Gotlieb, C.C.: 'An Efficient Algorithm for Graph Isomorphism', Journal of the Association for Computing Machinery, 17 (1) 51-64 (1970).

[Dö94] Dörr, Heiko: 'An Abstract Machine for the Execution of Graph Grammars', Proc. of the Poster Session of the International Conference on Compiler Construction, CC'94, Research Report Dept of Computer and Information Science, Linköping University, LiTH-IDA-R-94-11, pp.51-60.

[EKR91] Ehrig, Hartmut; Kreowski, Hans-Jorg; Rozenberg, Grzegorz (ed.): Graph-Grammars and Their Application to Computer Science, 4th Int. Workshop, Bremen, March 5-9, 1990, LNCS 532, Springer, Berlin, 1991.

[GaJo79] Garey, Michael R.; Johnson, David S.: 'Computers and Intractability', W.H. Freeman and Co., New York, 1979.

[Gött88] Göttler, Herbert: 'Graphgrammatiken in der Softwaretechnik', Informatik-Fachberichte 178, Springer, Berlin, 1988.

[Him89] Himsolt, Michael: 'Graphed: An interactive Graph Editor', in STACS 89, LNCS 349 Springer Verlag, Berlin, 1989.

[KlMa92] Klauck, Christoph; Mauss, Jakob: 'A Heuristic Driven Chart-Parser for Attributed Node Labelled Graph Grammars and its Application to Feature Recognition in CIM', Research Report, Deutsches Forschungszentrum fur Künstliche Intelligenz, Kaiserslautern/Saarbrücken, DFKI-RR-92-43, 1992.

[LoBe93] Löwe, Michael; Beyer, Martin: 'AGG - An Implementation of Algebraic Graph Rewriting' in Kirchner, Claude (ed.) Rewriting Techniques and Applications, Montreal, Canada, June 16-18, 1993, LNCS 690, Springer, Berlin, 1993, pp.451-456.

[Nag79] Nagl, Manfred: 'Graph-Grammatiken, Theorie, Implementierung, Anwendungen'; Vieweg, Braunschweig, 1979.

[Schü91] Schürr, Andreas: 'Operationales Spezifizieren mit programmierten Graphersetzungssystemen', Deutscher Universitäts-Verlag, Wiesbaden, 1991.

[Wit81] Witt, Kurt-Ulrich: 'On linearizing graphs', in Noltemeier, Hartmut (ed.) Graphtheoretic Concepts on Computer Science, WG '80, Bad Honnef, LNCS 100, Springer, Berlin, 1981, pp.32-41.

[Zün92] Zündorf, Albert: 'Implementation of the imperative/rule based language PROGRES', Aachener Informatik-Berichte Nr. 92-38, RWTH Fachgruppe Informatik, Aachen, 1992.

Efficient Algorithms
for a Mixed k-partition Problem
of Graphs without Specifying Bases

Koichi Wada*, Akinari Takaki and Kimio Kawaguchi

Nagoya Institute of Technology
Gokiso-cho, Syowa-ku, Nagoya 466, JAPAN
e-mail:(wada,akinari,kawaguch)@elcom.nitech.ac.jp

Abstract. This paper describes efficient algorithms for partitioning a k-edge-connected graph into k edge-disjoint connected subgraphs, each of which has a specified number of elements(vertices and edges). If each subgraph contains the specified element(called base), we call this problem the mixed k-partition problem with bases(called k-PART-WB), otherwise we call it the mixed k-partition problem without bases(called k-PART-WOB). In this paper, we show that k-PART-WB always has a solution for every k-edge-connected graph and we consider the problem without bases and we obtain the following results: (1)for any $k \geq 2$, k-PART-WOB can be solved in $O(|V|\sqrt{|V|\log_2|V|} + |E|)$ time for every 4-edge-connected graph $G = (V, E)$, (2)3-PART-WOB can be solved in $O(|V|^2)$ for every 2-edge-connected graph $G = (V, E)$ and (3)4-PART-WOB can be solved in $O(|E|^2)$ for every 3-edge-connected graph $G = (V, E)$.

1 Introduction

In this paper, we consider the following k-partition problem.

Input:

(1) an undirected graph $G = (V, E)$ with $n = |V|$ vertices and $m = |E|$ edges;

(2) $S \subseteq (V \cup E)(|S| \geq k)$;

(3) k distinct vertices and/or edges $a_i(1 \leq i \leq k) \in S$; and

(4) k natural numbers $n_1, n_2, \ldots, n_k$ such that $\sum_{i=1}^{k} n_i = |S|$.

Output: a partition $S_1 \cup S_2 \cup \ldots \cup S_k$ of the specified set S such that for each $i(1 \leq i \leq k)$

(a) $a_i \in S_i$;

(b) $|S_i| = n_i$; and

(c) there is a connected subgraph $G_i = (V_i, E_i)$ of G such that $S_i \subseteq (V_i \cup E_i)$ and $G_1, G_2, \ldots G_k$ are mutually edge-disjoint.

---

* Partially supported by the Grant-in-Aid of Scientific Research of the Ministry of Education, Science and Culture of Japan under Grant: (C)05680271 and the Okawa Institute of Information and Telecommunication(94-11).

The problem is called the mixed k-partition problem with respect to edge-disjointness and it is simply called the mixed k-partition problem unless confusion arises. Each a_i is called a base of the subgraph G_i and if all bases are not specified for the mixed k-partition problem, the problem is called the mixed k-partition problem without bases.

In the mixed k-partition problem, if $S = E$ then the problem corresponds to the k-edge-partition problem[6] and if $S \subseteq V$ then the problem corresponds to the k-vertex-partition problem with respect to edge-disjointness [15]. The mixed k-partition problem becomes the k-vertex-partition problem[6, 10] (the vertex-subset k-partition problem[14]) if $S = V$ ($S \subseteq V$) and the condition "edge-disjointness" in (c) is replaced by "vertex-disjointness".

It has been shown that the k-edge-partition problem and the k-vertex-partition problem with respect to edge-disjointness always have solutions for every k-edge-connected graph [6, 15] and the vertex-subset k-partition problem has a solution for every k-connected graph [14]. Although efficient algorithms are known for these problems provided that k is limited to 2 and 3 [12, 15], no polynomial algorithms are known so far as $k \geq 4$. In order to construct highly fault-tolerant routings in the surviving-route-graph model, it is necessary to solve efficiently the vertex-subset k-partition problem for k-connected graphs or the k-vertex-partition problem with respect to edge-disjointness for k-edge-connected graphs [8, 13]. Since it is shown that it is sufficient to solve these problems without bases in order to define such routings, we would like to obtain efficient solutions for these problems without bases for the cases that $k \geq 4$.

It has been also shown that the set of graphs for which the vertex-subset k-partition problem has a solution for any instance equals to the set of all k-connected graphs and that the set of graphs for which the k-edge-partition problem has a solution for any instance equals to the set of all k-edge-connected graphs [6, 15]. It is an interesting graph-theoretic question to reveal relationship between the number of partitions for these problems and the vertex- and/or edge-connectivity of input graphs for the cases without specifying bases.

In this paper, we show that the mixed k-partition problem with bases always has a solution for every k-edge-connected graph and it is computed efficiently for the cases that $k = 2$ and 3. Since the mixed k-partition problem includes the k-edge-partition problem as a special case, we can also show that the set of graphs for which the mixed k-partition problem has a solution for any instance equals to the set of all k-edge-connected graphs. Furthermore, we consider the mixed k-partition problem without bases and we obtain the following results:

1. For any $k \geq 2$, the mixed k-partition problem without bases can be solved in $O(|V|\sqrt{|V|}\log_2|V|+|E|)$ time for every 4-edge-connected graph $G = (V,E)$.
2. The mixed tripartition problem without bases can be solved in $O(|V|^2)$ time for every 2-edge-connected graph $G = (V,E)$.
3. The mixed 4-partition problem without bases can be solved in $O(|E|^2)$ time for every 3-edge-connected graph $G = (V,E)$.

For the mixed tripartition problem without bases, we show that some special cases are solved more efficiently. If the specified set S is a superset of V or E, it is solved in $O(|E| + min(n_1, n_2, n_3) \cdot |V|)$ time.

2 Preliminaries

2.1 Definitions

We deal with a connected undirected graph $G = (V, E)$ with a vertex set V and an edge set E. For a graph G, the vertex set is denoted by $V(G)$ and the edge set is denoted by $E(G)$. For a graph $G = (V, E)$ and a vertex subset V', the *induced subgraph* is denoted by $G[V']$. For two graphs $G = (V, E)$ and $G' = (V', E')$, the graph $(V \cup V', E \cup E')$ is denoted by $G \cup G'$. For a graph $G = (V, E)$ and a set E' of edges, the graph $(V, E - E')$ is denoted by $G - E'$, and if $E' = \{e\}$ then it is denoted by $G - e$.

A *cut-vertex* of G is a vertex whose removal disconnects G. A *bridge* of G is an edge whose removal disconnects G. A biconnected component of G is a maximal set of edges such that any two edges in the set lie on a common cycle. A *block* of G is a bridge or a biconnected component of G. An *Eulerian cycle* of a connected graph G is a cycle that traverses each edge of G exactly once, although it may visit a vertex more than once. We often treat a path, a cycle, a block and etc. as graphs. A graph G is *k-connected(k-edge-connected)* if there exist k internally node-disjoint(edge-disjoint) paths between every pair of distinct nodes in G. A k-connected graph $G = (V, E)$ is *minimal* if for any edge $e \in E$, $G - e$ is not k-connected. Usually 2-connected graphs are called *biconnected graphs* and 3-connected graphs are called *triconnected graphs*.

2.2 Graph Transformations

We define two kinds of graph transformations which are utilized in this paper to transform k-edge-connected graphs into k-connected graphs.

Let $k \geq 2$ and let $\ell = max(k - 2, 1)$. Given a graph $G = (V, E)$, define the graph $\varphi_k(G) = (\varphi(V), \varphi(E))$ as follows. For every vertex $v \in V$, there are ℓ vertices $\varphi(v_1), \varphi(v_2), \ldots, \varphi(v_\ell)$ in $\varphi(V)$. For every edge $e \in E$, there is a vertex $\varphi(e)$ in $\varphi(V)$.

The edge set $\varphi(E)$ is defined as follows: Let v be any vertex in V and $u_0, u_1, \ldots, u_{d-1}$ be the vertices adjacent to v. Let $e_i = (v, u_i)(0 \leq i \leq d - 1)$. Then there are edges $(\varphi(e_i), \varphi(e_{(i+1)mod\ d}))(0 \leq i \leq d-1)$ and $(\varphi(e_i), \varphi(v_j))(0 \leq i \leq d - 1, 1 \leq j \leq \ell)$ in $\varphi(E)$. Note that if $d = 2$, there is an edge $(\varphi(e_0), \varphi(e_1))$ in $\varphi(E)$.

From the definition $\varphi_k(G)$ has $\ell|V| + |E|$ vertices and $(2k - 2)|E|$ edges and it can be computed in $O(k(|V| + |E|))$.

Proposition 1. [5][2] *For any $k(\geq 2)$, G is k-edge-connected if and only if $\varphi_k(G)$ is k-connected.*

---

[2] In [5] the cases for $k \geq 3$ are shown and the case for $k = 2$ can be shown similarly.

Let $G' = (V', E')$ be a subgraph of $\varphi_k(G)$. A subgraph
$\varphi_k^{-1}(G') = (\varphi^{-1}(V'), \varphi^{-1}(E'))$ of $G = (V, E)$ is defined to be
$\varphi^{-1}(V') = \{v|\varphi(v_i) \in V' \text{ and } v \in V\} \cup \{\text{ endvertices of } e|\varphi(e) \in V' \text{ and } e \in E\}$
and
$\varphi^{-1}(E') = \{e|\varphi(e) \in V' \text{ and } e \in E\}$.

The subgraph $\varphi_k^{-1}(G')$ can be computed in $O(|E'|)$ time and it has the following properties.

Lemma 2. [16] *Let $k \geq 2$. (a)If a subgraph $G' = (V', E')$ of $\varphi_k(G)$ is connected, then the graph $\varphi_k^{-1}(G')$ is connected.*

(b)If subgraphs $G' = (V', E')$ and $G'' = (V'', E'')$ of $\varphi_k(G)$ are vertex-disjoint, then the graphs $\varphi_k^{-1}(G')$ and $\varphi_k^{-1}(G'')$ are edge-disjoint.

Given a 2-edge-connected graph $G = (V, E)$, define the graph $\psi_2(G) = (\psi(V), \psi(E))$ as follows. Let C be a set of cut-vertices in G. Let c be a cut-vertex in G such that ℓ blocks $B_1, B_2, \ldots, B_\ell$ contain c in common and let $b_{ij}(1 \leq j \leq d_i)$ denote the adjacent vertices to c in B_i.
$\psi(V) = V \cup \{c_i|1 \leq i \leq \ell - 1, c \in C\}$.
$\psi(E) = (E - \{(c, b_{id_i}), (c, b_{(i+1)1})|1 \leq i \leq \ell - 1, c \in C\})$
$\qquad \cup (\cup_{c \in C}(\{(b_{id_i}, c_i), (c_i, b_{(i+1)1})|1 \leq i \leq \ell - 1\})$

The vertices $c_i(1 \leq i \leq \ell - 1)$ are called the duplicate vertices for c. If G has p blocks, $\psi_2(G)$ has $|V| + p - 1 = O(|V|)$ vertices and $|E|$ edges and it can be computed in $O(|E|)$.

For $k \geq 3$, the graph $\psi_k(G) = (\psi(V), \psi(E))$ is defined as follows. For every vertex $v \in V$, there are $k - 2$ vertices $\psi(v_1), \psi(v_2), \ldots, \psi(v_{k-2})$ in $\psi(V)$. For every edge $e = (u, v) \in E$, there are two vertices $\psi(e_u)$ and $\psi(e_v)$ in $\psi(V)$. Therefore, for a vertex v of degree d, $d + k - 2$ vertices are created in $\psi_k(G)$. These vertices are called the duplicate vertices for v.

The edge set $\psi(E)$ is defined as follows: For every edge $e = (u, v) \in E$, there is an edge $(\psi(e_u), \psi(e_v)) \in \psi(E)$. Let v be any vertex in V and $u_0, u_1, \ldots, u_{d-1}$ be the vertices adjacent to v. Let $e_i = (v, u_i)(0 \leq i \leq d - 1)$. Then there are edges $(\psi((e_i)_v), \psi((e_{(i+1)\mod d})_v))((0 \leq i \leq d - 1)$ and $(\psi(e_i)_v), \psi(v_j))(0 \leq i \leq d - 1, 1 \leq j \leq k - 2)$ in $\psi(E)$.

From the definition $\psi_k(G)(k \geq 3)$ has $2|E| + (k-2)|V|$ vertices and $(2k-1)|E|$ edges and it can be computed in $O(k(|V| + |E|))$.

Lemma 3. [16] *For any $k(\geq 2)$, G is k-edge-connected if and only if $\psi_k(G)$ is k-connected.*

Similarly to $\varphi_k(G)$, an inverse graph of $\psi_k(G)(k \geq 2)$ is defined as follows:
Let $G' = (V', E')$ be a subgraph of $\psi_k(G)$. A subgraph
$\psi_k^{-1}(G') = (\psi^{-1}(V'), \psi^{-1}(E'))$ of $G = (V, E)$ is defined as follows.

Let U be V' in which duplicate vertices for v are changed into one original vertex v and let $G[U] = (U, F)$ denote the induced subgraph. Then
$\psi^{-1}(V') = U$,
$\psi^{-1}(E') = \{(u, v)|(u, v) \in F \text{ and } (u', v') \in E'$
$\qquad$ where u' and v' are some duplicate vertices for u and $v\}$.

The subgraph $\psi_k^{-1}(G')$ can be computed in $O(|E'|)$ time and it has the following properties.

Lemma 4. [16] *Let $k \geq 2$. (a)If a subgraph $G' = (V', E')$ of $\psi_k(G)$ is connected, then the graph $\psi_k^{-1}(G')$ is connected.*

(b)If subgraphs $G' = (V', E')$ and $G'' = (V'', E'')$ of $\psi_k(G)$ are edge-disjoint, then the graphs $\psi_k^{-1}(G')$ and $\psi_k^{-1}(G'')$ are edge-disjoint.

2.3 Nonseparating Ear Decomposition

An *ear decomposition* of a biconnected graph $G = (V, E)$ is a decomposition $G = P_0 \cup P_1 \cup \ldots \cup P_q$, where P_0 is a cycle and $P_i(1 \leq i \leq q)$ is a path whose end vertices are distinct and the vertices in common with $P_0 \cup \ldots \cup P_{i-1}$ are the end vertices. Each P_i is called an open ear. Note that $P_0 \cup P_1 \cup \ldots \cup P_i$ is biconnected for each $i(1 \leq i \leq q)$.

Given an ear decomposition $P_0 \cup P_1 \cup \ldots \cup P_q$ of G, let $V_i = V(P_0) \cup \ldots \cup V(P_i)$, let $G_i = G[V_i]$ and let $\overline{G_i} = G[V - V_i]$ for each $i(1 \leq i \leq q)$.

We say that $G = P_0 \cup P_1 \cup \ldots \cup P_q$ is an *ear decomposition through edge (a, b) and avoiding vertex c*, if the cycle P_0 contains the edge (a, b) and the last ear is of length 2 and has c as its only internal vertex.

An ear decomposition $P_0 \cup P_1 \cup \ldots \cup P_q$ of a graph G through edge (a, b) and avoiding vertex c is *nonseparating* if for all $i(1 \leq i < q)$, each graph $\overline{G_i}$ is connected and each internal vertex of the ear P_i has a neighbor in $\overline{G_i}$.

Proposition 5. [2] *For a triconnected graph $G = (V, E)$, any edge $(a, b) \in E$ and any vertex $c(\neq a, b) \in V$, a nonseparating ear decomposition $P_0 \cup P_1 \cup \ldots \cup P_q$ through edge (a, b) and avoiding vertex c can be constructed in $O(|V| \cdot |E|)$, where the path P_q has the vertex c as its only internal one. In particular, the cycle P_0 and each path $P_i(1 \leq i \leq q)$ can be constructed in $O(|E|)$.*

3 The Mixed k-partition Problem for k-edge-connected Graphs

Proposition 6. [15] *(a)For any integer $k(\geq 2)$, the vertex-subset k-partition problem has a solution for every k-connected graph.*

(b)For a biconnected graph $G = (V, E)$, the vertex-subset bipartition problem can be solved in $O(|E|)$ time.

(c)For a triconnected graph $G = (V, E)$, the vertex-subset tripartition problem can be solved in $O(|V|^2)$ time.

We obtain the next theorem by using Proposition 1, 6 and Lemma 2.

Theorem 7. [16] *For any integer $k(\geq 2)$, the mixed k-partition problem has a solution for every k-edge-connected graph.*

Since the transformations $\varphi_2(G)$ and $\varphi_3(G)$ and the inverse ones $\varphi_2^{-1}(G')$ and $\varphi_3^{-1}(G')$ can be computed in $O(|E|)$ time, the next theorem holds from Theorem 7 and Proposition 6(b) and (c). Note that $|V(\varphi_3(G))| = O(|E|)$.

Theorem 8. *(a)For a 2-edge-connected graph $G = (V, E)$, the mixed bipartition problem with bases can be solved in $O(|E|)$ time.*

(b)For a 3-edge-connected graph $G = (V, E)$, the mixed tripartition problem with bases can be solved in $O(|E|^2)$ time.

4 The Mixed k-partition Problem Without Bases

In order to make algorithms shown here efficient, we transform the mixed k-partition problem without bases into a modified one in which the input graph $G = (V, E)$ is converted to a spanning subgraph $G = (V, E')$(E' is desirable to be sparse) with a weight function $w : V \to N$, the k numbers $n_i(1 \le i \le k)$ are given such that $\Sigma_{i=1}^k n_i = |S| + \Sigma_{v \in V} w(v)$ and the output is mutually edge-disjoint subgraphs $G_i(1 \le i \le k)$ and a weight function $w_i : V(G_i) \to N(1 \le i \le k)$ such that $n_i = |(V(G_i) \cup E(G_i)) \cap S| + \Sigma_{v \in V(G_i)} w_i(v)(1 \le i \le k)$ and $\Sigma_{v \in V} w_i(v) = w(v)$ for every $v \in V$. If a vertex v is contained in both G_i and G_j, the weight $w(v)$ is divided into two weights $w_i(v)$ and $w_j(v)$ such that $w_i(v) + w_j(v) = w(v)$. For a weight function w, a vertex set U and an edge set F, we define $w(U \cup F)$ as $|U \cup F| + \Sigma_{u \in U} w(u)$.

The mixed k-partition problem without bases can be transformed into the modified one as follows. For the inputs $G = (V, E)$, $S(\subseteq V \cup E)$ and $n_i(1 \le i \le k)$, we construct a spanning subgraph $G' = (V, E')$, and $R(v)(v \in V)$ which is a set of edges in $S \cap (E - E')$ such that each $(u, v) \in S \cap (E - E')$ occurs once in either $R(u)$ or $R(v)$ and we define the weight function w as $w(v) = |R(v)|$. If this transformation takes $O(|E|)$ time and the modified version can be solved in $O(T_m)$ time for $G' = (V, E')$, $S \cap (V \cup E')$, w and $n_i(1 \le i \le k)$, the original one can be solved in $O(|E| + T_m)$ time. This transformation is effective for the mixed tri- and 4-partition problems shown in Section 5 and 6.

4.1 An Algorithm for the Mixed k-partition Problem Without Bases

If the input graph G includes an Eulerian cycle as a spanning subgraph, we can show that $G = (V, E)$ has a solution of the mixed k-partition problem without bases by using a similar method in [9].

Let $(x_0, x_1, \ldots, x_r)(r \ge |V|)$ be a spanning Eulerian cycle of G and let $R(v)(v \in V)$ be a set of edges that are not contained in the Eulerian cycle and are contained in S. In order to solve the original mixed k-partition problem without bases, it is sufficient to solve the modified one in which the input graph is the spanning Eulerian cycle of G with the weight function w as $w(v) = |R(v)|$.

Since for any $i(0 \le i \le r)$ two subgraphs $G_1 = (x_0, x_1, \ldots, x_i)$ and $G_2 = (x_i, x_{i+1}, \ldots, x_r)$ are connected and mutually edge-disjoint and arbitrary values can be assigned to $w_1(x)$ and $w_2(x)$, G can be partitioned into k mutually edge-disjoint subgraphs each of which are connected and contains the specified weight by using the Eulerian cycle and w.

Theorem 9. *Let $k \geq 2$. If $G = (V, E)$ has an Eulerian cycle as a spanning subgraph, the mixed k-partition problem without bases can be solved in $O(T_{ec}(G) + |E|)$ time, where $T_{ec}(G)$ is a computation time to find a spanning Eulerian cycle in G.*

It is known that if G is 4-edge-connected, then G has a spanning Eulerian cycle. A spanning Eulerian cycle G_{ec} for a 4-edge-connected graph G can be obtained as follows [9].

1. Since there are two edge-disjoint spanning trees for a 4-edge-connected graph, let T_1 and T_2 be such trees.
2. Let W be a set of vertices whose degree in T_1 are odd. (Note that $|W|$ is even.)
3. Let $(v_i, u_i)(1 \leq i \leq |W|/2)$ be $|W|/2$ pairs of vertices in W.
4. For each pair $(v_i, u_i)(1 \leq i \leq |W|/2)$, let P_i be a simple path between v_i and u_i in T_2 and let $E(P) = E(P_1) \cup \ldots \cup E(P_{|W|/2})$.
5. $G_{ec} = (V, E(T_1) \cup \{e | e \text{ appears in } E(P) \text{ odd times } \})$.

It is obvious that G_{ec} is a spanning Eulerian cycle of G from the fact G_{ec} is an Eulerian cycle iff every vertex in G_{ec} has even degree. For a 4-edge-connected graph two edge-disjoint spanning trees can be computed in $O(|V|\sqrt{|V|}\log_2 |V| + |E|)$ time [4]. Since other operations in the above procedure can be done in $O(|V|)$ time, the next corollary can be obtained.

Corollary 10. *Let $k \geq 2$. If $G = (V, E)$ is 4-edge-connected graph, the mixed k-partition problem without bases can be solved in $O(|V|\sqrt{|V|}\log_2 |V| + |E|)$ time.*

5 The Mixed Tripartition Problem Without Bases

In this section, we prove that the mixed tripartition problem without bases has a solution for every 2-edge-connected graph. We show an algorithm for the modified version in which the input graph is a minimal biconnected graph $G = (V, E)$ with a weight function $w : V \to N$ mentioned above. We can solve the mixed tripartition problem without bases for a biconnected graph with the algorithm. Using the algorithm for a biconnected graph as a subroutine and the graph transformation ψ_2, we solve the problem for a 2-edge-connected graph.

5.1 An Algorithm for a Minimal Biconnected Graph

Minimal biconnected graphs have the following property.

Proposition 11. [1] *Let $G = (V, E)$ be a minimal biconnected graph and let (x, y) be any edge in E. Then $G - (x, y)$ can be represented by blocks $B_i (0 \leq i \leq q - 1)$ and cutvertices $c_i (1 \leq q - 1)$ such that*

(1) B_i *is either a bridge or a minimal biconnected graph,*
(2) x *is contained in B_0 and y is contained in B_{q-1} and*
(3) *Only B_{i-1} and B_i contain c_i.*

In other words, $G - (x, y)$ is a linear structure of blocks.

An algorithm (called M-PART3-WOB) of the modified mixed tripartition problem without bases makes use of a linear structure of a minimal biconnected graph shown in Proposition 11. M-PART3-WOB has mainly three cases. Since each block B_i can be bipartitioned into two connected graphs B_{i1} and B_{i2} of proper weights such that $c_i \in B_{i1}$ and $c_{i+1} \in B_{i2}$, if there are no cases that one block is partitioned into three pieces, desired tripartition can be done. For the case that one block is partitioned into three pieces, M-PART3-WOB is called for the block which is a bridge or a minimal biconnected graph from Proposition 11 (1) recursively. In M-PART3-WOB, one vertex a_1 can be specified to be contained in one subgraph G_1. This enables us to call M-PART3-WOB recursively.

The algorithm M-PART3-WOB as follows: Its input graph is either a bridge or a minimal biconnected graph. In M-PART3-WOB, two functions M-PART2 and COUNT are used. M-PART2 solves the modified mixed bipartition problem for a biconnected graph(or a bridge) with a weight function $w : V \rightarrow N$ and returns two edge-disjoint subgraphs and weight functions w_1 and w_2. Using a similar algorithm shown in [15], we can show that the modified mixed bipartition problem can be solved for a biconnected graph(or a bridge) with a weight function $w : V \rightarrow N$. The other function $COUNT(B_i^j)$ computes $w(V(B_i^j) \cup E(B_i^j) - \{c_{j+1}\})$, where B_i^j denotes $B_i \cup \ldots \cup B_j$. In M-PART3-WOB the returned weight functions are not described since they can be easily filled out.

Algorithm M-PART3-WOB$(G = (V, E); a_1; n_1, n_2, n_3; S; w)$
    begin if G is a bridge **then** it can be easily solved **else begin**
    Let $e = (a_1, x) \in E$ for some $x \in V$
    and let $B_i (0 \leq i \leq q - 1)$ be the blocks of $G - e$
    and let $c_i (1 \leq i \leq q - 1)$ be cut-vertices between B_{i-1} and B_i and
    define B_q to be the graph consisting of only $e = (a_1, x)$
    and define $c_0 = a_1$ and $c_q = x$ for convenience;
    $B_i \cup B_{i+1} \cup \ldots \cup B_j$ is denoted by B_i^j;
    Let ℓ_{12} be the least number ℓ such that $n_1 \leq COUNT(B_0^\ell)$ and
    let ℓ_{23} be the least number ℓ such that $n_1 + n_2 \leq COUNT(B_0^\ell)$;
    if $\ell_{12} \neq \ell_{23}$ **then**
        begin
            $n_1' \leftarrow n_1 - COUNT(B_0^{\ell_{12}-1});$
            $n_2' \leftarrow n_1 + n_2 - COUNT(B_0^{\ell_{23}-1});$

$$(G_1', w_1'; G_2', w_2')$$
$$\leftarrow \text{M-PART2}(B_{\ell_{12}}; c_{\ell_{12}}, c_{\ell_{12}+1}; n_1', COUNT(B_{\ell_{12}}) - n_1';$$
$$S \cap (V(B_{\ell_{12}}) \cup E(B_{\ell_{12}})); w);$$
$$(G_2'', w_2''; G_3', w_3')$$
$$\leftarrow \text{M-PART2}(B_{\ell_{23}}; c_{\ell_{23}}, c_{\ell_{23}+1}; n_2', COUNT(B_{\ell_{23}}) - n_2';$$
$$S \cap (V(B_{\ell_{23}}) \cup E(B_{\ell_{23}}); w);$$
$$\textbf{return}(B_0^{\ell_{12}-1} \cup G_1', G_2' \cup B_{\ell_{12}+1}^{\ell_{23}-1} \cup G_2'', G_3' \cup B_{\ell_{23}+1}^q)$$
$\quad$ **end**

$\quad$ **else if** $\ell_{12} = \ell_{23}$ and $n_2 + n_3 \geq COUNT(B_{\ell_{12}})$ **then**

$\qquad$ **begin**

$\qquad$ Let ℓ_{31} be the least number ℓ such that $n_2 + n_3 \leq COUNT(B_{\ell_{12}}^{\ell})$;
$$n_3' \leftarrow (n_2 + n_3) - COUNT(B_{\ell_{12}}^{\ell_{31}-1});$$
$$(G_2, w_2'; G_3', w_3')$$
$$\leftarrow \text{M-PART2}(B_{\ell_{12}}; c_{\ell_{12}}, c_{\ell_{12}+1}; n_2, COUNT(B_{\ell_{12}}) - n_2;$$
$$S \cap (V(B_{\ell_{12}}) \cup E(B_{\ell_{12}})); w);$$
$$(G_3'', w_3''; G_1', w_1')$$
$$\leftarrow \text{M-PART2}(B_{\ell_{31}}; c_{\ell_{31}}, c_{\ell_{31}+1}; n_3', COUNT(B_{\ell_{31}}) - n_3';$$
$$S \cap (V(B_{\ell_{31}} \cup E(B_{\ell_{31}}))); w);$$
$$\textbf{return}(B_0^{\ell_{12}-1} \cup G_1' \cup B_{\ell_{31}+1}^q, G_2, G_3' \cup B_{\ell_{12}+1}^{\ell_{31}-1} \cup G_3'')$$
$\qquad$ **end**

$\quad$ **else if** $\ell_{12} = \ell_{23}$ and $n_2 + n_3 < COUNT(B_{\ell_{12}})$ **then**

$\qquad$ **begin**
$$n_1' \leftarrow COUNT(B_{\ell_{12}}) - (n_2 + n_3);$$
$$(G_1', w_1'; G_2, w_2; G_3, w_3) \leftarrow \text{M-PART3-WOB}$$
$$(B_{\ell_{12}}; c_{\ell_{12}+1}; n_1', n_2, n_3; S \cap (V(B_{\ell_{12}} \cup E(B_{\ell_{12}}))); w);$$
$$\textbf{return}(B_0^{\ell_{12}-1} \cup G_1' \cup B_{\ell_{12}+1}^q, G_2, G_3)$$
$\qquad$ **end end**

$\quad$ **end**

Theorem 12. [16] *For a bridge or a minimal biconnected graph $G = (V, E)$ the modified mixed tripartition problem without bases can be solved in $O(|E|^2) = O(|V|^2)$ time. If $S \supseteq V$ or $S \supseteq E$ then it can be solved in $O(min(n_1, n_2, n_3)|E|) = O(min(n_1, n_2, n_3)|V|)$ time.*

For a biconnected graph $G = (V, E)$, we construct a minimal biconnected graph $G_m = (V, E_m)$ and edge sets $R(v) = \{(v, u)|(v, u) \in S$ and $(v, u) \in E - E_m\}$ for each $v(\in V)$ such that each $e \in E - E_m$ occurs in all $R(v)$ just once. We define the weight function $w : V \rightarrow N$ to be $w(v) = |R(v)|(v \in V)$. Since a minimal biconnected graph can be computed in $O(|E|)$ time [7], $G_m, R(v)$ and w can be constructed in $O(|E|)$ time. For a biconnected graph G, the mixed tripartition problem without bases is solved as follows. We call M-PART3-WOB with G_m and w and let G_1, G_2, G_3 be subgraphs returned by M-PART3-WOB. We construct $G_i'(1 \leq i \leq 3)$ as adding G_i to proper vertices and edges for $v \in V(G_i)$ having non-zero weight $w_i(v)$, which are a solution for a biconnected graph and this process can be done in $O(|E|)$ time. We can also show that we solve the modified mixed tripartition problem without bases for a biconnected graph in $O(|V|^2)$ time. This algorithm is referred as M-PART3-WOB'.

Theorem 13. *For a biconnected graph $G = (V, E)$ the modified mixed tripartition problem without bases can be solved in $O(|V|^2)$ time. If $S \supseteq V$ or $S \supseteq E$ then it can be solved in $O(|E| + min(n_1, n_2, n_3)|V|)$).*

5.2 An Algorithm for a 2-edge-connected graph

The mixed tripartition problem without bases for a 2-edge-connected graph $G = (V, E)$ can be solved by using the algorithm for a biconnected graph and the graph transformation ψ_2. Since $\psi_2(G)$ is constructed in $O(|E|)$ time and $|V(\psi_2(G))| = O(|V|)$, it can be solved in $O(|V|^2)$ time. Furthermore, it can be solved in $O(|E| + min(n_1, n_2, n_3)|V|)$) time if $S \supseteq E$, because there is one-to-one correspondence between E and $E(\psi_2(G))$. Since there is no one-to-one correspondence between V and $V(\psi_2(G))$, if $S \supseteq V$, it is not solved in $O(|E| + min(n_1, n_2, n_3)|V|)$) by using ψ_2. However, by using the graph $\psi_2'(G)$ which is a modification of $\psi_2(G)$ and which has the vertex set V and has an edge $(b_{id_i}, b_{(i+1)1})$ instead of $(b_{id_i}, c_i), (c_i, b_{(i+1)1})$ in $\psi_2(G)$, it can be solved in $O(|E| + min(n_1, n_2, n_3)|V|)$) time for $S \supseteq V$. Lemmas 3 and 4 also hold for $\psi_2'(G)$ and there is one-to-one correspondence between V and $V(\psi_2'(G))$.

Theorem 14. *The mixed tripartition problem for 2-edge-connected graph $G = (V, E)$ can be solved in $O(|V|^2)$ time. In particular, if $S \supseteq V$ or $S \supseteq E$ then it can be solved in $O(|E| + min(n_1, n_2, n_3)|V|)$ time.*

It is not known so far whether there are 2-edge-connected graphs which cannot be 4-partitioned without bases or not. However under the assumption that one base can be specified to be contained in a specified subgraph,[3] we show that there is a 2-edge-connected graph(in fact a biconnected graph) which cannot be 4-partitioned without specifying bases. We can show an example of an instance which cannot be 4-partitioned under the assumption [16]. This assumption is reasonable because all algorithms shown in this paper satisfy the assumption. The next section will show that every 3-edge-connected graph can be 4-partitioned under the assumption.

6 The Mixed 4-partition Problem Without Bases

In this section, we prove that the mixed 4-partition problem without bases has a solution for every 3-edge-connected graph by using the algorithm M-PART3-WOB' which solves the modified version for a biconnected graph, a nonseparating ear decomposition for a triconnected graph and the graph transformation ψ_3 from a 3-edge-connected graph to a triconnected graph. Similarly to the case of

---

[3] Although in M-PART3-WOB one specified base is limited to a vertex, it may be an edge.

the mixed tripartition problem, we first show an algorithm of the modified problem for a sparse triconnected graph with a weight function $w : V \to N$ which can be used by solving the original mixed 4-partition problem for a triconnected graph and we extend the algorithm for a 3-edge-connected graph with the graph transformation ψ_3.

6.1 An Algorithm for a Triconnected Graph

An algorithm (denoted by M-PART4-WOB) of the modified mixed 4-partition problem for a sparse triconnected graph G is based on the following idea. Let $G' = (V'E')$ be a biconnected graph and let $P = (\{x_i | 0 \le i \le \ell\}, \{(x_i, x_{i+1}) | 0 \le i \le \ell-1\})$ be a path graph such that either x_0 or x_ℓ is contained in V'. By using the algorithm of M-PART3-WOB' we can solve the modified mixed tripartition problem for the graph $G' \cup P$, since one vertex(x_0 or x_ℓ) can be specified to be contained in one subgraph in M-PART3-WOB'. For a nonseparating ear decomposition $P_0 \cup P_1 \cup \ldots \cup P_q$ of the graph G through an edge (u, v) and avoiding a vertex a_1 for some edge (u, v) and a vertex a_1, we obtain the biconnected graph $P_0 \cup \ldots \cup P_{i-1}$ and the path graph P_i. The algorithm M-PART3-WOB' can be applied to the union of these graphs to obtain three edge-disjoint subgraphs each of which contains the specified weights. Moreover, the remaining graph $\overline{G_i}$ is connected and contains a_1. Therefore, we can solve the modified mixed 4-partition problem without bases for a triconnected graph so that one of edge-disjoint subgraphs contains the specified vertex a_1. For lack of space, the detailed description of M-PART4-WOB is omitted. It is shown in [16].

The original mixed 4-partition problem for a triconnected graph $G = (V, E)$ can be solved similarly to the case for a biconnected graph by constructing a sparse spanning subgraph $G' = (V, E')$ of G such that G' is triconnected and $|E'| = O(|V|)$ and edge sets $R(v) = \{(v, u) | (v, u) \in S$ and $(v, u) \in E - E'\}$ for each $v(\in V)$ such that each $e \in E - E'$ occurs in R just once. Since a sparse triconnected graph can be computed in $O(|E|)$ time [11], the next theorem can be obtained.

Theorem 15. *The mixed 4-partition problem without bases for a triconnected graph $G = (V, E)$ can be solved in $O(|V|^2)$. In particular, if $S \supseteq V$ or $S \supseteq E$ then it can be solved in $O(|E| + (|S| - max_{1 \le i \le 4}(n_i))|V|)$ time.*

6.2 An Algorithm for a 3-edge-connected Graph

The mixed 4-partition problem without bases for a 3-edge-connected graph can be solved by using the algorithm for a triconnected graph and the graph transformation ψ_3.

Theorem 16. *The mixed 4-partition problem without bases for a 3-edge-connected graph $G = (V, E)$ can be solved in $O(|E|^2)$.*

References

1. B.Bollobás: Extremal graph theory, Academic Press, 11-13 (1978).
2. J. Cherian and S.N. Maheshwari : "Finding nonseparating induced cycles and independent spanning trees in 3-connected graphs," *Journal of Algorithms*, 9, 507-537 (1988).
3. S.Even: Graph algorithms, Computer Science Press, Potomac, MD (1979).
4. H.N.Gabow and H.H.Westermann: "Forests frames and games: algorithms for matroid sums and applications," Algorithmica, 7, 5/6, 465-497 (1992).
5. Z. Galil and G.F. Italiano: "Reducing edge connectivity to vertex connectivity," *SIGACT NEWS*, 22, 1, 57-61 (1991).
6. E. Györi: "On division of connected subgraphs," in: Combinatorics(Proc. 5th Hungarian Combinational Coll., 1976, Keszthely) North-Holland, Amsterdam, 485-494 (1978).
7. X.Han, P.Kelsen, V.Ramachandran and R.Tarjan: "Computing minimal spanning subgraphs in linear time," Proc. of 3rd SODA, 146-156 (1992).
8. M.Imase and Y.Manabe: "Fault tolerant routings in a κ-connected network," *Information Processing Letters*, 28, 4, 171-175 (1988).
9. M.Junger, G.Reinelt and W.R.Pulleyblank: "On the partitioning the edges of graphs into connected subgraphs," *J. of Graph Theory*, 9, 539-549 (1985).
10. L. Lovász: "A homology theory for spanning trees of a graph," *Acta math. Acad. Sci. Hunger*, 30, 241-251 (1977).
11. H.Nagamochi and T.Ibaraki : "A linear-time algorithm for finding a sparse k-connected spanning subgraph of a k-connected graph," *Algorithmica*, 7, 5/6, 583-596 (1992).
12. H.Suzuki, N.Takahashi and T.Nishizeki: "A linear algorithm for bipartition of biconnected graphs," *Information Processing Letters*, 33, 5 (1990).
13. K.Wada, T.Shibuya, E.Shamoto and K.Kawaguchi: "A linear time (L, k)-edge-partition algorithm for connected graphs and fault-tolerant routings for k-edge-connected graphs," *Trans. of IEICE*, J75-D-I, 11, 993-1004 (1992).
14. K.Wada, K.Kawaguchi and N.Yokoyama: "On a generalized k-partition problem for graphs," Proc. of the 6th Karuizawa Workshop on Circuit and Systems, 243-248 (1993).
15. K.Wada and K.Kawaguchi: "Efficient algorithms for triconnected graphs and 3-edge-connected graphs," Proc. of the 19th International Workshop on Graph-Theoretic Concepts in Computer Science(WG'93), Lecture Notes in Computer Science, 790, 132-143(1994).
16. K.Wada, A.Takaki and K.Kawaguchi: "Efficient Algorithms for a Mixed k-partition Problem of Graphs without Specifying Bases," Kawaguchi Lab. Technical Report of ECE in Nagoya Institute of Technology, TR-01-94(1994).

Fugitive-Search Games on Graphs and Related Parameters*

Nick D. Dendris, Lefteris M. Kirousis, and Dimitris M. Thilikos

Department of Computer Engineering and Informatics
Patras University, Rio, 265 00 Patras, Greece, and
Computer Technology Institute
P.O. Box 1122, 261 10 Patras, Greece
E-mail: <dendris, kirousis, sedthilk>@cti.gr

Abstract. The goal of a fugitive-search game on a graph is to trap a fugitive that hides on the vertices of the graph by systematically placing searchers on the vertices. The fugitive is assumed to have complete knowledge of the graph and of the searchers' moves, but is restricted to move only along paths whose vertices are not guarded by searchers. The search number of the graph is the least number of searchers necessary to trap the fugitive. Variants of the fugitive-search game have been related to important graph parameters like treewidth and pathwidth. In this paper, we introduce a class of fugitive-search games where the searchers do not see the fugitive and the fugitive can only move just before a searcher is placed on the vertex it occupies. Letting the fugitive's speed (i.e. the maximum number of edges the fugitive can traverse at a single move) vary, we get different games. We show that if the speed of the fugitive is unbounded then the search number minus 1 is equal to the treewidth of the graph, while if the speed is 1 then the search number minus 1 is equal to the width, a polynomially computable graph parameter. We also show that in the above two cases, the search number remains the same even if we consider only these search strategies that at every step further restrict the fugitive's possible resorts (this monotonicity phenomenon is usually expressed as: "recontamination does not help"). Finally, we show that for any graph, if the length of any chordless cycle is bounded by a constant s ($s \geq 3$), then the treewidth of the graph plus 1 is equal to the search number for fugitive speed $s - 2$.

1 Introduction

The fugitive-search game was introduced by Parsons [26, 27] (see also [7]). In the original version of the game, the graph is thought of as a system of tunnels where an omniscient fugitive with unbounded speed is hidden. The object of the game is to trap the fugitive using searchers. A searcher can either be placed on an

---

* This research was partially supported by the European Union ESPRIT Basic Research Projects ALCOM II (contract no. 7141), GEPPCOM (contract no. 9072) and Insight II (contract no. 6019).

arbitrary vertex of the graph, or deleted from the graph, or slid along an edge. The fugitive cannot go through a vertex guarded by a searcher; it is trapped once there is no place to go. The searchers cannot see the fugitive. The fugitive being omniscient means that it a priori has complete knowledge of the graph and of the searchers' moves. It exploits this knowlegde to move to locations where it is harder to get trapped. The goal of the game is to trap the fugitive using the least possible number of searchers.

Megiddo et al. [24] showed that computing the search number is an NP-hard problem. The fact that it actually belongs to the class NP follows from an important result of LaPaugh [20] (see also [4]) stating that excluding search strategies which give to the fugitive the possibility to visit an already searched vertex does not increase the search number (i.e. allowing recontamination does not help to search the graph).

A variant of the game, called node-search, was introduced in [18]. In this variant searchers can only be placed on or removed from the vertices of the graph (no sliding is allowed). The fugitive resides on a vertex and moves from one vertex to another along unguarded paths (again, the fugitive is assumed to have unbounded speed and be omniscient; the searchers can be placed on any vertex, but they cannot see the fugitive). This variant has the monotonicity and NP-completeness properties of the original version ([18]) and the search number for it is equal to the interval thickness of the graph (i.e. the size of the smallest max-clique in any interval supergraph of G; see [17]) and therefore to the pathwidth of the graph plus 1 (see [25]).

Results relating search number to other graph parameters can be found in [11, 16, 25]. Franklin et al. [12] used this and similar versions of the fugitive-search game to model issues of privacy in distributed systems.

Seymour and Thomas [29] introduced still another variant of the fugitive-search game where, at every stage of the search, the searchers can see the component of the graph where the fugitive resides. They showed that the search number for this variant is equal to the treewidth plus 1 (for a survey of results related to treewidth see [5]). They also showed that the monotonicity property still holds (i.e. excluding search strategies that allow the fugitive to visit a searched vertex is of no help to the fugitive). They proved the monotonicity property by showing that if for a given number of searchers the fugitive has an escape strategy, then there is a nice escape strategy; i.e. there is a collection of sets of vertices that offer a resort to the fugitive, in the sense that the fugitive can always move from any such set of vertices to another one independently of the location of the searchers. The existence of such a resort is proved using ideas on obstruction sets (see [28]). Bienstock [3] gives a survey of the related results.

In this paper, we examine search games where the searchers are always assumed *not* to be able to see the fugitive. Again, the fugitive resides on vertices, moves along unguarded paths and is supposed to be omniscient. The searchers are systematically placed on the vertices with the goal to trap the fugitive. However, the mobility of the fugitive is restricted: we assume that the fugitive is *inert*, i.e., it only moves just before a searcher visits the vertex it occupies (given of course that there is a vertex that can be reached via an unguarded path; otherwise the fugitive is trapped). Formal definitions are given in the next section. We prove that this inert-fugitive search game has the monotonicity property (i.e., recontamination does not help) and that the corresponding search number is equal to the treewidth plus 1. In contrast, without the inertness restriction on the fugitive the corresponding search number (being equal to the node-search number [18]) is equal to the pathwidth plus 1.

We also examine search games where the fugitive, besides being inert, is further restricted to have speed a given number s, i.e. the number of unguarded edges it can traverse at each move is at most s (if $s = n - 1$, n is the number of vertices of the graph, we say that the speed is unbounded, since in this case the fugitive can traverse any unguarded path). We thus obtain a class of fugitive-search games parametrized in terms of the speed of the fugitive.

We show that if the speed is 1 then the monotonicity property holds (i.e. recontamination does not help) and moreover, the search number minus 1 is equal to the width (also known as linkage), a graph parameter studied in the context of the Constraint Satisfaction and Boolean Satisfiability problems (see e.g. [8, 9, 10, 13, 14, 21, 22]). Despite the etymological affinity, width is polynomially computable for arbitrary graphs, whereas treewidth and pathwidth are NP-complete. To define the width of a graph consider a layout of the graph and let the backdegree of a vertex v be the number of vertices earlier in the layout that are adjacent to v; the minimum over all layouts of the maximum backdegree of any vertex of the graph is the width of G. It is known that the width of G is equal to the maximum min-degree of any subgraph of G. Certain classes of graphs with bounded width are advantageous for applying backtracking, the classical method to solve the Constraint Satisfaction Problem (see [21]).

The above characterizations of the search numbers for fugitives with differing abilities, but with identical rules for searchers, offer a uniform game-theoretic approach to pathwidth, treewidth and width, three important graph parameters. Franklin et al. [12] introduce a very general frame for defining fugitive games by considering various mobility settings for both the searchers and the fugitive. However, our inertness restriction is not covered by their classification.

In the last section, we give an interesting from an algorithmic point of view result. We characterize the search number for an inert fugitive with a given speed in terms of an elimination ordering. Using this characterization, we prove that for any graph whose largest chordless cycle is at most $s + 2$, the treewidth plus 1 is equal to the search number for an inert fugitive with speed s. This is a new characterization of treewidth for graphs whose chordless cycles have bounded length.

2 Formal Definitions and Results on Pathwidth

A *search* on a graph $G = (V, E)$ is a sequence $S_0, \ldots, S_r$ of sets of vertices ($S_i \subseteq V, i = 0, \ldots, r$) such that: (i) $S_0 = \emptyset$, and (ii) for all $i = 1, \ldots, r$, the symmetric difference of the sets S_i and S_{i-1} has cardinality 1 (intuitively, either one searcher is added on or one searcher is deleted from the graph at each step of the search). The *search number of a search* is the maximum of the cardinalities of the sets $S_i, i = 0, \ldots, r$.

Let $S = \{S_i, i = 0, \ldots, r\}$ be a search. For $i = 0, \ldots, r$, we inductively define the set of *free locations* F_i for an *agile fugitive with unbounded speed* as follows: $F_0 = V$; For $i = 1, \ldots, r$, let $F_i = (F_{i-1} - S_i) \cup \{v \in V - S_i : \text{there is a path}$ from a vertex $u \in F_{i-1}$ to v whose vertices except u belong to $V - S_i\}$.

Intuitively, after the ith step of the search, the fugitive can be at *any* of the vertices of F_i. Being omniscient, after each step of the search, it chooses a most advantageous location in F_i. Intuitively also, $V - S_i$ is the set of unguarded vertices. The fugitive moves along unguarded paths to vertices that have possibly admitted a searcher in the past. The fugitive is agile in the sense that it has the ability to move whenever there appears an unguarded path that starts from its current location (also see below the definition of the inert-fugitive game). It is easy to see that the search game for an agile fugitive as defined here is exactly the same as the node-search game introduced in [18].

The set of free locations for an *inert fugitive with unbounded speed* is defined as follows: $F_0 = V$; for $i = 1, \ldots, r$, let $F_i = (F_{i-1} - S_i) \cup \{v \in V - S_i : \text{there is}$ a path from a vertex $u \in F_{i-1} \cap (S_i - S_{i-1})$ to v whose vertices except u belong to $V - S_i\}$. Intuitively, an inert fugitive is allowed to move only when a searcher is about to be placed on the vertex it occupies (this is so because the fugitive can move away from a vertex u only if $u \in F_{i-1} \cap (S_i - S_{i-1})$). Notice that for an inert fugitive, if $S_i \subseteq S_{i-1}$ or even if $S_i \subseteq V - F_{i-1}$, then $F_i = F_{i-1}$.

Finally, if n is the number of vertices of the graph and $1 \le s \le n - 1$ is an integer, the set of free locations for an *inert fugitive with speed* s is defined as follows: $F_0 = V$; for $i = 1, \ldots, r$, let $F_i = (F_{i-1} - S_i) \cup \{v \in V - S_i : \text{there is}$ a path of length at most s from a vertex $u \in F_{i-1} \cap (S_i - S_{i-1})$ to v whose vertices except u belong to $V - S_i\}$. Intuitively, an inert fugitive with speed s behaves exactly as an inert fugitive with unbounded speed except that it cannot traverse a path of lentgh more than s.

Given the type of the fugitive, a search S is defined to be *complete* if $F_r = \emptyset$. For each type of fugitive, the corresponding *search number of the graph* is the minimum search number over all searches which are complete with respect to this type of fugitive. Figure 1 depicts a graph whose search number varies depending on the type of the fugitive.

For all types of fugitives, a search is called *monotone* if $\forall i = 1, \ldots, r, F_i \subseteq F_{i-1}$. Notice that for a monotone search and for all types of fugitives, $F_i = F_{i-1} - S_i$. Intuitively, a search is monotone if it does not allow recontamination. For each type of fugitive, the corresponding *monotone search number of the graph* is the minimum search number over all monotone searches which are complete with respect to this type of fugitive.

It is known that (see [18, 20]):

Theorem 1. *For a graph G and for an agile fugitive with unbounded speed, the monotone search number of G is equal to the search number of G.*

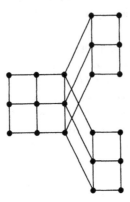

Fig. 1. An example graph which for an agile fugitive with unbounded speed has search number 6; for an inert fugitive with unbounded speed has search number 4; and for inert fugitive with speed 1 has search number 3.

Notice that in the case of an inert fugitive, a *monotone* search may entail re-insertion of a searcher on a vertex that has already been visited by a searcher (of course, by monotonicity, the fugitive cannot be on such a vertex). The monotonicity property for such searches guarantees only that the fugitive cannot visit an already searched vertex (and not that a searcher is never re-inserted on a vertex that has previously admitted a searcher). For example, to search the graph of Fig. 1 for a fugitive with unbounded speed using only 4 searchers, it is necessary, at some steps of the search, to re-insert searchers to already searched vertices. However, for an agile fugitive, it is not only known that recontamination does not help, but also that searcher re-insertion is unnecessary [17].

Finally, we mention (see [17, 25]):

Theorem 2. *The pathwidth of G incremented by 1 is equal to the search number of G for an agile fugitive with unbounded speed.*

3 Inert-Fugitive Game and Treewidth

In this section we show that the search number for an inert fugitive with unbounded speed is equal to the treewidth of the graph incremented by 1. We also show that for this type of fugitive, the monotone search number is equal to the (nonmonotone) search number. The proof of this result depends on the existence of *screens*, a notion introduced by Seymour and Thomas [29]. Screens are obstructions for graphs with small treewidth.

Treewidth has many equivalent chracterizations. One is in terms of *tree-decompositions* (see, e.g., [5]). Another is in terms of elimination orderings of graphs [1]. An *elimination ordering* of a graph $G = (V, E)$ is an ordering $\pi = (v_1, \ldots, v_n)$ of the vertices of G ($n = |V|$). The graphs generated during an elimination of the vertices of G according to π are defined to be: $G_1 = G$ and $G_{i+1} =$ the graph obtained from G_i by deleting the vertex v_i and adding new edges (if necessary) so that all pairs of neighbors of v_i in G_i are adjacent in G_{i+1}. Obviously, $G_{n+1} =$ empty graph. The *dimension of v_i* with respect to π is defined to be the degree of v_i in G_i. The *dimension of π* is the maximum dimension of any of the v_is, and finally the *elimination dimension* of G is the minimum dimension of any elimination ordering of G. The following result can be found in [1]:

Theorem 3. *The treewidth of a graph is equal to its elimination dimension.*

Given an elimination ordering $\pi = (v_1, \ldots, v_n)$ of G, it is convenient to define the *support* of a vertex v_i to be the set of vertices v_j with $j > i$ and connected to v_i by a path in G whose vertices except its endpoints v_i and v_j are earlier than v_i in π.

We will need the following easy technical lemma (the proof is omitted):

Lemma 4. *Let $\pi = (v_1, \ldots, v_n)$ be an elimination ordering of G. Then for every v_i, the support of v_i is equal to the set of neighbors of v_i in G_i and therefore the cardinality of the support of v_i is equal to the dimension of v_i with respect to π.*

We now prove the following:

Theorem 5. *The treewidth of a graph G plus 1 is equal to the monotone search number for an inert fugitive with unbounded speed.*

Proof. We first show that if the treewidth of G is k, then there is a complete monotone search for an inert fugitive with unbounded speed that uses at most $k + 1$ searchers.

By Theorem 3 there is an elimination ordering $\pi = (v_1, \ldots, v_n)$ of G which has dimension k. We define a complete monotone search with $k + 1$ searchers as follows: we traverse the vertices of G in the *reverse* order of the one defined by π, i.e. in the order $(v_n, \ldots, v_1)$; at each step of this traversal, we define how to add and delete searchers; by considering all these searchers' moves in succession, we get a search of the graph. Specifically, when at v_i, we first delete all the searchers (if any) that are on the graph, then we add searchers on the support of v_i and finally we add a searcher on v_i. The reader can easily verify, using Lemma 4, that the search thus defined satisfies the requirements. This completes the proof of the first direction.

We now prove that if there is a monotone complete search with $k+1$ searchers, then there is an elimination of dimension at most k. Order the vertices in terms of the search step that places a searcher on them for the first time; then reverse this order to get $\pi = (v_1, \ldots, v_n)$ (formally, during the search, a searcher visits v_i for the first time before a searcher visits v_j for the first time iff $i > j$). We claim that the elimination ordering thus defined has dimension at most k. Indeed, by Lemma 4, if a vertex v_i has dimension with respect to π strictly more than k, then also the support of v_i has cardinality strictly more than k. But then, since there are no more than $k+1$ searchers available, when visiting v_i with a searcher for the first time there would exist a vertex in the support of v_i not guarded by a searcher. This contradicts the monotonicity of the search. $\qquad\square$

The next step is to prove the monotonicity of the search game with an inert fugitive of unbounded speed. Crucial for the proof is the notion of screen introduced in [29]. Below we give the related definitions and then state the corresponding theorem.

Let $G = (V, E)$ be a graph and let $H_1, H_2 \subseteq V$. We say that H_1, H_2 *mutually touch* if $H_1 \cap H_2 \neq \emptyset$ or $\exists e = \{v_1, v_2\} \in E : v_1 \in H_1 \wedge v_2 \in H_2$.

Definition 6. A *screen* S is a collection $H_1, H_2, \ldots, H_r$ of connected subsets of V that are pairwise touching. A screen S has thickness $\geq k$ iff $\forall X \subseteq V$ with $|X| < k, \exists H_i : H_i \cap X = \emptyset$.

Theorem 7 (Seymour & Thomas). *Let $G = (V, E)$ be graph. If the treewidth of $G \geq k$ then G has a screen of thickness $\geq k + 1$.*

We now prove:

Theorem 8. *If G has a screen with thickness $\geq k+1$ then an inert fugitive with unbounded speed cannot be captured with $\leq k$ searchers (even by a nonmonotone search).*

Proof. Let $H_1, \ldots, H_r$ be a screen of G of thickness $\geq k + 1$. Then $\forall X \subseteq V$ if $|X| < k + 1$, then $\exists H_i : H_i \cap X = \emptyset$. We will now provide a strategy for the fugitive that allows him to avoid any search by $< k + 1$ searchers. Recall that the fugitive, being omniscient, knows in advance all the moves of the searchers. Initially, i.e., before any searchers appear on the graph, the fugitive arbitrarily selects some screen element H. Let v be the first vertex of H ever to be visited by a searcher. The fugitive chooses v as its very first location. Let X be the set of vertices occupied by a searcher immediately before v admits a searcher for the first time. Then, since $|X \cup \{v\}| < k + 1$, there is a screen element H' such that $H' \cap (X \cup \{v\}) = \emptyset$. Notice also that by definition, $X \cap H = \emptyset$. So when a searcher is placed on v, the fugitive can escape to any vertex in H'. This is so because H and H' are connected, mutually touching and moreover, just before putting a searcher on v, H and H' carry no searcher. The fugitive, being omniscient, chooses to go to that vertex of H' that will be visited first by a searcher after the current step of the search.

Repeating this procedure, it becomes clear that the fugitive can escape being captured by $\leq k$ searchers forever. $\qquad\square$

Now, from Theorems 5, 7 and 8, we get as immediate corollaries the following two theorems, our main results in this section:

Theorem 9. *For an inert fugitive with unbounded speed, the monotone search number of a graph G is equal to (nonmonotone) search number of G (i.e., recontamination does not help to search for such a fugitive).*

Theorem 10. *The treewidth of a graph G plus 1 is equal to its (nonmonotone) search number for an inert fugitive with unbounded speed.*

4 Unit-Speed Fugitive and Width

In this section, we examine the search game with inert fugitives that have speed 1. We prove that the search number in this case is equal to the width. We also prove that "recontamination does not help" to search for an inert fugitive with unit speed. The proof shows the existence of a very simple obstruction for small search number (for this type of fugitive).

We first give the related definitions.

A *layout* of a graph $G = (V, E)$ is an ordering $L = (v_1, v_2, \ldots, v_n)$ of the vertices of G ($n = |V|$).

The *width of a vertex* v with respect to a layout L is the number of vertices which are adjacent to v and precede v in the layout. The *width of a layout* L of G is the maximum width of any vertex in L. The *width of* G is the minimum width of any layout of G.

For clarity, let us mention that we use the term *min-degree of a subgraph H of G* to denote the minimum over all vertices v in H of the number of vertices in H which are adjacent to v.

The following theorem is proved in [13] (see also [23]).

Theorem 11. *The width of a graph is equal to the largest min-degree of any subgraph of G.*

Similarly to the proof of Theorem 5 we show that:

Theorem 12. *The monotone search number of G when searching for an inert fugitive with speed 1 equals its width plus one.*

Moreover:

Theorem 13. *If there is a subgraph H of G with min-degree k, then an inert fugitive with speed 1 cannot be captured using $\leq k$ searchers (even by a non-monotone search).*

Proof. If the fugitive chooses to reside in H, any attempt to capture it with $\leq k$ searchers (even allowing recontamination) will be futile. Indeed, whenever the search places a searcher on the vertex of H where the fugitive is hiding (call this vertex v) there will always exist a vertex u in H which is both unguarded and adjacent to v; the fugitive can escape to u. □

From the above theorems we easily get:

Theorem 14. *For an inert fugitive with speed 1, the monotone search number of a graph G is equal to the (nonmonotone) search number of G.*

Theorem 15. *The width of a graph plus 1 is equal to its search number for an inert fugitive with speed 1.*

5 Elimination Orderings — Treewidth of Graphs With Chordless Cycles of Bounded Length

In this section, we give a characterization of the monotone search number for an inert fugitive of a given arbitrary speed in terms of an elimination ordering of the graph. Using this result, we show that in the class of graphs whose largest chordless cycle has length at most $s + 2$, the treewidth plus 1 is equal to the monotone search number for an inert fugitive with speed s.

As mentioned in Sect. 3, an elimination ordering of a graph $G = (V, E)$ is an ordering $\pi = (v_1, \ldots, v_n)$ of the vertices of the graph $(n = |V|)$. Given an elimination ordering π and an integer s $(1 \leq s \leq n - 1)$, the graphs generated during an *s-elimination of the vertices of G* according to π are defined to be: $G_1 = G$; $V_{i+1} = V_i - \{v_i\}$ and E_{i+1} is the set of pairs $\{u, v\}$ such that $u, v \in V_{i+1}$ and there is a path in G that connects u with v, has length at most s and all its vertices except u and v are among $v_1, \ldots, v_i$. The *s-dimension of v_i with respect to π* is defined to be the degree of v_i in G_i. The *s-dimension of π* is defined to be the maximum s-dimension of any of the v_is with respect to π, and finally the *s-elimination dimension of G* is the minimum s-dimension of any elimination ordering of G.

Similarly with Theorem 5, one can show that:

Theorem 16. *For any s $(1 \leq s \leq n-1)$, the s-elimination dimension of G plus 1 is equal to its monotone search number for an inert fugitive with speed s.*

We now state the following characterization of treewidth for the class of graphs whose chordless cycles are of bounded length.

Theorem 17. *The treewidth of a graph with no chordless cycles of length $> s+2$ is equal to its s-elimination dimension.*

To prove the above theorem, we first give some definitions and certain lemmata.

For reasons of clarity, let us mention that we use the term *sub-tree* of given tree to refer to a connected subgraph of the tree.

The following is proved in [15] (see also [6]).

Lemma 18. *Given a tree $T = (I, F)$, let $\mathcal{T}$ be a class of sub-trees of T such that $\forall S, S' \in \mathcal{T}, S \cap S' \neq \emptyset$. Then, $\bigcap_{S \in \mathcal{T}} S \neq \emptyset$.*

Also, using the previous lemma the following is proved in [6].

Lemma 19. *Consider a tree-decomposition $(\{X_i, i \in I\}, T = (I, F))$ of a graph $G = (V, E)$. Then, for any clique K of G, $\exists i \in I : K \subseteq X_i$.*

Let $\pi = (v_1, \ldots, v_n)$ be an elimination ordering of a graph $G = (V(G), E(G))$. We define below a new procedure to eliminate the vertices of G according to π. We call this procedure *tree-elimination*:

procedure tree-elimination$(H = (V(H), E(H)) : V(H) \subseteq V(G))$

1. Let v be the first vertex of π that belongs to $V(H)$.

2. If there exist (nonempty) connected components, say $C_1, \ldots, C_r$, of the graph obtained from H by removing v and all its adjacent (in H) vertices, then do the following steps.

3. For each C_j, let ∂C_j be the set of vertices that are adjacent to v and, moreover, are adjacent to at least one vertex of C_j. Formally, $\partial C_j = \{u \in V(H) : \{u, v\} \in E(H) \text{ and } \exists w \in C_j : \{u, w\} \in E(H)\}$.

4. For each C_j, let $\bar{C}_j$ be the graph with set of vertices $V(C_j) \cup \partial C_j$ and set of edges the union of the following:

   (a) the set of edges induced from $E(H)$ on $V(C_j) \cup \partial C_j$.

   (b) the set of "new" edges necessary to make ∂C_j a clique.

5. Recursively apply tree-elimination on each $\bar{C}_j$, $j = 1, \ldots, r$.
   /*Notice that the graphs $\bar{C}_j$ are not necessarily pairwise disjoint.*/

Let $(H_l)_{l \in L}$ be the family of all graphs to which tree-elimination is recursively applied when we run tree-elimination on G according to π. Include G in this family (say $G = H_{l_0}$). Let v_{H_l} be the first vertex in π that belongs to $V(H_l)$. Notice that for two different H_ls, the corresponding v_{H_l}s may be equal. We give a *rooted* tree structure to the family $(H_l)_{l \in L}$ as follows: The root is $G = H_{l_0}$; H_l is the parent of H_m iff when the procedure tree-elimination is recursively applied on H_l, then H_m is one of the graphs $\bar{C}_j$ defined at Step 4 of this recursive call (notice that by this definition, the procedure tree-elimination will be subsequently recursively applied on H_m). We call this tree the *recursion tree*.

Let the *tree-dimension of* π be the maximum degree of any v_{H_l} with respect to H_l. Let the *tree-elimination dimension of* G be the minimum tree-dimension of any elimination ordering. Finally, given an edge $\{v, w\} \in E(H_l)$, let its weight with respect to H_l (notationally $\text{weight}_{H_l}(v, w)$) be the length of the shortest path in G that connects v to w and whose internal vertices lie to the left of v_{H_l} in the ordering π.

The following four lemmata, which, although non-trivial, are given here without proofs due to space limitation, outline the main steps in the proof of Theorem 17.

Lemma 20. *For any H_l, the treewidth of H_l is at most equal to the maximum of the degrees of $v_{H_m}s$ (the degrees are taken with respect to the corresponding $H_m s$), over all $H_m s$ that are either descendants of or coincide with H_l in the recursion tree. Therefore, the treewidth of G is at most equal to the tree-elimination dimension of G.*

Let now s be a constant positive integer. The following hold:

Lemma 21. *If G has no chordless cycle of length $> s + 2$, then no H_l contains a chordless cycle that has length 4 or more and whose edges have weights (with respect to H_l) with total sum $> s + 2$.*

Lemma 22. *If G has no chordless cycle of length $> s + 2$ then no H_l contains an edge of weight (in H_l) $> s$.*

Lemma 23. *If G has no chordless cycle of length $> s + 2$ then the tree-elimination dimension of G is $\leq$ the s-elimination dimension of G.*

Now, by Lemmata 20 and 23, we conclude that the treewidth of $G \leq$ the s-elimination dimension of G, which proves one direction of Theorem 17. The other direction follows immediately from Theorem 3 and the obvious fact that the elimination dimension of a graph is at least equal to its s-elimination dimension.

Therefore:

Theorem 24. *The treewidth of a graph with no chordless cycles of length $> s+2$ incremented by 1 is equal to its monotone search number for an inert fugitive with speed s.*

Acknowledgments

We thank Christos Papadimitriou and Paul Spirakis for their patience to hear us talk about fugitive searching and for their valuable suggestions. We thank Moti Yung, who during a short visit to Patras revived our interest for search games.

References

1. S. Arnborg: Efficient algorithms for combinatorial problems on graphs with bounded decomposability (A survey). BIT **25** (1985) 2–33

2. R. Anderson and E. Mayr: Parallelism and greedy algorithms. Advances in Computing Recearch **4** (1987) 17–38. See also: A P-complete problem and approximations to it. Technical Report, Dept. Computer Science, Stanford University, California (1984)

3. D. Bienstock: Graph searching, path-width, tree-width and related problems (A survey). DIMACS Series in Discrete Mathematics and Theoretical Computer Science **5** (1991) 33–49

4. D. Bienstock and P.D. Seymour: Monotonicity in graph searching. Journal of Algorithms **12** (1991) 239–245

5. H.L. Bodlaender: A tourist guide through treewidth. Acta Cyberbetica **11** (1993) 1–23

6. H.L. Bodlaender and R.H. Möhring: The pathwidth and treewidth of cographs. SIAM Journal on Discrete Mathematics **6** (1993) 181–188

7. R. Breisch: An intuitive approach to speleotopology. Southwestern Cavers (A publication of the Southwestern Region of the National Speleological Society) **VI** (1967) 72–78

8. R. Dechter: Constraint networks. In: S.C. Shapiro (ed.) Encyclopedia of Artificial Inteligence, Wiley, N.Y. (1992) 285–293

9. R. Dechter: Directional resolution: The Davis-Putnam procedure, revised. Working Notes AAAI Spring Symposioum on AI and NP-Hard Problems (1993) 29–35

10. R. Dechter and J. Pearl: Tree clustering for constraint networks. Artificial Inteligence **38** (1989) 353–366

11. J.A. Ellis, I.H. Sudborough, and J.S. Turner: The vertex separation number of a graph. Proceedings 1983 Allerton Conference on Comunication and Computing 224–233.

12. M. Franklin, Z. Galil, and M. Yung: Eavesdropping games: A graph-theoretic approach to privacy in distributed systems. Proceedings 34th Annual Symposioum on Foundations of Computer Science (FOCS) 1993, IEEE Computer Society Press (1993) 670–679

13. E.C. Freuder: A sufficient condition for backtrack-free search. Journal of the Association for Computing Machinery **29** (1982) 24–32

14. E.C. Freuder: A sufficient condition for backtrack-bounded search. Journal of the Association for Computing Machinery **32** (1985) 755–761

15. F. Gavril: The intersection graphs of subtrees in trees are exactly the chordal graphs. Journal of Combinatorial Theory (Series B) **16** (1974) 47–56

16. N.G. Kinnersley: The vertex separation number of a graph equals its path-width. Information Processing Letters **42** (1992) 345–350

17. L.M. Kirousis and C.H. Papadimitriou: Interval graphs and searching. Discrete Mathematics **55** (1985) 181–184

18. L.M. Kirousis and C.H. Papadimitriou: Searching and pebbling. Journal of Theoretical Computer Science **47** (1986) 205–218

19. L.M. Kirousis and D.M. Thilikos: The Linkage of a Graph. Technical Report TR 93.04.16, Computer Technology Institute, Patras, Greece (1993)

20. A.S. LaPaugh: Recontamination does not help to search a graph. Journal of the Association for Computing Machinery **40** (1993) 224–245

21. A. Mackworth: Constraint Satisfaction. In: S.C. Shapiro (ed.) Encyclopedia of Artificial Inteligence, Wiley, N.Y. (1992) 276–285

22. A. Mackworth and E. Freuder: The complexity of some polynomial network consistency algorithms for constraint satisfaction problems. Artificial Inteligence **25** (1985) 65–74

23. D.W. Matula: A min-max theorem for graphs with application to graph coloring. SIAM Rev. **10** (1968) 481–482

24. N. Megiddo, S.L. Hakimi, M.R. Garey, D.S. Johnson, and C.H. Papadimitriou: The complexity of searching a graph. Journal of the Association for Computing Machinery **35** (1988) 18–44

25. R.H. Möhring: Graph problems related to gate matrix layout and PLA folding. In: E. Mayr, H. Noltemeier, and M. Syslo (eds.) Computational Graph Theory, Computing Supplementum, **7** (1990) 17–51

26. T.D. Parsons: Pursuit-evasion in a graph. In: Y. Alavi and D.R. Lick (eds.) Theory and Applications of Graphs, Springer-Verlag (1976) 426–441

27. T.D. Parsons: The search number of a connected graph. In: Proceedings 9th Southeastern Conference on Combinatorics, Graph Theory, and Computing, Utilitas Mathematica, Winnipeg, Canada (1978), 549–554

28. N. Robertson and P.D. Seymour: Graph Minors III. Planar tree-width. Journal of Combinatorial Theory (Series B) **36** (1984) 49–64

29. P.D. Seymour and R. Thomas: Graph searching and a minimax theorem for tree-width. Journal of Combinatorial Theory (Series B) **58** (1993) 22–33

New Approximation Results on Graph Matching and Related Problems[*]

Yoji Kajitani[1], Jun Dong Cho[2] and Majid Sarrafzadeh[3]

[1] Japan Advanced Institute of Science and Technology, Hokuriku, Japan
[2] CAE, Samsung Electronics, Buchun Kyunggi-Do, Korea
[3] Northwestern University, Evanston, IL 60208

Abstract. For a graph G with e edges and n vertices, a *maximum cardinality matching* of G is a maximum subset M of edges such that no two edges of M are incident at a common vertex. The best known algorithm for solving the problem in general graphs requires $O(n^{5/2})$ time. We first propose an approximate maximum cardinality matching algorithm that runs in $O(e+n)$ sequential time yielding a matching of size at least $\frac{e}{n-1}$, improving the bound known before. For bipartite graphs, the algorithm yields a matching of size at least $\frac{2e}{n}$. The proposed algorithms are extremely simple, and the derived lowerbounds are existentially tight. Next, the proposed maximum cardinality matching algorithm is extended to the weighted case running in $O(e+n)$ time. The problem of approximate maximum matching has a number of applications, for example in, Vertex Cover, TSP, MAXCUT, and VLSI physical design problems.

1 Introduction

Let $G = (V, E), |V| = n$ and $|E| = e$, be an undirected graph (without loops and multiple edges) each of whose edges has a real-valued weight. A *matching* M is a subset of E such that no two edges of M are incident to a common vertex. A matching M is *maximal* if every remaining edge in $E - M$ has an endpoint in common with some member of M. A maximum cardinality (weighted) matching is a matching whose cardinality (edge weights) is maximum. We denote by $w(E)$ the total edge weights on the graph G.

There are a number of applications of matchings in such areas as crew scheduling, graph plotting, and heuristics and relaxations for hard problems defined on graphs, for example, the traveling salesman problem. Interest in these applications has in turn produced a large amount of efforts in the development of efficient implementations of matching algorithms. It is well known that even maximum cardinality matching in bipartite graphs requires $O(n^{5/2})$ time for its exact solution [19]. The best time bound in weighted bipartite graphs is currently $O(n(m + n \log n))$ [15]. Whereas, the current best algorithm for maximum weighted matching in general graphs requires $O(n(m \log \log \log_{\lceil m/n+1 \rceil} n + n \log n))$ time [12]. For planar graphs, an exact solution can be computed in

---

[*] This work has been supported in part by the National Science Foundation under Grant MIP-9207267.

$O(n^{3/2} \log n)$ time by a divide and conquer algorithm given in [22] using the $O(\sqrt{n})$ planar separator theorem [21]. For a more detailed history of the maximum matching problem in graphs see [13, 10, 24, 23, 28].

Approximation algorithms for NP-complete problems have been widely studied [25, 17, 7, 30, 5, 18], but rarely for problems known to be in P (polynomial time solvable) [28]. To approximate a problem known to be in P, we require that the relative approximation quality be "good", and that the algorithm runs "very fast". Avis [1] proposed a simple approximation algorithm on weighted graphs (referred to as GREEDY-W) that is to repeatedly match the two unmatched vertices with the heaviest edge, which runs in $O(n^2 \log n)$ time using sorting. GREEDY-W guarantees a matching of at least half the weight of the maximum weighted matching [2].

In this paper, we are interested in finding a linear-time approximation algorithm for the maximum cardinality matching problem which guarantees a "good" approximation quality. Although the matching problem can be solved exactly in polynomial time there are many situations, such as the control of plotters, where extremely fast algorithms are required so that the use of heuristics is appropriate. Suboptimal but conceptually simpler heuristics are easier to analyze and provide bounds on this expected value. For an earlier survey of heuristics on the matching problem see [2]. Recently, Hagin and Venkatesan [17] showed that any graph admits a matching of size $\lfloor e/n \rfloor$ in $O(e + n)$ sequential time. The lower-bound on matching size, $\lfloor e/n \rfloor$, has also been shown in [4]. For dense graphs, the bound exhibits a near-optimal solution. In this paper, we present a more efficient and simpler approximation algorithm for the problem. Our algorithm runs in $O(e + n)$ sequential time yielding a matching size of at least $\lceil e/(n - 1) \rceil$ (for weighted case, the algorithm runs in $O(e + n)$ time yielding the bound of at least $w(E)/(n - 1)$) for general graphs and $2e/n$ (for weighted case, $2w(E)/n$) for bipartite graphs, thus improving the bound known before. A summary of the main results is given in Table 1. In all cases, we improve upon previously known results; either the computing time, or the approximation bound, or the both.

This paper is organized as follows. In Sections 2 and 3, we will develop effective approximation bounds on both unweighted and weighted graphs and propose efficient approximate algorithms on both graphs independently. Also, we will extend the solutions by exploring the basic theorems in Section 4. Conclusion is presented in Section 5.

2 Approximation Algorithm for the Maximum Cardinality Matching Problem

Let us first define a number of terminologies as follows.

Definition 1. Let M be a matching in a graph G. An edge in M is *matched*; every edge not in M is *unmatched*. A vertex is *matched* if it is incident to a matching edge and *exposed* otherwise. An *alternating path* or *cycle* is a simple path or cycle whose edges are alternately matched and unmatched. For example,

Problems	Before		Ours	
	time	size	time	size
MCM (g)	$O(e+n)^{\dagger}$	$\lfloor e/n \rfloor^{\dagger}$	$O(e+n)$	$\lceil \frac{e}{n-1} \rceil$
MWM (g)	unknown	unknown	$O(e+n)$	$\frac{w_G}{n-1}$
MCM (b)	$O(e+n)^{\dagger}$	$\lfloor e/n \rfloor^{\dagger}$	$O(e+n)$	$\frac{2e}{n}$
MWM (b)	unknown	unknown	$O(e+n)$	$\frac{2w_G}{n}$

† : [17]. ‡ : [29].
MCM : maximum cardinality matching.
MWM : maximum weighted (both positive and negative) matching.
(g) : general graphs. (b) : bipartite graphs.

Table 1: A comparison between the previous results and ours

a path $v_0, v_1, \cdots v_k$ is said to be alternating if for all $i = 1, \cdots, k-1$, the edges v_{i-1}, v_i does not lie in M and the edges v_i, v_{i+1} lies in M. The length of an alternating path or cycle is the number of edges it contains. An alternating path is *augmenting* if both its ends are exposed vertices.

If M has an augmenting path then M is not of maximum size, since we can increase its size by one by interchanging matched edges with unmatched edges along the path. We call this process an *augmentation*.

Our algorithm is motivated by the following fundamental theorem due to Berge [3].

Theorem 2. *A matching is of maximum size (cardinality) if and only if it has no augmenting paths.*

We can construct a maximum size matching by beginning with an empty matching and repeatedly performing augmentation until there are no augmenting paths; this takes at most $n/2$ augmentations. In each stage a search for an augmenting path takes $O(e)$ time. Thus, the algorithm runs in $O(en)$ time. The best known algorithm using the above strategy is by Hopcroft and Karp for bipartite graphs [19] and Micali and Vazirani for general graphs [26]. They discovered a way to find many augmenting paths in one traversal of the graph. Their algorithm runs in $O(e\sqrt{n})$ time.

2.1 Algorithm Approximate Maximum Cardinality Matching

Let us consider the following approximation algorithm that leads to an efficient tradeoff between approximation quality and computing resources. An l-locally optimal cardinality matching (ℓ-$LOCM$) is a matching in which none of the paths of length less than or equal to ℓ in G is augmenting. The intuition behind the algorithm is to consider a set of augmenting paths, each of whose length is not more than a small constant $\ell \leq n$. We repeatedly perform augmentation to increase the matching size until local search fails to identify a way of improving the solution.

Input: a graph G and an integer ℓ
Output: Find an ℓ-$LOCM$ M in the graph G with the largest number of edges.
1. [Initialize] Partition the graph arbitrarily into two sets by finding an arbitrary maximal matching, i.e., one with a set of matched edges and another with a set of exposed nodes. Let M be the set of arbitrarily matched edges.
2. [Improvement along a given path of length ℓ] While an augmenting path of length $\ell' \leq \ell$ exists in the graph G, repeat the following operations. Search for an augmenting path with respect to M with the initial vertex v_0; if such a path $v_0, v_1, \cdots, v_{2i+1}, \cdots, v_{\ell'}$ has been found, delete from M the edges

$$(v_1, v_2), \cdots, (v_{2i-1}, v_{2i}), \cdots, (v_{\ell'-2}, v_{\ell'-1})$$

and add to M the edges

$$(v_0, v_1), \cdots, (v_{2i}, v_{2i+1}), \cdots, (v_{\ell'-1}, v_{\ell'})$$

That is, we construct a maximum size matching for ℓ-$LOCM$ by beginning with the initial matching and repeatedly performing augmentation until there are no augmenting paths of length less than or equal to ℓ. Now the question is how far from an optimum could such a matching be? In this paper, we obtain an upperbound on the size of matching by characterizing the above approximation algorithm. Especially for the case where $\ell = 3$, we show how such a local improvement algorithm could be implemented in linear time for unweighted graphs.

In the next section, we will show that an approximate maximum cardinality matching with $\ell = 3$ in an n-vertex e-edge graph can be found in $O(e + n)$ time yielding matching of size at least $\lceil e/(n-1) \rceil$ (when n is even). The underlying scheme will be effectively used to develop the lowerbound on weighted graphs.

2.2 Approximate Maximum Cardinality Matching Theorem

We observe an interesting property to find a lower bound on the matching size. We will use the following terminologies throughout the paper. Let us assume that we have a 3-LOCM M of size m. We partition the graph into two sets; T_{in} and T_{out} (refer to Figure 1). T_{in} contains $2m$ vertices of M. T_{out} contains s $(= n-2m)$ exposed nodes. There are no edges between vertices in T_{out}; otherwise, we could have added some of edges in T_{out} to M.

We are given a graph $G(V, E)$ with edge weight 1. G may not be complete. First assume that n is even. We provide an algorithm for finding a matching M that satisfies In this algorithm, M denotes the current matching. V_M denotes the set of end vertices of the elements of M. Let $m = |M|$. A vertex i is said to be an *outside vertex* if i is not in V_M.

- **Case 1: n is even:**

  - **1. Initialization:** Initially, set $M = \emptyset$.
  - **2. Simple addition:** While there is a pair (i, j) of outside vertices such that $(i, j) \in E$, $M \Leftarrow M \cup \{(i, j)\}$.
  - **3. Augmentation:** Find a pair (i, j) (called the *augmentable pair*) of outside vertices such that there exists an edge $(p, q) \in M$ with $(i, p), (j, q) \in E$. $M \Leftarrow M - \{(p, q)\} + \{(p, i), (q, j)\}$. If there is no such a pair, output M.

Lemma 3. *The resultant matching M satisfies $m \geq \lceil \frac{e}{n-1} \rceil$.*

Proof: (refer to Figure 1)

1. There is no augmenting path of length three in the resulting graph; otherwise we could have augmented some of edges between T_{in} and T_{out} to M in order to increase the size of matching.
2. There is no edge between the outside vertices.
3. Make all $n - 2m$ outside vertices into $\frac{n-2m}{2}$ pairs arbitrarily. The number of edges between one of such pairs and one of matching edges is at most 2, since otherwise, those outside vertices had been augmented. Therefore, the total number of edges between V_M and the outside vertices is at most $2\frac{m(n-2m)}{2} = m(n - 2m)$.
4. The possible maximal subgraph induced by V_M is a complete graph. The number of the edges in the complete graph is at most $\frac{2m(2m-1)}{2} = m(2m - 1)$.
5. Thus, $e \leq m(n - 2m) + m(2m - 1) = m(n - 1)$, i.e., $m \geq \lceil \frac{e}{n-1} \rceil$.

$\square$

The claim was implicitly assuming the number of exposed vertices is zero or more than one. If n is even, then $m \geq \lceil \frac{e}{n-1} \rceil$. However, if n is odd and G is a complete graph where the assumption always fails, then the claim is incorrect (refer to Example 1).

In general, when n is even and $e \leq (2m(2m - 1)/2 + (n - 2m)m)/\alpha$ for real value $\alpha > 1$, the lower bound can be improved to $\lceil \alpha e/(n - 1) \rceil$, where α is a measure of *sparsity* of the graph.

- **Case 2: n is odd:**

  For an arbitrary $v \in V$, let $G' = (V', E')$ whose vertex set is $V' = V - v$ and edge set is $E' = E - I_v$ (I_v: an edge set incident to vetex v). Let us denote M' a matching for G' and m' a lower bound on $|M'|$.

  Since the number of vertices in G' is even, we can apply Proposition 1. To obtain a tight lower bound, we select a vertex v such that $|I_v|$ is minimum over all vertices.

  Therefore,

$$m \geq m' \geq \frac{|E'|}{n - 2} = \frac{|E| - |I_v|}{n - 2} \tag{1}$$

Example 1 *Let $G = K_{2p+1}$. If we were to apply Proposition 1, it would be*

$$m \geq \lceil \frac{(2p+1)2p/2}{2p} \rceil = p + \lceil 1/2 \rceil$$

This is incorrect because $m = p$. The bound obtained for Case 2 claims correctly:

$$m \geq \frac{(2p+1)2p/2 - 2p}{2p - 1} = p.$$

Example 2 *Consider a maximum cardinality matching in a complete bipartite $K_{1,2p}$. Hence, for $p \geq 2$, both lower bounds give the same value, $\frac{w(E)}{n-1} = \frac{w(E) - w(I_{v'})}{n-2} = 1$.*

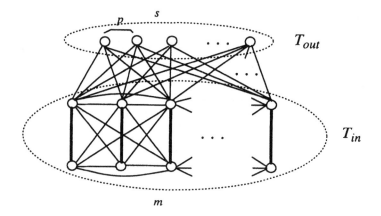

Fig. 1. A case where a matching is of size $\lceil \frac{e}{n-1} \rceil$ for unweighted cases

Theorem 4. *For a graph $G(V, E), |V| = n, -E- = e,$*

$$if \ n \ is \ even \ : \ |M| \geq \frac{e}{n - 1}$$

$$if \ n \ is \ odd \ : \ |M| \geq \frac{e - |I_{v'}|}{n - 2}.$$

2.3 Detailed Implementation

To obtain the lower bound in Theorem 4 we implement an algorithm as follows. Here we only consider the case where n is even. The case where n is odd can be trivially treated based on the algorithm. For the description of a matching, we will use the function $MATE$; if (u, w) is an edge of the matching, then $MATE(v) = w$ and $MATE(w) = v$. The function is undefined at exposed vertices. We also define by $EXP(v)$ a set of exposed vertices incident to a vertex v. We initialize $EXP(v_i) = \{v_j \mid (v_i, v_j) \in E\}$.

The given graph is represented by an adjacency list. Each header vertex $v_i, 1 \le i \le n$, in the list contains a list of vertices adjacent to itself denoted by N_i. Let $L_i = \{v_i\} \cup N_i$. Scanning through the adjacency list once, we initialize the set of matching edges induced by header vertices v_i and a vertex $v_j \in N_i$, for all $i \in V$. While looping through each L_i once again, if there is an exposed vertex v_j in L_i, then set $EXP(v_i) = EXP(v_i) \cup v_j$. Then, check if both $EXP(v_i)$ are not empty and $EXP(MATE(v_i))$ is not empty. If it is true, then we perform an augmentation. The time complexity of the proposed algorithm is as follows. The initialization step takes $O(n)$ time. Since a search for augmenting paths takes $O(e)$ time. Each augmentation can be done in a constant time, the algorithm runs in $O(e + n)$ time.

Procedure: Initialize (G, M, EXP)
Input: $G = (V, E)$;
Output: A partition with an arbitrary matching M and a set of exposed node for each vertex v_i, $EXP[v_i]$;
    begin
(1)   **for each** header $v_i \in V$
(2)   **if** $MATE[v_i] = 0$ **and** $\exists v_j \in N_{v_i}$ s.t. $MATE[v_j] = 0$
(3)   **then**
(4) $MATE[v_i] = v_j$; $MATE[v_j] = v_i$; $M = M \cup \{(v_i, v_j)\}$;
(5)   **for each** $v_j \in N_{v_i}$ s.t. $MATE[v_j] = 0$ **then**
(6) $EXP[v_i] = EXP[v_i] \cup v_j$;
    end

Algorithm 3-LOCM (G, M)
Input: $G = (V, E)$;
Output: A matching with $|M| = \lceil e/(n-1) \rceil$;
    begin
(1) $M = empty$;
(2)   **for each** $v_i \in V$
(3) $MATE[v_i] = 0$;
(4) $EXP[v_i] = 0$;
(5)   **call procedure** Initialize(G, M, EXP);
(6)   **for each** header $v_i \in V$
(7)   **for each** $v_j \in N_{v_i}$ s.t. $v_j \in EXP[v_i]$
(8)   **if** $EXP[MATE[v_i]] \ne \emptyset$ **and** $v_j \ne v_k \in EXP[MATE[v_i]]$
(9)   **then**
(10)   Perform an augmentation for Path $(j, i, MATE(v_i), k)$;
    end

Lemma 5. *The computational complexity of the algorithm is $O(e + n)$.*

Proof: When the simple addition stops, the total computation is $O(e+n)$. The augmentaion will be implemented as follows. Take a matching edge $e_k = (t_k, b_k)$ for $k = 1, \cdots, e_k$. We take an edge $\{t_k, b_k\}$ in M at a time and check if it is incident to more than one exposed vertices. If yes, we perform an augmentation; otherwise, increment k. Each augmentation requires a constant time. Hence, the total complexity is $O(e + n)$. $\square$

It is clear that after applying Algorithm 3-LOCM there is no augmenting path of length 3 in the resulting graph. Based on the above lemma along with Theorem 4, we have:

Theorem 6. *Given a graph $G = (V, E), |V| = n$ and $|E| = e$, there exists a matching of size at least $\lceil e/(n-1) \rceil$ which can be found in $O(e + n)$ time.*

Our algorithm improves the lowerbound on size of matching over the algorithm (that guarantees the size at least $\lfloor e/n \rfloor$) developed in [17].

3 Approximation Algorithm for the Maximum Weighted Matching Problem

Let G be a given edge-weighted graph. Associated with the maximum edge-weighted matching, we assume without loss of generality that G is a complete graph with non-negative edge weights and that a maximum weighted matching contains $\lfloor n/2 \rfloor$ edges (some of them may be weighted 0). For an edge set S, $w(S)$ denotes the sum of weights of the edges.

We also denote by $w(A, B)$ the total weights on the edges connecting vertices incident to set A with vertices incident to set B.

In this section, we will show that an approximate maximum weighted matching in an n-vertex e-edge graph can be found in $O(e + n)$ time yielding matching of size at least $w(E)/(n-1)$ when n is even and a similar lower bound when n is odd.

3.1 Approximate Maximum Weighted Matching Theorem

- **Case 1: n is even:** (refer to Figure 2)
  There is no exposed vertex. Let us consider K_4 (= a clique of size four) at a time. We label the two matched edges in K_4 by e_1 and e_2 and their weights by w_1 and w_2, respectively. The other edges in K_4 are labeled by $e_a, e_b, e_c,$ and e_d and their weights by w_a, w_b, w_c, and w_d, respectively. To have a maximum weighted matching in K_4, we should have chosen a matching satisfying the following inequalities:

$$w_1 + w_2 \geq w_a + w_d$$
$$w_1 + w_2 \geq w_b + w_c,$$

that is, $2(w_1 + w_2) \geq w(e_1, e_2)$. Applying the same condition for $(m-1)m/2$ matched-edge pairs in T_{in} (We shall call the condition *3-LOWM*), we have

$$w(M) + 2\Sigma_{\forall \ distinct \ pairs \ e_i, e_j \in M}(w(e_i) + w(e_j)) \geq w(E);$$

i.e.,

$$w(M) + 2(n/2 - 1)w(M) \geq w(E).$$

Thus,

$$w(M) \geq \frac{w(E)}{n-1}.$$

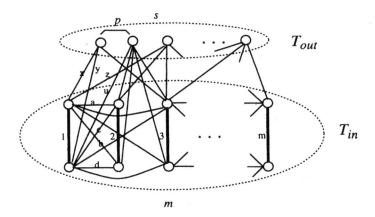

Fig. 2. A matching is of size at least $\frac{w(E)}{n-1}$ for weighted cases

- **Case 2: n is odd:**
  The lower bound obtained in Case 2 of Section 2.2 (unweightd case) can also be applied for weighted cases. Therefore,

$$w(M) \geq w(M') \geq \frac{w(E')}{n-2} = \frac{w(E) - w(I_v)}{n-2}$$

To see the difference between two lower bounds,

$$\frac{w(E)}{n-1} \geq \frac{w(E) - w(I_{v'})}{n-2}$$

if and only if

$$w(I_{v'}) \geq \frac{n}{2(n-1)}d_{av},$$

where $d_{av} = \frac{2w(E)}{n}$, the average of weights of incident edges over the vertices.

Theorem 7. *For a graph* $G(V, E), |V| = n$, *with arbitrary edge weights,*

$$if \quad n \quad is \quad even \quad : \quad w(M) \geq \frac{w(E)}{n - 1}$$

$$if \quad n \quad is \quad odd \quad : \quad w(M) \geq \frac{w(E) - w(I_{v'})}{n - 2}.$$

Theorem 7 also holds for negative weights.

3.2 Algorithm for Maximum Weighted Matching

We will show an algorithm to obtain a matching M that satisifies $w(M) \geq \frac{w(E)}{2|V_M|-1}$.

Note that the resultant matching does not necessarily satisfy the condition of 3-LOWM in Cazse 1 of Section 3.1 (See Example 3). Also, it is not trivial to find a fast (even $O(n^2)$ time) algorithm satisfying the condition which yields the lower bound obtained in Section 3.1. This motivated the results of this section.

Only graphs with even number of vertices will be considered. It is trivial to deduce the corresponding result for the graphs with odd number of vertices.

1. **initialization**:
   (a) Let the current matching be $M = \{e_1, e_2, \cdots, e_m\}$, $e_k = (t_k, b_k)$.
   (b) Start with an initial matching $M = \emptyset$.
   (c) Let (α, β) be any pair (edge) of outside vertices such that α, β are not in V_M.
   (d) Define the *swap-gains* $\delta(e_k)$ and $\delta'(e_k)$ of $e_k \in M$ associated with (α, β) as:

   $$\delta(e_k) = w(t_k, \alpha) + w(b_k, \beta) - w(e_k) - w(\alpha, \beta)$$

   and

   $$\delta'(e_k) = w(t_k, \beta) + w(b_k, \alpha) - w(e_k) - w(\alpha, \beta)$$

   Let us denote $\delta(e_r)$ by

   $$max\{\delta(e_1), \delta'(e_1), \cdots, \delta(e_m), \delta'(e_m)\}.$$

   (e) Define by a *SwapPositve* the case where the condition

   $$\delta(e_r) > 0$$

   is true.

2. **Main loop**: Continue until $V_M = V$ for any pair (α, β) of outside vertices;
   (a) **Simple addition**: If condition *SwapPositve* is not satisfied for $\forall e_k \in M$, then

   $$M \Leftarrow M \cup \{(\alpha, \beta)\}$$

   (b) **Augmentation**: Otherwise, if there exists an edge (t_r, b_r) in M satisfying the condition $\delta(e_r) > 0$, then

   $$M \Leftarrow M \cup \{(t_r, \alpha), (b_r, \beta)\} - \{e_r\}$$

Lemma 8. *If the current matching M satisfies $w(M) \geq \frac{w(E)}{2|V_M|-1}$ and SwapPositive is not true for (α, β), then the new matching $M' = M \cup \{(\alpha, \beta)\}$ satisfies $w(M') \geq \frac{w(E)}{2|V_M|-1}$.*

Proof: We denote the sum of weights of the edges between the vertices of V_M is denoted by $w(V_M)$. Let us assume by induction that the current matching satisfies $w(M) \geq \frac{w(E)}{2|V_M|-1}$, i.e.,

$$w(V_M) \leq (2m - 1)w(M)$$

. Since every swap-gain to (α, β) is non-positve,

$$\sum_{k=1}^{m} \delta(e_k) + \delta'(e_k) \leq 0.$$

The set of edges between the vertices of V_M and vertices of $\{\alpha, \beta\}$ is denoted as $X(M; \alpha, \beta)$. Then, the above is equivalent to

$$w(X(M; \alpha, \beta)) - 2w(M) - 2mw(\alpha, \beta) \leq 0.$$

Their sum is

$$w(V_M) + w(X(M; \alpha, \beta)) \leq (2m + 1)w(M) + 2mw(\alpha, \beta).$$

Therefore,

$$w(V_M) + w(X(M; \alpha, \beta)) + w(\alpha, \beta) \leq (2m + 1)(w(M) + w(\alpha, \beta)).$$

This satisfies the bound for M'. $\square$

Lemma 9. *If the current matching M satisfies and SwapPositive and $(w(M) \geq \frac{w(E)}{n-1}$ is true, then the new matching*

$$M' = M \cup \{(t_r, \alpha), (b_r, \beta)\} - \{e_r\}$$

satisfies $w(M) \geq \frac{w(E)}{n-1}$.

Proof: Since $\delta(e_k)$ is maximum,

$$\sum_{k=1}^{m} \delta(e_k) + \delta'(e_k) \leq 2m\delta(e_r)$$

or equivalently,

$$w(X(M; \alpha, \beta)) - 2w(M) - 2mw(\alpha, \beta) \leq 2m(w(t_r, \alpha) + w(b_r, \beta) - w(e_r) - w(\alpha, \beta)).$$

By assumption,

$$w(V_M) \leq (2m - 1)w(M).$$

Their sum leads to

$$w(V_M) + w(X(M; \alpha, \beta)) \leq (2m + 1)w(M) + 2m(w(t_r, \alpha) + w(b_r, \beta) - w(e_r))$$

or equivalently,

$$\begin{aligned}
&w(V_M) + w(X(M; \alpha, \beta)) + w(\alpha, \beta) \\
&\leq (2m + 1)(w(M) + w(t_r, \alpha) + w(b_r, \beta) - w(e_r)) \\
&\quad -(w(t_r, \alpha)) + w(b_r, \beta)) - w(e_r) - w(\alpha, \beta)).
\end{aligned}$$

By assumption of *SwapPositive*,

$$\delta(e_r) = w(t_r, \alpha) + w(b_r, \beta)) - w(e_r) - w(\alpha, \beta) > 0.$$

Therefore, we have

$$w(V_M) + w(X(M; \alpha, \beta)) + w(\alpha, \beta) < (2m+1)(w(M) + w(t_r, \alpha) + w(b_r, \beta) - w(e_r))$$

. $\square$

Lemma 10. *The computational complexity of the algorithm is $O(e + n)$.*

Proof: Consider a generic step: when testing an arbitrary edge (α, β) to be included in a current matching M. We check all neighbors of α and β. Among them, identify the neighbors in M, call them $NEIGHBOR$. When you compute the swap-gains between (α, β) and (t_k, b_k) you only choose those (t_k, b_k) that are in NEIGHBOR. This takes $O(d_\alpha + d_\beta)$ time, where d_α is the degree of α (similarly, d_β is degree of β). Summing over all steps, the algorithm takes $O(\sum d_i) = O(e)$ time for checking all neighbors and $O(n+e)$ time for initializing the data structure. $\square$

Example 3 *The following example shows that the result satisfies $w(M) \geq \frac{w(E)}{n-1}$, but not the condition of 3-LOWM. Let us consider a graph with $V = \{1, 2, \cdots, 8\}$ and $\{w(1, 2) = 3, w(1, 3) = 4, w(2, 4) = 1, w(3, 5) = 4, w(3, 4) = 3, w(4, 6) = 1, w(5, 6) = 3, w(5, 7) = 20, w(6, 8) = 30, w(7, 8) = 40, w(4, 7) = 40\}$. Other edges are all weighted 0.*

Suppose we have a matching $M = \{(1, 2), (3, 4), (5, 6)\}$ all generated by simple addition. Let $(\alpha, \beta) = (7, 8)$. Since the swap-gain of $(5, 6)$ is positive, we have $M = \{(1, 2), (3, 4), (5, 7), (6, 8)\}$ with $W(M) = 56$. We observe that 3-LOWM does not hold for a length 4 circuit consisting of $(3, 4), (5, 7) \in M$ and $(3, 5), (4, 7) \notin M$. If we were to apply augmetation here, then we could have had a better matching $M' = \{(1, 2), (3, 5), (4, 7), (6, 8)\}$ with $w(M') = 77$. However, even M satisfies $w(M) \geq \frac{w(E)}{n-1}$ as we see $56 \geq \frac{w(E)}{n-1} = \frac{149}{7} = 21.3..$

4 Extensions

Based on the proof of the previous theorems (we here assume that n is even for brevity), we also obtain upperbounds on both unweighted and weighted bipartite graphs. For brevity, in the remaining of the paper, we assume n is even. The case for odd number of vertices can be derived accordingly.

Corollary 11. *Given a bipartite graph* $G = (V, E), |V| = n$ *and* $|E| = e$, *there exists a matching of size at least* $2e/n$ *which can be found in* $O(e+n)$ *time. For weighted case, a matching is of weight* $2w(E)/n$ *and can be computed in* $O(n^2)$ *time.*

Proof: The proof is similar to Theorem 4. Since a bipartite graph does not contain an odd cycle, we consider either Equation 1 or Equation 2 for the first case and either Equation 3 or Equation 4 for the the second case. Thus, since $w_G = w(T_{in}, T_{in}) + w(T_{in}, T_{out}) \leq (m-1)w_M + w_M + (n-2m)w_M/2$, we prove $2w_G/n \leq w_M$. $\square$

The idea used for the proof of Theorem 4 can be extended to develop an i-LOM algorithm, for $i > 3$. For example, algorithm $3 - LOM$ can be effectively applied for the general case by recursively contracting those augmented paths into ones of length three. In general, in a local search heuristic, the more neighbors we search, the greater chance of finding a better solution. Thus, with higher orders, the solution may be further improved. However, here are some counter arguments.

The following fact verifies that the approximate lowerbounds obtained for both unweighted general and bipartite graphs are existentially tight.

Fact 1 (Existential Upper Bounds) *There exists graphs with n vertices and e edges such that the maximum matching is of size at most* $\frac{ne}{2(n-1)}$ *for trees,* $\frac{2e}{n}$ *for bipartite graphs,* $\frac{ne}{2(3n-6)}$ *for planar graph;* $\frac{e}{n-1}$ *otherwise.*

For example, building on Theorem 4, the lowerbound on matching size for 5-LOCM can be easily extended and is at least $\frac{e}{(n-1)}$, which is same as one for 3-LOCM.

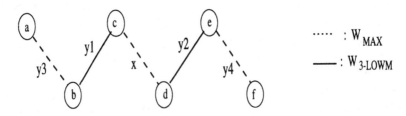

Fig. 3. Property 3-LOWM produces a matching of at least one-third the weight of the maximum weighted matching

For weighted case, we also have the following result on 3-LOWM.

Theorem 12. *3-LOWM produces a matching of at least one-half the weight of the maximum weighted matching.*

Proof: Refer to Figure 3. Let y_1 and y_2 be the weight of the first two matching edges that are selected by 3-LOWM. Let x be the weight of the non-matched edge such that $y_1 - x - y_2$ forms a path. Thus, we know that $w_x \leq w_{y_1} + w_{y_2}$. Now when all incident edges of the two matching edges are deleted, at most three edges of maximum matching may be removed (e.g., y_3, x and y_4). Let assume that those three edges are in maximum weighted matching. Then we have the following three inequalities:

$$w_x \leq w_{y_1} + w_{y_2} \tag{2}$$

$$w_{y_3} \leq w_{y_1} \tag{3}$$

$$w_{y_4} \leq w_{y_2} \tag{4}$$

When we sum up the above three inequalities, we have:

$$w_x + w_{y_3} + w_{y_4} \leq 2(w_{y_1} + w_{y_2})$$

. Therefore, total weights of the three edges in maximum weighted matching cannot exceed $2(y_1 + y_2)$. Since the argument may be repeated for each of the $n/2$ iterations of 3-LOWM, the theorem holds. $\square$

GREEDY-W [1] also provides a matching of at least half the weight of the maximum weighted matching. Therefore, a better heuristic for weighted graphs is to use GREEDY-W as an initialization step of the proposed algorithm to improve the solution. This strategy will guarantee a matching of at least (half $+ \beta (\geq 0)$) the weight of the maximum weighted matching. Here β is an improvement on a given graph after the initial matching.

For unweighted case, we can similarly prove:

Theorem 13. *The ℓ-LOCM, $\ell \geq 3$, produces a matching of at least $(\ell+1)/(\ell+2)$ the maximum cardinality matching.*

Thus, algorithm $3 - LOCM$ produces a matching of at least two-thirds the maximum cardinality matching. Whereas, similarly we can prove that GREEDY-C produces a matching of at least half the maximum cardinality matching.

5   Conclusion

The problem of approximate maximum matching has a number of applications, for example in, Vertex Cover [14], TSP [6], MAXCUT [5], and VLSI physical design problems [11, 20]. We have demonstrated iterative algorithms based on combinatorial techniques for approximating the maximum matching and related problems. We obtained linear time approximation algorithms for maximum cardinality matching by using the local optimality condition on path of length three. The proposed maximum cardinality matching algorithm was also extended to the weighted case. To the best our knowledge, the performance bounds achieved in this paper are the best known today. They are also existentially tight. Furthermore, the relative simplicity of the proposed algorithms and their computational economy are both keys to their practical applications. Also, two heuristics are proposed to improve the solutions of both 3-LOCM and 3-LOWM.

One natural open question is to ask whether Theorems 4 and 7 can be efficiently extended to handle the case for ℓ-LOM, where $\ell > 3$, and how to generalize Theorem 7 for ℓ-LOWM with the arbitrary length of ℓ.

Another open question is to find a lowerbound on the weighted matching when performing the combined strategy of GREEDY-W followed by 3-LOWM.

Euclidean maximum weighted matching problem is the problem of finding a matching with a maximum weight in a complete graph. The vertices are points in the Euclidean plane and the weight assigned to the corresponding edge is the Euclidean distance between two points. In this case, we expect much stronger results, for these weights satisfy the triangle inequality (i.e., geometry may help in matching).

6 Acknowledgment

We would like to thank Dr. Kunihiro Fujiyoshi of JAIST for his suggestions and Mr. Sunil Mishra for implementing the algorithms. The authors also thank Dr. David Williamson in MIT (dpwmson@theory.lcs.mit.edu) for pointing out a good survey on matching heuristics in [2]

References

1. D. Avis. "Worst Case Bounds for the Euclidean Matching Problem". In *International J. Comput. Math. Appl.*, 7:251–257, 1981.

2. D. Avis. "A Survey of Heuristics for the Weighted Matching Problem". In *Networks*, 13:475–493, 1983.

3. C. Berge. "Two Theorems in Graph Theory". In *Proc. Nat. Acad. Sci. 43*, pages 842–844, 1957.

4. B. Bollobás. *Extremal Graph Theory. London, NY; Academic Press*, page 58, 1978.

5. J. D. Cho, S. Raje, and M. Sarrafzadeh. "A Generalized Multi-Way Maxcut Partitioning". manuscript, *EECS Dept. Northwestern Univ.*, April 1993.

6. N. Christfides. "Worst-Case Analysis of a New Heuristic for the Travelling Salesman Problem". Technical Report, *Graduate School of Industrial Administration, Carnegie-Mellon University, Pittsburgh, PA*, 1976.

7. M. Fürer and B. Raghavachari. "Approximating the Minimum Degree Spanning Tree to within one from the Optimal Degree". In *Ann. ACM-SIAM Symp. on Discrete Algorithm*, volume 4, pages 317–324, 1992.

8. H. N. Gabow. "Implementation of Algorithm for Maximum Matching on Ninbipartite Graphs". *Ph.D. Thesis, Stanford University*, 1973.

9. H. N. Gabow. "An Efficient Implementation of Edmonds' Algorithm for Maximum Matching on Graphs". *Journal of ACM*, 23:221–234, 1976.

10. Z. Galil. "Efficient Algorithms for Finding Maximum Matching in Graphs". *Computing Surveys*, 18:23–38, 1987.

11. M. K. Goldberg, and M. Burstein. "Heuristic Improvement Technique for Bisection of VLSI Networks". *In Proc. of the IEEE International Conference on Computer Design* , 122–125, 1983.

12. H. N. Gabow, Z. Galil and T. H. Spencer. "Efficient Implementation of Graph Algorithms using Contraction". *In Proc. of the 25th Annual IEEE Symposium on Foundations of Computer Science*, 347–357, 1984.

13. M. R. Garey, D. S. Johnson, and L. Stockmeyer. "Some Simplified NP-Complete Graph Problems". *Theoretical Computer Science*, 1:237–267, 1976.

14. M. R. Garey, D. S. Johnson. "Computer and Intractability, A Guide to the Theory of NP-Completeness". *W. H. Freeman and Company, San Francisco, 1979.*

15. D. Goldfard. "Efficient Dual Simplex Algorithms for the Assignment Problem". *Math. Program.*, 33:187–203, 1985.

16. M. X. Goemans and D. P. Williamson. "A General Approximation Technique for Constrained Forest Problems". In *Ann. ACM-SIAM Symp. on Discrete Algorithm*, volume 4, pages 307–316, 1992.

17. D. J. Hagin and S. M. Venkatesan. "Approximation and Intractability Results for the Maximum Cut Problem and Its Variants". *IEEE Transactions on Computers*, 40(1):110–113, January 1991.

18. M. M. Halldórsson. "A Still Better Performance Guarantee for Approximate Graph Coloring". *Information Processing Letters*, 45:19–23, 1993.

19. J. E. Hopcroft and R. M. Karp. "$N^{5/2}$ algorithm for maximum matchings in bipartite graphs". *SIAM Journal on Computing*, 2:225–231, 1973.

20. T. Lengauer. "Combinatorial Algorithms for Integrated Circuit Layout". *Applicable Theory in Computer Science, John Wiley & Sons*, 273–274, 1990.

21. R. J. Lipton and R. E. Tarjan. "A Separator Theorem for Planar Graphs". *SIAM Journal on Applied Mathematics*, 36(2):177–189, April 1979.

22. R. J. Lipton and R. E. Tarjan. "Applications of a Planar Separator Theorem". *SIAM Journal on Computing*, 9:615–627, 1980.

23. L. Lovasz. "Matching Structure and the Matching lattice". *Journal of Combinatorial Theory (b)*, 43:187–222, 1987.

24. L. Lovasz and M. D. Plummer. *Matching Theory*, chapter "Chapter 3: Size and Structure of Maximum Matchings". Noth-Holland Mathematics Studies, Annals of Discrete Mathematics (29), 1986.

25. T. Mastumoto, N. Saigan, and K. Tsuji. "A New Efficient Approximation Algorithm for the Steiner Tree Problem in Rectilinear Graph ". In *International Symposium on Circuits and Systems*, pages 2874–2876. IEEE, 1990.

26. S. Micali and V. V. Vazirani. "An $O(\sqrt{|V|}|E|)$ Algorithm for Finding Maximum Matching in General Graphs". In *Proceedings of 21th Annual Symposium on the Foundations of Computer Science*, pages 17–27. IEEE, 1980.

27. T. Nishizeki. "Lower Bounds on the Cardinality of the Maximum Matchings of Graphs". In *Proceedings of 9th Southeastern Conference on Combinatorics, Graph Theory, and Computing*, 1978.

28. S. M. Venkatesan. "Approximation Algorithms for Weighted Matching". *Theoretical Computer Science*, 54:129–137, 1987.

29. P. M. B. Vitányi. "How Well Can A Graph Be n-Colored?". *Discrete Mathematics*, 34:69–80, 1981.

30. M. Yannakakis. "On the Approximation of Maximum Satisfiability". In *Proc. 3rd Annual ACM-SIAM Symp. on Discrete Algorithms*, pages 1–8, 1992.

New Lower Bounds and Hierarchy Results for Restricted Branching Programs

Detlef Sieling[*] and Ingo Wegener[*]

FB Informatik, LS II, Univ. Dortmund,
44221 Dortmund, Fed. Rep. of Germany

Abstract. Known lower bound techniques for depth restricted branching programs are not sensitive enough to lead to tight hierarchies. A new lower bound technique implies the separation of the classes of polynomial-size branching programs, where on each path k variables may be tested more than once and $k \leq (1 - \varepsilon)(n/3)^{1/3} / \log^{2/3} n$ for some $\varepsilon > 0$. Methods from communication complexity theory are adopted to separate the classes of polynomial-size ordered read k times branching programs, where $k = o(n^{1/2} / \log^2 n)$.

1 Restricted branching programs and known results

Branching programs are a well established computation model for discrete functions.

Definition 1. A branching program G for a function $f : A^n \to B$, where $A = \{0, \ldots, a-1\}$ and $B = \{0, \ldots, b-1\}$, is a directed acyclic graph with one source. The sink nodes are labeled by constants from B. The inner nodes are labeled by variables from $X = \{x_1, \ldots, x_n\}$ and have a outgoing edges labeled by the different elements from A. The computation path for the input $c \in A^n$ starts at the source. At an inner node with label x_i the outgoing edge with label c_i is chosen. The label of the sink which is finally reached has to equal $f(c)$. The size of G is equal to the number of its edges.

We discuss only explicitly defined functions (excluding diagonalization or counting methods for lower bounds) and with one exception we consider the Boolean case $a = b = 2$. The branching program size of Boolean functions f is known to be a measure for the space complexity of nonuniform Turing machines and known to lie between the circuit size of f and its $\{\wedge, \vee, \neg\}$-formula size (see e. g. Wegener [19]). Hence, only small polynomial size lower bounds can be proved, the largest one of size $n^2 / \log^2 n$ is still due to Nečiporuk [15].

Width-5 branching programs of polynomial size represent exactly the Boolean functions in NC^1 (Barrington [2]). Hence, exponential lower bounds are not known for width 5. But for some depth restricted models powerful lower bound techniques are known.

---

[*] Supported in part by DFG grant We 1066/7-1.

Definition 2. A branching program is called oblivious, if the node set can be partitioned into levels such that edges are leading from lower to higher levels and all inner nodes of one level are labeled by the same variable.

Alon and Maass [1] proved exponential lower bounds for some Boolean functions and oblivious branching programs with a linear number of levels. These lower bound techniques are too coarse to establish tight hierarchies, i.e. it cannot be proved (for explicitly defined functions) that the branching program size decreases from exponential to polynomial by increasing the number of levels a little bit.

Definition 3. A branching program is called read k times, if each variable is tested on each path at most k times.

Exponential lower bounds have been proved for read-once branching programs first independently by Žák [22] and Wegener [20]. For larger k we have exponential lower bounds by Okolnisch'kova [17] for $k \leq \varepsilon(\log n)/\log\log n$ and Borodin, Razborov and Smolensky [4] for $k \leq \varepsilon \log n$ and even nondeterministic branching programs. Again we do not obtain tight hierarchies. The considered functions are known to be computable in polynomial size only if $k \geq n^{1/2}$ (the functions of Okolnisch'kova [17]) or $k \geq n^2$ (the functions of Borodin, Razborov and Smolensky [4]).

Definition 4. A read k times branching program G is called ordered, if it respects a variable ordering π. I.e., there exists a partition of the node set of G into k layers such that edges do not lead from higher to lower layers, each layer is read-once and edges within each layer respect the variable ordering, i.e., $\pi(x_i) < \pi(x_j)$, if an edge leads from an x_i-node to an x_j-node.

Informally, it is allowed to test the variables in the given order and to repeat this k times. The lower bound method of Krause [11] leads to exponential lower bounds for $k = o(n/\log n)$ but it is also too coarse to allow tight hierarchy results.

We conclude that the existing lower bound techniques do not lead to tight hierarchies. In this paper new lower bound techniques are presented from which some tight hierarchies can be derived. In Section 2 an overview about the results and methods is given. Some details of the proofs are given in Section 3 and 4.

It is evident that lower bounds, lower bound techniques and hierarchy results contribute to the structure of branching program based complexity classes. But, moreover, some of the considered restricted branching program models can be used (and are used in practice) as data structure for Boolean functions.

In computer-aided design, symbolic verification, test pattern generation, symbolic simulation and logical synthesis one has to manipulate Boolean functions with software tools. Hence, a data structure or representation for Boolean functions is needed which allows a compact representation of many functions and efficient algorithms for a list of operations, among them satisfiability and equality test and the synthesis of representations by Boolean operators (see Wegener

[21]). The most often used representation (called ordered binary decision diagram) is in our notation the representation by ordered read-once branching programs. For this representation all operations have efficient algorithms (Bryant [5][6]) but already simple functions need exponential size (Bryant [7]). Gergov and Meinel [9] and Sieling and Wegener [18] have investigated how one can work with unordered read-once branching programs. Bollig, Sauerhoff, Sieling and Wegener [3] have proved that ordered read k times branching programs allow for constant k polynomial time algorithms for all important operations and may serve as data structure for Boolean functions. Without the order restriction the satisfiability problem is NP-hard already for read-twice branching programs.

2 New results

Ordered and unordered read-once branching programs are the most restricted models of branching programs representing all Boolean functions. Starting from these models we discuss two types of hierarchies. First we relax the restrictions as slightly as possible by allowing that on each path k variables are tested more than once. Then we investigate the situation where we increase the number of layers. The hierarchy problem for unordered read k times branching programs remains open. To be precise we define the considered complexity classes.

Definition 5. i) $C_1(k)$ is the class of Boolean functions computable by polynomial-size $BP_1(k)$s, i.e. ordered read-twice branching programs, where the length of the paths in the second layer is bounded by k.
ii) $C_2(k)$ is the class of Boolean functions computable by polynomial-size $BP_2(k)$s, i.e. branching programs, where on each path at most k variables are tested more than once.
iii) $C_3^{int}(k)$ and $C_3^{Bool}(k)$ are the classes of functions $f_n : \{0, \ldots, n-1\}^n \to \{0, \ldots, n-1\}$ and $f_n : \{0,1\}^n \to \{0,1\}$ resp. computable by polynomial-size $BP_3(k)$s, i.e. ordered read k times branching programs.

For the $C_1(k)$- and the $C_2(k)$-hierarchy we use the same functions. The idea is to force polynomial-size branching programs to test k variables for a second time. The variables are partitioned to k groups. Each group selects one variable in such a way that after the test of many variables each variable has still the chance of being chosen. The output is the parity of all selected variables.

Definition 6. The Boolean function $f_n^k : \{0,1\}^n \to \{0,1\}$ is defined on the set of variables $X = \{x_0, \ldots, x_{n-1}\}$. Let m be the largest number, where $mk\lceil \log n \rceil \leq n$. We partition the variables to k groups each consisting of m numbers of bit length $\lceil \log n \rceil$. Let $s(j)$ be the sum (mod n, if n is odd, and mod$(n-1)$ else) of the numbers of the j-th group. Then $f_n^k(x_0, \ldots, x_{n-1}) = x_{s(1)} \oplus \ldots \oplus x_{s(k)}$.

Theorem 7. *(Upper bounds for f_n^k.)*
i) *The function f_n^k can be represented by a $BP_1(k)$ of size $O(n^{k+1})$.*
ii) *The function f_n^k can be represented by a $BP_2(k)$ of size $O(n^2)$.*

Theorem 8. *(Lower bounds for f_n^k.)*

Each $BP_1(k-1)$ or $BP_2(k-1)$ for f_n^k has size $2^{\Omega\left(\frac{n}{k^2 \log n} - 3k \log n - 2k\right)}$.

Theorem 9. *(Hierarchy results.)*

i) $C_1(k-1) \subsetneq C_1(k)$, if $k = O(\log^{1-\varepsilon} n)$ for some $\varepsilon > 0$.

ii) $C_2(k-1) \subsetneq C_2(k)$, if $k \leq (1-\varepsilon)(n/3)^{1/3}/\log^{2/3} n$ for some $\varepsilon > 0$.

The hierarchy theorem follows directly from Theorem 7 and Theorem 8, if one applies the usual padding method for part i). The simple proof of Theorem 7 is given in Section 3.

For the proof of Theorem 8 a new lower bound technique is needed, which is precise enough to allow the hierarchy result (see Section 3). Here we give an outline of the proof technique. Lower bounds for read-once branching programs can be proved with the cut-and-paste technique (Wegener [20]). One proves that paths which split early cannot be merged after a small number of tests, since otherwise the read-once branching program works wrong for some input. Now some variables may be tested again and again and a more careful analysis is necessary.

The proof argues about consistent paths, i.e. paths, where no variable is tested with different results. For some appropriately chosen number u we mark on each consistent path the first node, where the number of different tested variables reaches u. The parameter u is chosen large enough such that the number of marked nodes may be large. But u is also chosen small enough such that not too many paths, which split before the marked node, may be merged again before the marked node. The lower bound is proved for the number of marked nodes. The main part is a proof of an upper bound on the number of paths with the same marked node v. The set of paths p from the source to the marked node v is partitioned according to the set of variables tested on p up to v. By combinatorial methods it is proved that there cannot be too many different sets of tested variables, since only k variables may be tested again below v. Then the paths with a fixed set X' of tested variables are investigated. The properties of $BP_2(k)$s imply necessary conditions for those assignments of the variables in X' which lead to v. These conditions can be stated as a system of Boolean equations. Finally, the number of solutions of this system of Boolean equations can be estimated by algebraical methods.

For the hierarchy result for ordered read k times branching programs we investigate directed bipartite graphs, whose nodes have outdegree 1. The node set is $U \cup V \cup W$, where $U = \{u\}$, $V = \{v_0, \ldots, v_{n-1}\}$, $W = \{w_0, \ldots, w_{n-1}\}$. The edges from nodes in $U \cup W$ lead to nodes in V and the edges from nodes in V lead to nodes in W. With $p = (p_0 = u, p_1, \ldots, p_{2k+1})$ we denote the unique path of length $2k+1$ starting at u. The edge leaving u, v_i or w_j is described by the variable z, x_i and y_j resp. The pointer jumping function $\mathrm{PJ}_{k,n}^{\mathrm{int}}$ outputs the index j, if $p_{2k+1} = v_j$, and works on $2n+1$ variables $z, x_0, \ldots, x_{n-1}, y_0, \ldots, y_{n-1}$ with values in $\{0, \ldots, n-1\}$ describing the index of the node reached by the corresponding edge. In order to obtain a Boolean variant $\mathrm{PJ}_{k,n}^{\mathrm{Bool}}$, if $n = 2^l$ and $l \in \mathbb{N}$, each variable is replaced by $\log n$ Boolean variables, e.g. x_i by

$x_{i,\log n-1}, \ldots, x_{i,0}$. Furthermore, n Boolean variables $c_0, \ldots, c_{n-1}$ describing a coloring $c : V \to \{0, 1\}$ of the nodes in V are added. The output of $\mathrm{PJ}_{k,n}^{\mathrm{Bool}}$ is $c(p_{2k+1})$, where p_{2k+1} is defined as before.

Theorem 10. *(Upper bounds for* $\mathrm{PJ}_{k,n}^{\mathrm{int}}$ *and* $\mathrm{PJ}_{k,n}^{\mathrm{Bool}}$.*)*
The functions $\mathrm{PJ}_{k,n}^{\mathrm{int}}$ *and* $\mathrm{PJ}_{k,n}^{\mathrm{Bool}}$ *can be computed by* $BP_3(k)s$ *of size* $O(kn^2)$.

Theorem 11. *(Lower bounds for* $\mathrm{PJ}_{k,n}^{\mathrm{int}}$ *and* $\mathrm{PJ}_{k,n}^{\mathrm{Bool}}$.*)*
i) *Each* $BP_3(k-1)$ *for* $\mathrm{PJ}_{k,n}^{\mathrm{int}}$ *has size* $2^{\Omega(n/k)}/n$.
ii) *Each* $BP_3(k-1)$ *for* $\mathrm{PJ}_{k,n}^{\mathrm{Bool}}$ *has size* $2^{\Omega(n^{1/2}/k)}/n$.

Theorem 12. *(Hierarchy results.)*
i) $C_3^{\mathrm{int}}(k-1) \subsetneqq C_3^{\mathrm{int}}(k)$, *if* $k = o(n/\log n)$.
ii) $C_3^{\mathrm{Bool}}(k-1) \subsetneqq C_3^{\mathrm{Bool}}(k)$, *if* $k = o(n^{1/2}/\log^2 n)$.

The lower bound proofs adopt methods from communication complexity (in particular Nisan and Wigderson [16], but also Duris, Galil and Schnitger [8], Halstenberg and Reischuk [10], Lam and Ruzzo [12] and McGeoch [14]), see Section 4.

3 Read-once branching programs with k additional tests

In this section we prove Theorem 7 and Theorem 8. W.l.o.g. we assume that n is odd.

Sketch of Proof of Theorem 7. We describe an ordered read-once branching program P_j computing $s(j)$. The depth of P_j is bounded by $m\lceil\log n\rceil \le n/k$, since $s(j)$ depends only on $m\lceil\log n\rceil$ variables. A variable x_r at position $\mathrm{pos}(r)$ in its binary number contributes $x_r 2^{\mathrm{pos}(r)} \bmod n$ to the sum $s(j)$. Hence, width n is sufficient for each level to store the different partial sums mod n. The size of P_j is $O(n^2/k)$.

In the $BP_1(k)$ for f_n^k we arrange copies of $P_1, \ldots, P_k$ in a complete n-ary tree of depth k. At the j-th level of this tree the number $s(j)$ is computed and the vector $(s(1), \ldots, s(j))$ is stored, because paths with different values for $s(j)$ are never joined. In this part of the branching program each variable is tested only once on each path. For each value of the vector $(s(1), \ldots, s(k))$ we reach a different sink. Now we replace the sink for $(s(1), \ldots, s(k))$ by a branching program that computes $x_{s(1)} \oplus \ldots \oplus x_{s(k)}$.

In the $BP_2(k)$ we may repeat tests not only at the end of each path. We perform the test of $x_{s(j)}$ immediately after the computation of $s(j)$, i.e. we replace the i-th sink of P_j by a test of x_i and obtain a branching program P_j^* that computes for the j-th block the value $x_{s(j)}$. Then P_1^* and two copies of each P_j^*, $j > 1$, are sufficient for the computation of $x_{s(1)} \oplus \ldots \oplus x_{s(k)}$. $\square$

For the proof of Theorem 8 the following lemmas are useful. Lemma 13 states that for $s(1), \ldots, s(k)$ arbitrary values are possible, even if a large number of input bits has been replaced by constants. We recall that m is the number of binary numbers in each group.

Lemma 13. *Let* $t(1), \ldots, t(k) \in \{0, \ldots, n-1\}$. *If in the input* $x = (x_0, \ldots, x_{n-1})$ *at most* $(m-1)$ *bits are replaced by arbitrary constants, there is an assignment to the remaining bits so that* $s(l) = t(l)$ *for all* $l \in \{1, \ldots, k\}$.

Proof. If at most $m-1$ bits are replaced by constants, there is in each group some binary number in which no bit has been replaced. For each group we can replace all bits outside this number by arbitrary constants and then we can choose a suitable value for this number in order to get $s(l) = t(l)$. $\square$

Let v be a node in a branching program. For each consistent path leading from the source to v we count how many different variables are tested on this path before v is reached. We denote the largest of these numbers by $L(v)$ and the smallest by $S(v)$. Since we consider only consistent paths, null-chains (i. e. inconsistent paths) can affect neither $S(v)$ nor $L(v)$. The following lemma implies that for the nodes in the top part of a $BP_1(k-1)$ or $BP_2(k-1)$ that computes f_n^k the values $S(v)$ and $L(v)$ cannot differ too much. We present this lemma without proof.

Lemma 14. *Let a* $BP_2(k-1)$ *for* f_n^k *be given and let* $u^* = m - 2k - 2$. *Let* $u \le u^*$ *and let* v *be a node in the branching program that is reachable from the source via a consistent path* P, *on which* u *variables* $x_{i(1)}, \ldots, x_{i(u)}$ *are tested. Then on each other path* Q *from the source to* v *at most* $(k-1)$ *variables not contained in* $\{x_{i(1)}, \ldots, x_{i(u)}\}$ *are tested.*
In particular, if $S(v) \le u^*$, *then* $S(v) \ge L(v) - k + 1$.

Sketch of Proof of Theorem 8. Let a $BP_2(k-1)$ for f_n^k be given and let $u^* = m - 2k - 2$. In the given branching program we mark all nodes v for which $L(v) \le u^*$ and $S(v) \ge u^* - 2k + 2$. By Lemma 13 each consistent path contains at least $m - 1 \ge u^*$ different variables. Therefore, it contains a node v such that $u^* - k + 1$ different variables are tested before. By Lemma 14 this node becomes marked. We also know that on each consistent path at least $u^* - 2k + 2$ variables are tested before a marked node is reached. Therefore, there are at least $2^{u^* - 2k + 2}$ consistent paths from the source to all marked nodes. In the following we prove that the number of consistent paths from the source to a single marked node is bounded by $O(n^{3k-2} 2^{u^*(\frac{k-1}{k})})$. Hence, the number of marked nodes is at least

$$\Omega\left(\frac{2^{u^* - 2k + 2}}{n^{3k-2} 2^{u^*(\frac{k-1}{k})}}\right) = 2^{\Omega\left(\frac{n}{k^2 \log n} - 3k \log n - 2k\right)}.$$

Claim. Let v be a node in a $BP_2(k-1)$ for f_n^k and let $u^* = m - 2k - 2$. If $L(v) \le u^*$, then the number of consistent paths leading from the source to v is bounded by $O(n^{3k-2} 2^{u^*(\frac{k-1}{k})})$.

Sketch of Proof. Let T be the decision tree for f_n^k such that for all inputs the sequence of tested variables is the same as in the given $BP_2(k-1)$. Let V^* be the set of nodes in T representing the given node v and being reached in T on a consistent path. We partition the set of paths leading from the source of T to some node $v^* \in V^*$ into sets $P(j)$, $1 \leq j \leq A_v$, of paths on which exactly the same variables are tested. The claim is proved by proving an upper bound $O(n^{2k-2})$ for A_v and an upper bound $O(n^k 2^{u^*(\frac{k-1}{k})})$ for the size of the sets $P(j)$. By Lemma 14 the set of variables tested on different consistent paths to some node $v^* \in V^*$ cannot differ to much. This implies the upper bound on A_v.

For the proof of the upper bound $O(n^k 2^{u^*(\frac{k-1}{k})})$ on the size of the sets $P(j)$ let us consider some set $P(j)$ and let $U = \{x_{i(1)}, \ldots, x_{i(u)}\}$ be the set of variables tested on the paths in $P(j)$. We know that $u \leq u^*$, since $L(v) \leq u^*$. Since the subtrees whose sources are in V^* are isomorphic, it is possible to merge all nodes in V^* which belong to paths in $P(j)$. Let v^* be the resulting node and T^* the decision tree whose source is v^*. On each path in T^* at most $k-1$ variables contained in U are tested. We rearrange T^* in such a way that the U-variables are tested at the end of each path. Let $Y := X - U$. Perform on the decision tree successively the following operations for each $x^* \in Y$:

- Create a new source node labeled by x^*. The successors of this node are two copies of the previous decision tree.
- Eliminate redundant tests and nonreachable nodes and edges.

In the second step all nodes labeled by x^* except the new source node are removed. The new decision tree computes the same function as the old one. Before the rearrangement on each path in the decision tree at most $k-1$ U-variables are tested. The same holds afterwards, because only tests of $x^* \in Y$ are inserted. Now the tests of the U-variables are the last tests on each path. These tests are arranged in small decision trees of depth $k-1$ in the bottom part of the decision tree with root v^*. In the following we examine which functions have to be computed by these small decision trees.

Each path from the source to v^* defines an assignment to $x_{i(1)}, \ldots, x_{i(u)}$. We call $(s^*(1), \ldots, s^*(k))$ the value of this partial assignment, if after assigning 0 to all other variables we get $s^*(i)$ as the sum mod n of the numbers in the i-th group ($i \in \{1, \ldots, k\}$). Similarly we can define the value of an assignment to variables in Y. Then we get $(s(1), \ldots, s(k))$ as the sum of the value of the assignment to the variables in U and the value of the assignment to the variables in Y. We derive an upper bound for the number of those paths leading from the source to v^* for which the values of the partial assignments to $x_{i(1)}, \ldots, x_{i(u)}$ are equal to a fixed vector $(s^*(1), \ldots, s^*(k))$. We multiply this upper bound by n^k in order to obtain the upper bound for the number of all paths leading to v^*.

Now we fix $(s^*(1), \ldots, s^*(k))$ and consider only assignments to $x_{i(1)}, \ldots, x_{i(u)}$ with value $(s^*(1), \ldots, s^*(k))$. Lemma 13 implies that we can choose for $(s(1), \ldots, s(k))$ every value in $\{0, \ldots, n-1\}^k$ and that we can assign suitable values to the variables in Y in order to get the chosen value $s(i)$ as the sum of the numbers in the i-th group. On the other hand this assignment to the variables in Y determines a path starting at v^* and leading to one of the small decision trees in the bottom part of the decision tree with root v^*.

We assign values to the variables in Y so that $s(1) = i(1)$, $s(2) = i(2), \ldots,$ $s(k) = i(k)$. Then the value of f_n^k is $x_{i(1)} \oplus \ldots \oplus x_{i(k)}$. According to the assignments of the variables in Y we reach one of the small decision trees in the bottom part which computes a function $g_1(x_{i(1)}, \ldots, x_{i(u)})$. This is also the value that the branching program computes for the chosen assignment to the variables in Y and each assignment to $(x_{i(1)}, \ldots, x_{i(u)})$ with value $(s^*(1), \ldots, s^*(k))$ that defines a path from the source to v^*. Since $g_1(x_{i(1)}, \ldots, x_{i(u)})$ is computed by a decision tree of depth $k-1$, it is different from $x_{i(1)} \oplus \ldots \oplus x_{i(k)}$. Among the assignments to $x_{i(1)}, \ldots, x_{i(u)}$ with value $(s^*(1), \ldots, s^*(k))$ only those may define paths leading to v^*, for which the equation $x_{i(1)} \oplus \ldots \oplus x_{i(k)} \oplus g_1(x_{i(1)}, \ldots, x_{i(u)}) = 0$ holds.

We can derive more equations by choosing assignments to the variables in Y for which $s(1) = i(lk + 1)$, $s(2) = i(lk + 2), \ldots,$ $s(k) = i((l+1)k)$, where $l = 1, \ldots, t-1$ and $t := u/k$ (w.l.o.g. an integer). Therefore, all of the following equations have to be satisfied by assignments to $x_{i(1)}, \ldots, x_{i(u)}$ with value $(s^*(1), \ldots, s^*(k))$ which define paths leading to v^*.

$$x_{i(1)} \qquad \oplus \ldots \oplus x_{i(k)} \oplus g_1(x_{i(1)}, \ldots, x_{i(u)}) = 0$$
$$\vdots \tag{1}$$
$$x_{i((t-1)k+1)} \oplus \ldots \oplus x_{i(tk)} \oplus g_t(x_{i(1)}, \ldots, x_{i(u)}) = 0$$

The function g_j is the function computed by the decision tree of depth $k - 1$ which is reached for the corresponding assignment to the variables in Y.

The number of solutions of the system of equations (1) is an upper bound for the number of paths with value $(s^*(1), \ldots, s^*(k))$ leading to v^*. It suffices to prove the upper bound $2^{u^*(\frac{k-1}{k})}$ for this number. Since there are n^k possible values for $(s^*(1), \ldots, s^*(k))$, the number of paths leading to v^* is bounded by $n^k 2^{u^*(\frac{k-1}{k})}$. This implies the desired upper bound for the number of consistent paths leading to v and completes the proof of Theorem 8.

Let $G_j(x_{i(1)}, \ldots, x_{i(tk)})$ denote the left-hand side of the j-th equation of (1), i.e. $G_j(x_{i(1)}, \ldots, x_{i(tk)}) := x_{i((j-1)k+1)} \oplus \ldots \oplus x_{i(jk)} \oplus g_j(x_{i(1)}, \ldots, x_{i(u)})$. Let N_w, $w \in \{0,1\}^t$, denote the number of assignments to $x_{i(1)}, \ldots, x_{i(tk)}$, for which $(G_1(x_{i(1)}, \ldots, x_{i(tk)}), \ldots, G_t(x_{i(1)}, \ldots, x_{i(tk)})) = w$. The number of assignments satisfying all equations in (1) is $N_{(0,\ldots,0)}$. We show $N_w = 2^{tk-t} = 2^{u(\frac{k-1}{k})}$ not only for $w = (0, \ldots, 0)$, but even for all $w \in \{0,1\}^t$.

Since there are 2^{tk} assignments to $x_{i(1)}, \ldots, x_{i(tk)}$, we get the equation

$$\sum_{w \in \{0,1\}^t} N_w = 2^{tk}. \tag{2}$$

Now fix some set $J \subseteq \{1, \ldots, t\}$, $J \neq \emptyset$. It can be shown that the numbers of assignments to $x_{i(1)}, \ldots, x_{i(tk)}$ for which $\bigoplus_{j \in J} G_j(x_{i(1)}, \ldots, x_{i(tk)}) = 0$ and $\bigoplus_{j \in J} G_j(x_{i(1)}, \ldots, x_{i(tk)}) = 1$ is the same. This implies

$$\sum_{w \mid \bigoplus_{j \in J} w_j = 0} N_w - \sum_{w \mid \bigoplus_{j \in J} w_j = 1} N_w = 0. \tag{3}$$

Since there are $2^t - 1$ choices for the set J, we get $2^t - 1$ such equations. Together with equation (2) we obtain a system of 2^t linear equations with 2^t variables $N_w, w \in \{0,1\}^t$. It is easy to check that $N_w = 2^{tk-t}$ for all $w \in \{0,1\}^t$ satisfies all linear equations. Therefore, it suffices to prove that this is the unique solution. This follows from the fact that the rank of the matrix of coefficients is maximal because it is a Sylvester-matrix (see e.g. MacWilliams and Sloane [13]).

4   Ordered read k times branching programs

The upper bounds of Theorem 10 have a simple proof. At the source (level 0) z ist tested. On level i, $1 \le i \le 2k$, we have n nodes labeled by the variables $x_0, \ldots, x_{n-1}$, if i is odd, and by $y_0, \ldots, y_{n-1}$, if i is even. On level $2k+1$ we have n sinks labeled $0, \ldots, n-1$. Edges from level i lead to level $i+1$. If the edge label is j, the edge reaches the node with label x_j, y_j or j. This branching program represents $PJ_{k,n}^{int}$ with size $O(kn^2)$. It is ordered read k times for the variable ordering $z, x_0, \ldots, x_{n-1}, y_0, \ldots, y_{n-1}$. For $PJ_{k,n}^{Bool}$ the inner nodes are replaced by binary decision trees for the Boolean variables replacing the considered integer variable and the sink with label j is replaced by a test of the color variable c_j. The a-successor of this test is the sink with label $a \in \{0,1\}$.

For the proof of the lower bounds of Theorem 11 the following strategy is used. We start with an ordered read $(k-1)$ times branching program G representing $PJ_{k,n}^{int}$ or $PJ_{k,n}^{Bool}$ with size s. Based on the ordering of the variables used by G a communication game for a suitable subfunction of the pointer jumping function is discussed. The branching program G leads to an upper bound on the minimal protocol length l_{min} for the communication game which depends on s. Adopting the methods of Nisan and Wigderson [16] a lower bound on l_{min} is proved. From these two bounds the lower bound on s is derived.

First we consider $PJ_{k,n}^{int}$. It is always possible to assume that z is tested only at the source of G. This may change the variable ordering but the size of G is increased at most by the factor n, since we may consider disjoint copies of G as successors of the source. The tests of z in these copies can be eliminated. Let L be the list representing the ordering of the x- and y-variables used by G. We break L in the middle into L_A and L_B. If L_A contains at least $n/2$ x-variables, player A of our communication game obtains z and the x-variables of L_A and player B the y-variables of L_B. Otherwise player A obtains z and the y-variables of L_A and player B the x-variables of L_B. Let $V' \subseteq V$ and $W' \subseteq W$ be the sets of nodes v_i resp. w_j such that x_i resp. y_j is given to some player. The set of inputs is restricted to those graphs, where the edges from V' (resp. $U \cup W'$) reach nodes in W' (resp. V'). All edges leaving $V - V'$ (resp. $W - W'$) reach w_0 (resp. v_0). The players A and B have to evaluate $PJ_{k,n}^{int}$ in $2k - 2$ rounds of communication, where A writes in the first round.

Using G we obtain a communication protocol of length $(2k - 2)\lceil \log s \rceil + \lceil \log n \rceil$. The i-th layer of G is partitioned into two sublayers such that player A knows the variables in the first sublayer and player B knows the variables in the second one. The computation path starts at the source. If a player knows at which node the computation path reaches one of his sublayers, it computes and writes the number of the first successor node on the computation path not belonging to his sublayers. The number of the sink reached by the computation path is written not later than in the $(2k-2)$-th round. The bound on the protocol length follows, since A may write in the first round, which copy of G is used by the computation path. Afterwards the numbers of the nodes in the considered copy of G can be described with $\lceil \log s \rceil$ bits.

The lower bound $\varepsilon n - (2k-2)\lceil \log n \rceil$ for some $\varepsilon > 0$ on the protocol length follows from Theorem 8 in Nisan and Wigderson [16]. The input restriction is not crucial, since V' and W' contain at least $n/2$ nodes each. The lower bound works for $2k-1$ rounds of communication and even for $2k$ rounds of communication, if player A gets y-variables. Hence, $\log s = \Omega(n/k) - \log n$ by combining the bounds.

Nisan and Wigderson [16] have shown that their lower bound holds even if a single bit of the binary representation of p_{2k+1} has to be computed. This is for our communication game not necessarily true. The nodes in V' may have the same, e. g., last bit. This is the reason why we have introduced the color variables for the Boolean variant $\mathrm{PJ}_{k,n}^{\mathrm{Bool}}$. The other problem is that the Boolean variables $x_{j,\log n-1}, \ldots, x_{j,0}$ replacing the integer variable x_j may be distributed arbitrarily over the variable ordering used by G.

Now we give more details of the proof of the lower bound in the Boolean case. Let G be an ordered read $(k-1)$ times branching program representing $\mathrm{PJ}_{k,n}^{\mathrm{Bool}}$ with size s. Again it can be assumed w.l.o.g. that $z_{\log n-1}, \ldots, z_0$ are tested only at the top of G. Let L be the list representing the ordering of the x- and y-variables used by G. For each i we mark in L the $(\log n)/2$-th Boolean variable $x_{i,\cdot}$ and the same for the $y_{i,\cdot}$-variables. Now we break L into L_A and L_B. The breakpoint is the n-th marked variable. If L_A contains at least $n/2$ marked x-variables, player A of our communication game obtains the z-variables and the $x_{i,\cdot}$-variables of list L_A, where the marked $x_{i,\cdot}$-variable belongs to L_A, and player B obtains the $y_{j,\cdot}$-variables of list L_B, where the marked $y_{j,\cdot}$-variable belongs to L_B. In the other case L_A contains at least $n/2$ marked y-variables and player A gets the z-variables and some y-variables chosen in a similar way as in the first case the x-variables. In the same way player B gets some x-variables. Let $V' \subseteq V$ and $W' \subseteq W$ be the sets of nodes v_i resp. w_j such that some $x_{i,\cdot}$- resp. $y_{j,\cdot}$-variables are given to some player.

The variables not given to some player are now fixed. Variables belonging to nodes in $V - V'$ or $W - W'$ are set to 0. Let $v_i \in V'$ (the case $w_j \in W'$ is handled similarly). The $x_{i,\cdot}$-variables not given to some player are fixed in such a way that at least $n^{1/2}/2$ nodes in W' are reachable by an edge from v_i. A simple counting argument shows that this is possible. The color variables are fixed in such a way that for each $w_j \in W'$ at least a third of the nodes in V' reachable by an edge from w_j has color 0 and at least a third has color 1. The existence of such a coloring follows by probabilistic arguments. The probability that a random coloring has not the desired property for some node $w_j \in W'$ is exponentially small. The claim follows, since the number of nodes in W' is linear.

The players have to evaluate $PJ_{k,n}^{\text{Bool}}$ in $2k - 2$ rounds of communication, where A writes in the first round. The ordered read $(k - 1)$ times branching program G leads to a legal protocol of length $(2k - 2)\lceil \log s \rceil + \log n$.

We want to prove the existence of some $\varepsilon > 0$ such that protocols of length $\varepsilon n^{1/2} - (2k - 2)\log n$ are too short to solve our communication game. We discuss how this bound follows with the methods of Nisan and Wigderson [16]. We assume that the input is chosen uniformly at random from all graphs considered in the communication game, where, moreover, the edges from $U \cup W'$ lead to V' and the edges from V' lead to W'. Player A knows p_1 and in Case 1 of the definition of the communication game, where he obtains x-variables, also p_2. It is assumed that the message written in the t-th round is completed in Case 1 by p_{t+1} and in Case 2 by p_t. Then the notion of nice nodes can be adopted from Nisan and Wigderson [16], if one replaces v_{t-1} by p_{t+1} (Case 1) or p_t (Case 2). The proof of the Main Lemma of Nisan and Wigderson [16] works, since we discuss protocols whose length is bounded by $\varepsilon n^{1/2} - (2k - 2)\log n$. This assumption is necessary, since the nodes have only at least $n^{1/2}/2$ possible successors. It follows that the information about p_{2k+1} is bounded above by a constant δ after $2k - 1$ rounds of communication (even after $2k$ rounds of communication in Case 2). The constant δ can be made arbitrarily small by choosing ε small enough. But by the choice of the coloring of the nodes the random variable describing the color of p_{2k+1}, if p_{2k} is known, takes each of the two possible values with probability at least $1/3$. If the protocol is legal, the color of p_{2k+1} is known after $2k - 2$ rounds of communication and the information about p_{2k+1} is larger than 0.9. Altogether, we have proved the existence of an $\varepsilon > 0$ such that protocols of length $\varepsilon n^{1/2} - (2k - 2)\log n$ fail on some inputs.

We remark that the lower bounds of Theorem 11 even hold for ordered read k times branching programs, where in the k-th layer only the c-, x- and z-variables may be tested. The hierarchy results of Theorem 12 follow easily from Theorem 10 and Theorem 11. For the Boolean variant we have to pay attention to the fact that $PJ_{k,n}^{\text{Bool}}$ is defined on $\Theta(n \log n)$ variables.

Acknowledgement

The work on the hierarchy for ordered read k times branching programs was initiated by a discussion between Noam Nisan and the second author during the Leibniz complexity theory workshop in Jerusalem (May 1993).

References

1. Alon, N. and Maass, W.: Meanders and their applications in lower bound arguments. Journal of Computer and System Sciences **37** (1988) 118–129
2. Barrington, D.: Bounded-width polynomial-size branching programs recognize exactly those languages in NC^1. In 18th Symp. on Theory of Comp. (1986) 1–5
3. Bollig, B., Sauerhoff, M., Sieling, D. and Wegener, I.: Read k times ordered binary decision diagrams — efficient algorithms in the presence of null-chains. Preprint Univ. Dortmund (1993)

4. Borodin, A., Razborov, A. and Smolensky, R.: On lower bounds for read-k-times branching programs. Computational Complexity **3** (1993) 1–18
5. Bryant, R.E.: Symbolic manipulation of Boolean functions using a graphical representation. In 22nd Design Automation Conference (1985) 688–694
6. Bryant, R.E.: Graph-based algorithms for Boolean function manipulation. IEEE Transactions on Computers **35** (1986) 677–691
7. Bryant, R.E.: On the complexity of VLSI implementations and graph representations of Boolean functions with application to integer multiplication. IEEE Transactions on Computers **40** (1991) 205–213
8. Duris, P., Galil, Z. and Schnitger, G.: Lower bounds of communication complexity. In 16th Symp. on Theory of Comp. (1984) 81–91
9. Gergov, J. and Meinel, C.: Frontiers of feasible and probabilistic feasible Boolean manipulation with branching programs. In 10th Symp. on Theoretical Aspects of Computer Science (1993) 576–585
10. Halstenberg, B. and Reischuk, R.: On different modes of communication. In 20th Symp. on Theory of Comp. (1988) 162–172
11. Krause, M.: Lower bounds for depth-restricted branching programs. Information and Computation **91** (1991) 1–14
12. Lam, T. and Ruzzo, L.: Results on communication complexity classes. In 4th Structures in Complexity Theory Conference (1989) 148–157
13. MacWilliams, F.J. and Sloane, N.J.A.: The Theory of Error-Correcting Codes. North-Holland Publishing Company (1977)
14. McGeoch, L.A.: A strong separation between k and $k-1$ round communication complexity for a constructive language. Techn. Rep. CMU-CS-86-157, Carnegie Mellon University, Pittsburg, PA (1986)
15. Nečiporuk, È.I.: A Boolean function. Soviet Mathematics Doklady **7(4)** (1966) 999–1000
16. Nisan, N. and Wigderson, A.: Rounds in communication complexity revisited. SIAM Journal on Computing **22** (1993) 211–219
17. Okolnisch'kova, E.A.: Lower bounds on the complexity of realization of characteristic functions of binary codes by branching programs. Diskretnii Analiz **51** (1991) 61–83 (in Russian)
18. Sieling, D. and Wegener, I.: Graph driven BDD's — A new data structure for Boolean functions. Theoretical Computer Science (to appear)
19. Wegener, I.: The complexity of Boolean functions. Wiley (1987)
20. Wegener, I.: On the complexity of branching programs and decision trees for clique functions. Journal of the ACM **35(2)** (1988) 461–471
21. Wegener, I.: Efficient data structures for Boolean functions. Discrete Mathematics (Special Volume on "Trends in Discrete Mathematics", to appear)
22. Žák, S.: An exponential lower bound for one-time-only branching programs. In 11th Symp. on Mathematical Foundations of Computer Science (1984) 562–566

On-line Algorithms for Satisfiability Problems with Uncertainty *

Roberto Giaccio[1]

Dipartimento di Informatica e Sistemistica,
Università di Roma "La Sapienza",
via Salaria 113, I-00198 Roma, Italy.

Abstract. In this paper the problem of the on-line satisfiability of a Horn formula with uncertainty is addressed; we show how to represent a significant class of formulae by weighted directed hypergraphs and we present two algorithms that solve the on-line SAT problem and find a minimal interpretation for the formula working on the dynamic hypergraph representation. These algorithms make increasing assumptions on the formula and we will find that the second one solves the on-line SAT problem with a total time linear in the size of the formula, matching the optimal result for boolean Horn formulae.

1 Introduction

Working on a knowledge base over a realistically large domain necessarily involves some kind of uncertainty to be taken into account; such uncertainty can derive from different sources; for instance, we may know that some facts have some probability to be true, that a fact is true for a given fraction of the population or that a rule has some degree of applicability.

A first classification of the approaches to uncertainty is numerical vs symbolic approaches: numerical approaches extend the set of allowed certainty values from the boolean set; symbolic approaches maintain the boolean truth but try to elicit the exceptions from the set of rules and relax the monotonicity requirement of the classical logic.

Among the numerical approaches, another classification can be done based on the number of values that express the certainty of a fact: the most important one-valued approaches are the probabilistic method based on the Bayesian rule [12, 18, 19], the confirmation theory [24], used in MYCIN [6], also founded on probability theory and the fuzzy logic, by Zadeh [11, 25], derived from the fuzzy sets theory, that introduces some generalizations of the boolean functions. Another one-valued approach is the proposal by Bonissone et al. [5] that also uses triangular norms and conorms as general replacements to the AND, OR and NOT boolean function. The two-valued approaches express at the same

---

* Work partially supported by the ESPRIT II Basic Resarch Actions Program Project no. 7141 "ALCOM II", and by the Italian MURST National Project "Algoritmi, Modelli di Calcolo e Strutture Informative".

time the amount of certainty of a fact and the amount of certainty of its nega-
tion. Two valued approaches includes the Evidence Theory, by Dempster and
Shafer [23], Evidential Reasoning [15] and Evidence Space [22]. In Necessity
and Possibility Theory, by Zadeh [26], each fact has a corresponding normalized
possibility distribution, i.e., an infinite number of truth values, each with its
possibility. Symbolic approaches include, among others, Nonmonotonic logic by
McDermott and Doyle [9, 10], Default Logic by Reiter [21] and Circumscription
[7]; the major benefit of the latter is its capability to distinguish between certain
facts and conjectures, which allows to maintain a certain degree of monotonicity in presence of conflicting evidences. The references listed here have the only
purpose to introduce to the main research fields of the uncertainty. More recent
items can be found in the various conference proceedings related to uncertainty.

Another classification in knowledge bases is extensional vs intensional approaches: extensional approaches, also known as rule-based, attach a certainty
value to each fact and compute the certainty of a formula as a function of the
certainties of its subformulas.

In this work we will limit our attention on rule-based, one-valued knowledge
systems; furthermore our knowledge base will be a set of Horn clauses. Given
a Horn formula with uncertainty F, we will show several algorithms to check
the satisfiability of F and, if F is satisfiable, to find the interpretation I such
that F is true under I and I is minimal under some minimality constraints.
All the algorithms explained represent F by a weighted directed hypergraph H;
furthermore F is built incrementally, i.e., adding one Horn clause at a time
starting from an empty formula.

According to the previous statements, this paper is structured as follows:
first, in Sect. 2, we will define formally Horn formulae with uncertainty and
the related concepts of satisfiability and minimal interpretation; we will then
limit our attention to a subset of all the possible Horn formulae with uncertainty. In Sect. 3 we will define weighted directed hypergraphs and hyperpaths;
subsequently, in Sect. 4, we will show how we can map a Horn formula with uncertainty to a weighted directed hypergraph and we will find some results that
relate the truth of a propositional symbol to the weight of a hyperpath; in particular we will prove that a Horn formula with uncertainty is satisfiable if and
only if on the corresponding hypergraph the maximum weight of a hyperpath
from node $TRUE$ to node $FALSE$ is zero; furthermore we will prove that given
a satisfiable Horn formula with uncertainty, the minimal interpretation is the
one that assigns to each propositional symbol P_i a certainty factor equal to the
maximum weight of an hyperpath from node $TRUE$ to the node P_i. In Sect. 5
we will show two algorithms to check the satisfiability and find the minimal interpretation of a Horn formula with uncertainty F; for each algorithm we will
prove the correctness and analyze the time complexity; in particular it will be
shown that the second algorithm solves the SAT problem for Horn formulae with
uncertainty in time linear in the length of the formula, extending to logics with
uncertainty previous results [2] on propositional calculus; finally, in Sect. 6, we
will point out some open problems that deserve further investigation.

2 Horn Formulae with Uncertainty

The definitions below are from [17]; some of them have been modified or added for a better understanding and a greater analogy with the propositional calculus; the use of the terms *certainty factor* [17] to denote a truth value and *certainty function* and *combination function* to denote the functions that combine different certainty factors belongs to Confirmation Theory [24].

A *certainty domain* D is a set of numbers in $[0, 1]$ containing 0 and 1; a certainty domain contains all the possible certainty factors in a given logic with uncertainty; for example, in the propositional calculus the certainty domain is the set $\{0, 1\}$, and in some logics with uncertainty the certainty domain is the real interval $[0, 1]$; a *certainty factor* is an element of a certainty domain; the symbols *true* and *false* are also used for the certainty factors 0 and 1 respectively; a *certainty variable* is a variable with values in a certainty domain. A *Horn clause with uncertainty* C_j is a clause with the following structure:

$$P_i : g_j (\alpha_1, \ldots, \alpha_{N_j}) \leftarrow P_1 : \alpha_1 \wedge \cdots \wedge P_{N_j} : \alpha_{N_j}$$

where $P_1, \ldots, P_{N_j}$ are propositional symbols, $\alpha_1, \ldots, \alpha_{N_j}$ are certainty variables and g_j is a certainty function; a *certainty function* is a function $g_j : D^{N_j} \to D$, where D is a certainty domain, which combines the certainty factors of the premises of the clause and returns the certainty factor of the conclusion given by that clause; a certainty function has the following properties:

$$g_j (x_1, \ldots, x_{N_j}) \leq g_j (y_1, \ldots, y_{N_j}) \Leftarrow x_i \leq y_i, i = 1, \ldots, N_j \qquad (1)$$

$$g_j (x_1, \ldots, x_N) \leq x_i, i = 1, \ldots, N_j \ . \qquad (2)$$

Property (1) means that a greater certainty of the premises cannot yield a smaller certainty of the conclusion; Property (2) that a conclusion cannot have a certainty larger than its premises.

A *Horn formula with uncertainty* F is a conjunction of Horn clauses with uncertainty $C_1 \wedge \cdots \wedge C_M$; given a Horn formula with uncertainty, it is possible to partition its clauses in N sets X_i, one for each propositional symbol P_i, where X_i is the set $\{C_1, \ldots, C_{M_i}\}$ of the clauses having the same conclusion P_i; for each set X_i there is a *combination function* $f_i : D^{M_i} \to D$, which combines the certainty factors of the conclusions of the clauses in X_i and returns the certainty factor of P_i, i.e., the certainty factor obtained from all the clauses implicating P_i; a combination function has the following properties:

$$f_i (x_1, \ldots, x_{M_i}) \leq f_i (y_1, \ldots, y_{M_i}) \Leftarrow x_j \leq y_j, j = 1, \ldots, M_i \qquad (3)$$

$$f_i (x_1, \ldots, x_{M_i}) \geq x_j, j = 1, \ldots, M_i \ . \qquad (4)$$

Property (3) means that greater evidences cannot yield a smaller overall support to a fact; Property (4) that a fact cannot have a support lower than its evidences.

The above properties of certainty and combination functions are a subset of the properties that Kifer [17] shows to be desiderable for these functions; we choose these particular properties because they are sufficient for the correctness of the algorithms presented below. Also, the properties of the certainty and combination functions hold for the *fuzzy-and* and *fuzzy-or* of the fuzzy logic or the triangular norms and conorms in [5].

For instance, in fuzzy logic we replace the set $\{true, false\}$ with the set $[0 \ldots 1]$ and define the *fuzzy-and,fuzzy-or* and *fuzzy-not* functions as follows:

$$x \wedge y \equiv \min(x, y) \qquad x \vee y \equiv \max(x, y) \qquad not(x) \equiv 1 - x .$$

As a consequence the Horn formula $(P_i \leftarrow P_1 \wedge P_2) \wedge (P_i \leftarrow P_3 \wedge P_4)$ can be written as

$$P_i \leftarrow (P_1 \wedge P_2) \vee (P_3 \wedge P_4) = P_i \leftarrow \max(\min(P_1, P_2), \min(P_3, P_4))$$

where the *min* function composes the premises of a single clause, and the *max* function composes the clauses with the same conclusion.

On the other side, in probabilistic systems, giving that all the propositions are independent, the following definitions are often assumed:

$$x \wedge y \equiv x \cdot y \qquad x \vee y \equiv x + y - x \cdot y \qquad not(x) \equiv 1 - x$$

and the previous formula gives

$$P_i \leftarrow (P_1 \wedge P_2) \vee (P_3 \wedge P_4) = P_i \leftarrow (P_1 \cdot P_2) + (P_3 \cdot P_4) - (P_1 \cdot P_2) \cdot (P_3 \cdot P_4)$$

where the product function composes the premises of a single clause, and the $f(x, y) = (x + y - x \cdot y)$ function composes the clauses with the same conclusion. A well known example using these functions is MYCIN.

An *interpretation I* is a function $I : A \rightarrow D$, where A is the set of all propositional symbols and D is a certainty domain; a Horn clause with uncertainty

$$C_j \equiv P_i : g_j (\alpha_1, \ldots, \alpha_{N_j}) \leftarrow P_1 : \alpha_1 \wedge \cdots \wedge P_{N_j} : \alpha_{N_j}$$

is said to be *true under I* if

$$I(P_i) \geq g_j (I(P_1), \ldots, I(P_{N_j})) . \tag{5}$$

A Horn formula with uncertainty F is said to be *true under I* if for each propositional symbol P_i in F the following condition holds:

$$I(P_i) \geq f_i (\gamma_1, \ldots, \gamma_{M_j}) \tag{6}$$

where $\gamma_j = g_j (I(P_1), \ldots, I(P_{N_j}))$ is the value of of the certainty function $g_j (\alpha_1, \ldots, \alpha_{N_j})$ under interpretation I.

Theorem 1. *If a Horn formula F is true under the interpretation I then all its clauses are true under I.*

Proof. See [16]. □

In the following we will assume $f_i(x_1, \ldots, x_{M_i}) \equiv \max(x_1, \ldots, x_{M_i})$ for each propositional symbol P_i. The max combination function is very useful for an efficient on-line handling of the formula; with this assumption, the certainty of a fact after the introduction of a new clause will be updated only once, because the chain of updates starting from a fact will never reach it again; this behavior is stronger than the *Finite Termination Property* cited in [17], because the number of updates is linear in the number of the facts whose certainty has been modified by the new clause, not only finite. Moreover, since we have no cycles, we can easily express bidirectional dependencies, a structure that different assumptions, like those in MYCIN, make very difficult to handle; in fact, if we define the combination function as $f(x, y) \equiv x + y - x \cdot y$ it's easy to see that a new clause can cause an infinite number of cyclic updates.

Theorem 2. *If all the clauses of a Horn formula F are true under the interpretation I and for each propositional symbol P_i the combination function is defined as follows: $f_i(x_1, \ldots, x_{M_i}) \equiv \max(x_1, \ldots, x_{M_i})$ then F is true under I.*

Proof. See [16]. □

A Horn formula with uncertainty F is *satisfiable* if there is at least one interpretation I such that F is true under I.

Theorem 3. *If there exists an interpretation I that satisfy F such that*

$$\forall I', \forall P \in A : F \text{ is true under } I' \text{ and } I(P) \leq I'(P) \tag{7}$$

then I is unique.

Proof. See [16]. □

The interpretation I is the *minimal interpretation* for F. In the following we will see that all the presented algorithms find the minimal interpretation if and only if the formula is satisfiable.

The *length* of a Horn formula with uncertainty F is the length of the string representing F, and will be denoted with $Length(F)$.

3 Directed Hypergraphs

The following definitions concerning directed hypergraphs are from [1] and are consistent with the more general definitions given in [14].

A *directed hypergraph* H is a pair $\langle N, E \rangle$, where N is a non empty set of *nodes* and E is a set of hyperedges; a *hyperedge* e is an ordered pair $\langle T, h \rangle$, with $T \subseteq N$, $T \neq \emptyset$ and $h \in N$; h and T are called the *head* and the *tail* of the hyperedge e and will be denoted with $Head(e)$ and $Tail(e)$ respectively.

The *forward star* of a node n, also $FStar(n)$, is the set of hyperedges $\{e : n \in Tail(e)\}$, and the *backward star* of a node n, also $BStar(n)$, is the set of hyperedges $\{e : n = Head(e)\}$.

A set of nodes S who is the tail of at least one hyperedge is called a *source set*; given a hypergraph H, its *source area* is the sum of the cardinalities of all the source sets of H; the source area of H will be denoted with $Area(H)$; The *size* of a hypergraph H is the value $Size(H) = \sum_{e_j \in E} |Tail(e_j)|$; in the literature a different kind of size is also used: a hypergraph H' is derived from H where the m_i hyperedges with the same source set s are represented by a single hyperedge from s to a dummy node c, the *compound node*, and m_i hyperedges from c; this size will be called $Minsize(H) = Size(H')$, since it is the size of a hypergraph H' equivalent to H which is "minimal" under a simple minimality requirement; it's easy to see that $Minsize(H) = Area(H) + |E|$.

Given a hypergraph H, a *hyperpath* $\Pi_{s,t} \equiv \langle N_{\Pi_{s,t}}, E_{\Pi_{s,t}} \rangle$ from node s to node t on H, with $N_{\Pi_{s,t}} \subseteq N$ and $E_{\Pi_{s,t}} \subseteq E$, is a hypergraph such that for each node $n \in N_{\Pi_{s,t}}$ one of the following conditions is true:

$$n \equiv s \tag{8}$$

$$\exists e \equiv \langle X, n \rangle \in E_{\Pi_{s,t}} : \forall x \in X, x \in N_{\Pi_{s,t}} . \tag{9}$$

Furthermore, the following conditions hold:

$$\forall H' \equiv \langle N_{\Pi_{s,t}}, E' \rangle, E' \subset E_{\Pi_{s,t}} H' \text{ is not a hyperpath from } s \text{ to } t \tag{10}$$

$$\forall n \in N_{\Pi_{s,t}} - t : FStar(n) \neq \emptyset . \tag{11}$$

A node s is said to be *reachable* from node t on H if exists a hyperpath $\Pi_{t,s}$ on H; by (8) and (9) all the nodes in $\Pi_{s,t}$ are reachable from s; by (10) $\Pi_{s,t}$ is minimal and by (11) node t is the only node without outgoing hyperedges. The *predecessor* of a node n on the hyperpath Π, denoted with $Pred(n)$, is the hyperedge $e \in BStar(n) \cup E_\Pi$.

A *weighted directed hypergraph* H is a triad $\langle N, E, W \rangle$ where N is a non empty set of nodes, E is a set of hyperedges and W is a *weighting function* which assigns a weight to all the nodes of every hyperpath $\Pi_{s,t}$ on H; given a hyperpath $\Pi_{s,t}$ on H and a node $n \in N_{\Pi_{s,t}}$, $W_{\Pi_{s,t}}(n)$ is the *weight of the node* n and $W_{\Pi_{s,t}}(t)$ is the *weight of the hyperpath* $\Pi_{s,t}$; given a node n, $W_{MAX\Pi_{s,n}}$ is the the *maximum weight* of a hyperpath $\Pi_{s,n}$ and also the maximum weight of n; if no hyperpath $\Pi_{s,n}$ exists then we assume $W_{MAX\Pi_{s,n}} = 0$. In the following we will consider only weighted hypergraphs whose weighting functions have the form:

$$W_{\Pi_{s,t}(n)} \equiv \begin{cases} g_e\left(W_{\Pi_{s,t}}(i), i \in Tail(e)\right) & \text{if } n \neq s, \text{ with } e \equiv Pred(n) \\ w(s) & \text{otherwise} \end{cases}$$

where g_e is a function $g_e : \{W_{\Pi_{s,t}}(i), i \in Tail(e)\} \to \Re$ related to hyperedge e and $w(s)$ is a function $w : N \to \Re$ which assigns a real value to the source node s of each hyperpath $\Pi_{s,t}$.

4 Horn Formulae and Directed Hypergraphs

The following definitions and theorems extend some known results [2] on the relationships between Horn formulae of propositional calculus and directed hypergraphs to Horn formulae with uncertainty.

We can map a Horn formula with uncertainty to a weighted directed hypergraph with the following procedure:

- for each propositional symbol P_i in F the node P_i is defined;
- if the propositional symbols $TRUE$ and $FALSE$ are not in F then the nodes $TRUE$ and $FALSE$ are defined;
- for each Horn clause with uncertainty

$$C_j \equiv P_i : g_j\left(\alpha_1, \ldots, \alpha_{N_j}\right) \leftarrow P_1 : \alpha_1 \wedge \cdots \wedge P_{N_j} : \alpha_{N_j}$$

the hyperedge $e_j \equiv \left\langle \{P_1, \ldots, P_{N_j}\}, P_i \right\rangle$ is defined on H with the related function $g_j\left(W\left(P_1\right), \ldots, W\left(P_{N_j}\right)\right)$;
- for each Horn clause with uncertainty

$$C_j \equiv P_i : \tau$$

the hyperedge $e_j \equiv \langle TRUE, P_i \rangle$ is defined on H with the related function $g_j\left(W\left(TRUE\right)\right) \equiv \tau \cdot W\left(TRUE\right)$;
- for each Horn clause with uncertainty

$$C_j \equiv \leftarrow P_1 : \alpha_1 \wedge \cdots \wedge P_{N_j} : \alpha_{N_j}$$

the hyperedge $e_j \equiv \left\langle \{P_1, \ldots, P_{N_j}\}, FALSE \right\rangle$ is defined on H with the related function $g_j\left(W\left(P_1\right), \ldots, W\left(P_{N_j}\right)\right)$;
- the function w is defined as

$$w\left(P\right) = \begin{cases} 1 \text{ if } P \equiv TRUE \\ 0 \text{ otherwise} \end{cases}.$$

Theorem 4. *Let F be a Horn formula with uncertainty and H the corresponding hypergraph; if there exists a hyperpath Π_{TRUE,P_i} on H with weight $W_{\Pi_{TRUE,P_i}}\left(P_i\right)$ and F is true under the interpretation I then*

$$I\left(P_i\right) \geq W_{\Pi_{TRUE,P_i}}\left(P_i\right) .$$

Proof. See [16]. □

Theorem 5. *Let F be a Horn formula with uncertainty and H the corresponding hypergraph; if $W_{MAX\Pi_{TRUE,FALSE}} = 0$ then F is true under the interpretation*

$$I\left(P_i\right) = W_{MAX\Pi_{TRUE,P_i}} . \tag{12}$$

Proof. See [16]. □

Corollary 6. *A Horn formula F is satisfiable if and only if*

$$W_{MAX\,\Pi_{TRUE,FALSE}} = 0 \ .$$

Proof. If $W_{MAX\,\Pi_{TRUE,FALSE}} > 0$ then, by Theorem 4, for each interpretation I such that F is true under I we have $I\,(FALSE) > 0$, which is not possible; if $W_{MAX\,\Pi_{TRUE,FALSE}} = 0$ then, by Theorem 5, F is true under the Interpretation (12) and so F is satisfiable. □

Corollary 7. *If a Horn formula F is satisfiable then the interpretation (12) is the minimal interpretation.*

Proof. If F is satisfiable then, by Corollary 6, $W_{MAX\,\Pi_{TRUE,FALSE}} = 0$ and, by Theorem 5, F is true under I; by Theorem 4, for each interpretation I' under which F is true $I'\,(P_i) \geq W_{MAX\,\Pi_{TRUE,P_i}}$ holds so I is the minimal interpretation. □

5 On-line Algorithms for Satisfiability Problems with Uncertainty

The problem of on-line evaluation of boolean Horn formulae has been studied by Ausiello and Italiano [2], where it's shown that the on-line requirement doesn't increase the overall time complexity of the SAT problem. The on-line SAT problem with uncertainty has been studied in many areas related to knowledge representation, expert systems and uncertain reasoning, but with more attention to modelling and tractability aspects. The main contribution of this paper is to show the existence of efficient algorithms for this problem; we will show two algorithms: the first one, *Assert1*, solves the SAT problem making the only assumption that the combination function is the function *max*. The *Assert2* algorithm assumes that the certainty domain is finite and that the certainty function is a function $g'\,(min\,(\cdot))$, where g' is computable in constant time; in this case the overall time complexity is linear in the size of the formula times the cardinality of the certainty domain. This result shows that, under some conditions, the on-line SAT problem remains linear in the size of the formula even in case of uncertainty; the constant factor expresses the increasing number of intermediate states at the growing of the certainty domain.

5.1 The Assert1 Algorithm

The *Assert1* algorithm checks the satisfiability of a Horn formula with uncertainty F incrementally, i.e., adding one Horn clause a time starting from an empty formula; the algorithm maps the formula to a hypergraph and add one hyperedge to the hypergraph when a clause is added; then it recompute the maximum weights of the nodes of the hypergraph starting from the values known before the hyperedge was added; the maximum weights of the nodes of the hypergraph define the minimal interpretation as in (12).

In the following $UpdateG(e : Hyperedge)$ is assumed to be a procedure that computes the new value of $g_e\,(\cdot)$ when the certainty factor of a node $n \in Tail\,(e)$ changes. We will use the following variables:

$$G : \quad \textbf{array } [Hyperedge] \textbf{ of } CertaintyFactor$$
$$W : \quad \textbf{array } [Node] \textbf{ of } CertaintyFactor$$
$$Pred : \textbf{array } [Node] \textbf{ of } Hyperedge$$

$G[e]$ is the actual value of the function $g_e\,(\cdot)$, $W[n]$ is the actual value of the maximum weight of the node n and $Pred[n]$ is the predecessor of node n on the maximum weight hyperpath $\Pi_{TRUE,n}$. The arrays G and W are initialized to 0 except for $W[TRUE]$, which is initialized to 1. Furthermore we will use $Nodes$ to denote an implementation-dependent set of nodes.

procedure $Assert1(e : Hyperedge)$
    var
        $Q : Nodes$
begin
    $Q := \emptyset$
    $UpdateG(e)$
    $n := Head(e)$
    if $G[e] > W[n]$ **then begin**
        $W[n] := G[e]$
        $Pred[n] := e$
        $Insert(Q, n)$
    end
    while $Q \neq \emptyset$ **do begin**
        $RemoveMax(Q, n)$
        for each $Hyperedge\ j \in FStar[n]$ **do begin**
            $i := Head[j]$
            $UpdateG(j)$
            if $G[j] > W[i]$ **then begin**
                $W[i] := G[j]$
                $Pred[i] := j$
                $Insert(Q, i)$
            end
        end
    end
end

Now we will show that $Assert1$ is correct and we will find its time complexity: first we will show that $Assert1$ stops and we will find its time complexity; then we will see that when $Assert1$ stops we can check if the formula built up to now is satisfiable, and find the minimal interpretation for it.

In the following N will be the number of nodes P_i whose maximum weight $W_{MAX \Pi_{TRUE,P_i}}$ changes during the execution of $Assert1$; M will be the number of hyperedges e_j whose related function $g_e\,(\cdot)$ changes value during the execution of $Assert1$; F will be the maximum cardinality of $FStar\,(P_i)$ of the nodes P_i whose maximum weight $W_{MAX \Pi_{TRUE,P_i}}$ changes during the execution of *Assert1*; T will be the maximum cardinality of $Tail(e_j)$ of the hyperedges e_j whose related function $g_e(\cdot)$ changes value during the execution of *Assert1*.

Lemma 8. *If the node P_k is removed from Q immediately after the node P_i then*

$$W[P_k] \le W[P_i] \ .$$

Proof. See [16]. □

Corollary 9. *Each node P_i is removed from Q at most once.*

Proof. Suppose that node P_i has been removed from Q when $W[P_i] = a$; P_i will be inserted again into Q if and only if $W[P_i]$ changes from b to c with $c > b \ge a$; so, when P_i will be removed again from Q, we will have

$$W[P_i] = d \ge c > a$$

but, by Lemma 8, this is not possible. □

Corollary 10. *The maximum size of Q is the number N of nodes P_i whose maximum weight $W[P_i]$ changes during Assert1.*

Theorem 11. *The Assert1 algorithm has a time complexity*

$$O(N \cdot cost(RemoveMax) + N \cdot F(cost(UpdateG) + cost(Insert))) \ . \quad (13)$$

Proof. The **while ... do** loop is executed, by Corollary 10, at most N times; inside the **while ... do** loop, the **for ... each** loop is executed at most $|FStar(P_i)|$ times for each node P_i removed from Q, so *UpdateG* is computed at most $N \cdot F$ times and *Insert* is executed at most $N \cdot F$ times. □

We have $cost(RemoveMax) = O(N)$, $cost(Insert) = O(1)$ if the set Q is represented by an unordered array, so the time complexity of *Assert1* is $O(N^2 + N \cdot F cost(UpdateG))$; else, if a heap is used to represent Q, we have $cost(RemoveMax) = O(\log N)$ and $cost(Insert) = O(\log N)$, so the time complexity of *Assert1* is $O(N \cdot F(cost(UpdateG) + \log N))$.

If $cost(UpdateG) = O(1)$, as when the certainty of a conclusion is the product of the certainty of the premises, the time complexity is $O(N^2 + N \cdot F)$ using an array and $O(N \cdot F \log N)$ using a heap. If $cost(UpdateG) = O(T)$, as when the certainty of a conclusion is a function computable in time linear in the number of the premises, the time complexity is $O(N^2 + N \cdot F \cdot T)$ using an array and $O(N \cdot F(T + \log N))$ using a heap. If the certainty of a conclusion is the minimum of the certainty of the premises, we can efficiently compute *UpdateG* using a heap for each hyperedge; we have $cost(UpdateG) = O(\log T)$ and the time complexity becomes $O(N^2 + N \cdot F \log T)$ using an array and $O(N \cdot F(\log T + \log N))$ using a heap.

Lemma 12. *When Assert1 stops for each node $P_i \in N$ the following condition holds:*

$$\text{if } W[P_i] > 0 \text{ then exists a hyperpath } \Pi_{TRUE,P_i} \text{ with} \tag{14}$$
$$W_{\Pi_{TRUE,P_i}} = W[P_i] .$$

Proof. See [16]. □

Lemma 13. *When Assert1 stops for each hyperpath Π_{TRUE,P_i} holds*

$$W_{\Pi_{TRUE,P_i}} \leq W[P_i] .$$

Proof. See [16]. □

If F is the Horn formula with uncertainty formed by the clauses added until *Assert1* stops, we have the following results:

Corollary 14. *When Assert1 stops the interpretation: $I(P_i) = W[P_i]$ is the minimal interpretation.*

Corollary 15. *When Assert1 stops F is satisfiable if and only if*

$$W[FALSE] = 0 .$$

If the certainty domain is finite, i.e., if there are L certainty factors, we can enhance the time complexity of the *Assert1* algorithm structuring the set Q as an array of size L and using an auxiliary variable l equal to the maximum weight of the nodes in Q; at each moment $Q[h]$ contains the nodes P_i to be extracted whose weights $W[P_i]$ are equal to h.

After a node P_i is removed from Q two possibilities arise:

- the set $Q[l]$ isn't empty: l is left unchanged;
- the set $Q[l]$ is empty: l is decremented until $Q[l]$ isn't empty; during N remotions l can change at most L times, so the total cost of N remotions is $O(\max(N, L))$.

The time complexity becomes $O(L + N \cdot F \cdot cost(UpdateG))$ since the initialization of Q has a cost $O(L)$, the total cost of *Remove* is $O(N)$ and $cost(Insert) = O(1)$. This feature will be used in the next algorithm.

5.2   The Assert2 Algorithm

A significant class of certainty functions is the class

$$g_j(x_1, \ldots, x_{N_j}) = g'_j(\min(x_1, \ldots, x_{N_j}))$$

where g' is computable in constant time; for this class, several improvements can be made to the *Assert1* algorithm. For each hyperedge e_j we can use a heap of size at most T ordered on the weights of the nodes $P_i \in Tail(e_j)$; in this way we have $cost(UpdateG) = O(\log T)$, and the time complexity is $O(L + N \cdot F \log T)$ Alternatively, for each hyperedge e_j we can group the nodes $P_i \in Tail(e_j)$ with the same weight w and use an array or a heap of size at most L ordered on the weights of these sets; in this way we have $cost\,(UpdateG) = O\,(L)$ with the array and $cost\,(UpdateG) = O\,(\log L)$ with the heap, and the time complexity becomes $O\,(N \cdot F \cdot L)$ using an array and $O\,(L + N \cdot F \log L)$ using a heap. If we use an array NW of size L for each hyperedge, storing in $NW[e, l]$ the cardinality of the set $\{P_i : P_i \in Tail\,(e), W[P_i] = l\}$ we have relevant results regarding the total time complexity, i.e., the time needed to dynamically insert all the clauses of a given formula F; the following variables are used:

$$NW : \mathbf{array}\;[Hypededge, CertaintyFactor]\;\mathbf{of}\;Integer$$
$$MIN : \mathbf{array}\;[Hyperedge]\;\mathbf{of}\;CertaintyFactor$$
$$oldW : \mathbf{array}\;[Node]\;\mathbf{of}\;CertaintyFactor$$

The array $oldW$ is initialized like the array W; the $UpdateG$ procedure is the following:

```
procedure UpdateG(e : Hyperedge; oldW, oldW : CertaintyFactor)
begin
        Decrement(NW[e, oldW])
        Increment(NW[e, newW])
        while NW[e, MIN[e]] = 0 do
              Increment(MIN[e])
        G[e] = G'(MIN[e])
end
```

Finally, this is the $Assert2$ algorithm:

```
procedure Assert2(e : Hyperedge)
        var
                Q : array [CertaintyFactor] of Nodes
begin
        for each CertaintyFactor l do begin
              Q[l] := ∅
              NW[e, l] = 0
        end
        MIN[e] := 1
        for each Node n ∈ Tail[e] do begin
              Increment(NW[e, W[n]])
              MIN[e] := Min(MIN[e], W[n])
        end
        G[e] := G'(MIN[e])
        l := 0
        n := Head(e)
        if G[e] > W[n] then begin
              W[n] := G[e]
              Pred[n] := e
              l := W[n]
              Insert(Q[l], n)
```

```
      end
      while Q[l] ≠ ∅ do begin
            Remove(Q[l], n)
            while l > 0 and Q[l] = ∅ do
                  Decrement(l)
            for each Hyperedge j ∈ FStar[n] do begin
                  i := Head[j]
                  UpdateG(j, oldW[n], W[n])
                  if G[j] > W[i] then begin
                        oldl := W[i]
                        W[i] := G[j]
                        Pred[i] := j
                        Remove(Q[oldl], i)
                        Insert(Q[l], i)
                  end
            end
            oldW[n] := W[n]
      end
end
```

From the previous discussions, and since the cost of the initialization phase is $O(L + T)$, it's easy to see that the time complexity of the *Assert2* algorithm is $O(N \cdot F \cdot L)$; furthermore the following result holds:

Theorem 16. *The total time complexity of the Assert2 algorithm for M insertions, starting from an empty formula, is $O(L \cdot N_{tot})$, where N_{tot} is $\sum_{e_j \in E} |Tail(e_j)|$.*

Proof. The total cost of the initialization phase is $O(L \cdot M + N_{tot})$; a node P_i is removed from Q at most L times, since it is inserted in Q only if $W[P_i]$ changes, and it can change at most L times; so the total cost of *Remove* is $O(L \cdot N)$; the variable l is decremented at most L times for each hyperedge inserted, so the total cost of the *Decrement(l)* instruction is $O(L \cdot M)$; the *UpdateG* function is called at most $L \cdot N_{tot}$ times, since the **for ... each** loop is executed at most $|FStar(P_i)|$ times for each node P_i, $\sum_{P_i \in N} |FStar(P_i)| = \sum_{e_j \in E} |Tail(e_j)|$ and a node can be removed from Q at most L times; in the *UpdateG* function, the inner loop is executed at most $L \cdot M$ times; hence the amortized time complexity of the *Assert2* algorithm is

$$O(L \cdot M + N_{tot} + L \cdot N + L \cdot M + L \cdot N_{tot} + L \cdot M) =$$

$$= O(L \cdot M + L \cdot N_{tot}) = O(L \cdot N_{tot})$$

since $N_{tot} \geq M$. □

If a formula F has M clauses, since $N_{tot} = Size(H) = Length(F)$, the following corollary holds:

Corollary 17. *The Assert2 algorithm checks the satisfiability of a Horn formula with uncertainty in time* $O(L \cdot Length(F))$.

This result extends to the logics with uncertainty a similar result in [2]; an approach often used in the literature is to work on the hypergraph H', expressing the complexity of the algorithms as a function of $Minsize(H)$; in this case, since $Minsize(H) \leq Size(H)$, the time complexity of the $Assert2$ algorithm improves, but we have also to maintain the minimal representation of H; the main approach is to impose an order on the nodes and a lexicographic order on the source sets, using a balanced search tree T whose leaves are the source sets; when inserting a hyperedge e, we first search $Tail(e)$ in T; if there is no such source set, we insert the source set $s = Tail(e)$ in T, create the compound node c and the hyperedges $\langle s, c \rangle$ and $\langle c, Head(e) \rangle$; if there is a source set $s = Tail(e)$ in T, we just insert the hyperedge $\langle c, Head(e) \rangle$. If in the formula F there are N_S different source sets, the total cost of maintaining H' is $O(M \log N_S)$ and the following corollary holds:

Corollary 18. *The Assert2 algorithm checks the satisfiability of a Horn formula with uncertainty in a time* $O(L \cdot Minsize(H) + M \log N_S)$.

Note that if $N_S = \Theta(M)$, as with sparse hypergraphs or with small source sets, we have

$$Minsize(H) = O(Size(H)) = O(Length(F))$$

and the time complexity working with H' is $O(L \cdot Length(F) + M \cdot \log M)$.

6 Conclusion and Future Work

We have shown a method to represent a significant class of Horn formulae with uncertainty using weighted directed hypergraphs; this class has the combination function defined as the maximum of the arguments. We presented two algorithms that solve the SAT problem with uncertainty making increasing assumptions on the formula. We found that the second algorithm solves the on-line SAT problem in a time linear in the size of the formula, matching the optimal result given in [2] and showing that the greater number of allowed certainty factors with respect to the boolean set doesn't prevent the on-line problem to have the same overall time bounds of the off-line one.

More work can be done; for instance, we can add clauses until we obtain an unsatisfiable formula; if we associate to each clause a fixed reliability value, it would be interesting to determine some set of clauses whose deletion makes the formula satisfiable and which is minimum with respect to some function of the reliability of the clauses. It's easy to see that in the case of boolean Horn formulae, if we want to minimize the number of deleted clauses we have the MAX-Horn-SAT problem, which is known to be NP-complete. Under uncertainty conditions, if we want to minimize the sum of the reliability of the deleted clauses the problem is still NP-complete, being very easy to reduce it to the MAX-Horn-SAT problem giving unit reliability to each clause, but in this case there are other and more significant functions to minimize instead of the sum. An interesting function to minimize is the maximum reliability of the deleted clauses, because it is both semantically reasonable and polinomially solvable; it would be interesting to find efficient algorithms that determine such set of clauses.

Since many of the results in this paper derive from the tight relationship between minimal weight hyperpath problems on weighted directed hypergraphs and satisfiability problems on boolean Horn formulae with uncertainty, further investigations can also be made on the relations between flow and cut problems on weighted directed hypergraphs and the corresponding problems on boolean Horn formulae with uncertainty.

7 Acknowledgments

I wish to thank Giorgio Ausiello; without his invaluable help and encouragement this paper couldn't have been written. I am also grateful to Umberto Nanni whose precise criticism and suggestions allowed me to improve this work.

References

1. Ausiello, G., D'Atri, A., Sacca, D.: Minimal representation of directed hypergraphs. SIAM Journal on Computing **2** (1986) 418–431
2. Ausiello, G., Italiano, G.F.: On-line algorithms for polinomially solvable satisfiability problems. Journal of Logic Programming **10** (1991) 69–90
3. Ausiello, G., Italiano, G. F., Nanni, U.: Dynamic maintenance of directed hypergraphs. Theoretical Computer Science **72** (1990) 97–117
4. Ausiello, G., Italiano, G. F., Nanni, U.: Optimal traversal of directed hypergraphs. ICSI Technical Report TR-92-073 (September 1992)
5. Bonissone, P. P., Decker, K. S.: Selecting uncertainty calculi and granularity: an experiment in trading-off precision and complexity. Proceedings of the Workshop on Uncertainty and Probability in Artificial Intelligence, University of California, Los Angeles, August 14-16 (1985) 57–66,
6. Buchanan, B. G., Shortliffe, E.H. : Rule-based expert systems. Addison-Wesley (1984)
7. McCarthy, J.: Circumscription – a form of non-monotonic reasoning. Artif. Intell. **13** (1980) 27–40
8. Cambini, R., Gallo, G., Scutella, M. G.: Minimum cost flows on hypergraphs. TR-1/92 Dipartimento di Informatica, Università di Pisa (1992)
9. McDermott, D., Doyle, J.: Nonmonotonic logic I. JACM **29** (1982) 33–57
10. McDermott, D.: Nonmonotonic logic II. Artif. Intell. **13** (1980) 41–72
11. Dubois, D., Prade, H.: Fuzzy sets and systems: theory and applications. Academic Press (1980)
12. Duda, R. O., Hart, P. E., Nilsson, N. J.: Subjective bayesian methods for rule-based inference systems. AFIPS Conference Proceedings, New York (1976) 1074–1982
13. Encyclopedia of artificial intelligence John Wiley & Sons, Inc. (1987)

14. Gallo, G., Longo, G., Nguyen, S., Pallottino, S.: Directed hypergraphs and applications. Discrete Applied Mathematics **42** (1993) 177–201
15. Garvey, T. D., Lovrance, J. D., Fishler, M. A.: An inference technique for integrating knowledge from disparate sources. Proceedings of the Seventh International Joint Conference on Artificial Intelligence, Vancouver, B.C. (1981) 319–325
16. Giaccio, R.: On-line algorithms for satisfiability formulae with uncertainty. TR-2/94 Dipartimento di Informatica e Sistemistica, Università degli studi di Roma "La Sapienza".
17. Kifer, M., Li, A.: On the semantics of rule-based expert systems with uncertainty. International Conference on Database Theory 1988, LNCS vol. 326, Springer-Verlag (September 1988) 186–202
18. Pearl, J.: Reverend bayes on inference engine: a distributed hierarchical approach. Proceedings of the Second National Conference on Artificial Intelligence, Pittsburgh, PA (1982) 133–136
19. Pearl, J.: Probabilistic reasoning in intelligent systems: network of plausible inference. Morgan Kaufmann Publishers (1988)
20. Pretolani, D.: Satisfiability and hypergraphs. TD 12/93, Dipartimento di Informatica, Università di Pisa (1993)
21. Reiter, R.: A logic for default reasoning. Artif. Intell. **13** (1980) 81–132
22. Rollinger, C. R.: How to represent evidence: aspects of uncertainty reasoning. Proceedings of the Eighth International Joint Conference on Artificial Intelligence, Karlsruhe, FRG (1983) 358–361
23. Shafer, G.: A mathematical theory of evidence. Princeton University Press (1976)
24. Shortliffe, E. H., Buchanan, B. G.: A model of inexact reasoning in medicine. Math. Biosci. **23** (1975) 351–379
25. Zadeh, L. A. : Fuzzy sets. Inf. Ctrl. **8** (1965) 338–353
26. Zadeh, L. A.: Fuzzy sets as a basys for a theory of possibility. Fuzzy Sets Sys. **1** (1978) 3–28

NC Algorithms for Antidirected Hamiltonian Paths and Cycles in Tournaments

(Extended Abstract)

E. Bampis[1,2], Y. Manoussakis[1], and I. Milis[1]*

[1] LRI, Bât 490, Université de Paris Sud, 91405 Orsay Cedex, France
[2] LIVE, Université d'Evry, Bd des Coquibus, 91025 Evry Cedex, France

Abstract

Two classical theorems about tournaments state that a tournament with no less than eight vertices admits an antidirected Hamiltonian path and an even cardinality tournament with no less than sixteen vertices admits an antidirected Hamiltonian cycle. Sequential algorithms for finding such a path as well as a cycle follow directly from the proofs of the theorems. Unfortunately, these proofs are inherently sequential and can not be exploited in a parallel context. In this paper we propose new proofs leading to efficient parallel algorithms.

1 Introduction

An antidirected path (ADP) in a digraph is a simple path, every two adjacent arcs of which, have opposite orientations i.e. no two consecutive arcs of the path form a directed path. An antidirected Hamiltonian path (ADHP) in a digraph is a simple antidirected path containing all the vertices. Similarly, we define an antidirected Hamiltonian cycle (ADHC). The problem of finding an ADHP in an arbitrary digraph is trivially NP-complete.

A tournament is an oriented complete graph. In [7] and [11] it has been shown that all tournaments have an ADHP, except exactly three tournaments of sizes 3, 5 and 7, respectively. It has been shown, also, that each tournament with no less than sixteen vertices has an ADHC [9] [12] [14]. The proofs of these results imply efficient sequential algorithms for finding ADHPs and ADHCs of complexities $O(n)$ and $O(n^2)$, respectively (n is the number of vertices of the tournament). Such an algorithm for ADHPs is presented in the more general context of [8].

Although many results have been appeared in the literature on NC algorithms for (directed) Hamiltonian paths and cycles in tournaments [1] [4] [13], there are no analog results for the antidirected case. Unfortunately, the existing efficient algorithms for constructing ADHPs and ADHCs are implied by inductive proofs and are inherently sequential.

In this paper we present NC algorithms for both problems. The development of these algorithms is based on new proofs for the existence of ADHPs and

---

* Fellow of the E.E.C. program Human Capital and Mobility.

ADHCs. Our algorithms are in the well known CRCW PRAM model, where simultaneous reading is allowed while for simultaneous writing processors are required to write the same value. The complexity of the algorithms presented are $O(\log n)$ time, $O(n/\log n)$ processors for finding ADHPs and $O(\log^2 n)$ time using $O(n^2/\log n)$ processors for finding ADHCs. Thus the first algorithm is optimal and the second optimal over a factor of $\log n$, in respect with the sequential complexities of the problems.

2 Terminology and Preliminaries

Throughout this paper, T_n denotes a tournament with n vertices. A trivial, but useful fact is that any induced subgraph, of a tournament is also a tournament. If v, w are vertices of a tournament T_n then we say that v dominates w if the arc (v, w) exists and denote this relation by $v \to w$. Note that since the directions of the arcs are arbitrary the domination relation is not necessarily transitive. By $\Gamma^-(v)$ and $\Gamma^+(v)$ we denote the sets of vertices which, respectively, dominate and are dominated by the vertex v. The indegree and the outdegree of v are defined as $|\Gamma^-(v)|$ and $|\Gamma^+(v)|$ and are denoted by $d^-(v)$ and $d^+(v)$, respectively.

Following [11], we say that a vertex v is a starting vertex (resp. an ending vertex) of an antidirected path, if the path is of the form $v \to v_1 \leftarrow \ldots$ (resp. $v \leftarrow v_1 \to \ldots$). If v is both a starting and an ending point we say that v is a **double point**. It follows directly from this definition that one of the extremities of an antidirected path with odd number of vertices, $v_1 \to v_2 \leftarrow \ldots \leftarrow v_k$, $k = 2r + 1$, is a double point. To see that, it is enough to examine the arc between its extremities, i.e. if $v_k \to v_1$ (resp. $v_k \leftarrow v_1$) then the double point is the vertex v_1 (resp. v_k).

We say that a tournament is *transitive* if the domination relation is transitive. Clearly the vertices of a transitive tournament is linearly ordered by the domination relation and there is a unique such order, in which $i \to j$ if and only if $i < j$. In the following a transitive tournament, with n vertices is denoted by TT_n and its ordered vertex set by $\{1, 2, \ldots, n\}$.

Transitive tournaments will be very helpful for the problems examined in this paper because of their following interesting property [11]:

Lemma 1. [11] *Let TT_n be a transitive tournament with the set $\{1, 2, \ldots, n\}$ as vertex set.*
(i) If n is even, then TT_n contains an ADHP starting from i and ending to j unless either $i = n$ or $j = 1$ or $\{i, j\} = \{1, 2\}$ or $\{i, j\} = \{n - 1, n\}$.
(ii) If n is odd, then TT_n contains an ADHP with starting (ending) vertices i, j unless either $i = n (= 1)$ or $j = n (= 1)$ or $\{i, j\} = \{n - 2, n - 1\} (= \{2, 3\})$.

Although M. Rosenfeld in [11] was not interested on the complexity of finding the ADHPs in Lemma 1, it is easy to show that, given the ordering of the vertices of a transitive tournament, no searching is needed to find these paths, i.e. it takes $O(1)$ sequential time, while the rank of the vertices in the ADHPs can be computed in parallel in $O(\log n)$ time using $O(n)$ processors [6].

The next Lemma of D. Soroker [13] is, also, used in the sequel.

Lemma 2. [13]. *Any tournament T_n contains a vertex u which dominates, and is dominated by at least $\lfloor \frac{n}{4} \rfloor$ vertices.*

3 Antidirected Hamiltonian Path

In this section we prove an NC algorithm for finding an ADHP in a tournament T_n. The following lemma helps to exploit the divide and conquer approach in order to find such a path.

Lemma 3. *The following operations on ADPs can be done in constant time.*
(i) Add a vertex to an ADP with odd number of vertices.
(ii) Connect two vertex-disjoint ADPs, X and Y, both of odd number of vertices, in a new one with vertex set $V(X) \cup X(Y)$.
(iii) Add two vertices to an ADP with even number of vertices.

Proof.
(i) Let x_1 be the double point of an ADHP, $X \in T_n$ and u be a vertex, $u \in T_n - X$. It is enough to examine the orientation of the arc between x_1 and u.

(ii) Let x_1 and y_1 be the double points of the ADHP's X and Y respectively. It is enough to examine the orientation of the arc between x_1 and x_2.

(iii) Let $X = x_1 \to x_2 \leftarrow \ldots \leftarrow x_{k-1} \to x_k$, $k = 2p$, an antidirected path with even number of vertices in T_n and two vertices $u_1, u_2 \in T_n - X$. If u_1 (or u_2) $\leftarrow x_1$ (resp. u_1 (or u_2) $\to x_k$), then replace X by $u_1(u_2) \leftarrow x_1 \to x_2 \leftarrow \ldots \leftarrow x_{k-1} \to x_k$ (resp. $x_1 \to x_2 \leftarrow \ldots \leftarrow x_{k-1} \to x_k \leftarrow u_1(u_2)$). Moreover, if u_1 (or u_2) $\to x_2$ (resp. u_1 (or u_2) $\leftarrow x_{k-1}$), then replace X by $x_1 \leftarrow u_1(u_2) \to x_2 \leftarrow \ldots \leftarrow x_{k-1} \to x_k$ (resp. $x_1 \to x_2 \leftarrow \ldots \leftarrow x_{k-1} \to u_1(u_2) \leftarrow x_k$). If only one vertex of u_1, u_2 is added to X, then the path formed so far has odd number of vertices, that is, it contains a double point. To add the remaining vertex it is enough to examine the orientation of the arc between the double point and the remaining vertex. Otherwise (if no vertex is added to X so far) assume without loss of generality, that $u_1 \to u_2$. If $x_1 \to x_{k-1}$, then replace X by $x_k \to u_2 \leftarrow u_1 \to x_{k-1} \leftarrow x_1 \to x_2 \leftarrow \ldots \to x_{k-2}$, else replace X by $x_k \to u_2 \leftarrow u_1 \to x_1 \leftarrow x_{k-1} \to \ldots \to x_2$.

It is clear, that all these operations take constant time, since the orientation of a constant number of arcs (2, 10 and 3 in cases (i), (ii) and (iii) respectively) must be examined. $\square$

However, it is not obvious yet how Lemma 3 can be used in order to divide T_n and to combine the obtained in each part ADHPs.

In the case of **even** cardinality tournaments we can handle this problem as it is shown in the following procedure:

procedure FIND-EVEN-ADHP(T_n, n even)

   (1) **If** $|T_n| \leq 2$ **then return** an ADHP of T_n.

   (2) Split T into two subtournaments T_{n_1} and T_{n_2}
             of roughly equal even cardinalities such that $n_1 + n_2 = n$.

   (3) In parallel, find ADHPs X_1=FIND-EVEN-ADHP(T_{n_1}, n_1 even)
             and X_2=FIND-EVEN-ADHP(T_{n_2}, n_2 even).

   (4) $X_1 = X_1 - \{u_1\}$, $X_2 = X_2 - \{u_2\}$. {u_1 and u_2 are extremities of
             the ADHPs X_1 and X_2 respectively }

   (5) Use Lemma 3(ii) to connect paths X_1 and X_2 in a new path X.

   (6) Add, by Lemma 3(iii) vertices u_1, u_2 to X and **return** an ADHP of T_n.

end FIND-EVEN-ADHP.

Lemma 4. *For any tournament T_n, n even, procedure FIND-EVEN-ADHP obtains an ADHP in $O(\log n)$ time using $O(n/\log n)$ processors.*

Proof. The depth of the recursion tree of the above procedure is $O(\log n)$ (Step (3)) and it can be implemented using $O(n)$ processors. Steps (5) and (6) can be implemented in constant time, by Lemma 3, and the same is obvious for Step (4). The rank of the vertices in the ADHPs can be computed in $O(\log n)$ time using $O(n)$ processors [6]. Thus, by using Brent's principle [5], the parallel complexity of FIND-EVEN-ADHP procedure becomes $O(\log n)$ on $O(n/\log n)$ processors. □

Unfortunately, in the case of **odd** cardinality tournaments a similar approach does not work. This is due to parity reasons and we proceed in a different way.

First, we find a transitive subtournament of T_n consisting of 5 vertices, i.e. a TT_5, as following: It is known that every tournament contains a vertex with $d^+(x_1) \geq \frac{n}{2}$. If we repeat this argument on the tournament induced by $\Gamma^+(x)$ and so on we find a sequence of vertices $x_1, x_2, \ldots, x_t$ which clearly form a transitive tournament. From this procedure it is obvious that any tournament contains a transitive subtournament TT_t with $t \geq \lfloor \log n \rfloor + 1$ and therefore every tournament T_n, $n \geq 16$, contains a TT_5. By applying the above procedure on any subtournament of T_n induced by 16 of its vertices a TT_5 can be found in constant time.

Next, we consider the subtournament $T_k = T_n - TT_5$, in which we can find an ADHP using FIND-EVEN-ADHP, since k is even. Now, using Lemma 1, we can prove the following:

Lemma 5. *Let T_n, $n > 16$, be a tournament of odd cardinality. Given a TT_5 of T_n and an ADHP in $T_k = T_n - TT_5$, an ADHP in T_n can be constructed in constant time.*

Proof. Let $\{1, 2, 3, 4, 5\}$ be a TT_5 in T_n and $x_1 \rightarrow x_2 \leftarrow \ldots \leftarrow x_{k-1} \rightarrow x_k$, $k = 2s$ be an ADHP in $T_n - TT_5$. We test if one of the extremities of this path, x_1 or x_k, can be inserted into TT_5 in such a way that preserves the transitivity, i.e. to form a TT_6. Each of these vertices can not be inserted in TT_5 if and only if there are two vertices i and j in TT_5, $i < j$, such that $j \rightarrow x_1(x_k)$ and

$x_1(x_k) \rightarrow i$. It is clear that this test can be done in constant time, since TT_5 is a constant size transitive subtournament. Next, we consider two cases:

Case (i): No extremity can be inserted in TT_5. If for some vertex i of TT_5, $1 \leq i \leq 4$, $i \rightarrow x_k$, then, by Lemma 1, there is an antidirected path in TT_5 starting from i and the desired path is $x_1 \rightarrow x_2 \leftarrow \ldots \leftarrow x_{k-1} \rightarrow x_k \leftarrow i \rightarrow$[ADHP in TT_5 starting from i], for some i such that $1 \leq i \leq 4$. Similarly, if for some vertex i of TT_5 $2 \leq i \leq 5$, $i \leftarrow x_1$, then, by Lemma 1, there is an antidirected path in TT_5 ending at i and the desired path is $x_k \leftarrow x_{k-1} \rightarrow \ldots \rightarrow x_2 \leftarrow x_1 \rightarrow i \leftarrow$ [ADHP in TT_5 ending at i], for some i such that $2 \leq i \leq 5$. If none of these is the case, then $x_1 \rightarrow 1$, for otherwise x_1 can be inserted into TT_5, a contradiction. Then the desired path is $x_{k-1} \rightarrow \ldots \rightarrow x_2 \leftarrow x_1 \rightarrow 1 \leftarrow x_k \rightarrow i \leftarrow$[ADHP in $TT_5 - \{1\}$ ending at i], $3 \leq i \leq 5$.

Case (ii): An extremity can be inserted in TT_5. Assume that x_k can be inserted into TT_5, the proof being similar in the case of x_1. Now, we consider a TT_6 and the antidirected path $x_1 \rightarrow x_2 \leftarrow \ldots \leftarrow x_{k-1}$ in $T_n - TT_6$, which now contains an odd number of vertices. Let us assume, without loss of generality, that x_1 is the double point of this path. It is enough to consider an arc between x_1 and some vertex i of TT_6, $2 \leq i \leq 5$. If $x_1 \rightarrow i$ then the path is [ADHP in TT_6 ending at i] $\rightarrow i \leftarrow x_1 \rightarrow x_2 \leftarrow \ldots \leftarrow x_{k-1}$ else the path is $x_2 \leftarrow \ldots \leftarrow x_{k-1} \rightarrow x_1 \leftarrow i \rightarrow$[ADHP in TT_6 starting from i].

Since we consider a constant size transitive subtournament of T_n, i.e. a TT_5, in both cases the direction of a constant number of arcs is examined and the corresponding ADHPs can be found in constant time. $\square$

Given Lemma 5, it is clear that the complexity of finding an ADHP in a tournament of odd cardinality is determined by the complexity of the procedure FIND-EVEN-ADHP. As a consequence of this fact and Lemma 4 we obtain the next theorem.

Theorem 1. *For any tournament T_n, $n \geq 16$, an ADHP can be found in $O(\log n)$ time using $O(n/\log n)$ processors.*

Note that the complexity of our parallel algorithm for finding ADHPs is optimal over the best known sequential one of complexity $O(n)$.

4 Antidirected Hamiltonian Cycle

A restricted antidirected Hamiltonian path of a tournament T_n, is an ADHP with a specified extremity, either the first or the last vertex, not both. Given a specified extremity, say x, we denote such an ADHP as x-ADHP. Notice that we are not interested if the specified extremity is a starting or an ending point.

In [11] M. Rosenfeld has proved that for any vertex x of a tournament T_n, $n \geq 9$, there is an x-ADHP. From a careful reading of the proof of Theorem 3 in [11] it is not hard to see that its arguments can be easily implemented in parallel. In particular, if n is even, then it is enough to find an ADHP in $T_n - \{x\}$. Since this ADHP has an odd number of vertices it is trivial to construct a x-ADHP.

If n is odd then we test first if x is an internal vertex in some $TT_4 \subset T_n$. Such a test can be done in $O(\log n)$ time using $O(n^2/\log n)$ processors. If this is the case we find an ADHP in $T_n - TT_4$ and then we can easily construct a x-ADHP. Otherwise, T_n has a special structure which implies directly a x-ADHP. Hence, taking into account Theorem 1, we have the following corollary.

Corollary 1. *For any tournament T_n, $n \geq 9$, and x-ADHP can be found in $O(\log n)$ time using $O(n^2/\log n)$ processors.*

Let us now consider the problem of finding an ADHC in tournaments. Obviously, such a cycle exists only in even cardinality tournaments. It is known that any tournament T_n, n even, $n \geq 16$, has an ADHC. This result was initially proved in [14] for $n \geq 50$. For $n \geq 16$, if T_n contains a TT_6, then a proof can be found in [12], otherwise the proof is given in [9]. We point out, here, that all these proofs use a maximal transitive subtournament of T_n.

It is known that the problem of finding a maximum transitive subtournament is NP-complete [3]. On the other hand a maximal, transitive subtournament can be found by a greedy sequential algorithm of complexity $O(n^2)$. Unfortunately, in our knowledge the problem of finding efficiently in parallel a maximal transitive subtournament of a tournament is open. Consequently, a parallel algorithm for the ADHC problem can not be based in these proofs.

In what follows we give a new proof for the existence of ADHCs. Instead of the maximal transitive subtournament our proof uses a **nice transitive subtournament** which is defined as following:

Definition 1. *A transitive subtournament $TT_p \subset T_n$ is said to be **nice** if for each vertex $x \in T_n - TT_p$ there is an arc $x \rightarrow i$ for some $i \in TT_p$.*

To find a **nice** transitive subtournament we use Lemma 2 of D. Soroker [13]. We find a vertex $u_1 \in T_n$ whose both in-degree and out-degree are at least $\lfloor \frac{n}{4} \rfloor$. We repeat the same argument on the tournament induced by $\Gamma^+(u_1)$ and so on. Thus, we find a sequence of vertices $u_1, u_2, \ldots \ldots, u_p$ which clearly form a transitive tournament. Furthermore for each vertex $x \in T_n - TT_p$ we know that for some $u_i \in TT_p$, we have $x \in \Gamma^-(u_i)$. Therefore, for each vertex $x \in T_n - TT_p$ there is an arc $x \rightarrow u_i$ for some $u_i \in TT_p$, and thus, TT_p is a nice transitive subtournament.

Lemma 6. *For any tournament, T_n, a nice transitive subtournament, $TT_p \subset T_n$, with $p \geq \lfloor \frac{\log n}{2} \rfloor + 1$ vertices can be found in $O(\log^2 n)$ time using $O(n^2/\log n)$ processors.*

Proof. Since, by Lemma 2, both $\Gamma^+(u)$ and $\Gamma^-(u)$ have at least $\lfloor \frac{n}{4} \rfloor$ vertices it follows that $\Gamma^+(u) \geq \lfloor \frac{n}{4} \rfloor$ and $\Gamma^+(u) \leq \lfloor \frac{3n}{4} \rfloor$. At each step we add a vertex to TT_p, and we finally obtain that any tournament contains a nice transitive subtournament with $p \geq \lfloor \frac{\log n}{2} \rfloor + 1$ vertices.

The procedure described above terminates in $O(\log n)$ steps and at each step a vertex $u \in T_n$ whose in-degree and out-degree are at least $\lfloor \frac{n}{4} \rfloor$ must be found. Such a vertex can be found in $O(\log n)$ time using $O(n^2)$ processors. The rank of the vertices in TT_p can be computed in $O(\log n)$ time using $O(n)$ processors [6].

Therefore, the complexity of the procedure becomes $O(\log^2 n)$ time using $O(n^2)$ processors. By applying the well known Brent's principle [5] we can reduce the number of processors to $O(n^2/\log n)$. $\square$

Lemma 7. *Let* T_n *be a tournament,* $n \geq 256$, n *even. Given a nice transitive subtournament* TT_p *of* T_n *and an ADHP in* $T_k = T_n - TT_p$, *an ADHC in* T_n *can be constructed in* $O(\log n)$ *time using* $O(n^2/\log n)$ *processors.*

Proof. Let $TT_p = \{1, 2, \ldots, p\}$ be a nice transitive subtournament of T_n. We distinguish between two cases depending on the parity of p.

(i) If p is odd, then $k = n - p$ is odd. Let $x_1 \leftarrow x_2 \rightarrow \ldots \leftarrow x_{k-1} \rightarrow x_k$, be an ADHP of $T_n - TT_p$ and let x_1 be its double point. We assume without loss of generality that there are at least two vertices, say i, j in TT_p dominating x_1 (if this is not the case we inverse each arc of T_n and study the new tournament). If for some vertex $u \in \{1, 2, \ldots, p - 1\}$, $u \rightarrow x_k$ then the desired cycle is $i \rightarrow x_1 \leftarrow x_2 \rightarrow \ldots \rightarrow x_k \leftarrow u \rightarrow$ [ADHP in TT_p with starting vertices u and i] $\leftarrow i$. Otherwise we consider the vertex x_{k-1}. If for some vertex $v \in \{1, 2, \ldots, p - 1\}$, $x_{k-1} \rightarrow v$ then the desired cycle is $i \rightarrow x_1 \leftarrow x_2 \rightarrow \ldots \leftarrow x_{k-1} \rightarrow v \leftarrow x_k \rightarrow u \leftarrow$ [ADHP in $TT_p - v$ ending at u and starting from i] $\leftarrow i$. If this is not the case, then $x_{k-1} \rightarrow p$, since TT_p is nice. Then, we consider the path $x_1 \leftarrow x_2 \rightarrow \ldots \leftarrow x_{k-1} \rightarrow p$ and the transitive subtournament $TT'_p = \{x_k, 1, 2, \ldots, p - 1\}$. The cycle now is $i \rightarrow x_1 \leftarrow x_2 \rightarrow \ldots \leftarrow x_{k-1} \rightarrow p \leftarrow t \leftarrow$ [ADHP in TT'_p with starting vertices t and i] $\leftarrow i$, where $t \in TT'_p$ and $t \neq p - 1$.

The above arguments are valid for $p \geq 5$. We know that for a nice transitive subtournament $p \geq \lfloor \frac{\log n}{2} \rfloor + 1$, that is $n \geq 256$. These arguments can be implemented in $O(1)$ time using $O(\log n)$ processors, since the cardinality of TT_p is $O(\log n)$.

(ii) If p is even, we examine the vertices in $T_n - TT_p$. If there is a vertex, y, which can be inserted in TT_p, then we consider the transitive subtournament $TT_{p+1} = TT_p \cup \{y\}$, which remain nice and therefore we are in the case (i). If no such a vertex exist, then TT_p is maximal and the proof is given by M. Rosenfeld in [11]. For his proof a restricted x-ADHP must be constructed and this dominates the implementation complexity, that is, by Corollary 1, $O(\log n)$ time using $O(n^2/\log n)$ processors.

Combining the two cases we obtain that this Lemma can be implemented in $O(\log n)$ time using $O(n^2/\log n)$ processors. $\square$.

Hence, combining Lemma 6 (construction of a nice TT_p) Lemma 7 and Theorem 1 (construction of an ADHP) we obtain:

Theorem 2. *For any tournament* T_n, $n \geq 256$, *an ADHC can be found in* $O(\log^2 n)$ *time using* $O(n^2/\log n)$ *processors.*

5 Concluding Remarks

We have shown that the problems of finding an ADHP and an ADHC in tournaments are both in NC. By this work some interesting unsolved questions are addressed:

1) Does the maximal transitive subtournament problem belong to NC? If this is true, then employing our results for ADHPs and the proofs given by M. Rosenfeld in [12] we can find ADHCs in tournaments with no less than 16 vertices. An interesting generalization of this problem is that of finding a maximal acyclic subdigraph of a given digraph or equivalently a minimal feedback vertex set in a digraph.

2) What is the complexity of finding, if there exists, a doubly restricted ADHP in a tournament, i.e. an ADHP whose both extremities are specified? In other words, what is the complexity (both sequential and parallel) of testing the antidirected Hamiltonian connectedness of a tournament? A similar question for the (directed) doubly restricted Hamiltonian path has been stated by D. Soroker in [13]. Recently, J. Bang-Jensen, Y. Manoussakis and C.Thomassen [2], answered Soroker's question in the affirmative by presenting a polynomial sequential algorithm, while it is not known if the problem is in NC.

References

1. E. Bampis, M. El Haddad, Y. Manoussakis and M. Santha, *A parallel reduction of Hamiltonian cycle to Hamiltonian path in tournaments*, PARLE '93, Lect. Notes in Comp. Sc. 694 (1993) 553-560.

2. J. Bang-Jensen, Y. Manoussakis and C. Thomassen, *A polynomial algorithm for Hamiltonian-connectedness in semicomplete graphs*, Journal of Algorithms 13 (1992) 114-127.

3. J. Bang-Jensen and C. Thomassen, *A polynomial algorithm for the 2-path problem for semicomplete digraphs*, SIAM J. Discr. Math. (1992) 366-376.

4. A. Bar-Noy and J. Naor, *Sorting, minimal feedback sets and Hamiltonian paths in tournaments*, SIAM J. Discr. Math. 3 (1990) 7-20.

5. R. Brent, *The parallel evaluation of general arithmetic expressions*, J. ACM 21 (1974) 201-206.

6. R. Cole and U. Vishkin, *Approximate and exact parallel scheduling with applications to list tree and graph problems*, In Proc. 27th FOCS (1986) 478-491.

7. B. Grünbaum, *Antidirected Hamiltonian paths in tournaments*, J. Combin. Theory (B) 11 (1971) 249-257.

8. P. Hell and M. Rosenfeld, *The complexity of finding generalized paths in tournaments*, Journal of Algorithms 4 (1983) 303-309.

9. V. Petrovic, *Antidirected Hamiltonian circuits in tournaments*, In Proc. 4th Yogoslavian Seminar of Graph Theory, Novi Sad, 1983.

10. K. B. Reid and E. T. Parker, *Disproof of a conjecture of Erdös and Moser*, J. Combin. Theory (B) 9 (1970) 93-99.

11. M. Rosenfeld, *Antidirected Hamiltonian paths in tournaments*, J. Combin. Theory (B) 12 (1972) 93-99.

12. M. Rosenfeld, *Antidirected Hamiltonian circuits in tournaments*, J. Combin. Theory (B) 16 (1974) 234-242.

13. D. Soroker, *Fast parallel algorithms for finding Hamiltonian paths and cycles in a tournament*, Journal of Algorithms 9 (1988) 276-286.

14. C. Thomassen, *Antidirected Hamiltonian circuits and paths in tournaments*, Math. Ann. 201 (1973) 231-238.

Directed Path Graph Isomorphism

(Extended Abstract)

Luitpold Babel[1], Ilia Ponomarenko[2] and Gottfried Tinhofer[1]

[1] Institut für Mathematik, Technische Universität München,
80290 München, Germany
[2] POMI, Fontanka 27, 191011 Sankt-Petersburg, Russia

Abstract. This paper deals with the isomorphism problem of directed path graphs and rooted directed path graphs. Both graph classes belong to the class of chordal graphs, and for both classes the relative complexity of the isomorphism problem is yet unknown. We prove that deciding isomorphism of directed path graphs is isomorphism complete, whereas for rooted directed path graphs we present a polynomial-time isomorphism algorithm.

1 Introduction

Let $G = (V, E)$ be a simple undirected graph, V its node set, E its edge set, $n = |V|$, $m = |E|$. For $v \in V$ and $W \subseteq V$ let $N_W(v)$ be the set of neighbours of v that belong to the subset W of V. We write $N(v)$ instead of $N_V(v)$. The subgraph induced by W is denoted by $G(W)$. $G-W$ means the subgraph $G(V-W)$ induced by the complement of W. $E(W)$ denotes the set of edges in G with both end nodes in W. Two graphs $G = (V, E)$ and $G' = (V', E')$ are *isomorphic* iff there is a one-to-one mapping $\psi : V \to V'$ which preserves the adjacency relation between nodes, i.e. which satisfies the condition $uv \in E$ iff $\psi(u)\psi(v) \in E'$ for every pair of nodes u and v. ψ is called an *isomorphism* of G and G'. If $G = G'$, then an isomorphism is called an *automorphism* of G.

A node v is called *simplicial* iff $N(v)$ induces a complete subgraph of G. A *clique* is a complete subgraph $G(W)$ such that for no proper superset W' of W the subgraph $G(W')$ is also complete. If v is a simplicial node and $W = \{v\} \cup N(v)$ then $G(W)$ is a clique, the only clique of G that contains v. A graph G is called *chordal* iff every induced subgraph $G(W)$ has at least one simplicial node.

Assume that for a graph $G = (V, E)$ there is a tree T and a set $\mathbf{T} = \{T_v | v \in V\}$ of subtrees of T such that $T_u \cap T_v \neq \emptyset$ iff $uv \in E$. In this case, $(T, \mathbf{T})$ is called a *tree model* for G. It is well known that a graph has a tree model iff it is chordal (see [9], [11]).

A tree model for a chordal graph G is called a *clique model* if the node set of T is the set Cl of cliques in G and $c \in T_v$ is equivalent to $v \in c$ for all $c \in Cl$ and all $v \in V$. It is known that every chordal graph has even a clique model.

In what follows we restrict our considerations to the class of chordal graphs. Furthermore, we assume that any graph G is given by its clique model $(T, \mathbf{T})$.

Graphs with a clique model $(T, \mathbf{P})$ where each $P_v \in \mathbf{P}$ is a path of T are called *undirected path graphs*. If T is a directed tree, i.e. if each edge of T has a fixed orientation, and if all paths in $\mathbf{P}$ are directed paths, then G is called *directed path graph*. If T can be chosen to be a rooted tree such that all edges are oriented downwards, i.e. in the direction from the root r to the leaves, then G is called *rooted directed path graph* (*rdp-graph* for short). All graph properties mentionned here can be recognized in polynomial time. A recognition algorithm for rdp-graphs is presented and discussed in [10]. A tree model for an rdp-graph may be changed to become a clique model using a straight forward procedure also explained in [10]. A review on path graphs is given in [20].

Undirected trees are trivial examples of rdp-graphs. Other examples are interval graphs (for more details see [11]). Let $\mathcal{T}$ be the class of trees, $\mathcal{I}$ the class of interval graphs, $\mathcal{RDP}$ the class of rooted directed path graphs, $\mathcal{DP}$ the class of directed path graphs, $\mathcal{UP}$ the class of undirected path graphs and $\mathcal{Ch}$ the class of chordal graphs. Then we have

$$\mathcal{T} \subset \mathcal{RDP} \quad \text{and} \quad \mathcal{I} \subset \mathcal{RDP} \subset \mathcal{DP} \subset \mathcal{UP} \subset \mathcal{Ch}.$$

All inclusions are strict. Figure 1a shows an example for an undirected path graph which is not a directed path graph. Figure 1b shows an example for a directed path graph which is not a rooted directed path graph. In both cases a corresponding clique tree is shown. It is obvious how the system of paths is to be chosen in order to get the desired clique model.

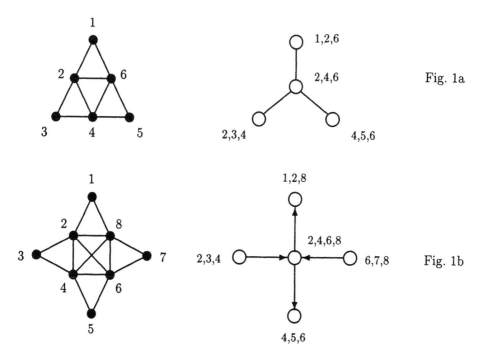

Fig. 1a

Fig. 1b

It is known that isomorphism testing is *isomorphism complete* (i.e. polynomial time equivalent to graph isomorphism) for some special classes of graphs including regular graphs, bipartite graphs, line graphs, rooted acyclic digraphs, comparability graphs, split graphs, k-trees (for unbounded k) and undirected path graphs. For a review on isomorphism complete graph classes see [3].

On the other side it has been shown that the isomorphism problem is solvable in polynomial time when restricted to special graph classes, e.g. to graphs of bounded degrees [18], planar graphs [8], [12], trees (folklore), interval graphs [4], cographs [5], P_4-sparse and P_4-extendible graphs [13], [14], permutation graphs [6], k-trees (for fixed k) [16], transitive series parallel digraphs [17], graphs of bounded genus [7], graphs of bounded eigenvalue multiplicity [1] and graphs of bounded Hadwiger number [21]. In this latter paper a set of additional isomorphism complete problems is discussed, too. A comprehensive survey on the state of art in graph isomorphism problems until 1985 has been given in [15].

There is still a gap within the chordal graphs between the subclasses of interval graphs (which can be interpreted as rdp-graphs with a clique model where the tree T is a path) and undirected path graphs. Deciding isomorphism of interval graphs is a polynomial problem whereas deciding isomorphism of undirected path graphs is isomorphism complete. The relative complexity of the isomorphism problem for directed path graphs and for rooted directed path graphs is yet unsettled. It is the aim of this paper to close this gap. The upper part of the gap is relatively easy to close, this is done in the following Theorem 1. The remaining part is done in the rest of the paper. It needs a detailed study of rdp-graphs. For all proofs see [2].

Theorem 1. *Isomorphism testing for directed path graphs is isomorphism complete.*

2 Simplicial Partitions of Chordal Graphs

Let a chordal graph $G = (V, E)$ be given. An ordered partition $V = S_1 \cup S_2 \cup \ldots \cup S_q$ of V is called *simplicial* iff for $1 \leq i \leq q$ all the nodes in S_i are simplicial nodes of the subgraph $G(S_i \cup S_{i+1} \cup \ldots \cup S_q)$. It is well known that a graph has a simplicial partition iff it is a chordal graph. If the simplicial partition is total, i.e. if $|S_i| = 1$, $1 \leq i \leq q$, then it is called a *perfect elimination scheme*. Given a chordal graph G a perfect elimination scheme can be computed within time $O(n + m)$ ([11]). Let $S_i = \{v_i\}$, $1 \leq i \leq n$, be such an elimination scheme. W.l.o.g. we may assume that G has no isolated nodes. For arbitrary i let $t(i)$ be the smallest index belonging to a neighbour of v_i. A node v_i is a simplicial node of G iff either $t(i) > i$ or $N(v_i) = N(v_{t(i)}) \cap \{v_{t(i)+1}, \ldots, v_n\}$. Based on this criterion we may compute the coarsest simplicial partition of G in time at most $O(n^3)$.

A graph G together with a specified simplicial partition $S = \{S_1, S_2, \ldots, S_q\}$ will be called a *layered graph*. It will be denoted by (G, S). The triple (n, m, q) will be called the *size* of (G, S). The sets S_i are called the *cells* of (G, S). Two cells S_i and S_j are called *adjacent* iff at least one node of S_i is adjacent to at least one node in S_j, they are called *totally adjacent* iff all nodes in S_i are adjacent to all nodes in S_j. A cell is called *complete* iff it induces a complete subgraph.

Let (G, S) be a layered graph with simplicial partition $S = \{S_1, ..., S_q\}$. For arbitrary two cells A and B we shall write $A \prec B$ iff $A = S_i, B = S_j, j > i$, and there is a sequence $i = i_0 < i_1 ... < i_k = j$ such that S_{i_s} and $S_{i_{s+1}}$ are adjacent for $0 \leq s \leq k - 1$.

Now, let A be an arbitrary cell of S. The following statements follow trivially from the simpliciality of S:

(2.1) The subgraph $G(A)$ is a union of complete graphs, the components of $G(A)$. Let their node sets be $A_1, ..., A_p, p = p(A)$.

(2.2) No node $x \in A$ has neighbours in different components of a cell $B, A \prec B$.

(2.3) If x and y belong to the same component of $G(A)$ then $N_B(x) = N_B(y)$ for every cell $B, B \succ A$.

(2.4) $S - A$ is a simplicial partition for $G - A$.

The following simple statements, too, follow immediately from the definition of layered graphs:

(2.5) If G is connected then $G(S_q)$ is connected, i.e. $G(S_q)$ is complete.

(2.6) If S is the coarsest simplicial partition, i.e. if S_i equals the set of all simplicial nodes in $G(S_i \cup ... \cup S_q)$, then $G' = G - S_q$ is not connected.

(2.7) If $G(S_q)$ is not complete then G is not connected. In this case, let W be the node set of some component of G'. Then $S \cap W = \{S_1 \cap W, S_2 \cap W, ..., S_{q-1} \cap W\}$ is a simplicial partition of $G(W)$ (empty sets $S_i \cap W$ being removed).

A simplicial partition S of G is called *regular* iff for any $A, B \in S$ and for any $x, x' \in A$ we have $|N_B(x)| = |N_B(x')|$. With other words, in a regular simplicial partition every node x in a cell A has the same number of neighbours in any cell B ($B = A$ included). For a regular simplicial partition S and arbitrary $A, B \in S$ define $n_A = |A|$ and $d(A, B) = |N_B(x)|$, $x \in A$ arbitrary.

Note that, in general, simpliciality does not follow from regularity. Simpliciality is a notion for ordered partitions whereas regularity is independent of any ordering on the set of cells. The following property holds in general only for regular simplicial partitions:

(2.8) Let W and W' be the node sets of different components of G and assume $A \cap W \neq \emptyset$ and $A \cap W' \neq \emptyset$ for some $A \in S$. Then $B \cap W \neq \emptyset$ iff $B \cap W' \neq \emptyset$ for all $B \in S$.

A given partition S can be refined to become a regular partition in the following obvious way. For every cell S_i and every $x \in S_i$ compute the vector $d(x) = (d_1(x), ..., d_q(x))$ where $d_j(x)$ means the number of neighbours of x which belong to S_j, $1 \leq j \leq q$. If $d(x)$ is not constant on S_i then sort the vectors $d(x)$, $x \in S_i$, lexicographically. Assume the result is a sequence $d_1 \succ d_2 \succ ... \succ d_r$. Define $S_{ij} = \{x \in S_i \mid d(x) = d_j\}$, $1 \leq j \leq r$. This gives an ordered partition of S_i. Replace in S the cell S_i by the sequence of cells $S_{i1}, ..., S_{ir}$, in this order, define $S'_k = S_k$ for $1 \leq k \leq i - 1$, $S'_{i+k-1} = S_{ik}$ for $1 \leq k \leq r$, and $S'_{i+r+k-1} = S_{i+k}$ for $1 \leq k \leq q-i$ and update $q \leftarrow q+r-1$. The resulting ordered partition is a refinement of S. Repeat this process with different cells as long as possible. The process will stop when no cell S_i can be partitioned further by the described procedure. Call the resulting ordered partition $S*$. Obviously, $S*$ is regular and if S was simplicial then $S*$ will be simplicial, too. $S*$ is the coarsest regular partition which is finer than S. Moreover, $S*$ is canonically ordered in the following sense: Any automorphism of G which preserves S preserves also $S*$ (see also the definition of S-automorphism at the end of this section).

A given partition S can also easily be refined to become a simplicial partition. Let $Y = (Y_1, \ldots, Y_\rho)$ be the coarsest simplicial partition of G. Sort the set $J = \{(s,t) \mid Y_s \cap S_t \neq \emptyset\}$ lexicographically and find the mapping $j : J \longrightarrow \{1, \ldots, Q\}$, $Q = |J|$, such that $j(s,t) < j(s',t')$ iff $(s,t) \prec (s',t')$. Define

$$\tilde{S} = (\tilde{S}_1, \ldots, \tilde{S}_Q); \quad \tilde{S}_{j(s,t)} = Y_s \cap S_t.$$

Obviously, $\tilde{S}$ is the coarsest simplicial partition which is finer than S. Finally, $\tilde{S}*$ is the coarsest regular simplicial partition which is finer than S. Since all components of the degree vectors $d(x)$ are bounded by $n-1$ we may use bucket sort as a sorting routine. Thus, given S, the refinement $S*$ can be found in time $O(n^3)$.

Regular partitions of the node set of a graph play an important role in the algorithmic approach to graph isomorphisms. They have been broadly used for designing practical isomorphism algorithms. In a series of papers [22], [23], [24], [25] such partitions have been investigated in great detail. It turned out in this work that the set of regular partitions can be associated in a canonical way to the polytope of doubly stochastic matrices which commute with the adjacency matrix of the graph.

A graph G together with a regular simplicial partition S is called a *regularly layered graph*. The following lemma is of considerable interest in studying regularly layered graphs.

Lemma 2. *Let G be regularly layered with partition S, and let $A \in S$ be arbitrary. Then any two nodes x and x' in A belong to the same number of cliques of any given size.*

Corollary 3. *Let $(T, \mathbf{T})$ be a clique model for G and let S be a regular simplicial partition. Then for an arbitrary cell $A \in S$ and for any two nodes $x, x' \in A$ we have $|T_x| = |T_{x'}|$, i.e. T_x and $T_{x'}$ contain the same number of cliques. If G is an rdp-graph then the paths P_x and $P_{x'}$ corresponding to x and x' have equal length.*

To each regularly layered graph (G, S) there is a corresponding matrix *degree* (G, S) of dimension $q \times q$ defined by

$$degree(G, S)_{ij} = d(S_i, S_j), \quad 1 \leq i, j \leq q.$$

It will be called the *degree matrix* of (G, S).

Lemma 4. *Let degree(G, S) be given and assume that G is connected. Then we can find from degree(G, S)*
(a) the node numbers n_A of $G(A)$, $A \in S$, and therefore also the total node number n.
(b) the number $p(A)$ of components of each $G(A)$.
If G is not connected, then degree(G, S) has a decomposition into $t \geq 1$ irreducible blocks. Each component $G(W)$ is a regularly layered graph with partition $S_W = S \cap W$. The components of G are partitioned into t classes such that the matrices degree$(G(W), S_W)$ are equal for all components within a fixed class. The classes are in one-to-one correspondence with the blocks of degree(G, S), these blocks are the degree matrices for the components in the corresponding class as well as for the union of all components in the class.

The subgraphs $G_{I_1}, G_{I_2}, ..., G_{I_t}$ of a disconnected graph G which according to the lemma correspond to the blocks of $degree(G, S)$ will be called the *blocks* of G. Note that the blocks are not necessarily connected subgraphs.

Again, consider an arbitrary graph (G, S) where $S = \{S_1, S_2, ..., S_q\}$ is a regular simplicial partition of G. We construct an auxiliary graph $\hat{G}(S)$ in the following way. Let $\hat{V} = \{v_A \mid A \in S\}$ be a set of vertices representing the cells of S and define $\hat{E} = \{v_A v_B \mid A \text{ adjacent but not totally adjacent to } B\}$. The graph $\hat{G}(S)$ is called the *cell graph* of (G, S). (G, S) is called *irreducible* iff $\hat{G}(S)$ is connected, otherwise it is called *reducible*. Clearly, if (G, S) is irreducible, then G is connected.

Assume that (G, S) is reducible and let $\hat{G}_k = (\hat{V}_k, \hat{E}_k), 1 \leq k \leq r$, be the components of $\hat{G}(S)$. Define

$$L_k = \{i \mid v_{S_i} \in \hat{V}_k\}$$
$$\hat{S}_k = \{S_i \mid i \in L_k\}$$
$$U_k = \bigcup_{i \in L_k} S_i.$$

Then, $(G(U_k), \hat{S}_k)$ is a regularly layered subgraph of (G, S).

An automorphism ψ of G which preserves the (not necessarily simplicial or regular) partition S, i.e. which satisfies $\psi(S_i) = S_i$, $1 \leq i \leq q$, is called S-*automorphism* of G, or automorphism of (G, S). The group of all S-automorphisms of G is denoted by $aut(G, S)$. Clearly, if S is the coarsest regular simplicial partition for G, then $aut(G, S) = aut(G)$.

Analoguously, an isomorphism ϕ of two layered graphs (G, S) and $(H, \mathcal{R})$ is called an $S - \mathcal{R}$-isomorphism iff it satisfies $\phi(S_i) = R_i$, $1 \leq i \leq q$. Clearly, $degree(G, S) = degree(H, \mathcal{R})$ is a necessary condition for the existence of an $S - \mathcal{R}$-isomorphism of (G, S) and $(H, \mathcal{R})$.

Note that if (G, S) is reducible with layered subgraphs $(G(U_k), \hat{S}_k)$ as defined above, then

$$aut(G, S) = \bigotimes_{k=1}^{r} aut(G(U_k), \hat{S}_k).$$

It can easily be shown (see [2]) that checking reducibility of a regularly layered connected graph and finding the graphs $(G(U_k), \hat{S}_k)$ requires time $O(n^2)$.

3   Thinning of rdp-graphs

From now on, we deal with layered rdp-graphs (G, S) only.

An edge $e = uv$ is called *simplicial* if the subgraph induced by $N(u) \cap N(v)$ is complete. Obviously a simplicial edge is contained in exactly one clique.

Lemma 5. *Each clique of a connected rdp-graph has a simplicial edge.*

Virtually, the last lemma is true even for directed path graphs. We know how to prove this claim. However, it is false for undirected path graphs, and therefore it is false for chordal graphs in general.

Let K be a complete subgraph of G and $H_1, H_2, ..., H_p$ the components of $G - K$. A component H_i will be called *comprehensive* with respect to K iff there is a node v in H_i which is adjacent to all nodes of K. The node v will also be called *comprehensive*. Note that if K is a union of cells then in the case of a regularly layered graph (G, S), if a cell S_i contains comprehensive nodes, then all of its nodes are comprehensive. In the latter case S_i is called a *comprehensive cell*. Let $S = \{S_1, S_2, ..., S_q\}$ be a regular simplicial partition of G. The *core* of a connected rdp-graph G is denoted by $Core(G)$ and defined in the following procedure:

Procedure CORE(G, S);
(1) $Core \leftarrow S_q$;
(2) **if** there is exactly one comprehensive component with respect to
    $Core$ **then** find the comprehensive cell S_k with largest index k **else** STOP;
(3) $Core \leftarrow Core \cup S_k$; **goto** (2).

Since G was assumed to be connected we know that S_q is complete. It is easy to see that each S_k added to $Core$ in step (3) of the procedure is a complete cell. Therefore, $Core$ induces a complete subgraph of G.

In the following the term *comprehensive* always refers to $Core = Core(G)$ of the graph G at hand.

If (G, S) has comprehensive components then due to the construction of $Core$ it has at least two of them. The following lemma tells us that an irreducible graph has at most two comprehensive components.

Lemma 6. *If (G, S) has more than two comprehensive components then it is reducible.*

Our goal is to settle the isomorphism (automorphism) problem for rdp-graphs by reducing it to the isomorphism (automorphism) problem for smaller parts of the graphs under consideration or for some thinned out versions of them. The last lemma combined with the remark at the end of the last section shows that in the following we can assume that (G, S) has 0 or 2 comprehensive components. We are now ready to state the main result of this section.

Theorem 7. *Let (G, S) be a connected regularly layered rdp-graph. Then there exists a nonempty set $F \subseteq E$ such that $G' = (V, E - F)$ is an rdp-graph and each S-automorphism of G is an S-automorphism of G' and vice versa.*

We shall write $T(G, S)$ for the layered graph which is thinned out as stated in the theorem. This is done by applying one of three possible types of thinning operations which depend on the number of comprehensive components and the kind of the neighbourhood of the nodes of $Core$ (for more details see [2]).

Corollary 8. *Let (G, S) and $(H, \mathcal{R})$ be irreducible regularly layered rdp-graphs such that S_i is a core, respectively comprehensive, cell in (G, S) iff R_i is a core, respectively comprehensive, cell in $(H, \mathcal{R})$. Let $(\tilde{G}, \tilde{S}*)$ and $(\tilde{H}, \tilde{\mathcal{R}}*)$ be the regularly layered rdp-graphs resulting from thinning and making the partitions simplicial and regular again. Then each $\tilde{S} * -\tilde{\mathcal{R}}*$-isomorphism of $\tilde{G}$ and $\tilde{H}$ is an $S - \mathcal{R}$- isomorphism of G and H, and vice versa.*

Finally, we are able to prove:

Theorem 9. *Let (G, S) be a regularly layered connected rdp-graph. The core and the set of comprehensive nodes of G and the type of thinning operation applicable to (G, S) can be computed from $degree(G, S)$ in time $O(n^3)$. Consequently, $T(G, S)$ can be found in time $O(n^3)$.*

Furthermore, if (G, S) and $(H, \mathcal{R})$ have the same degree matrix, then S_i is a core cell (a comprehensive cell, respectively) iff R_i is a core cell (a comprehensive cell, respectively).

4 An Isomorphism Test for rdp-graphs

Let $\mathcal{G} = (G, S)$ and $\mathcal{H} = (H, \mathcal{R})$ be two graphs of equal size (n, m, q). We are now ready to formulate a simple algorithm for deciding whether $\mathcal{G}$ and $\mathcal{H}$ are isomorphic as layered graphs or not, i.e. whether there is an isomorphism $\psi : \mathcal{G} \xrightarrow{\ *\ } \mathcal{H}$ satisfying $\psi(S_i) = R_i$ for $1 \leq i \leq q$. The algorithm needs time $O(n^6)$. This settles the complexity question of the isomorphism problem for rdp-graphs.

The algorithm which will be described below works as follows:

In a first step the algorithm replaces the given partitions S and $\mathcal{R}$ by their refinements $\tilde{S}*$ and $\tilde{\mathcal{R}}*$, respectively, and computes $degree(\mathcal{G})$ and $degree(\mathcal{H})$. If the degree matrices are not identical, then the algorithm stops in state 'false', otherwise it continues working.

If $\mathcal{G}$ and $\mathcal{H}$ are not connected let $\mathcal{G}_i$, $1 \leq i \leq s$, and $\mathcal{H}_j$, $1 \leq j \leq t$, be the components of $\mathcal{G}$ and $\mathcal{H}$, respectively. If $s = t$, then the algorithm, by calling itself recursively, tests all pairs $\mathcal{G}_i$ and $\mathcal{H}_j$ with $size(\mathcal{G}_i) = size(\mathcal{H}_j)$ for isomorphism. If by this procedure a permutation ϕ of $\{1, 2, \ldots, s\}$ can be found such that $\mathcal{G}_i$ is isomorphic to $\mathcal{H}_{\phi(i)}$, then the algorithm stops in state 'true', otherwise it stops in state 'false'.

If $\mathcal{G}$ and $\mathcal{H}$ are connected then let D be a copy of the common degree matrix $degree(\mathcal{G})$ of $\mathcal{G}$ and $\mathcal{H}$. The algorithm calls a subroutine which will be called RE-DUCE. This routine changes D by reducing all entries D_{ij} to 0 which correspond to totally adjacent cells S_i and S_j. Afterwards the algorithm finds the blocks of the reduced matrix D. $\mathcal{G}$ and $\mathcal{H}$ are reducible iff D is a block-diagonal matrix, i.e. $D = \bigoplus_{k=1}^{r} D_k$ with $r > 1$. If reducibility occurs then let $\hat{\mathcal{G}}_k = (\hat{Y}_k, \hat{S}_k)$ and $\hat{\mathcal{H}}_k = (\hat{Z}_k, \hat{\mathcal{R}}_k)$, $1 \leq k \leq r$, be the subgraphs of $\mathcal{G}$ and $\mathcal{H}$, respectively, which correspond to the components of $\hat{G}$ and $\hat{H}$ (i.e. if U_k is the union of all cells in $\hat{S}_k$ then $\hat{Y}_k = G(U_k)$, and $\hat{Z}_k$ has an analoguous meaning). Again calling itself recursively, the algorithm tests for all k, $1 \leq k \leq r$, whether $\hat{\mathcal{G}}_k$ is isomorphic to $\hat{\mathcal{H}}_k$ or not. If the answer is negative for some k, then the algorithm stops in state 'false', otherwise it stops in state 'true'.

If $\mathcal{G}$ and $\mathcal{H}$ turn out to be irreducible then the algorithm computes the thinned versions $T(G, \mathcal{S})$ and $T(H, \mathcal{R})$ and tests whether these two graphs are isomorphic or not. In the positive case the algorithm stops in state 'true', otherwise in state 'false'.

Here is now a more formal (but not too formal) description of the algorithm.

Algorithm ISOMORPH($\mathcal{G}, \mathcal{H}, Boole$);
Input: Two layered rdp-graphs $\mathcal{G} = (G, \mathcal{S})$ and $\mathcal{H} = (H, \mathcal{R})$ of equal size (n, m, q).
Output: The value of the Boolean variable $Boole$.

(1) $\mathcal{S} \leftarrow \tilde{\mathcal{S}}*$; $\mathcal{R} \leftarrow \tilde{\mathcal{R}}*$;
 $D \leftarrow degree(G, \mathcal{S})$; $D_1 \leftarrow degree(H, \mathcal{R})$;
 if $D \neq D_1$ **then**
 begin $Boole \leftarrow' false'$; **goto** (6); **end**;

(2) **compute** the components $\mathcal{G}_i = (G_i, \mathcal{S}_i)$, $1 \leq i \leq s$, of $\mathcal{G}$;
 compute the components $\mathcal{H}_j = (H_j, \mathcal{R}_j)$, $1 \leq j \leq t$, of $\mathcal{H}$;
 if $s \neq t$ **then**
 begin $Boole \leftarrow' false'$; **goto** (6); **end**;
 if $s = 1$ **then goto** (4);

(3) **compute** $size(\mathcal{G}_i)$ and $size(\mathcal{H}_i)$, $1 \leq i \leq s$;
 $J \leftarrow \{1, 2, \ldots, s\}$; $z \leftarrow 0$;
 for $1 \leq i \leq s$ **do**
 begin
 for $j \in J$ **do**
 if $size(\mathcal{G}_i) = size(\mathcal{H}_j)$ **then**
  begin ISOMORPH($\mathcal{G}_i, \mathcal{H}_j, B$);
  if $B =' true'$ **then begin** $J \leftarrow J - \{j\}$; $z \leftarrow z + 1$; **goto** (*); **end**;
  end;

(*) **end**;
if $z < s$ **then** $Boole \leftarrow' false'$ **else** $Boole \leftarrow' true'$;
goto (6);

(4) $D \leftarrow \text{REDUCE}(D)$;
compute the blocks D_k, $1 \le k \le r$, of D;
if $r = 1$ **then goto** (5);
compute $\hat{\mathcal{G}}_k = (\hat{Y}_k, \hat{S}_k)$ and $\hat{\mathcal{H}}_k = (\hat{Z}_k, \hat{\mathcal{R}}_k)$, $1 \le k \le r$;
for $1 \le k \le r$ **do**
   begin $\text{ISOMORPH}(\hat{\mathcal{G}}_k, \hat{\mathcal{H}}_k, B)$;
   if $B =' false'$ **then begin** $Boole \leftarrow' false'$; **goto** (6); **end**;
   end;
$Boole \leftarrow' true'$; **goto** (6);

(5) **compute** $\mathcal{T}(\mathcal{G}) = \mathcal{T}(G, \mathcal{S})$ and $\mathcal{T}(\mathcal{H}) = \mathcal{T}(H, \mathcal{R})$;
$\mathcal{G} \leftarrow \mathcal{T}(\mathcal{G})$; $\mathcal{H} \leftarrow \mathcal{T}(\mathcal{H})$;
goto (1);

(6) STOP.

The correctness of steps (1) - (4) is evident. The correctness of step (5) follows from Theorem 7 and the Corollary. We still háve to prove the polynomiality of the algorithm. This is done using the following lemma.

Lemma 10. *Assume that $\mathcal{A}$ is an algorithm which compares two graphs G and H of equal node number $n > 1$ according to the following strategy. In a first phase $\mathcal{A}$ dissects G and H into at least two smaller parts $G_1, \ldots, G_s$ and $H_1, \ldots, H_t$. If $s \ne t$ then $\mathcal{A}$ stops sending a failure message whereas otherwise, in a second phase, the algorithm calls itself recursively to compare all pairs of graphs G_i and H_j for all i and j such that G_i and H_j have the same number of nodes. Assume that for $n = 1$ the algorithm compares G and H directly and needs time at most D' and that for larger values of n the time needed to perform Phase 1 is bounded from above by Cn^p. Then there exists a constant D such that the time needed by $\mathcal{A}$ to compare two graphs of node number n is bounded by Dn^{p+1}.*

Now, let us apply this lemma to Algorithm ISOMORPH. Phase 1 of this algorithm consists of computing $\tilde{\mathcal{S}}*$ and $\tilde{\mathcal{R}}*$ followed by a test for connectivity, a test for reducibility and eventually an application of a thinning operation. All this can be done in time $O(n^3)$. However, the result of a thinning operation is not necessarily a disconnected or reducible graph. Thus, Phase 1 can consist of a series of thinning operations. The number of thinning operations needed to partition the graphs into smaller parts is bounded by $O(n^2)$. This means that our upper bound for the amount of time needed to perform Phase 1 is Cn^5. Hence, Algorithm ISOMORPH works in time at most $O(n^6)$.

5 Conclusions

We have shown in Theorem 1 that deciding graph isomorphism for directed path graphs is isomorphism complete. In the case of rooted directed path graphs we have presented a decomposition method which is based on Theorem 7 and which allows to reduce the isomorphism problem for large graphs to a sequence of isomorphism problems for smaller parts of them. The result is a polynomial-time algorithm for deciding rooted directed path graph isomorphism. The method presented is not extendible to superclasses of rdp-graphs. One of the main reasons is that Lemma 5 does not hold for general chordal graphs. Another reason is that we cannot ensure that the thinning operations considered in Section 3, if applicable at all to general chordal graphs which may happen occasionally, preserve the type of the graphs under consideration.

Our results close a gap in the literature concerning the relative complexity of isomorphism problems for various subclasses of chordal graphs.

References

1. L. Babai, D.Y. Grigor'ev and D.M. Mount, Isomorphism of graphs with bounded eigenvalue multiplicity, in "Proc. 14th ACM STOC" (1982), 310-324

2. L. Babel, I.N. Ponomarenko, G. Tinhofer, The isomorphism problem for directed path graphs and for rooted directed path graphs, *Report No. TUM-M9401, Institute of Mathematics, Technical University of Munich (1994), submitted*

3. K.S. Booth and C.J. Colbourn, Problems polynomially equivalent to graph isomorphism, *Report No. CS-77-04, Computer Science Department, University of Waterloo (1979)*

4. K.S. Booth and G.S. Lueker, A linear-time algorithm for deciding interval graph isomorphism, *J. of ACM 26 (1979), 183-195*

5. D.G. Corneil, H. Lerchs and L. Stewart Burlingham, Complement reducible graphs, *Discrete Appl. Math. 3 (1981), 163-174*

6. G.J. Colbourn, On testing isomorphism of permutation graphs, *Networks 11 (1981), 13-21*

7. I.S. Filotti and J.N. Mayer, A polynomial-time algorithm for determining the isomorphism of graphs of fixed genus, in "Proc. 12th ACM Symposium on Theory of Computing", 1980, 235-243

8. M. Fontet, A linear algorithm for testing isomorphism of planar graphs, in "Proc. 3rd Colloquium on Automata, Languages and Programming", 1976, 411-423

9. F. Gavril, The intersection graphs of subtrees in trees are exactly the chordal graphs, *J. Combinat. Theory B16 (1974), 47-56*

10. F. Gavril, A recognition algorithm for the intersection graphs of directed paths in directed trees, *Discrete Math. 13 (1975), 237-249*

11. M.C. Golumbic, Algorithmic graph theory and perfect graphs, *Academic Press, New York, 1980*

12. J.E. Hopcroft and J.K. Wong, Linear-time algorithm for isomorphism of planar graphs, in "Proc. 6th ACM Symp. Theory of Computing, 1974, 172-184

13. R. Jamison and S. Olariu, On a unique tree representation for P_4-extendible graphs, *Discrete Applied Mathematics 34 (1991), 151-164*

14. R. Jamison and S. Olariu, A unique tree representation for P_4-sparse graphs, *Discrete Applied Mathematics, 35 (1992), 115-129*

15. D.S. Johnson, The NP-completeness column: an ongoing guide, *Journal of Algorithms 6 (1985), 434-451*

16. M.M. Klawe, M.M. Corneil and A. Proskurowski, Isomorphism testing in hook-up graphs, *SIAM J. of Algebraic and Discrete Methods 3 (1982), 260-274*

17. E.L. Lawler, Graphical algorithms and their complexity, *Math. Centre Tracts 81 (1976), 3-32*

18. E.M. Luks, Isomorphism of graphs of bounded valence can be tested in polynomial time, *J. Comput. Syst. Sci. 25 (1982), 42-65*

19. G.L. Miller, Isomorphism testing for graphs of bounded genus, in "Proc. 12th ACM Symposium on Theory of Computing", 1980, 225-235

20. C.L. Monma and V.K. Wei, Intersection graphs of paths in a tree, *Journal of Combinatorial Theory B41 (1986), 141-181*

21. I.N. Ponomarenko, On the isomorphism problem for classes of graphs closed under contractions, in: "Zapiski Nauchnykh Seminarov LOMI", 174 (1988), 147-177 (Russian)

22. H. Schreck and G. Tinhofer, A note on certain subpolytopes of the assignment polytope associated with circulant graphs, *Linear Algebra and Its Applications 111 (1988), 125-134*

23. G. Tinhofer, Graph isomorphism and theorems of Birkhoff type, *Computing 36 (1986), 285-300*

24. G. Tinhofer, Strong tree-cographs are Birkhoff graphs, *Discrete Applied Mathematics 22 (1988/89), 275-288*

25. G. Tinhofer, A note on compact graphs, *Discrete Applied Mathematics 30 (1991), 253-264*

List of Participants

Prof. Dr. Thomas Andreae
Universität Hamburg
Mathematisches Seminar
Bundesstr. 55
20146 Hamburg, Germany
andreae@math.uni-hamburg.de

Dr. Luitpold Babel
Institut für Mathematik
TU München
80290 München, Germany
babel@statistik.tu-muenchen.de

Stefan Baumann
Institut für Mathematik
TU München
80290 München, Germany

Arne Bayer
Fakultät für Informatik
Universität der Bundeswehr
D-85577 Neubiberg, Germany
bayer@informatik.unibw-muenchen.de

Dr. Hans Bodlaender
Department of Computer Science
Utrecht University
P.O. Box 80.089
3508 TB Utrecht, The Netherlands
hans@cs.ruu.nl

Prof. Dr. Andreas Brandtstädt
FB Mathematik FG Informatik I
Gerhard-Mercator-Universität-GH-
Duisburg
47048 Duisburg, Germany
ab@marvin.uni-duisburg.de

Dr. Shiva Chaudhuri
MPI Informatik
Im Stadtwald
66123 Saarbrücken, Germany
shiva@mpi-sb.mpg.de

Zhi-Zhong Chen
Department of Mathematical Sciences
Tokyo Denki University
Hatoyama, Saitama 350-03, Japan
chen@r.dendai.ac.jp
chen@info.mie-u.ac.jp

Jun-Dong Cho, Ph.D.
CAE, Samsung Electronics Co., Ltd.
82-3 Dodang-Dong, Buchun
Kyunggi-Do, Korea
bscad@saitgw.sait.samsung.co.kr

Dr. Elias Dahlhaus
Basser Dept. of Computer Science
University of Sydney
NSW 2006, Australia
dahlhaus@cs.su.oz.au

Nick Dendris
Patras University
Department of Computer Engineering
and Informatics
Rio, 265 00 Patras, Greece
dendris@cti.gr

Heiko Dörr
Institut für Informatik
Freie Universität Berlin
Takustr. 9
14195 Berlin, Germany
doerr@inf.fu-berlin.de

Prof. Dr. Jürgen Ebert
Universität Koblenz-Landau
Institut für Informatik
Rheinau 1
56075 Koblenz, Germany
ebert@informatik.uni-koblenz.de

Angelika Franzke
Universität Koblenz-Landau
Institut für Informatik
Rheinau 1
56075 Koblenz, Germany
franzke@informatik.uni-koblenz.de

Dipl.-Math. Renate Garbe
University of Twente
Postbus 217
7500 AE Enschede, The Netherlands
garbe@math.utwente.ne

Dr. Roberto Giacco
Dip. di Informatica e Sistemistica
Universita' di Roma "La Sapienza"
Via Salaria 113
00198 Roma, Italy
giacco@athena.dis.uniroma1.it

Dr. M.J. Golin
Department of Computer Science
Hong Kong University of Science
and Technology
Clear Water Bay, Kowloon
Hong Kong
golin@cs.ust.hk

Torben Hagerup
MPI-Saarbrücken
Im Stadtwald
66111 Saarbrücken, Germany
torben@mpi-sb.mpg.de

Wolfram Kahl
Fakultät für Informatik
Universität der Bundeswehr
D-85577 Neubiberg, Germany
kahl@informatik.unibw-muenchen.de

Prof. Kimio Kawaguchi
Dept. of Electrical and
Comp. Engineering
Nagoya Institute of Technology
Gokiso-cho Showa-ku Nagoya 466,
Japan
kawaguch@phaser.elcom.nitech.ac.jp

Peter Kempf
Fakultät für Informatik
Universität der Bundeswehr
D-85577 Neubiberg, Germany
kempf@informatik.unibw-muenchen.de

Prof. Lefteris Kirousis
Patras University
Department of Computer Engineering
and Informatics
Rio, 265 00 Patras, Greece
kirousis@cti.gr

Irene Knödel
Institut für Informatik
Breitwiesenstr. 20
70565 Stuttgart, Germany
knoedel@informatik.uni-stuttgart.de

Dr. Walter Knödel
Institut für Informatik
Breitwiesenstr. 20
70565 Stuttgart, Germany
knoedel@informatik.uni-stuttgart.de

Dr. Ton Kloks
Department of Mathematics
and Computer Science
Eindhoven University of Technology
P.O. Box 513
5600 MB Eindhoven, The Netherlands
ton@win.tue.nl

Dr. Jan Kratochvíl
KA MFF UK
Sokolovská 83
18600 Praha 8, Czech Republic
kratoch@earn.cvut.cz

Prof. Dr. Jan van Leeuwen
Department of Computer Science
Utrecht University
Padualaan 14
3584 CH Utrecht, The Netherlands
jan@cs.ruu.nl

Prof. Yannis Manoussakis
L.R.I., Bât 490
Université Paris-Sud
91405 Orsay Cedex, France
yannis@lri.fr

Prof. Alberto Marchetti-Spaccamela
Universita' di Roma "La Sapienza"
Dept. Informatica e Sistemistica
Via Salaria 113
00198 Roma, Italy
marchetti@aquila.infn.it
marchetti@irmiasi.rm.cnr.it

Prof. Ernst W. Mayr
Lehrstuhl Effiziente Algorithmen
Institut für Informatik
TU München
80290 München, Germany
mayr@informatik.tu-muenchen.de

Prof. Dr. Rolf Möhring
Technische Universität Berlin
Fachbereich Mathematik
Straße des 17. Juni 135
10623 Berlin, Germany
moehring@math.tu-berlin.de

Dr. Haiko Müller
Friedrich-Schiller-Universität Jena
Fakultät für Mathematik und
Informatik
Leutragraben 1, UHH 17.OG
07740 Jena, Germany
hm@minet.uni-jena.de

Prof. Dr.-Ing. Manfred Nagl
Lehrstuhl Informatik III
RWTH Aachen
52056 Aachen, Germany
nagl@rwthi3.informatik.rwth-
aachen.de

Dipl.-Math. Valeska Naumann
TU Berlin
FB Mathematik, Sekretariat MA 6-1
Straße des 17.Juni 136
10623 Berlin, Germany
naumann@math.tu-berlin.de

Prof. Stephan Olariu
Department of Computer Science
Old Dominion University
Nordfolk, Virginia 23529, USA
olariu@cs.odu.edu

Prof. Dr. Francesco Parisi-Presicce
Dip. di Scienze dell' Informazione
Universita' di Roma La Sapienza
Via Salaria 113
00198 Roma, Italy
parisi@dsi.uniroma1.it

Dipl.-Math Andreas Parra Asensio
TU Berlin
FB Mathematik, Sekretariat MA 6-1
Straße des 17.Juni 136
10623 Berlin, Germany
parra@math.tu-berlin.de

Prof. Andrzej Proskurowski
University of Oregon
Computer Science Department
Eugene, OR 97403, USA
andrzej@cs.uoregon.edu

Siddharthan Ramachandramurthi
Department of Computer Science
University of Tennessee
Knoxville, TN 37996-1301, USA
siddhart@cs.utk.edu

Prof. Dr. Rüdiger Reischuk
Medizinische Universität Lübeck
Wallstr. 40
23560 Lübeck, Germany
reischuk@informatik.mu-luebeck.de

Dr. Petra Scheffler
TU Berlin
FB Mathematik, Sekretariat MA 6-1
Straße des 17.Juni 136
10623 Berlin, Germany
scheffle@math.tu-berlin.de

Franz Schmalhofer
Fakultät für Informatik
Universität der Bundeswehr
D-85577 Neubiberg, Germany
schmalhofer@informatik.unibw-
muenchen.de

Dr. Andy Schürr
Lehrstuhl für Informatik III
RWTH Aachen
52074 Aachen, Germany
andy@i3.informatik.rwth-aachen.de

Prof. Dr. Gunther Schmidt
Fakultät für Informatik
Universität der Bundeswehr
D-85577 Neubiberg, Germany
schmidt@informatik.unibw-
muenchen.de

Gerald Schreiber
TU Hamburg-Harburg
Technische Informatik I
Harburger Schlossstr. 20
21071 Hamburg, Germany
g-schreiber@tu-harburg.d400.de

Detlev Sieling
Universität Dortmund
FB Informatik, LS II
44221 Dortmund, Germany
sieling@ls2.informatik.uni-
dortmund.de

Dipl.-Inf. Konstantin Skodinis
University of Passau
94030 Passau, Germany
skodinis@fmi.uni-passau.de

Dr. Ondrej Sýkora
Institute for Informatics
Slovak Academy of Sciences
Dubravska' 9
84235 Bratislava, Slovak Republic
sykora@savba.sk

Prof. Dr. Maciej M. Syslo
Institute of Computer Science
University of Wroclaw
51151 Wroclaw, Poland
syslo@ii.uni.wroc.pl

Dr. Jan Telle
University of Oregon
Computer Science Department
Eugene, OR 97403, USA
telle@cs.uoregon.edu

Prof. Gottfried Tinhofer
Institut für Mathematik
TU München
80290 München, Germany
gottin@statistik.tu-muenchen.de

Dimitris Thilikos
Patras University
Department of Computer Engineering
and Informatics
Rio, 265 00 Patras, Greece
sedthilk@cti.gr

Luca Trevisan
Dip. di Scienze dell' Informazione
Via Salaria 113 (III piano)
00198 Roma, Italy
trevisan@encore.dsi.uniroma1.it

Prof. Dr. Volker Turau
FH Giessen-Friedberg
Fachbereich MND
Wilhelm-Leuschner Str. 13
61169 Friedberg, Germany
turau@courbet.fh-friedberg.de

Prof. Dr. Zsolt Tuza
Computer and Automation Institute
Hungarian Academy of Sciences
H-1111 Budapest, Kende u. 13-17
Hungary
h684tuz

List of Authors

The 20 WG-Volumes

1. PAPE, U.: *Graphen-Sprachen und Algorithmen auf Graphen*. 1. Fachtagung Graphentheoret. Konzepte der Informatik, Hanser, Munich, 1976, 236 pages, ISBN 3-446-12215 X.
2. NOLTEMEIER, H.: *Graphen, Algorithmen, Datenstrukturen*. Proc. WG 76, Graphtheoretic Concepts in Computer Science, Hanser, Munich, 1977, 336 pages, ISBN 3-446-12330 4.
3. MÜHLBACHER, J.: *Datenstrukturen, Graphen, Algorithmen*. Proc. WG 77, Hanser, Munich, 1978, 368 pages, ISBN 3-446-12526 3.
4. NAGL, M., SCHNEIDER, H.-J.: *Graphs, Data Structures, Algorithms*. Proc. WG 78, Hanser, Munich, 1979, 320 pages, ISBN 3-446-12748 3.
5. PAPE, U.: *Discrete Structures and Algorithms*. Proc. WG 79, Hanser, Munich, 1980, 270 pages, ISBN 3-446-13135 3.
6. NOLTEMEIER, H.: *Graphtheoretic Concepts in Computer Science*. Proc. WG 80, Lect. Notes Comput. Sci. 100, Springer, Berlin, 1981, 403 pages, ISBN 0-387-10291 4.
7. MÜHLBACHER, J.: *Proc. of the 7th Conf. Graphtheoretic Concepts in Computer Science (WG 81)*. Hanser, Munich, 1982, 355 pages, ISBN 3-446-13538 3.
8. SCHNEIDER, H.-J., GÖTTLER, H.: *Proc. of the 8th Conf. Graphtheoretic Concepts in Computer Science (WG 82)*. Hanser, Munich, 1983, 280 pages, ISBN 3-446-13778 5.
9. NAGL, M., PERL, J.: *Proc. WG 83, Workshop on Graphtheoretic Concepts in Computer Science*. Trauner, Linz, 1984, 397 pages, ISBN 3-853-20311 6.
10. PAPE, U.: *Proc. WG 84, Workshop on Graphtheoretic Concepts in Computer Science*. Trauner, Linz, 1985, 381 pages, ISBN 3-853-20334 5.
11. NOLTEMEIER, H.: *Graphtheoretic Concepts in Computer Science*. Proc. WG 85, Trauner, Linz, 1986, 443 pages, ISBN 3-853-20357 4.
12. TINHOFER, G., SCHMIDT, G.: *Graph-Theoretic Concepts in Computer Science*. Proc. WG 86, Lect. Notes Comput. Sci. 246, Springer, Berlin, 1987, 305 pages, ISBN 0-387-17218 1.
13. GÖTTLER, H., SCHNEIDER, H.-J.: *Graph-Theoretic Concepts in Computer Science*. Proc. WG 87, Lect. Notes Comput. Sci. 314, Springer, Berlin, 1988, 254 pages, ISBN 0-387-19422 3.
14. VAN LEEUWEN, J.: *Graph-Theoretic Concepts in Computer Science*. Proc. WG 88, Lect. Notes Comput. Sci. 344, Springer, Berlin, 1989, 457 pages, ISBN 0-387-50728 0.
15. NAGL, M.: *Graph-Theoretic Concepts in Computer Science*. Proc. WG 89, Lect. Notes Comput. Sci. 411, Springer, Berlin, 1990, 374 pages, ISBN 0-387-52292 1.
16. MÖHRING, M.: *Graph-Theoretic Concepts in Computer Science*. Proc. WG 90, Lect. Notes Comput. Sci. 484, Springer, Berlin, 1991, 360 pages, ISBN 0-387-53832 1.

17. SCHMIDT, G., BERGHAMMER, R.: *Graph-Theoretic Concepts in Computer Science.* Proc. WG 91, Lect. Notes Comput. Sci. 570, Springer, Berlin, 1992, 253 pages, ISBN 0-387-55121 2.

18. MAYR, E.W.: *Graph-Theoretic Concepts in Computer Science.* Proc. WG 92, Lect. Notes Comput. Sci. 657, Springer, Berlin, 1993, 350 pages, ISBN 0-387-56402 0.

19. VAN LEEUWEN, J.: *Graph-Theoretic Concepts in Computer Science.* Proc. WG 93, Lect. Notes Comput. Sci. 790, Springer, Berlin, 1994, 431 pages, ISBN 0-387-57889 4.

20. MAYR, E.W., SCHMIDT, G., TINHOFER, G.: *Graph-Theoretic Concepts in Computer Science.* Proc. WG 94, Lect. Notes Comput. Sci., Springer, Berlin, 1995

Springer-Verlag
and the Environment

We at Springer-Verlag firmly believe that an international science publisher has a special obligation to the environment, and our corporate policies consistently reflect this conviction.

We also expect our business partners – paper mills, printers, packaging manufacturers, etc. – to commit themselves to using environmentally friendly materials and production processes.

The paper in this book is made from low- or no-chlorine pulp and is acid free, in conformance with international standards for paper permanency.

Lecture Notes in Computer Science

For information about Vols. 1–822
please contact your bookseller or Springer-Verlag

Vol. 859: T. F. Melham, J. Camilleri (Eds.), Higher Order Logic Theorem Proving and Its Applications. Proceedings, 1994. IX, 470 pages. 1994.

Vol. 860: W. L. Zagler, G. Busby, R. R. Wagner (Eds.), Computers for Handicapped Persons. Proceedings, 1994. XX, 625 pages. 1994.

Vol: 861: B. Nebel, L. Dreschler-Fischer (Eds.), KI-94: Advances in Artificial Intelligence. Proceedings, 1994. IX, 401 pages. 1994. (Subseries LNAI).

Vol. 862: R. C. Carrasco, J. Oncina (Eds.), Grammatical Inference and Applications. Proceedings, 1994. VIII, 290 pages. 1994. (Subseries LNAI).

Vol. 863: H. Langmaack, W.-P. de Roever, J. Vytopil (Eds.), Formal Techniques in Real-Time and Fault-Tolerant Systems. Proceedings, 1994. XIV, 787 pages. 1994.

Vol. 864: B. Le Charlier (Ed.), Static Analysis. Proceedings, 1994. XII, 465 pages. 1994.

Vol. 865: T. C. Fogarty (Ed.), Evolutionary Computing. Proceedings, 1994. XII, 332 pages. 1994.

Vol. 866: Y. Davidor, H.-P. Schwefel, R. Männer (Eds.), Parallel Problem Solving from Nature - PPSN III. Proceedings, 1994. XV, 642 pages. 1994.

Vol 867: L. Steels, G. Schreiber, W. Van de Velde (Eds.), A Future for Knowledge Acquisition. Proceedings, 1994. XII, 414 pages. 1994. (Subseries LNAI).

Vol. 868: R. Steinmetz (Ed.), Multimedia: Advanced Teleservices and High-Speed Communication Architectures. Proceedings, 1994. IX, 451 pages. 1994.

Vol. 869: Z. W. Raś, Zemankova (Eds.), Methodologies for Intelligent Systems. Proceedings, 1994. X, 613 pages. 1994. (Subseries LNAI).

Vol. 870: J. S. Greenfield, Distributed Programming Paradigms with Cryptography Applications. XI, 182 pages. 1994.

Vol. 871: J. P. Lee, G. G. Grinstein (Eds.), Database Issues for Data Visualization. Proceedings, 1993. XIV, 229 pages. 1994.

Vol. 872: S Arikawa, K. P. Jantke (Eds.), Algorithmic Learning Theory. Proceedings, 1994. XIV, 575 pages. 1994.

Vol. 873: M. Naftalin, T. Denvir, M. Bertran (Eds.), FME '94: Industrial Benefit of Formal Methods. Proceedings, 1994. XI, 723 pages. 1994.

Vol. 874: A. Borning (Ed.), Principles and Practice of Constraint Programming. Proceedings, 1994. IX, 361 pages. 1994.

Vol. 875: D. Gollmann (Ed.), Computer Security – ESORICS 94. Proceedings, 1994. XI, 469 pages. 1994.

Vol. 876: B. Blumenthal, J. Gornostaev, C. Unger (Eds.), Human-Computer Interaction. Proceedings, 1994. IX, 239 pages. 1994.

Vol. 877: L. M. Adleman, M.-D. Huang (Eds.), Algorithmic Number Theory. Proceedings, 1994. IX, 323 pages. 1994.

Vol. 878: T. Ishida; Parallel, Distributed and Multiagent Production Systems. XVII, 166 pages. 1994. (Subseries LNAI).

Vol. 879: J. Dongarra, J. Waśniewski (Eds.), Parallel Scientific Computing. Proceedings, 1994. XI, 566 pages. 1994.

Vol. 880: P. S. Thiagarajan (Ed.), Foundations of Software Technology and Theoretical Computer Science. Proceedings, 1994. XI, 451 pages. 1994.

Vol. 881: P. Loucopoulos (Ed.), Entity-Relationship Approach – ER'94. Proceedings, 1994. XIII, 579 pages. 1994.

Vol. 882: D. Hutchison, A. Danthine, H. Leopold, G. Coulson (Eds.), Multimedia Transport and Teleservices. Proceedings, 1994. XI, 380 pages. 1994.

Vol. 883: L. Fribourg, F. Turini (Eds.), Logic Program Synthesis and Transformation – Meta-Programming in Logic. Proceedings, 1994. IX, 451 pages. 1994.

Vol. 884: J. Nievergelt, T. Roos, H.-J. Schek, P. Widmayer (Eds.), IGIS '94: Geographic Information Systems. Proceedings, 1994. VIII, 292 pages. 19944.

Vol. 885: R. C. Veltkamp, Closed Objects Boundaries from Scattered Points. VIII, 144 pages. 1994.

Vol. 886: M. M. Veloso, Planning and Learning by Analogical Reasoning. XIII, 181 pages. 1994. (Subseries LNAI).

Vol. 887: M. Toussaint (Ed.), Ada in Europe. Proceedings, 1994. XII, 521 pages. 1994.

Vol. 888: S. A. Andersson (Ed.), Analysis of Dynamical and Cognitive Systems. Proceedings, 1993. VII, 260 pages. 1995.

Vol. 889: H. P. Lubich, Towards a CSCW Framework for Scientific Cooperation in Europe. X, 268 pages. 1995.

Vol. 890: M. J. Wooldridge, N. R. Jennings (Eds.), Intelligent Agents. Proceedings, 1994. VIII, 407 pages. 1995. (Subseries LNAI).

Vol. 891: C. Lewerentz, T. Lindner (Eds.), Formal Development of Reactive Systems. XI, 394 pages. 1995.

Vol. 892: K. Pingali, U. Banerjee, D. Gelernter, A. Nicolau, D. Padua (Eds.), Languages and Compilers for Parallel Computing. Proceedings, 1994. XI, 496 pages. 1995.

Vol. 893: G. Gottlob, M. Y. Vardi (Eds.), Database Theory – ICDT '95. Proceedings, 1995. XI, 454 pages. 1995.

Vol. 894: R. Tamassia, I. G. Tollis (Eds.), Graph Drawing. Proceedings, 1994. X, 471 pages. 1995.

Vol. 895: R. L. Ibrahim (Ed.), Software Engineering Education. Proceedings, 1995. XII, 449 pages. 1995.

Vol. 896: R. M. Taylor, J. Coutaz (Eds.), Software Engineering and Human-Computer Interaction. Proceedings, 1994. X, 281 pages. 1995.

Vol. 898: P. Steffens (Ed.), Machine Translation and the Lexicon. Proceedings, 1993. X, 251 pages. 1995. (Subseries LNAI).

Vol. 899: W. Banzhaf, F. H. Eeckman (Eds.), Evolution and Biocomputation. VII, 277 pages. 1995.

Vol. 900: E. W. Mayr, C. Puech (Eds.), STACS 95. Proceedings, 1995. XIII, 654 pages. 1995.

Vol. 901: R. Kumar, T. Kropf (Eds.), Theorem Provers in Circuit Design. Proceedings, 1994. VIII, 303 pages. 1995.

Vol. 902: M. Dezani-Ciancaglini, G. Plotkin (Eds.), Typed Lambda Calculi and Applications. Proceedings, 1995. VIII, 443 pages. 1995.

Vol. 903: E. W. Mayr, G. Schmidt, G. Tinhofer (Eds.), Graph-Theoretic Concepts in Computer Science. Proceedings, 1994. IX, 414 pages. 1995.